Operator Theory: Advances and Applications
Vol. 131

Convolution Operators and Factorization of Almost Periodic Matrix Functions

Albrecht Böttcher
Yuri I. Karlovich
Ilya M. Spitkovsky

Springer Basel AG

Authors:

Albrecht Böttcher
Faculty of Mathematics
Technical University Chemnitz
09107 Chemnitz
Germany
e-mail: aboettch@mathematik.tu-chemnitz.de

Yuri I. Karlovich
Department of Mathematics
CINVESTAV of the I.P.N.
P.O. Box 14-740
07000 Mexico D.F.
Mexico
e-mail: karlovic@math.cinvestav.mx

Ilya M. Spitkovsky
Department of Mathematics
College of William and Mary
P.O. Box 8795
Williamsburg, VA 23187-8795
USA
e-mail: ilya@math.wm.edu

2000 Mathematics Subject Classification 47A68, 47B35

A CIP catalogue record for this book is available from the
Library of Congress, Washington D.C., USA

Deutsche Bibliothek Cataloging-in-Publication Data

Böttcher, Albrecht:
Convolution operators and factorization of almost periodic matrix functions
/ Albrecht Böttcher ; Yuri I. Karlovich ; Ilya M. Spitkovsky. - Basel ;
Boston ; Berlin : Birkhäuser, 2002
(Operator theory ; Vol. 131)
ISBN 978-3-0348-9457-9 ISBN 978-3-0348-8152-4 (eBook)
DOI 10.1007/978-3-0348-8152-4

ISBN 978-3-0348-9457-9

Originally published by Birkhäuser Verlag in 2002
Softcover reprint of the hardcover 1st edition 2002
Member of the BertelsmannSpringer Publishing Group
Printed on acid-free paper produced from chlorine-free pulp. TCF ∞
Cover design: Heinz Hiltbrunner, Basel

ISBN 978-3-0348-9457-9

Contents

Preface

Many problems of the engineering sciences, physics, and mathematics lead to convolution equations and their various modifications. Convolution equations on a half-line can be studied by having recourse to the methods and results of the theory of Toeplitz and Wiener-Hopf operators. Convolutions by integrable kernels have continuous symbols and the Cauchy singular integral operator is the most prominent example of a convolution operator with a piecewise continuous symbol. The Fredholm theory of Toeplitz and Wiener-Hopf operators with continuous and piecewise continuous (matrix) symbols is well presented in a series of classical and recent monographs.

Symbols beyond piecewise continuous symbols have discontinuities of oscillating type. Such symbols emerge very naturally. For example, difference operators are nothing but convolution operators with almost periodic symbols: the operator defined by $(A\varphi)(x) = \sum a_k \varphi(x - \lambda_k)$ is the convolution operator with the symbol $f(x) = \sum a_k e^{i\lambda_k x}$. Moreover, a convolution operator on a finite interval is, in a sense, equivalent to a convolution operator on the half-line whose symbol is a 2×2 oscillating matrix function: consideration of the convolution operator with the symbol $f(x)$ on the interval $(0, \lambda)$ leads to the convolution operator with the matrix symbol

$$G_f^{(\lambda)}(x) = \begin{pmatrix} e^{-i\lambda x} & 0 \\ f(x) & e^{i\lambda x} \end{pmatrix}$$

on the half-line $(0, \infty)$. Notice that $G_f^{(\lambda)}(x)$ is oscillating even if $f(x)$ is continuous. We finally mention that convolution operators with oscillating symbols have properties that are not shared by operators with continuous or piecewise continuous symbols. Some basic phenomena of Toeplitz and Wiener-Hopf theory stay in the dark when working with piecewise continuous symbols but come to light after passage to, say, semi-almost periodic symbols. One illustration of this statement is as follows. Let a_n be functions that converge uniformly to some function a. If the functions a_n are continuous or piecewise continuous, then the spectra of the Wiener-Hopf operators with the symbols a_n converge in the Hausdorff metric to the spectrum of the Wiener-Hopf operator with the symbol a. Surprisingly, this need not be true in case the functions a_n are semi-almost periodic.

Convolution equations on the half-line can be solved once a Wiener-Hopf factorization of the symbol is available. In this book we investigate several factorizations of almost periodic matrix functions. For instance, in case the function $f(x)$ is almost periodic, we are interested in representing $G_f^{(\lambda)}(x)$ in the form

$$G_f^{(\lambda)}(x) = G_-(x) \begin{pmatrix} e^{i\mu_1 x} & 0 \\ 0 & e^{i\mu_2 x} \end{pmatrix} G_+(x),$$

where μ_1, μ_2 are real numbers and $G_\pm(x)$ are invertible almost periodic matrix functions such that the Bohr-Fourier spectra of $G_+^{\pm 1}(x)$ and $G_-^{\pm 1}(x)$ are contained in $[0, \infty)$ and $(-\infty, 0]$, respectively. Such a factorization is called an *AP* factorization. The numbers μ_1, μ_2 are referred to as the almost periodic indices of $G_f^{(\lambda)}(x)$. Sole knowledge of the almost periodic indices tells us much about the properties of the original convolution operator. If, in addition, the factors $G_-(x)$ and $G_+(x)$ are at our disposal, we have nearly complete command of the convolution equation.

The notion of *AP* factorization was introduced in the eighties. During the last twenty years, *AP* factorization has become a relatively independent business. Nowadays one knows large classes of symbols $f(x)$ for which an *AP* factorization of $G_f^{(\lambda)}(x)$ can be constructed explicitly. In the other direction, one has discovered classes of *very nice* symbols $f(x)$ such that $G_f^{(\lambda)}(x)$ has definitely *no AP* factorization. The border between these two extreme situations remains a rugged crest with many open questions and mysteries.

The purpose of this book is to elucidate the close relationship between convolution equations and *AP* factorization on the one hand and to acquaint the reader with what is known about the *AP* factorization of several concrete classes of matrix functions on the other. Our emphasis is both on the results and on the methods of *AP* factorization. The majority of the results we will quote are accompanied with full proofs. Moreover, many results are taken from the periodicals and are cited with full proofs here for the first time. We suppose that knowledge of the basics of functional, real, complex, and harmonic analysis will enable the interested and motivated reader to go through the bulk of the book without serious difficulty.

A look at the table of contents provides an idea of what this book is actually about. We have split the large amount of material into many small pieces, so that the book has 23 chapters. We believe that this facilitates getting in the matter. Moreover, each chapter has its own flavor: some of them are introductory surveys, others are devoted to delightful stories, and part of the chapters are forced marches. The pictures on the first pages of the chapters show the ranges of different semi-almost periodic functions and are explained in Chapter 3.

It should not be hidden that, despite many genuine achievements made during the last two decades, the edifice of *AP* factorization is still incomplete and that our original optimism to finish this edifice in due time has deceased over the years. We nevertheless think the time is ripe for this book. It is the first attempt of comprehensively presenting the core of the theory of *AP* factorization and its applications, it is a certain account of what is going on in this field, and we hope the book can serve as some kind of a guide for further research into the subject.

Acknowledgements. Our joint work over the last years and, in particular, the work on this book was generously supported by NATO Collaborative Research Grant CRG 950332. This support stimulated and enabled us to work together regularly, and without this support the present book would certainly not have been written. We wish to express our sincere gratitude for this grant.

Karlovich was also supported by JNICT grant PRAXIS XXI/BCC/4355/94 (Portugal) and CONACYT grant Cátedra Patrimonial, Nivel II, No. 990017-Ex (México), and Spitkovsky also acknowledges support by NSF grants DMS-9401848, DMS-9800704, and DMS-9988579.

Last but not least, we are greatly indebted to Sylvia Böttcher for producing the LaTeX masters of the book with her usual patience, reliability, and precision.

Chemnitz, Mexico City, Williamsburg, Summer 2001 The authors

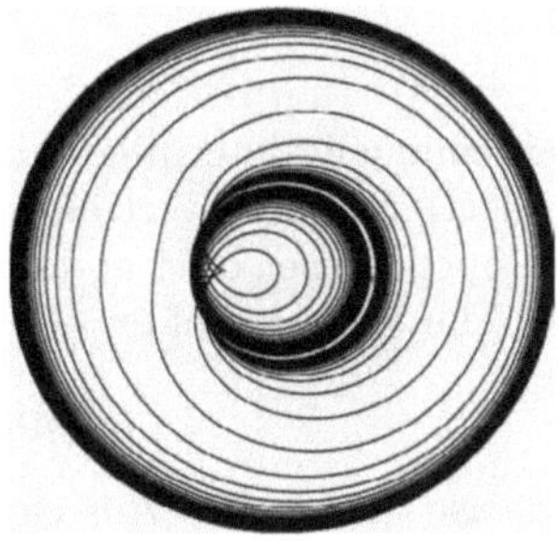

Chapter 1

Convolution Operators and Their Symbols

In this chapter we introduce convolution operators on the line, on a half-line, and on finite intervals. The Fourier transform of the convolution kernel is called the symbol of the operator. Shift operators are convolutions by the shifted Dirac delta function and lead to almost periodic symbols. The symbol of the Cauchy singular integral operator is piecewise continuous. The class of semi-almost periodic symbols emerges naturally when considering the algebra generated by shift operators and the Cauchy singular integral operator. Moreover, convolution operators with continuous and piecewise continuous symbols on finite intervals can be transformed into convolution operators on a half-line with 2×2*-matrix valued symbols whose entries are almost periodic functions and semi-almost periodic (or even piecewise-almost periodic) functions, respectively.*

1.1 Banach and C^*-Algebras

The purpose of this section is to record a few definitions and results on Banach algebras and to fix some notation.

A *Banach algebra* is a complex Banach space A with an associative and distributive multiplication that satisfies $\|ab\| \le \|a\|\|b\|$ for all $a, b \in A$. A Banach algebra is said to be *unital* if it has a unit element. The unit element is frequently also called the identity and is usually denoted by $e, 1$, or I. We always require that $\|e\| = \|1\| = \|I\| = 1$.

Let A be a Banach algebra with the unit element e. An element $a \in A$ is said to be *invertible* (in A) if there is a $b \in A$ such that $ab = ba = e$. If it exists, this element

b is uniquely determined; it is denoted by a^{-1} and is called the *inverse* of A. The collection of all invertible elements of A is denoted by GA. We remark that GA is an open subset of A and that GA is a group with respect to the multiplication of A. The *spectrum* of an element $a \in A$ is the set

$$\mathrm{sp}_A\, a := \{\lambda \in \mathbf{C} : a - \lambda e \notin GA\}.$$

In case the algebra A is clear from the context, we abbreviate $\mathrm{sp}_A\, a$ to $\mathrm{sp}\, a$. Note that $\mathrm{sp}\, a$ is always a nonempty compact subset of $\mathbf{C}$ which is contained in the disk $\{\lambda \in \mathbf{C} : |\lambda| \le \|a\|\}$.

Example 1.1. If X is a Banach space, then the set $\mathcal{B}(X)$ of all bounded linear operators on X is a unital Banach algebra under the natural algebraic operations and the operator norm. The unit element is the identity operator I. By the spectrum of an operator $A \in \mathcal{B}(X)$ we always understand the set $\mathrm{sp}_{\mathcal{B}(X)} A$. □

Example 1.2. For $1 \le p \le \infty$, we denote by $L^p(\mathbf{R})$ the usual Lebesgue spaces on the real line $\mathbf{R}$. The norm in $L^p(\mathbf{R})$ is denoted by $\|\cdot\|_p$. The space $L^\infty(\mathbf{R})$ is a unital Banach algebra under pointwise algebraic operations and the norm $\|\cdot\|_\infty$. Although the elements of $L^p(\mathbf{R})$ are cosets of functions, we simply refer to them as functions. A function $a \in L^\infty(\mathbf{R})$ is invertible in $L^\infty(\mathbf{R})$ if and only if $\operatorname{ess\,inf}_{x \in \mathbf{R}} |a(x)| > 0$. The spectrum $\mathrm{sp}_{L^\infty(\mathbf{R})}\, a$ of $a \in L^\infty(\mathbf{R})$ is called the *essential range* of a and is denoted by $\mathcal{R}(a)$. □

An element $a \in A$ of a unital Banach algebra A is said to be *left* (resp. *right*) *invertible* if there is a $b \in A$ such that $ba = e$ (resp. $ab = e$).

A Banach algebra A is called *commutative* if $ab = ba$ for all $a, b \in A$. Given a commutative Banach algebra A with identity element e, we denote by G_0A the connected component of GA which contains e. It is well known that

$$G_0A = \{\exp(b) : b \in A\} \tag{1.1}$$

where $\exp(b) := \sum_{n=0}^{\infty} b^n/n!$.

Let A be a Banach algebra. For a subset $E \subset A$, we denote by $\mathrm{alg}_A\, E$ or simply by $\mathrm{alg}\, E$ the smallest closed subalgebra of A that contains E, i.e., the intersection of all closed subalgebras of A containing E. Equivalently,

$$\mathrm{alg}\, E = \mathrm{clos}\Big\{ \sum_j \gamma_j \prod_k a_{jk} : \gamma_j \in \mathbf{C},\ a_{jk} \in E \Big\},$$

the sum and the (ordered) products finite; of course, here clos stands for the closure in A. Given subsets $E, F, \ldots \subset A$ and elements $a, b, \ldots \in A$, we put

$$\mathrm{alg}\,(E, F, \ldots, a, b, \ldots) := \mathrm{alg}\big(E \cup F \cup \ldots \cup \{a, b, \ldots\}\big).$$

Let now A and B be two Banach algebras. A bounded linear operator $\varphi : A \to B$ is called a *Banach algebra homomorphism* if $\varphi(ab) = \varphi(a)\varphi(b)$ for all $a, b \in$

A. Bijective Banach algebra homomorphisms are referred to as *Banach algebra isomorphisms*.

An especially nice class of Banach algebras is the class of C^*-algebras. A map $a \mapsto a^*$ of a Banach algebra A into itself is called an *involution* if

$$(a^*)^* = a, \ (\lambda a)^* = \overline{\lambda} a^*, \ (a+b)^* = a^* + b^*, \ (ab)^* = b^* a^*$$

for all $\lambda \in \mathbf{C}$ and all $a, b \in A$. A Banach algebra A with an involution $a \mapsto a^*$ is said to be a *C^*-algebra* if $\|aa^*\| = \|a\|^2$ for all $a \in A$.

Example 1.3. If H is a Hilbert space, then $\mathcal{B}(H)$ is a C^*-algebra with passage to the adjoint as involution. The Banach algebra $L^\infty(\mathbf{R})$ is a C^*-algebra under the involution $a^*(x) := \overline{a(x)}$, where the bar denotes passage to the complex conjugate. A subalgebra $\mathcal{B}$ of a C^*-algebra is called selfadjoint if $b^* \in \mathcal{B}$ for every $b \in \mathcal{B}$. Closed selfadjoint subalgebras of C^*-algebras are obviously again C^*-algebras. □

Let A be a Banach algebra with unit element e and let B be a closed subalgebra of A which contains e. Then, for $b \in B$, the spectrum $\operatorname{sp}_B b$ may be larger than $\operatorname{sp}_A b$. If this does not happen, that is, if $\operatorname{sp}_B b = \operatorname{sp}_A b$ for every $b \in B$, then B is said to be *inverse closed* in A. Clearly, inverse closedness of B in A is equivalent to saying that every $b \in B$ which is invertible in A is automatically invertible in the smaller algebra B.

Proposition 1.4. (a) *If A is a unital C^*-algebra and B is a C^*-subalgebra of A which contains the identity element of A, then B is inverse closed in A.*

(b) *If A is a unital C^*-algebra and $a \in A$, then*

$$\begin{aligned} a \textit{ is left-invertible} &\iff a^*a \textit{ is invertible};\\ a \textit{ is right-invertible} &\iff aa^* \textit{ is invertible}. \end{aligned}$$

A Banach algebra homomorphism $\varphi : A \to B$ between two C^*-algebras A and B is called a *C^*-algebra homomorphism* if $\varphi(a^*) = (\varphi(a))^*$ for all $a \in A$. A *C^*-algebra isomorphism* is a bijective C^*-algebra homomorphism. Here are some very useful properties of C^*-algebra homomorphisms.

Proposition 1.5. *Let A and B be C^*-algebras and let $\varphi : A \to B$ be a C^*-algebra homomorphism.*

(a) *We have $\|\varphi(a)\| \le \|a\|$ for all $a \in A$.*

(b) *If φ is injective, then φ is an isometry, that is, $\|\varphi(a)\| = \|a\|$ for all $a \in A$.*

(c) *The image $\varphi(A)$ is a closed subset and thus a C^*-subalgebra of B.*

(d) *If φ is injective, then $\operatorname{sp}_A a = \operatorname{sp}_B \varphi(a)$ for all $a \in A$.*

Let A be a unital commutative Banach algebra. Denote by M the set of all nonzero Banach algebra homomorphisms $m : A \to \mathbf{C}$. The set M can be shown to be a compact Hausdorff space in the weak $*$-topology of the dual space A^* and is referred to as the *maximal ideal space* of A. The map

$$\Gamma : A \to C(M), \quad (\Gamma a)(m) := m(a)$$

is a Banach algebra homomorphism and is called the *Gelfand map.*

Theorem 1.6 (Gelfand). *Let A be a unital commutative Banach algebra and let $\Gamma : A \to C(M)$ be the Gelfand map. Then an element $a \in A$ is invertible in A if and only if $(\Gamma a)(m) \neq 0$ for all $m \in M$.*

Theorem 1.7 (Gelfand and Naimark). (a) *If A is a commutative unital C^*-algebra, then the Gelfand map is a C^*-algebra isomorphism of A onto $C(M)$.*

(b) *If A is any C^*-algebra, then there exist a Hilbert space H and an injective C^*-algebra homomorphism $\varphi : A \to \mathcal{B}(H)$. In other words, every C^*-algebra can be identified with a C^*-subalgebra of $\mathcal{B}(H)$ for some Hilbert space H.*

1.2 Convolution Operators on the Line

A convolution operator A is formally given by the formula

$$(Af)(t) = \int k(t-s)f(s)\,ds.$$

The function k is in this context referred to as the *convolution kernel* of A.

We first consider convolution operators on the Hilbert space $L^2(\mathbf{R})$:

$$(Af)(t) = \int_{\mathbf{R}} k(t-s)f(s)\,ds, \quad t \in \mathbf{R}. \tag{1.2}$$

Let $F : L^2(\mathbf{R}) \to L^2(\mathbf{R})$ denote the *Fourier transform,*

$$(Ff)(x) := \hat{f}(x) := \int_{\mathbf{R}} f(t)e^{itx}\,dt, \quad x \in \mathbf{R}, \tag{1.3}$$

and let $F^{-1} : L^2(\mathbf{R}) \to L^2(\mathbf{R})$ be the inverse of F,

$$(F^{-1}g)(t) = \frac{1}{2\pi}\int_{\mathbf{R}} g(x)e^{-itx}\,dx, \quad t \in \mathbf{R}.$$

The operator (1.2) can formally be written in the form

$$A = F^{-1}\hat{k}F \tag{1.4}$$

or equivalently, $(Af)\hat{}\,(x) = \hat{k}(x)\hat{f}(x)$, $x \in \mathbf{R}$. Thus, the Fourier transform takes convolution by k to multiplication by $\hat{k}$. The function $\hat{k}$ is called the *symbol* of the operator (1.2), (1.4).

Given a function $a \in L^\infty(\mathbf{R})$, we denote by aI the operator of multiplication by a on $L^2(\mathbf{R})$:

$$aI : L^2(\mathbf{R}) \to L^2(\mathbf{R}),\ g \mapsto ag.$$

Obviously, aI is bounded on $L^2(\mathbf{R})$ and $\|aI\| \le \|a\|_\infty$. It is not difficult to show that actually $\|aI\| = \|a\|_\infty$. In case aI follows another operator, X say, we omit the I and abbreviate aIX to aX. For $a \in L^\infty(\mathbf{R})$, we define the operator $W^0(a)$ by

$$W^0(a) : L^2(\mathbf{R}) \to L^2(\mathbf{R}),\ f \mapsto F^{-1}aFf.$$

Since $\|aI\| = \|a\|_\infty$ and $(2\pi)^{-1/2}F$ is an isometry of $L^2(\mathbf{R})$, it follows that $\|W^0(a)\| = \|a\|_\infty$. Due to (1.2) and (1.4), the operator $W^0(a)$ is referred to as the *convolution operator with the symbol* a. Clearly, the operator given by (1.2) or (1.4) is nothing but $W^0(\hat{k})$.

Proposition 1.8. *The map*

$$W^0 : L^\infty(\mathbf{R}) \to \mathcal{B}(L^2(\mathbf{R})),\ a \mapsto W^0(a)$$

is an isometric C^*-algebra homomorphism. If $a \in L^\infty(\mathbf{R})$ then $\operatorname{sp} W^0(a) = \mathcal{R}(a)$, *and if $a \in GL^\infty(\mathbf{R})$ then the inverse of the operator $W^0(a)$ is the operator $W^0(a^{-1})$.*

Proof. It is immediate from the definition of $W^0(a)$ that W^0 is a C^*-algebra homomorphism. That W^0 is an isometry was shown above and can also be deduced from Proposition 1.5(b) and the fact that W^0 is obviously injective. Since W^0 is a C^*-algebra isomorphism of $L^\infty(\mathbf{R})$ onto the image $W^0(L^\infty(\mathbf{R}))$, we obtain from Example 1.2 and Proposition 1.5(c) and 1.4(a) (or simply from Proposition 1.5(d)) that

$$\operatorname{sp} W^0(a) := \operatorname{sp}_{\mathcal{B}(L^2(\mathbf{R}))} W^0(a) = \operatorname{sp}_{W^0(L^\infty(\mathbf{R}))} W^0(a) = \operatorname{sp}_{L^\infty(\mathbf{R})} a =: \mathcal{R}(a).$$

Finally, it is clear that $(W^0(a))^{-1} = W^0(a^{-1})$ whenever $a \in GL^\infty(\mathbf{R})$. □

Thus, the problem of finding the spectrum of the convolution operator (1.2) amounts to finding the essential range of its symbol.

Example 1.9. If $k \in L^1(\mathbf{R})$, then $\hat{k}$ is a continuous function which vanishes at infinity and we have

$$\int_{\mathbf{R}} k(t-s)f(s)\,ds = (W^0(\hat{k})f)(t), \quad t \in \mathbf{R}.$$

If $\gamma \in \mathbf{C}$, then clearly $\gamma I = W^0(\gamma)$. Notice that γI may be interpreted as convolution by $\gamma\delta(t)$ where $\delta(t)$ is the Dirac delta function. Hence, the operator A given by

$$(Af)(t) = \gamma f(t) + \int_{\mathbf{R}} k(t-s)f(s)\,ds, \quad t \in \mathbf{R},$$

with $\gamma \in \mathbf{C}$ and $k \in L^1(\mathbf{R})$ has the symbol $\gamma+\hat{k}(x)$. This is a continuous function on $\mathbf{R}$ whose limits at $\pm\infty$ exist and are both equal to γ. Consequently, by Proposition 1.8,

$$\text{sp}\, W^0(\gamma+\hat{k}) = \{\gamma + \hat{k}(x) : x \in \mathbf{R}\} \cup \{\gamma\}. \qquad \square$$

The sets

$$\mathcal{A}_0 := \{W^0(\hat{k}) : k \in L^1(\mathbf{R})\}, \ \mathcal{A} := \{W^0(\gamma+\hat{k}) : \gamma \in \mathbf{C},\ k \in L^1(\mathbf{R})\}$$

are algebras, but they are not closed in $\mathcal{B}(L^2(\mathbf{R}))$. Let us denote by $\mathcal{C}_0$ and $\mathcal{C}$, respectively, their closures in $\mathcal{B}(L^2(\mathbf{R}))$. Clearly, $\mathcal{C}_0$ and $\mathcal{C}$ are C^*-subalgebras of $\mathcal{B}(L^2(\mathbf{R}))$. These two C^*-algebras admit a very simple description.

Let $C(\mathbf{R})$ denote the set of all continuous (complex-valued) functions on $\mathbf{R}$ and let $BC(\mathbf{R}) := C(\mathbf{R}) \cap L^\infty(\mathbf{R})$ stand for the bounded continuous functions on $\mathbf{R}$. We denote by $C(\dot{\mathbf{R}})$ the functions $a \in BC(\mathbf{R})$ for which the two limits

$$a(-\infty) := \lim_{x\to-\infty} a(x), \quad a(+\infty) := \lim_{x\to+\infty} a(x)$$

exist and coincide; the common value of these two limits will be denoted by $a(\infty)$. Finally, let $C_0(\dot{\mathbf{R}})$ be the collection of all $a \in C(\dot{\mathbf{R}})$ for which $a(\infty) = 0$. Clearly, $BC(\mathbf{R})$, $C(\dot{\mathbf{R}})$, and $C_0(\dot{\mathbf{R}})$ are C^*-subalgebras of $L^\infty(\mathbf{R})$.

Taking into account that infinitely differentiable functions with compact support are dense in $C_0(\dot{\mathbf{R}})$ and that their inverse Fourier transforms lie in $L^1(\mathbf{R})$, it is readily seen that

$$\mathcal{C}_0 := \text{clos}\, \mathcal{A}_0 = W^0(C_0(\dot{\mathbf{R}})), \quad \mathcal{C} := \text{clos}\, \mathcal{A} = W^0(C(\dot{\mathbf{R}})).$$

Thus, we have identified $\mathcal{C}_0$ and $\mathcal{C}$ as the convolution operators with symbols in $C_0(\dot{\mathbf{R}})$ and $C(\dot{\mathbf{R}})$, respectively. Note that $C(\dot{\mathbf{R}})$ decomposes into the direct sum $C(\dot{\mathbf{R}}) = \mathbf{C}\dot{+}C_0(\dot{\mathbf{R}})$, where $\mathbf{C}$ here stands for the constant (complex-valued) functions.

1.3 Almost Periodic Symbols

We begin by considering a simple but important class of convolution operators on $L^2(\mathbf{R})$.

Example 1.10. Fix $\lambda \in \mathbf{R}$ and let $U_\lambda \in \mathcal{B}(L^2(\mathbf{R}))$ be the *shift operator* defined by

$$(U_\lambda f)(t) := f(t-\lambda), \quad t \in \mathbf{R}.$$

Since

$$(U_\lambda f)\hat{}(x) = \int_{\mathbf{R}} f(t-\lambda)e^{itx}\,dt = \int_{\mathbf{R}} f(s)e^{i(s+\lambda)x}\,ds = e^{i\lambda x}\hat{f}(x),$$

we see that $U_\lambda = W^0(e_\lambda)$ with $e_\lambda(x) := e^{i\lambda x}$. Thus, U_λ is the convolution operator with the symbol e_λ. Using the Dirac delta function $\delta(t)$, we have

$$(U_\lambda f)(t) = \int_{\mathbf{R}} \delta(t-\lambda-s)f(s)\,ds, \quad t \in \mathbf{R},$$

which shows that U_λ is convolution by the kernel $\delta(t-\lambda)$. From Proposition 1.8 we deduce that $\operatorname{sp} U_\lambda = \mathbf{T} := \{z \in \mathbf{C} : |z| = 1\}$ whenever $\lambda \neq 0$. □

The study of algebras of shift operators leads to almost periodic functions.

Definition 1.11. The smallest closed subalgebra of $L^\infty(\mathbf{R})$ that contains all the functions e_λ ($\lambda \in \mathbf{R}$) is denoted by AP and called the algebra of (uniformly) *almost periodic functions*:

$$AP := \operatorname{alg}_{L^\infty(\mathbf{R})}\{e_\lambda : \lambda \in \mathbf{R}\}, \quad e_\lambda(x) := e^{i\lambda x}.$$ □

A function $p : \mathbf{R} \to \mathbf{C}$ which can be represented as a finite sum of the form $p = \sum r_j e_{\lambda_j}$ with $r_j \in \mathbf{C}$ and $\lambda_j \in \mathbf{R}$ is called an *almost periodic polynomial.* The collection of all almost periodic polynomials is denoted by AP^0. Obviously, AP is the closure of AP^0 in $L^\infty(\mathbf{R})$. It is clear that AP is a C^*-subalgebra of $L^\infty(\mathbf{R})$. Figures 1.1 and 1.2 show the behavior of two almost periodic polynomials.

Figure 1.1: The image of $10\,e^{ix} + (3+5i)e^{2ix} - 3i\,e^{\pi i x} + 2e^{5ix}$ for x in $[-6,6]$, $[-100,100]$, $[-300,300]$.

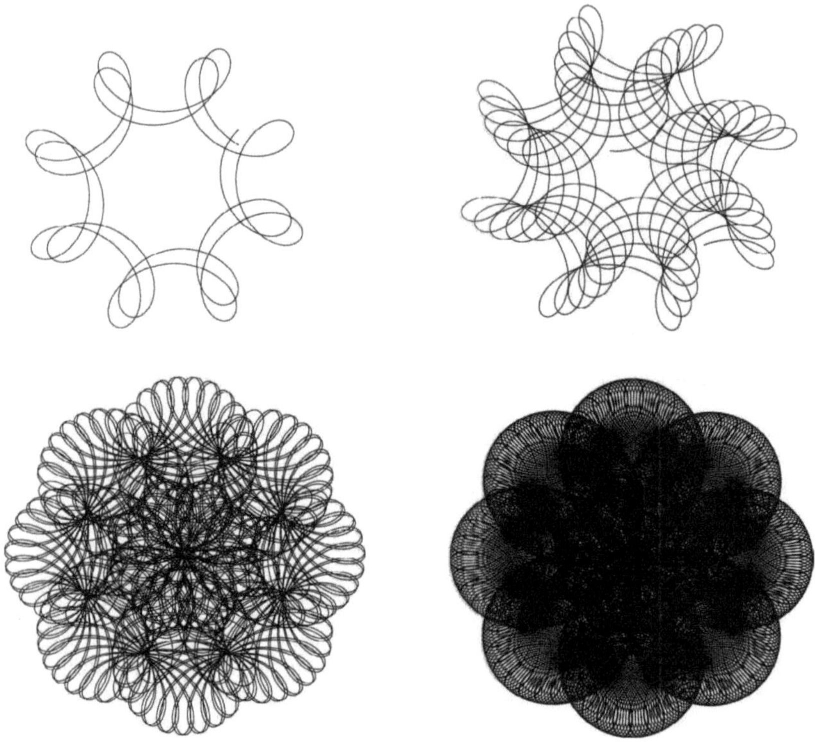

Figure 1.2: The image of $4\,e^{-7ix} + 10\,e^{ix} + 6i\,e^{\pi i x/3}$ for x in $[-6,6]$, $[-20,20]$, $[-120,120]$, $[-400,400]$.

The smallest closed subalgebra of $\mathcal{B}(L^2(\mathbf{R}))$ which contains all shift operators U_λ can now be characterized as the image of AP under the map W^0:

$$\text{alg}_{\mathcal{B}(L^2(\mathbf{R}))}\{U_\lambda : \lambda \in \mathbf{R}\} = W^0(AP).$$

1.4 Kronecker's Theorem

Many properties of almost periodic functions are eventually based on Kronecker's theorem.

A finite collection $\lambda_1, \ldots, \lambda_m$ of real numbers is said to be rationally dependent if there are $k_1, \ldots, k_m \in \mathbf{Z}$ such that $k_1\lambda_1 + \ldots + k_m\lambda_m \in \mathbf{Z}$ and not all $k_1, \ldots, k_m$ are zero. Otherwise, that is, if $k_1\lambda_1 + \ldots + k_m\lambda_m \in \mathbf{Z}$ with $k_1, \ldots, k_m \in \mathbf{Z}$ is only possible for $k_1 = \ldots = k_m = 0$, the numbers $\lambda_1, \ldots, \lambda_m$ are called *rationally independent.*

Theorem 1.12 (Kronecker). *Let $\lambda_1, \ldots, \lambda_m \in \mathbf{R}$ and consider the map*

$$\varphi : \mathbf{N} \to \mathbf{T}^m,\ l \mapsto \left(e^{2\pi i \lambda_1 l}, \ldots, e^{2\pi i \lambda_m l}\right).$$

If the numbers $\lambda_1, \ldots, \lambda_m$ are rationally independent, then $\varphi(\mathbf{N})$ is a dense subset of $\mathbf{T}^m$.

Proof (after H. Weyl). We identify the torus $\mathbf{T}^m$ with $[0, 2\pi)^m$ in the natural manner. Thus, for $l \in \mathbf{N}$, we think of $\varphi(l)$ as the point

$$\left(2\pi\lambda_1 l (\operatorname{mod} 2\pi), \ldots, 2\pi\lambda_m l (\operatorname{mod} 2\pi)\right) \in [0, 2\pi)^m.$$

Let U be an open subset of $\mathbf{T}^m$ and denote by $U(N)$ the number of points of the collection $\varphi(1), \ldots, \varphi(N)$ that lie in U. Let χ_U be the characteristic function of U. We prove that

$$\frac{U(N)}{N} \to \frac{1}{(2\pi)^m}|U| := \frac{1}{(2\pi)^m}\int_{[0,2\pi)^m} \chi_U(x)\, dx \tag{1.5}$$

as $N \to \infty$, which clearly implies that $U(N) \geq 1$ for sufficiently large N and thus shows that U contains a point of $\varphi(\mathbf{N})$. Obviously, (1.5) is equivalent to the relation

$$\lim_{N\to\infty} \frac{1}{N}\sum_{l=1}^{N} \chi_U(\varphi(l)) = \frac{1}{(2\pi)^m}\int_{[0,2\pi)^m} \chi_U(x)\, dx. \tag{1.6}$$

There are $f, g \in C(\mathbf{T}^m)$ such that $f \leq \chi_U \leq g$ on $\mathbf{T}^m$ and such that the L^1 norm $\|f - g\|_1$ is as small as desired. This easily implies that (1.6) holds whenever

$$\lim_{N\to\infty} \frac{1}{N}\sum_{l=1}^{N} f(\varphi(l)) = \frac{1}{(2\pi)^m}\int_{[0,2\pi)^m} f(x)\, dx \tag{1.7}$$

for every $f \in C(\mathbf{T}^m)$. Since the trigonometric polynomials are dense in $C(\mathbf{T}^m)$, (1.7) will follow for all $f \in C(\mathbf{T}^m)$ once we have shown that (1.7) is true for trigonometric polynomials. If f is constant, then (1.7) is trivial. So suppose

$$f(x_1, \ldots, x_m) = e^{ik_1x_1} \ldots e^{ik_mx_m}$$

where $k_1, \ldots, k_m \in \mathbf{Z}$ and not all $k_1, \ldots, k_m$ are zero. Then the right-hand side of (1.7) equals zero, and because $k_1\lambda_1 + \ldots + k_m\lambda_m \notin \mathbf{Z}$, we get

$$\begin{aligned} \frac{1}{N}\sum_{l=1}^{N} f(\varphi(l)) &= \frac{1}{N}\sum_{l=1}^{N} e^{2\pi i k_1\lambda_1 l} \ldots e^{2\pi i k_m \lambda_m l} \\ &= \frac{1}{N}\,\frac{e^{2\pi i(k_1\lambda_1+\ldots+k_m\lambda_m)} - e^{2\pi i(k_1\lambda_1+\ldots+k_m\lambda_m)(N+1)}}{1 - e^{2\pi i(k_1\lambda_1+\ldots+k_m\lambda_m)}}, \end{aligned}$$

which converges indeed to zero as $N \to \infty$. □

One says that a finite collection $\lambda_1, \ldots, \lambda_m$ of real numbers is *linearly independent over* $\mathbf{Z}$ if the equality $k_1\lambda_1 + \ldots + k_m\lambda_m = 0$ with $k_1, \ldots, k_m \in \mathbf{Z}$ implies that $k_1 = \ldots = k_m = 0$. Clearly, if $\lambda_1, \ldots, \lambda_m$ are rationally independent, then they are linearly independent over $\mathbf{Z}$. The reverse implication is not true: for example, a single number λ is linearly independent over $\mathbf{Z}$ if and only if $\lambda \neq 0$, while λ is rationally independent if and only if it is irrational. It is easily seen that $\lambda_1, \ldots, \lambda_m$ are rationally independent exactly if $\lambda_1, \ldots, \lambda_m, 1$ are linearly independent over $\mathbf{Z}$.

Corollary 1.13. *Let $\lambda_1, \ldots, \lambda_m \in \mathbf{R}$, $T \in \mathbf{R}$, and let ψ be the map*

$$\psi : (T, \infty) \to \mathbf{T}^m, \; x \mapsto \left(e^{i\lambda_1 x}, \ldots, e^{i\lambda_m x}\right).$$

If the numbers $\lambda_1, \ldots, \lambda_m$ are linearly independent over $\mathbf{Z}$, then $\psi((T, \infty))$ is a dense subset of $\mathbf{T}^m$ (see Figure 1.3).

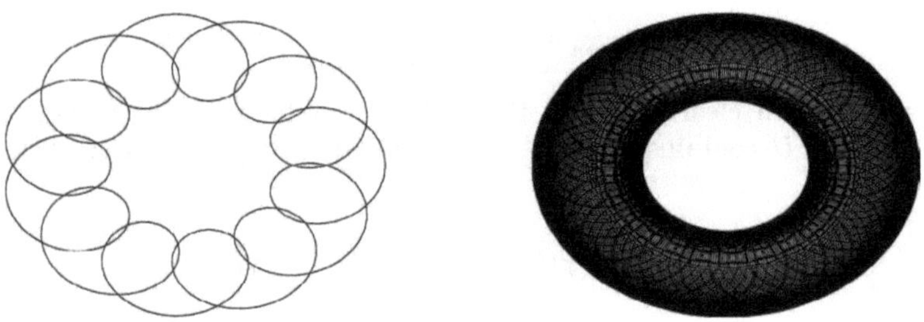

Figure 1.3: The plots show the image of $e^{ix} + 0.4\, e^{i\lambda x}$ as x ranges from -100 to 100 for $\lambda = 12$ (left) and $\lambda = 4\pi$ (right), thus indicating that the map $x \mapsto (e^{ix}, e^{12ix})$ traces out a closed curve on the torus, while the map $x \mapsto (e^{ix}, e^{4\pi ix})$ covers the torus densely. In the pictures, the horizontal scale differs from the vertical scale in order to make circles become ellipses.

Proof. Let $\mathbf{Q}$ denote the field of rational numbers. The set

$$M := \left\{r_1\lambda_1 + \ldots + r_m\lambda_m : (r_1, \ldots, r_m) \in \mathbf{Q}^m\right\}$$

is at most countable. Consequently, there exists a positive real number $\mu \notin M$. We claim that $\lambda_1/\mu, \ldots, \lambda_m/\mu$ are rationally independent. To see this, let

$$k_1(\lambda_1/\mu) + \ldots + k_m(\lambda_m/\mu) = k$$

with $k_1, \ldots, k_m,\; k \in \mathbf{Z}$. If $k = 0$, then $k_1 = \ldots = k_m = 0$ because $\lambda_1, \ldots, \lambda_m$ are linearly independent over $\mathbf{Z}$. If $k \neq 0$, we obtain $\mu = \sum r_j\lambda_j$ with $r_j = k_j/k$ and thus $\mu \in M$, which is impossible. This proves our claim. Applying Theorem 1.12 to the map

$$\varphi : \mathbf{N} \to \mathbf{T}^m, \; l \mapsto \left(e^{2\pi i(\lambda_1/\mu)l}, \ldots, e^{2\pi i(\lambda_m/\mu)l}\right),$$

we obtain that $\psi(2\pi\mu^{-1}\mathbf{N}) = \varphi(\mathbf{N})$ is dense in $\mathbf{T}^m$. □

Lemma 1.14. *Given any finite collection $\lambda_1, \ldots, \lambda_n$ of real numbers, there exists a finite collection $\beta_1, \ldots, \beta_m$ of $m \le n$ real numbers which are linearly independent over $\mathbf{Z}$ such that*

$$\lambda_j = k_1^{(j)}\beta_1 + \ldots + k_m^{(j)}\beta_m \quad \textit{with } k_1^{(j)}, \ldots, k_m^{(j)} \in \mathbf{Z} \tag{1.8}$$

for $j \in \{1, \ldots, n\}$.

Proof. If $n = 1$, we can take $m = 1$ and $\beta_1 = \lambda_1$ for $\lambda_1 \neq 0$ and $\beta_1 = 1$ for $\lambda_1 = 0$. Suppose now we have proved the assertion for $n \ge 1$. We prove it for a collection of $n + 1$ real numbers $\lambda_1, \ldots, \lambda_{n+1}$. If the $\lambda_1, \ldots, \lambda_{n+1}$ are linearly independent over $\mathbf{Z}$, there is nothing to prove. So assume there are $l_1, \ldots, l_{n+1} \in \mathbf{Z}$ such that

$$l_1\lambda_1 + \ldots + l_n\lambda_n + l_{n+1}\lambda_{n+1} = 0$$

and, for example, $l_{n+1} \neq 0$. We then have

$$\begin{aligned} \lambda_j &= -l_{n+1}(-\lambda_j/l_{n+1}) \text{ for } j \in \{1, \ldots, n\}, \\ \lambda_{n+1} &= l_1(-\lambda_1/l_{n+1}) + \ldots + l_n(-\lambda_n/l_{n+1}). \end{aligned}$$

By assumption, the numbers $-\lambda_j/l_{n+1}$ ($j = 1, \ldots, n$) are integral linear combinations of numbers which are linearly independent over $\mathbf{Z}$. Hence, $\lambda_1, \ldots, \lambda_n, \lambda_{n+1}$ are also integral linear combinations of these numbers. □

We remark that the lemma is not valid with linear independence over $\mathbf{Z}$ replaced by rational independence: there are no rationally independent $\beta_1, \ldots, \beta_m$ such that $1 = k_1\beta_1 + \ldots + k_m\beta_m$ with $k_1, \ldots, k_m \in \mathbf{Z}$.

Here now is a first simple consequence of Kronecker's theorem.

Theorem 1.15. *Let $a \in AP$ and $\mu \in \mathcal{R}(a)$. If $\varepsilon > 0$, then each set (T, ∞) (and also each set $(-\infty, T)$) contains a point x such that $|a(x) - \mu| < \varepsilon$. In particular,*

$$\|a\|_\infty = \limsup_{x\to+\infty} |a(x)| = \limsup_{x\to-\infty} |a(x)|. \tag{1.9}$$

Proof. Pick an $x_0 \in \mathbf{R}$ such that $|a(x_0) - \mu| < \varepsilon/4$ and choose an almost periodic polynomial $p = \sum r_j e_{\lambda_j}$ such that $\|a - p\|_\infty < \varepsilon/4$. Since

$$\begin{aligned} |a(x) - \mu| &\le |a(x) - p(x)| + |p(x) - p(x_0)| + |p(x_0) - a(x_0)| + |a(x_0) - \mu| \\ &\le \varepsilon/4 + |p(x) - p(x_0)| + \varepsilon/4 + \varepsilon/4, \end{aligned}$$

it suffices to prove that there is an $x \in (T, \infty)$ such that $|p(x) - p(x_0)| < \varepsilon/4$.

By Lemma 1.14, there are $\beta_1, \dots, \beta_m$ which satisfy (1.8) and are linearly independent over $\mathbf{Z}$. We have

$$p(x) = \sum_{j=1}^{n} r_j e^{i\lambda_j x} = \sum_{j=1}^{n} r_j \left(e^{i\beta_1 x}\right)^{k_1^{(j)}} \dots \left(e^{i\beta_m x}\right)^{k_m^{(j)}},$$

and by Corollary 1.13, every open neighborhood of $\left(e^{i\beta_1 x_0}, \dots, e^{i\beta_m x_0}\right) \in \mathbf{T}^m$ contains a point $(e^{i\beta_1 x}, \dots, e^{i\beta_m x})$ with $x \in (T, \infty)$. Hence, we can find an x in (T, ∞) such that $|p(x) - p(x_0)| < \varepsilon/4$. Replacing $a(x)$ by $a(-x)$, we obtain the assertion for $(-\infty, T)$. □

We now return to convolution operators. Consider the C^*-algebra

$$\operatorname{alg}_{\mathcal{B}(L^2(\mathbf{R}))}\Big(\{U_\lambda : \lambda \in \mathbf{R}\},\ \{W^0(\hat{k}) : k \in L^1(\mathbf{R})\}\Big), \tag{1.10}$$

i.e., the smallest closed subalgebra of $\mathcal{B}(L^2(\mathbf{R}))$ which contains all shift operators U_λ and all convolution operators $W^0(\hat{k})$ with kernels in $L^1(\mathbf{R})$. Note that $\gamma I = \gamma U_0$, so that this algebra automatically contains the operators $W^0(\gamma + \hat{k})$ with $\gamma \in \mathbf{C}$ and $k \in L^1(\mathbf{R})$. By the definition of AP and by what was said in Section 1.2, the algebra (1.10) equals

$$W^0\Big(\operatorname{alg}_{L^\infty(\mathbf{R})}(AP, C(\dot{\mathbf{R}}))\Big) =: W^0\Big(\operatorname{alg}(AP, C(\dot{\mathbf{R}}))\Big). \tag{1.11}$$

Theorem 1.16. *The algebra* $\operatorname{alg}(AP, C(\dot{\mathbf{R}}))$ *decomposes into the direct sum*

$$\operatorname{alg}(AP, C(\dot{\mathbf{R}})) = AP \dot{+} C_0(\dot{\mathbf{R}}),$$

and the projection of $\operatorname{alg}(AP, C(\dot{\mathbf{R}}))$ *onto* AP *parallel to* $C_0(\dot{\mathbf{R}})$ *is a* C^**-algebra homomorphism.*

Proof. If $a, b \in AP$ and $f, g \in C_0(\dot{\mathbf{R}})$, then

$$(a + f)(b + g) = ab + ag + fb + fg, \tag{1.12}$$

and it is clear that $ab \in AP$ and $ag + fb + fg \in C_0(\dot{\mathbf{R}})$. Hence

$$\operatorname{alg}(AP, C(\dot{\mathbf{R}})) = \operatorname{clos}(AP + C_0(\dot{\mathbf{R}})). \tag{1.13}$$

We claim that $AP + C_0(\dot{\mathbf{R}})$ is closed. To see this, let $a_n \in AP$ and $f_n \in C_0(\dot{\mathbf{R}})$, and suppose $\{a_n + f_n\}$ is a Cauchy sequence in $L^\infty(\mathbf{R})$. From (1.9) we get

$$\begin{aligned} \|a_n - a_m\|_\infty &= \limsup_{x \to +\infty} |a_n(x) - a_m(x)| \\ &= \limsup_{x \to +\infty} |a_n(x) + f_n(x) - a_m(x) - f_m(x)| \le \|(a_n + f_n) - (a_m + f_m)\|_\infty, \end{aligned}$$

and hence $\{a_n\}$ itself is a Cauchy sequence. It follows that there is an $a \in AP$ such that $\|a - a_n\|_\infty \to 0$ as $n \to \infty$. This implies that $f_n = (a_n + f_n) - a_n$ also converges in $L^\infty(\mathbf{R})$ and thus has a limit $f \in C_0(\dot{\mathbf{R}})$. Consequently, $AP + C_0(\dot{\mathbf{R}})$ is closed. From (1.13) we therefore obtain

$$\operatorname{alg}\big(AP, C(\dot{\mathbf{R}})\big) = AP + C_0(\dot{\mathbf{R}}). \tag{1.14}$$

If $a \in AP$, then $\|a\|_\infty = \limsup_{x\to\infty} |a(x)|$ by (1.9), and the upper limit is zero if, at the same time, $a \in C_0(\dot{\mathbf{R}})$. Hence $AP \cap C_0(\dot{\mathbf{R}}) = \{0\}$, which shows that the sum in (1.14) is direct.

Finally, if $a \in AP$ and $f \in C_0(\dot{\mathbf{R}})$ then

$$\|a\|_\infty = \limsup_{x\to\infty} |a(x)| = \limsup_{x\to\infty} |a(x) + f(x)| \le \|a + f\|_\infty$$

again by virtue of (1.9). Therefore the projection Π of $AP \dot{+} C_0(\dot{\mathbf{R}})$ onto AP parallel to $C_0(\dot{\mathbf{R}})$ is bounded. From (1.12) and the obvious equality $\overline{\Pi a} = \Pi\bar{a}$, we see that Π is a C^*-algebra homomorphism. □

Combining Proposition 1.8 and Theorem 1.16 we arrive at the conclusion that the algebra (1.10) (= the algebra (1.11)) is $W^0(AP) \dot{+} W^0(C_0(\dot{\mathbf{R}}))$.

The images of two functions in $\operatorname{alg}\big(AP, C(\dot{\mathbf{R}})\big) = AP \dot{+} C_0(\dot{\mathbf{R}})$ are shown in Figure 1.4

Figure 1.4: Addition of $i/(x+i)$ (left) and $1/(1+x^2)$ (right) to almost periodic functions.

1.5 Semi-Almost Periodic Symbols

Many interesting convolutions have piecewise continuous symbols.

Example 1.17. The *Cauchy singular integral operator* S on $L^2(\mathbf{R})$ is formally given by

$$(Sf)(t) := \frac{1}{\pi i} \int_{\mathbf{R}} \frac{f(s)}{s-t}\, ds, \quad t \in \mathbf{R},$$

the integral understood in the principal value sense:

$$\int_{\mathbf{R}} \frac{f(s)}{s-t}\, ds := \lim_{\varepsilon \to 0} \int_{|s-t| \geq \varepsilon} \frac{f(s)}{s-t}\, ds.$$

Classical results say that if $f \in L^2(\mathbf{R})$, then Sf exists almost everywhere on $\mathbf{R}$ and that the map $f \mapsto Sf$ is a bounded operator on $L^2(\mathbf{R})$.

The convolution kernel of S is $k(t) = -1/(\pi i t)$. For $0 < \varepsilon < \delta$, we have

$$\int_{(-\delta,\delta)\setminus(-\varepsilon,\varepsilon)} k(t) e^{ixt}\, dt = -\frac{1}{\pi i}\left(\int_{-\delta}^{-\varepsilon} \frac{e^{ixt}}{t}\, dt + \int_{\varepsilon}^{\delta} \frac{e^{ixt}}{t}\, dt \right)$$

$$= -\frac{1}{\pi i} \int_{\varepsilon}^{\delta} \frac{e^{ixt} - e^{-ixt}}{t}\, dt = -\frac{2}{\pi} \int_{\varepsilon}^{\delta} \frac{\sin(xt)}{t}\, dt,$$

and as $\varepsilon \to 0$ and $\delta \to \infty$, this converges to

$$-\frac{2}{\pi} \int_0^{\infty} \frac{\sin(xt)}{t}\, dt = -\frac{2}{\pi} \operatorname{sgn} x \int_0^{\infty} \frac{\sin t}{t}\, dt = -\operatorname{sgn} x$$

where $\operatorname{sgn} x$ is the piecewise continuous function

$$\operatorname{sgn} x := \begin{cases} -1 & \text{if} \quad x < 0 \\ 0 & \text{if} \quad x = 0 \\ 1 & \text{if} \quad x > 0 \end{cases}.$$

This implies that $S = W^0(-\operatorname{sgn} x)$; notice that here and in similar situations $-\operatorname{sgn} x$ stands for the function $x \mapsto -\operatorname{sgn} x$ and not for the value of this function at the point x. □

Example 1.18. Let T be convolution by $k(t) = \sin t/(\pi t)$,

$$(Tf)(t) := \frac{1}{\pi} \int_{\mathbf{R}} \frac{\sin(t-s)}{t-s} f(s)\, ds, \quad t \in \mathbf{R}.$$

We have $\sin t/(\pi t) = (e^{it} - e^{-it})/(2\pi i t)$, and after defining S_α ($\alpha \in \mathbf{R}$) on $L^2(\mathbf{R})$ by

$$(S_\alpha f)(t) := \frac{1}{\pi i} \int_{\mathbf{R}} \frac{e^{i\alpha(s-t)}}{s-t} f(s)\, ds, \quad t \in \mathbf{R},$$

we can therefore write $T = (S_{-1} - S_1)/2$. From Example 1.17 we infer that

$$S_\alpha = e_{-\alpha} S e_\alpha I = e_{-\alpha} W^0(-\operatorname{sgn} x) e_\alpha I = W^0\big(-\operatorname{sgn}(x-\alpha)\big)$$

Consequently,

$$T = W^0\Big(-\frac{1}{2}\operatorname{sgn}(x+1) + \frac{1}{2}\operatorname{sgn}(x-1)\Big) = W^0(-\chi_{(-1,1)}),$$

where $\chi_{(-1,1)}$ is the characteristic function of the interval $(-1,1)$. □

Example 1.19. Consider convolution by the kernel

$$k(t) = \frac{1}{\pi i}\frac{1}{e^{t/2} - e^{-t/2}} = \frac{1}{2\pi i}\frac{1}{\sinh(t/2)},$$

that is, let

$$(Rf)(t) := \frac{1}{2\pi i}\int_{\mathbf{R}} \frac{f(s)}{\sinh((t-s)/2)}\, ds, \quad t \in \mathbf{R},$$

the integral again understood in the Cauchy principal value sense. Although this is not obvious at the present moment, we remark that the operator R plays a key role in the theory of Wiener-Hopf operators with piecewise continuous symbols (see Example 1.24). Clearly,

$$\begin{aligned}\int_{(-\delta,\delta)\setminus(-\varepsilon,\varepsilon)} k(t)e^{ixt}\, dt &= \frac{1}{2\pi i}\Big(\int_{-\delta}^{-\varepsilon}\frac{e^{ixt}}{\sinh(t/2)}\, dt + \int_{\varepsilon}^{\delta}\frac{e^{ixt}}{\sinh(t/2)}\, dt\Big)\\ &= \frac{1}{2\pi i}\int_{\varepsilon}^{\delta}\frac{e^{ixt} - e^{-ixt}}{\sinh(t/2)}\, dt = \frac{1}{\pi}\int_{\varepsilon}^{\delta}\frac{\sin(xt)}{\sinh(t/2)}\, dt,\end{aligned}$$

and as $\varepsilon \to 0$ and $\delta \to \infty$, this has the limit

$$\frac{1}{\pi}\int_0^\infty \frac{\sin(xt)}{\sinh(t/2)}\, dt = \tanh(\pi x) = \frac{e^{2\pi x} - 1}{e^{2\pi x} + 1}.$$

It results that $R = W^0(\tanh(\pi x))$. Note that $\tanh(\pi x)$ is continuous on $\mathbf{R}$, but that

$$\lim_{x\to-\infty}\tanh(\pi x) = -1, \qquad \lim_{x\to+\infty}\tanh(\pi x) = +1,$$

which means that $\tanh(\pi x)$ has a jump at infinity. □

From Proposition 1.8 we get the spectra of the operators introduced in the preceding three examples:

$$\operatorname{sp} S = \{-1,1\}, \quad \operatorname{sp} T = \{-1,0\}, \quad \operatorname{sp} R = [-1,1].$$

We denote by $PC := PC(\dot{\mathbf{R}})$ the C^*-algebra of all bounded piecewise continuous functions on $\dot{\mathbf{R}}$. By definition, $a \in PC$ if and only if $a \in L^\infty(\mathbf{R})$ and the one-sided limits

$$a(x_0 - 0) := \lim_{x \to x_0 - 0} a(x), \quad a(x_0 + 0) := \lim_{x \to x_0 + 0} a(x)$$

exist for each $x_0 \in \dot{\mathbf{R}} := \mathbf{R} \cup \{\infty\}$; by convention,

$$\begin{aligned} a(\infty - 0) &:= a(+\infty) := \lim_{x \to +\infty} a(x), \\ a(\infty + 0) &:= a(-\infty) := \lim_{x \to -\infty} a(x). \end{aligned}$$

We remark that the set of all piecewise constant functions with at most finitely many jumps is dense in PC. Also notice that a function $a \in PC$ has at most countably many jumps and that for each $\delta > 0$ the set

$$\{x \in \dot{\mathbf{R}} : |a(x+0) - a(x-0)| > \delta\}$$

is finite. Finally, we put

$$C(\overline{\mathbf{R}}) := C(\mathbf{R}) \cap PC, \quad PC_0 := \{a \in PC : a(\pm\infty) = 0\}.$$

Clearly, $C(\overline{\mathbf{R}})$ and PC_0 are C^*-subalgebras of PC.
The symbol of the operator R considered in Example 1.19 belongs to $C(\overline{\mathbf{R}})$. With S_α as in Example 1.18, we have

$$\begin{aligned} &\mathrm{alg}_{\mathcal{B}(L^2(\mathbf{R}))}\{S_\alpha : \alpha \in \mathbf{R}\} \\ &= \mathrm{alg}_{\mathcal{B}(L^2(\mathbf{R}))}\Big(\{S_\alpha : \alpha \in \mathbf{R}\} \cup \{W^0(\hat{k}) : k \in L^1(\mathbf{R})\}\Big) \\ &= W^0(PC). \end{aligned} \tag{1.15}$$

Adding the shift operators U_λ to the latter algebra we arrive at convolution operators with semi-almost periodic symbols.

Definition 1.20. The C^*-algebra SAP of all *semi-almost periodic functions* on $\mathbf{R}$ is defined as the smallest closed subalgebra of $L^\infty(\mathbf{R})$ that contains AP and $C(\overline{\mathbf{R}})$:

$$SAP = \mathrm{alg}_{L^\infty(\mathbf{R})}\big(AP, C(\overline{\mathbf{R}})\big). \qquad \square$$

The following theorem reveals the structure of functions in SAP.

Theorem 1.21 (Sarason). *Let $u \in C(\overline{\mathbf{R}})$ be any function for which $u(-\infty) = 0$ and $u(+\infty) = 1$. If $a \in SAP$, then there exist $a_l, a_r \in AP$ and $a_0 \in C_0(\dot{\mathbf{R}})$ such that*

$$a = (1-u)a_l + ua_r + a_0. \tag{1.16}$$

The functions a_l, a_r are uniquely determined by a and independent of the particular choice of u. The maps

$$a \mapsto a_l, \ a \mapsto a_r \tag{1.17}$$

are C^*-algebra homomorphisms of SAP onto AP.

Proof. If $a \in AP$, then $a = (1-u)a + ua$ is a representation of a in the form (1.16). For $f \in C(\overline{\mathbf{R}})$, the function $f_0 = f - (1-u)f(-\infty) - uf(+\infty)$ belongs to $C_0(\dot{\mathbf{R}})$. Thus,

$$f = (1-u)f(-\infty) + uf(+\infty) + f_0$$

is a representation of f in the desired form. Let $a_l, a_r, b_l, b_r \in AP$ and $a_0, b_0 \in C_0(\dot{\mathbf{R}})$. We have

$$\begin{aligned} &\big((1-u)a_l + ua_r + a_0\big)\big((1-u)b_l + ub_r + b_0\big) \\ &= (1-u)^2 a_l b_l + u^2 a_r b_r + c_0 \end{aligned} \tag{1.18}$$

with some $c_0 \in C_0(\dot{\mathbf{R}})$, and as $(1-u)^2 - (1-u)$ and $u^2 - u$ are in $C_0(\dot{\mathbf{R}})$, it follows that (1.18) is of the form

$$(1-u)a_l b_l + ua_r b_r + d_0 \tag{1.19}$$

with $d_0 \in C_0(\dot{\mathbf{R}})$. Hence, $SAP = \text{alg}\,(AP, C(\overline{\mathbf{R}}))$ is the closure of the set

$$\big\{(1-u)a_l + ua_r + a_0 : a_l, a_r \in AP, \ a_0 \in C_0(\dot{\mathbf{R}})\big\}. \tag{1.20}$$

For $a_l, a_r \in AP$ and $a_0 \in C_0(\dot{\mathbf{R}})$, we deduce from (1.9) that

$$\begin{aligned} \|a_l\|_\infty &= \limsup_{x \to -\infty} |a_l(x)| \\ &= \limsup_{x \to -\infty} \big|(1-u(x))a_l(x) + u(x)a_r(x) + a_0(x)\big| \\ &\leq \big\|(1-u)a_l + ua_r + a_0\big\|_\infty \end{aligned} \tag{1.21}$$

and, analogously,

$$\|a_r\|_\infty \leq \big\|(1-u)a_l + ua_r + a_0\big\|_\infty. \tag{1.22}$$

Consequently, if $\{(1-u)a_l^{(n)} + ua_r^{(n)} + a_0^{(n)}\}$ is a Cauchy sequence in the set (1.20), then $\{a_l^{(n)}\}$ and $\{a_r^{(n)}\}$ are Cauchy sequences in AP and $\{a_0^{(n)}\}$ is a Cauchy sequence in $C_0(\dot{\mathbf{R}})$. This shows that the set (1.20) is closed and therefore coincides with SAP. In particular, every function in SAP is of the form (1.16).

From (1.21) and (1.22) we infer that the representation (1.16) is unique. Combining (1.18), (1.19), (1.21), (1.22) it is easily seen that the maps (1.17) are C^*-algebra homomorphisms. □

For $a \in SAP$, the functions $a_l \in AP$ and $a_r \in AP$ given by (1.16) are referred to as the *almost periodic representatives* of a at $-\infty$ and $+\infty$, respectively. The subscripts l and r stand for "left" $(-\infty)$ and "right" $(+\infty)$. Figures 1.5 and 1.6 indicate the behavior of two functions in SAP. The pictures on the first pages of the chapters are also the images of SAP functions.

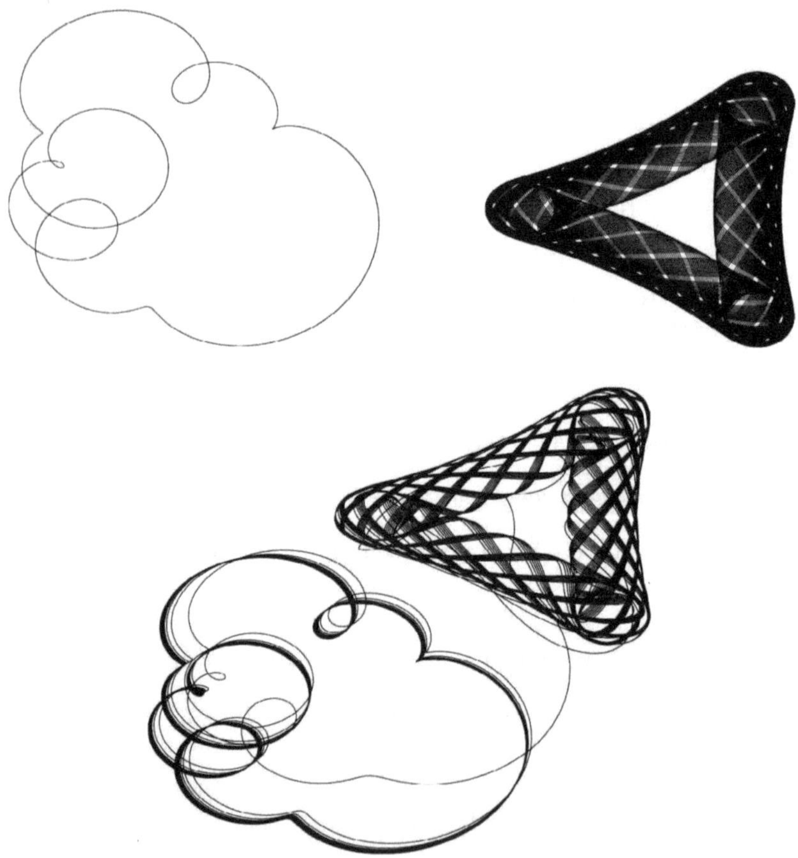

Figure 1.5: The range of $a_l(x) = 10\,e^{ix} + (6+i)e^{2ix} + 4ie^{4ix} - 3\,e^{8ix}$ for x between -900 and -300 (upper left), of $a_r(x) = 22 + 22i - 4\,e^{-2ix} + 10\,e^{ix} + 3i\,e^{\pi ix}$ for x between 300 and 900 (upper right), and of $a(x) = (1-u(x))a_l(x) + u(x)a_r(x)$ for x between -300 and 300 (bottom); here $u(x) = 1/2 + (1/\pi)\arctan x$.

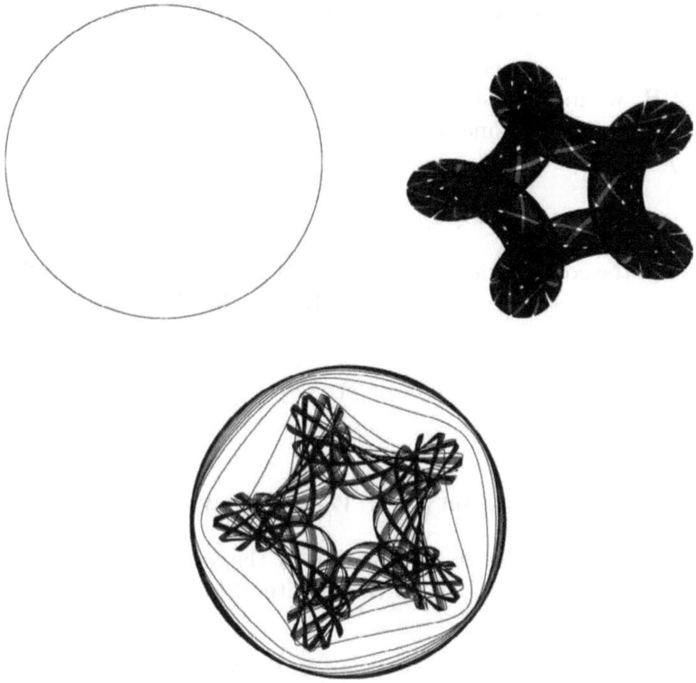

Figure 1.6: The range of $a_l(x) = 20\, e^{ix}$ for x between -900 and -300 (upper left), of $a_r(x) = -4\, e^{-4ix} + 10\, e^{ix} + 3i\, e^{\pi ix}$ for x between 300 and 900 (upper right), and of $a(x) = (1 - u(x))a_l(x) + u(x)a_r(x)$ for x between -300 and 300 (bottom); here $u(x) = 1/2 + (1/\pi)\arctan(x/5)$.

Taking into account (1.15), we obtain

$$\begin{aligned}&\operatorname{alg}\Big(\{U_\lambda : \lambda \in \mathbf{R}\},\ \{S_\alpha : \alpha \in \mathbf{R}\},\ \{W^0(\hat{k}) : k \in L^1(\mathbf{R})\}\Big)\\ &= W^0\big(\operatorname{alg}(AP, PC)\big).\end{aligned}$$

Since $PC = C(\overline{\mathbf{R}}) + PC_0$ and $AP \cdot PC_0 \subset PC_0$, we have

$$\operatorname{alg}(AP, PC) = SAP + PC_0. \tag{1.23}$$

Note that $SAP \cap PC_0 = C_0(\dot{\mathbf{R}})$, so that the sum in (1.23) is not direct. A decomposition of the algebra (1.23) into a direct sum is

$$SAP + PC_0 = (1-u)AP \dotplus uAP \dotplus PC_0,$$

but we remark that neither $(1-u)AP$ nor uAP is an algebra.

1.6 Convolution Operators on a Half-Line

Let $\mathbf{R}_+ := (0, \infty)$ be the positive half-line. On extending functions in $L^2(\mathbf{R}_+)$ by zero to all of $\mathbf{R}$, we may consider $L^2(\mathbf{R}_+)$ as a subspace of $L^2(\mathbf{R})$. We denote by χ_+ both the characteristic function of $\mathbf{R}_+$ and the canonical projection of $L^2(\mathbf{R})$ onto $L^2(\mathbf{R}_+)$:

$$\chi_+ : L^2(\mathbf{R}) \to L^2(\mathbf{R}_+), \; f \mapsto \chi_+ f.$$

For $a \in L^\infty(\mathbf{R})$, the compression of the convolution operator $W^0(a) \in \mathcal{B}(L^2(\mathbf{R}))$ to $L^2(\mathbf{R}_+)$ is denoted by $W(a)$ and called the *Wiener-Hopf operator with the symbol* a. Thus,

$$W(a) : L^2(\mathbf{R}_+) \to L^2(\mathbf{R}_+), \; f \mapsto \chi_+ W^0(a) f.$$

Frequently one writes

$$W(a) = \chi_+ W^0(a) | \operatorname{Im} \chi_+ \quad \text{or} \quad W(a) = \chi_+ W^0(a) \chi_+ | \operatorname{Im} \chi_+.$$

We remark that by a simple transformation of the variables convolution operators on an arbitrary half-line can be reduced to Wiener-Hopf operators, i.e., to convolution operators on $\mathbf{R}_+$.

Example 1.22. If $a = \gamma + \hat{k}$ with $\gamma \in \mathbf{C}$ and $k \in L^1(\mathbf{R})$, then $W(a)$ acts on $L^2(\mathbf{R}_+)$ by the rule

$$(W(a)f)(t) = \gamma f(t) + \int_{\mathbf{R}_+} k(t-s) f(s)\, ds, \quad t \in \mathbf{R}_+. \qquad \square$$

Example 1.23. Let $e_\lambda(x) = e^{i\lambda x}$ and let $U_\lambda = W^0(e_\lambda)$ be the shift operator introduced in Example 1.10. For the operator $V_\lambda := \chi_+ U_\lambda \chi_+ | \operatorname{Im} \chi_+ = W(e_\lambda)$ we have

$$(V_\lambda f)(t) = f(t - \lambda) \quad \text{for all } t \in \mathbf{R}_+ \qquad \text{if } \lambda < 0,$$

$$(V_\lambda f)(t) = \begin{cases} 0 & \text{for } t \in (0, \lambda) \\ f(t-\lambda) & \text{for } t \in (\lambda, \infty) \end{cases} \qquad \text{if } \lambda > 0.$$

Clearly, $V_0 = I$ is the identity operator. $\square$

The map $W : L^\infty(\mathbf{R}) \to \mathcal{B}(L^2(\mathbf{R}_+))$, $a \mapsto W(a)$ is obviously linear and continuous. One can show that W is even an isometry, i.e.,

$$\|W(a)\| = \|a\|_\infty \quad \text{for all } a \in L^\infty(\mathbf{R}).$$

However, unlike the map W^0, the map W is not an algebra homomorphism: for example, $W(e_1)W(e_{-1}) = V_1 V_{-1}$ is multiplication by the characteristic function of $(1,\infty)$ and is therefore not equal to $W(e_1 e_{-1}) = I$. This lack of multiplicativity makes the spectral theory of Wiener-Hopf operators less trivial but much more interesting than the spectral theory of convolution operators on the whole line. Later we will establish invertibility criteria for Wiener-Hopf operators with symbols in $SAP + PC_0$. At the present moment we confine ourselves to an (important) example in which the spectrum can be determined due to some happy circumstance.

Example 1.24. Let S_+ be the Cauchy singular integral operator on $L^2(\mathbf{R}_+)$:

$$(S_+ f)(t) := \frac{1}{\pi i}\int_{\mathbf{R}_+} \frac{f(s)}{s-t}\,ds, \quad t \in \mathbf{R}_+,$$

the integral again understood in the principal value sense. From Example 1.17 we infer that $S_+ = W(-\operatorname{sgn} x)$. It can be verified straightforwardly that the map

$$\Phi : L^2(\mathbf{R}_+) \to L^2(\mathbf{R}),\ (\Phi f)(t) := e^{-t/2} f(e^{-t})$$

is an isometric isomorphism. We have

$$(\Phi^{-1} g)(s) = s^{-1/2} g(-\log s),\ s > 0,$$

and hence

$$(\Phi S_+ \Phi^{-1} f)(t) = \frac{1}{\pi i}\int_{\mathbf{R}_+} \frac{e^{-t/2} s^{-1/2} f(-\log s)}{s - e^{-t}}\,ds$$

$$= \frac{1}{\pi i}\int_{\mathbf{R}} \frac{e^{-t/2} e^{y/2} f(y)}{e^{-y} - e^{-t}} e^{-y}\,dy = \frac{1}{2\pi i}\int_{\mathbf{R}} \frac{f(y)}{\sinh((t-y)/2)}\,dy.$$

Thus, we arrive at the convolution operator studied in Example 1.19:

$$\Phi S_+ \Phi^{-1} = W^0\big(\tanh(\pi x)\big).$$

Consequently, $\operatorname{sp} S_+ = \operatorname{sp} W(-\operatorname{sgn} x) = [-1,1]$. □

1.7 Convolution Operators on Finite Intervals

The purpose of this section is to show that the theory of convolution operators on finite intervals is equivalent to the theory of Wiener-Hopf operators with 2×2-matrix symbols.

Let $\lambda \in (0,\infty)$ and let $\chi_{(0,\lambda)}$ be the characteristic function of the interval $(0,\lambda)$. We identify $L^2(0,\lambda)$ with $\chi_{(0,\lambda)} L^2(\mathbf{R}_+)$ and consider $L^2(0,\lambda)$ as a subspace of $L^2(\mathbf{R}_+)$. For $a \in L^\infty(\mathbf{R})$, the *(finite) convolution operator* $W_\lambda(a)$ is defined by

$$W_\lambda(a) : L^2(0,\lambda) \to L^2(0,\lambda),\ f \mapsto \chi_{(0,\lambda)} W(a) f.$$

Thus, $W_\lambda(a)$ is the compression of the Wiener-Hopf operator $W(a)$ to $L^2(0,\lambda)$. Of course, we may also think of $W_\lambda(a)$ as the compression of $W^0(a)$ to $L^2(0,\lambda)$. If $I \subset \mathbf{R}$ is any interval of length λ, then the compression of $W^0(a)$ to $L^2(I)$ can be transformed to $W_\lambda(a)$ by means of a simple change of variables.

Given a measurable subset E of $\mathbf{R}$, we denote by $L^2_N(E)$ the direct sum of N copies of $L^2(E)$. Let $L^\infty_{N\times N}(\mathbf{R})$ stand for the algebra of all matrix functions $a : \mathbf{R} \to \mathbf{C}^{N\times N}$ with entries in $L^\infty(\mathbf{R})$. If $\mathcal{A}$ is a subalgebra of $L^\infty(\mathbf{R})$, we let $\mathcal{A}_{N\times N}$ or $[\mathcal{A}]_{N\times N}$ denote the matrix functions $a : \mathbf{R} \to \mathbf{C}^{N\times N}$ whose entries belong to $\mathcal{A}$.

For $a = (a_{jk})_{j,k=1}^N \in L^\infty_{N\times N}(\mathbf{R})$, the multiplication operator aI, the convolution operator $W^0(a)$, the Wiener-Hopf operator $W(a)$, and the finite convolution operator $W_\lambda(a)$ are defined in the natural manner:

$$\begin{aligned}
aI \ :& \quad L^2_N(\mathbf{R}) \to L^2_N(\mathbf{R}), && (f_j)_{j=1}^N \mapsto \Big(\textstyle\sum_{k=1}^N a_{jk} f_k\Big)_{j=1}^N,\\
W^0(a) :& \quad L^2_N(\mathbf{R}) \to L^2_N(\mathbf{R}), && (f_j)_{j=1}^N \mapsto \Big(\textstyle\sum_{k=1}^N W^0(a_{jk}) f_k\Big)_{j=1}^N,\\
W(a) :& \quad L^2_N(\mathbf{R}_+) \to L^2_N(\mathbf{R}_+), && (f_j)_{j=1}^N \mapsto \Big(\textstyle\sum_{k=1}^N W(a_{jk}) f_k\Big)_{j=1}^N,\\
W_\lambda(a) :& \quad L^2_N(0,\lambda) \to L^2_N(0,\lambda); && (f_j)_{j=1}^N \mapsto \Big(\textstyle\sum_{k=1}^N W_\lambda(a_{jk}) f_k\Big)_{j=1}^N.
\end{aligned}$$

In other terms, we think of $L^2_N(E)$ as a space of columns and of the operator as an $N \times N$ operator matrix acting on this column space.

Let X be a Banach space and $A \in \mathcal{B}(X)$. The *kernel* $\operatorname{Ker} A$ and the *image* (= *range*) $\operatorname{Im} A$ are defined by

$$\operatorname{Ker} A := \{x \in X : Ax = 0\}, \quad \operatorname{Im} A := \{Ax : x \in X\}.$$

Both $\operatorname{Ker} A$ and $\operatorname{Im} A$ are linear subspaces of X, and $\operatorname{Ker} A$ is always closed. The operator A is said to be *normally solvable* if $\operatorname{Im} A$ is a closed subspace of X. If A is normally solvable, then the *cokernel* $\operatorname{Coker} A$ is defined as

$$\operatorname{Coker} A := X/\operatorname{Im} A.$$

For a normally solvable operator A, we put

$$n(A) := \dim \operatorname{Ker} A, \quad d(A) := \dim \operatorname{Coker} A. \tag{1.24}$$

Notice that $n(A)$ and $d(A)$ lie in $\{0, 1, 2, \ldots\} \cup \{\infty\}$. We also remark that

$$n(A) = d(A^*), \quad d(A) = n(A^*)$$

where $A^* \in \mathcal{B}(X^*)$ is the adjoint operator of A. Two operators $A \in \mathcal{B}(X)$ and $B \in \mathcal{B}(Y)$ are said to be *strongly Φ-equivalent* if either none of them is normally solvable or both A and B are normally solvable and $n(A) = n(B)$, $d(A) = d(B)$.

Here is the result of this section.

Theorem 1.25. *If $a \in L^\infty_{N\times N}(\mathbf{R})$ and $\lambda \in (0,\infty)$, then $W_\lambda(a) \in \mathcal{B}(L^2_N(0,\lambda))$ is strongly Φ-equivalent to the operator $W(H_a) \in \mathcal{B}(L^2_{2N}(\mathbf{R}_+))$ where H_a is the matrix function in $L^\infty_{2N\times 2N}(\mathbf{R})$ given by*

$$H_a(x) = \begin{pmatrix} e^{-i\lambda x}I & 0 \\ a(x) & e^{i\lambda x}I \end{pmatrix}, \quad x \in \mathbf{R},$$

I standing for the $N \times N$ identity matrix.

Proof. Define P_λ on $L^2_N(\mathbf{R}_+)$ by $P_\lambda := \chi_{(0,\lambda)}I$, put $Q_\lambda := I - P_\lambda$, and let $V_\lambda = W(e_\lambda I)$, $V_{-\lambda} = W(e_{-\lambda}I)$. It can be checked directly that $Q_\lambda = V_\lambda V_{-\lambda}$. The operator $W_\lambda(a) = P_\lambda W(a) P_\lambda | \mathrm{Im} P_\lambda$ is obviously strongly Φ-equivalent to the operator $P_\lambda W(a) P_\lambda + Q_\lambda$. Since

$$P_\lambda W(a) P_\lambda + Q_\lambda = \big(W(a)P_\lambda + Q_\lambda\big)\big(I - Q_\lambda W(a) P_\lambda\big)$$

and $I - Q_\lambda W(a) P_\lambda$ is invertible, the inverse being $I + Q_\lambda W(a) P_\lambda$, it follows that $W_\lambda(a)$ is strongly Φ-equivalent to

$$B_\lambda := W(a)P_\lambda + Q_\lambda = W(a) - \big(W(a) - I\big)V_\lambda V_{-\lambda}.$$

The identity

$$\begin{pmatrix} I & C \\ B & A \end{pmatrix} = \begin{pmatrix} I & 0 \\ B & I \end{pmatrix}\begin{pmatrix} I & 0 \\ 0 & A - BC \end{pmatrix}\begin{pmatrix} I & C \\ 0 & I \end{pmatrix}$$

reveals that $A - BC$ is strongly Φ-equivalent to $\begin{pmatrix} I & C \\ B & A \end{pmatrix}$. Consequently, B_λ is strongly Φ-equivalent to

$$\begin{pmatrix} I & V_{-\lambda} \\ (W(a) - I)V_\lambda & W(a) \end{pmatrix},$$

and as the latter matrix equals

$$\begin{pmatrix} V_{-\lambda} & 0 \\ W(a) & V_\lambda \end{pmatrix}\begin{pmatrix} V_\lambda & I \\ -I & 0 \end{pmatrix} = W(H_a)\begin{pmatrix} V_\lambda & I \\ -I & 0 \end{pmatrix},$$

we arrive at the conclusion that $W_\lambda(a)$ is strongly Φ-equivalent to $W(H_a)$. □

Theorem 1.25 (for $N = 1$) shows that the consideration of convolution operators on finite intervals with symbols in $C(\dot{\mathbf{R}})$ automatically leads to Wiener-Hopf operators with symbols in

$$\big[AP + C(\dot{\mathbf{R}})\big]_{2\times 2} = \big[AP \dot{+} C_0(\dot{\mathbf{R}})\big]_{2\times 2},$$

while finite convolution operators with symbols in PC result in Wiener-Hopf operators with symbols in

$$\left[\text{alg}\,(PC, AP)\right]_{2\times 2} = [SAP + PC_0]_{2\times 2}.$$

Notes

The results on Banach and C^*-algebras we have cited are all in the standard texts on the topic, for example, in [62], [64], [66], [93], [103], [160]. Treatises on almost periodic functions include the books [60], [153], [154], [171]. The text [109] by Hlawka, Schoissengaier, and Taschner contains a delightful chapter on Kronecker's theorem. The proof of Kronecker's theorem given here is from this text.

The pioneering work on Wiener-Hopf equations goes back to Wiener and Hopf [231]. Widom's survey article [229] is a beautiful introduction to the basic ideas of Wiener and Hopf. It was Krein and Gohberg [87], [94], [143] who realized the Banach algebraic background behind Wiener-Hopf operators and Wiener-Hopf factorization and thus laid the foundation of the present-day edifice of Wiener-Hopf theory.

Convolution operators with symbols in $AP + C_0(\dot{\mathbf{R}})$ were first studied by Gohberg and Feldman [90], [91], [92] and Coburn and Douglas [59]. The problem of developing a theory that comprises both piecewise continuous and almost periodic symbols was suggested to Donald Sarason by Israel Gohberg and led to the creation of semi-almost periodic functions. Sarason [197] worked out the corresponding Fredholm theory of Wiener-Hopf operators and, among other things, he established Theorem 1.21.

That convolution operators on an interval are equivalent to Wiener-Hopf operators with $\mathbf{C}^{2\times 2}$-valued symbols has been known for a long time. This idea probably first appeared in the work by Ganin [81]. See also Novokshenov [166] and Paltsev [170]. In the latter paper, Paltsev writes: "However, the matrix coefficient of this problem, though it is smooth and nonsingular on the real axis, does not have a limit as $x \to \pm\infty$ because it contains oscillating elements of the form $\exp(\pm i2Tx)$; and therefore the theory of the Riemann-Hilbert problem, as developed at present, is not immediately applicable to this problem".

Theorem 1.25 can be generalized to convolution type equations, or systems of such equations, over the union of several intervals. The resulting matrix functions are then of higher dimensions, and their non-zero (block) entries are all located on the main diagonal and one additional row. This transformation was done in a direct way in [133], [211] and was discussed from the point of view of "equivalence after extension" in [55]. Furthermore, in [21] the abstract approach of [147] was extended to a setting embracing the case of several intervals.

Another procedure for passing from finite intervals to the half-line can be found in [77]. It also makes use of [76], [147], but it leads to non-triangular block matrix symbols of the form

$$\begin{pmatrix} -e_{-\lambda}\hat{k} & -I+\hat{k} \\ I+\hat{k} & -e_{\lambda}\hat{k} \end{pmatrix}.$$

The elegant proof of Theorem 1.25 given here is due to Krupnik and Feldman [147] (but see also Lemma 4.1 of [76]). In fact, Theorem 1.25 shows that the operators $W_\lambda(a)$ and $W(H_a)$ are matricially coupled in the sense of Bart-Gohberg-Kaashoek [14] and Bart-Tsekanovskii [15].

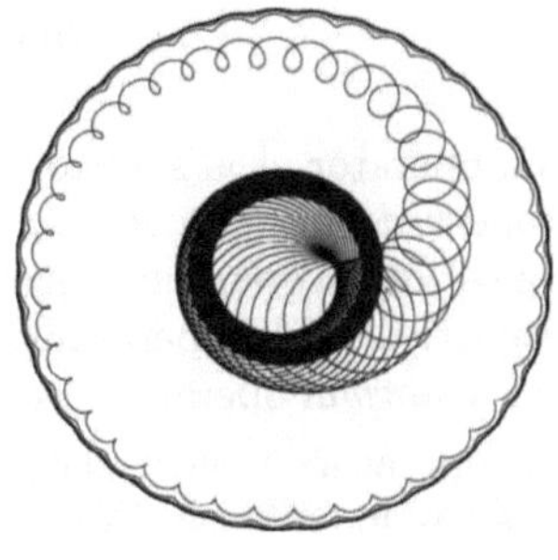

Chapter 2

Introduction to Scalar Wiener-Hopf Operators

Wiener-Hopf operators are convolution operators on a half-line. There is a fascinating interplay between the operator theoretic properties of Wiener-Hopf equations and the topological characteristics of their symbol. This connection is visually perfect for continuous and piecewise continuous symbols. Symbols in $C+H^\infty$*, which will repeatedly emerge in the forthcoming chapters when studying semi-almost periodic symbols, require some more machinery. After proving Bohr's theorem, the notion of the mean motion of an invertible almost periodic function is available. Having the concept of the mean motion, we can establish Fredholm and invertibility criteria for Wiener-Hopf operators with almost periodic symbols. In summary, this chapter contains the Fredholm theory of scalar Wiener-Hopf operators with symbols in* $C+H^\infty$*,* PC*, and* AP*.*

Operators with symbols in $C+H^\infty$ *or* PC *are "typically Fredholm", while those with symbols in* AP *are "typically semi-Fredholm". Thus, we begin with an outline of semi-Fredholm operators.*

2.1 Fredholm and Semi-Fredholm Operators

Let X be a Banach space and let $A \in \mathcal{B}(X)$. Suppose that A is *normally solvable*, which means that the image (= range) $\operatorname{Im} A$ is a closed subset of X. The operator A is called *Fredholm* if $n(A)$ and $d(A)$ defined by (1.24) are finite, and in that case the *index* of A is defined as

$$\operatorname{Ind} A := n(A) - d(A).$$

The operator A is said to be *properly n-normal* if $n(A) < \infty$ and $d(A) = \infty$ and *properly d-normal* if $d(A) < \infty$ and $n(A) = \infty$. We denote by $\Phi(X)$, $\Phi_n(X)$, $\Phi_d(X)$ the set of all Fredholm, properly n-normal, properly d-normal operators in $\mathcal{B}(X)$, respectively. The operators in $\Phi(X) \cup \Phi_n(X)$ and $\Phi(X) \cup \Phi_d(X)$ are called *n-normal* and *d-normal* operators, respectively.

The operators in $\Phi(X) \cup \Phi_n(X) \cup \Phi_d(X)$ are referred to as *semi-Fredholm operators*, those in $\Phi_n(X) \cup \Phi_d(X)$ are called *properly semi-Fredholm*. We put

$$\operatorname{Ind} A = \begin{cases} -\infty & \text{for} \quad A \in \Phi_n(X), \\ +\infty & \text{for} \quad A \in \Phi_d(X). \end{cases}$$

We now record a few well-known properties of semi-Fredholm operators. Let $\mathcal{K}(X)$ denote the collection of all compact operators on X. Notice that $\mathcal{K}(X)$ is a closed two-sided ideal of $\mathcal{B}(X)$. One can show that $A \in \Phi(X)$ if and only if the coset $A + \mathcal{K}(X)$ is invertible in the quotient algebra $\mathcal{B}(X)/\mathcal{K}(X)$. If $B + \mathcal{K}(X)$ is the inverse of $A + \mathcal{K}(X)$, then B is called a *regularizer* of A. We define the *essential spectrum* $\mathrm{sp}_{\mathrm{ess}} A$ of $A \in \mathcal{B}(X)$ as the spectrum of $A + \mathcal{K}(X)$ in $\mathcal{B}(X)/\mathcal{K}(X)$, i.e.,

$$\mathrm{sp}_{\mathrm{ess}} A := \{\lambda \in \mathbf{C} : A - \lambda I \ \text{ is not Fredholm}\}.$$

Clearly, $\mathrm{sp}_{\mathrm{ess}}\, A$ is a compact nonempty subset of $\mathrm{sp}\, A$.

An operator $A \in \mathcal{B}(X)$ is said to be *left-Fredholm* (resp. *right-Fredholm*) if there is a $B \in \mathcal{B}(X)$ such that $BA - I$ (resp. $AB - I$) is compact. One can show that A is left-Fredholm if and only if $A \in \Phi(X) \cup \Phi_n(X)$ and $\operatorname{Im} A$ is a complementable subspace of X. Analogously, A is right-Fredholm if and only if $A \in \Phi(X) \cup \Phi_d(X)$ and $\operatorname{Ker} A$ is a complementable subspace of X.

The operator A is called *left-invertible* (resp. *right-invertible*) if there exists a $B \in \mathcal{B}(X)$ such that $BA = I$ (resp. $AB = I$). It can be shown that A is left-invertible if and only if A is injective, normally solvable, and $\operatorname{Im} A$ is a complementable subspace of X. In the same vein, A is right-invertible if and only if A is surjective and the (always closed) space $\operatorname{Ker} A$ is complementable in X.

Since every closed subspace of a Hilbert space is complementable (for example, by its orthogonal complement), things are simpler for Hilbert space operators. Let H be a Hilbert space and $A \in \mathcal{B}(H)$. We then have

$$\begin{aligned} A \in \Phi(H) \cup \Phi_n(H) &\Longleftrightarrow A \ \text{ is left-Fredholm} \\ &\Longleftrightarrow A^*A \ \text{ is Fredholm} \end{aligned} \tag{2.1}$$

$$\begin{aligned} A \in \Phi(H) \cup \Phi_d(H) &\Longleftrightarrow A \ \text{ is right-Fredholm} \\ &\Longleftrightarrow AA^* \ \text{ is Fredholm} \end{aligned} \tag{2.2}$$

(recall Proposition 1.4 (b)). Furthermore,

$$\begin{aligned} &A \ \text{ is left-invertible} \Longleftrightarrow A \ \text{ is normally solvable and } \ n(A) = 0, \\ &A \ \text{ is right-invertible} \Longleftrightarrow A \ \text{ is normally solvable and } \ d(A) = 0. \end{aligned}$$

Here are some properties of the index of semi-Fredholm operators. In what follows, X is a Banach space.

Theorem 2.1. *The sets* $\Phi(X)$, $\Phi_n(X)$, $\Phi_d(X)$ *are open subsets of* $\mathcal{B}(X)$ *and the function*

$$\mathrm{Ind} : \Phi(X) \cup \Phi_n(X) \cup \Phi_d(X) \to \mathbf{Z} \cup \{-\infty, +\infty\}$$

is constant on the connected components of $\Phi(X) \cup \Phi_n(X) \cup \Phi_d(X)$. *Moreover, if* A *is semi-Fredholm, then there exists an* $\varepsilon > 0$ *such that*

$$n(A + C) \leq n(A), \quad d(A + C) \leq d(A)$$

whenever $C \in \mathcal{B}(X)$ *and* $\|C\| < \varepsilon$.

Theorem 2.2. *If* $A \in \Phi(X)$ *(resp.* $\Phi_n(X)$, $\Phi_d(X)$*) and* $K \in \mathcal{K}(X)$, *then* $A + K \in \Phi(X)$ *(resp.* $\Phi_n(X)$, $\Phi_d(X)$*) and* $\mathrm{Ind}(A + K) = \mathrm{Ind}\, A$.

Theorem 2.3 (Atkinson). *If* $A, B \in \Phi(X) \cup \Phi_n(X)$ *(resp.* $\Phi(X) \cup \Phi_d(X)$*), then* $AB \in \Phi(X) \cup \Phi_n(X)$ *(resp.* $\Phi(X) \cup \Phi_d(X)$*) and* $\mathrm{Ind}\, AB = \mathrm{Ind}\, A + \mathrm{Ind}\, B$.

Theorem 2.4. *Let* $A^* \in \mathcal{B}(X^*)$ *be the adjoint operator of* $A \in \mathcal{B}(X)$. *Then* A *is normally solvable if and only if so is* A^*. *Furthermore, if* A *and* A^* *are normally solvable, then*

$$\begin{aligned}
&A \in \Phi(X) \Longleftrightarrow A^* \in \Phi(X^*),\\
&A \in \Phi_n(X) \Longleftrightarrow A^* \in \Phi_d(X^*),\ A \in \Phi_d(X) \Longleftrightarrow A^* \in \Phi_n(X^*),\\
&n(A^*) = d(A),\ d(A^*) = n(A),\ \mathrm{Ind}\, A^* = -\mathrm{Ind}\, A.
\end{aligned}$$

2.2 Two Basic Theorems

In this section we cite some well-known general results on Wiener-Hopf operators. Unless otherwise stated, all operators considered in the rest of this chapter are regarded as acting on the Hilbert space $L^2(\mathbf{R}_+)$.

Theorem 2.5 (Coburn, Simonenko). *If* $a \in L^\infty(\mathbf{R})$ *is not identically zero then* $\mathrm{Ker}\, W(a) = \{0\}$ *or* $\mathrm{Im}\, W(a)$ *is dense in* $L^2(\mathbf{R}_+)$.

This beautiful theorem has far-reaching consequences for the spectral theory of scalar Wiener-Hopf operators. Here are some of them.

Corollary 2.6. (a) *Except for the zero operator* ($a = 0$), *a normally solvable Wiener-Hopf operator is automatically semi-Fredholm.*

(b) *If* $a \in L^\infty(\mathbf{R})$ *and* $W(a)$ *is a semi-Fredholm operator, then* $n(W(a)) = 0$ *or* $d(W(a)) = 0$. *Equivalently,*

$$n(W(a)) = \max\{0, \operatorname{Ind} W(a)\}, \quad d(W(a)) = \max\{0, -\operatorname{Ind} W(a)\}.$$

(c) *If* $a \in L^\infty(\mathbf{R})$ *and* $W(a)$ *is semi-Fredholm, then*

$$\begin{aligned} &W(a) \text{ is left-invertible} \Longleftrightarrow \operatorname{Ind} W(a) \le 0,\\ &W(a) \text{ is right-invertible} \Longleftrightarrow \operatorname{Ind} W(a) \ge 0,\\ &W(a) \text{ is invertible} \Longleftrightarrow \operatorname{Ind} W(a) = 0.\end{aligned}$$

The following theorem provides useful estimates for the spectrum and the essential spectrum of Wiener-Hopf operators. We denote by $\operatorname{conv} E$ the convex hull of a set $E \subset \mathbf{C}$.

Theorem 2.7. *If* $a \in L^\infty(\mathbf{R})$, *then both* $\operatorname{sp} W(a)$ *and* $\operatorname{sp_{ess}} W(a)$ *are connected sets and*

$$\mathcal{R}(a) \subset \operatorname{sp_{ess}} W(a) \subset \operatorname{sp} W(a) \subset \operatorname{conv} \mathcal{R}(a).$$

Moreover, if $a \in L^\infty(\mathbf{R})$ *and* $W(a)$ *is semi-Fredholm, then* $a \in GL^\infty(\mathbf{R})$.

For further referencing, we also mention the following fact.

Corollary 2.8. *Let* A *be a* C^**-subalgebra of* $L^\infty(\mathbf{R})$. *If* $a \in A \setminus \{0\}$ *and* $W(a)$ *is normally solvable, then* $a \in GA$.

Proof. Immediate from Corollary 2.6(a), Theorem 2.7, and Proposition 1.4(a). □

Since $(W(a))^* = W(\overline{a})$, a Wiener-Hopf operator is selfadjoint if and only if its symbol is real-valued. For such symbols, Theorem 2.7 immediately gives the following.

Corollary 2.9. *If* $a \in L^\infty(\mathbf{R})$ *is real-valued, then*

$$\operatorname{sp_{ess}} W(a) = \operatorname{sp} W(a) = \operatorname{conv} \mathcal{R}(a) = [\operatorname{ess\,inf} a, \operatorname{ess\,sup} a]. \qquad \square$$

In particular: $\operatorname{sp_{ess}} S_+ = \operatorname{sp} S_+ = [-1, +1]$ (recall Example 1.24).

2.3 Hankel Operators

Let χ_+ and χ_- denote the characteristic functions of $\mathbf{R}_+ := (0, \infty)$ and $\mathbf{R}_- := (-\infty, 0)$, respectively. For $f \in L^2(\mathbf{R}_-) \cup L^2(\mathbf{R}_+)$, define Jf by $(Jf)(t) := f(-t)$. Thus, J is an isometric isomorphism of $L^2(\mathbf{R}_\pm)$ onto $L^2(\mathbf{R}_\mp)$. Obviously, $J^2 = I$. Given $a \in L^\infty(\mathbf{R})$, the two *Hankel operators* $H(a)$ and $H(\tilde{a})$ are defined by

$$\begin{aligned} H(a) &: L^2(\mathbf{R}_+) \to L^2(\mathbf{R}_+), \; f \mapsto \chi_+ W^0(a) \chi_- J,\\ H(\tilde{a}) &: L^2(\mathbf{R}_+) \to L^2(\mathbf{R}_+), \; f \mapsto J\chi_- W^0(a) \chi_+ I.\end{aligned}$$

If $a = \gamma + \hat{k} \in \mathbf{C} + FL^1(\mathbf{R})$, then $H(a)$ and $H(\tilde{a})$ are easily seen to act by the rules

$$\begin{aligned}\big(H(a)f\big)(t) &= \int_{\mathbf{R}_+} k(t+s)f(s)\,ds, \quad t \in \mathbf{R}_+, \\ \big(H(\tilde{a})f\big)(t) &= \int_{\mathbf{R}_+} k(-t-s)f(s)\,ds, \quad t \in \mathbf{R}_+.\end{aligned} \tag{2.3}$$

For $a \in L^\infty(\mathbf{R})$, define $\tilde{a} \in L^\infty(\mathbf{R})$ by $\tilde{a}(x) := a(-x)$. It can be verified directly that

$$\chi_+ W^0(\tilde{a})\chi_- J = J\chi_- W^0(a)\chi_+ I.$$

Thus, we can also regard $H(\tilde{a})$ as the operator $H(b)$ with $b(x) := a(-x)$.

We know from Section 1.6 that the map $W : L^\infty(\mathbf{R}) \to \mathcal{B}(L^2(\mathbf{R}_+))$ is not multiplicative. The following simple proposition relates $W(ab)$ to $W(a)W(b)$ via Hankel operators and is a key result to the understanding of the nature of Wiener-Hopf operators.

Proposition 2.10. *If $a, b \in L^\infty(\mathbf{R})$, then*

$$W(ab) = W(a)W(b) + H(a)H(\tilde{b}), \tag{2.4}$$

$$H(ab) = W(a)H(b) + H(a)W(\tilde{b}). \tag{2.5}$$

Proof. Since $1 = \chi_-^2 + \chi_+^2$ and $J^2 = I$, we have

$$\begin{aligned}W(ab) &= \chi_+ W^0(ab)\chi_+ = \chi_+ W^0(a)W^0(b)\chi_+ \\ &= \chi_+ W^0(a)\chi_+\chi_+ W^0(b)\chi_+ + \chi_+ W^0(a)\chi_-\chi_- W^0(b)\chi_+ \\ &= W(a)W(b) + H(a)H(\tilde{b}), \\ H(ab) &= \chi_+ W^0(ab)\chi_- J = \chi_+ W^0(a)W^0(b)\chi_- J \\ &= \chi_+ W^0(a)\chi_+\chi_+ W^0(b)\chi_- J + \chi_+ W^0(a)\chi_-\chi_- W^0(b)\chi_- J \\ &= \chi_+ W^0(a)\chi_+\chi_+ W^0(b)\chi_- J + \chi_+ W^0(a)\chi_- J J\chi_- W^0(b)\chi_- J \\ &= W(a)H(b) + H(a)W(\tilde{b}).\end{aligned}$$

□

Proposition 2.11. *If $c \in C(\dot{\mathbf{R}})$, then the Hankel operators $H(c)$ and $H(\tilde{c})$ are compact.*

Proof. Let us show that $H(c)$ is compact. The symbol c can be uniformly approximated by symbols $\gamma + \hat{k} \in \mathbf{C} + FL^1(\mathbf{R})$. For such symbols, we have the representation (2.3). The kernel $k(t)$ $(t > 0)$ can be approximated in the L^1 norm by continuous functions with compact support, and these functions can be uniformly approximated by functions of the form $e^{-t}p(t)$ $(t > 0)$, where p is a polynomial.

Since, for $n \in \mathbf{Z}_+$, the operator

$$f(t) \mapsto \int_{\mathbf{R}_+} e^{-t-s}(t+s)^n f(s)\, ds, \quad t > 0,$$

has finite rank, we arrive at the assertion. □

Corollary 2.12. *If $a, b \in L^\infty(\mathbf{R})$ and at least one of the functions a, b belongs to $C(\dot{\mathbf{R}})$ then $W(ab) - W(a)W(b)$ is compact.*

Proof. Combine Propositions 2.10 and 2.11. □

By employing a continuous partition of the identity, one can sharpen Corollary 2.12 as follows.

Theorem 2.13. *If $a, b \in L^\infty(\mathbf{R})$ and if at each point $x_0 \in \dot{\mathbf{R}}$ at least one of the functions a and b is continuous, then $W(ab) - W(a)W(b)$ is compact.*

We conclude this section with Simonenko's localization principle. Two functions $a, b \in L^\infty(\mathbf{R})$ are said to be locally equivalent at a point $x_0 \in \dot{\mathbf{R}}$ if

$$\inf\Big\{ \|(a-b)\varphi\|_\infty : \varphi \in C(\dot{\mathbf{R}}),\ \varphi(x_0) = 1 \Big\} = 0.$$

Theorem 2.14 (Simonenko). *Let $a \in L^\infty(\mathbf{R})$ and suppose that for each $x_0 \in \dot{\mathbf{R}}$ there is an $a_{x_0} \in L^\infty(\mathbf{R})$ such that a is locally equivalent to a_{x_0} at x_0 and $W(a_{x_0})$ is Fredholm (resp. left-Fredholm, right-Fredholm). Then $W(a)$ itself is Fredholm (resp. left-Fredholm, right-Fredholm).*

Once Corollary 2.9 (for $a \in C(\dot{\mathbf{R}})$) and Corollary 2.12 are at our disposal, this theorem can be proved by having recourse to one of the so-called local principles for (abstract) Banach algebras, for example, to the local principle of Gohberg and Krupnik or the local principle of Allan and Douglas.

2.4 Continuous Symbols

In the following sections, we associate a real number ind a and an integer wind a with functions a belonging to several subsets of $GL^\infty(\mathbf{R})$. The number ind a is referred to as the *Cauchy index* of a, while wind a is called the *winding number* of a. In a sense, ind a is $1/(2\pi)$ times the increment of the argument of $a(x)$ as x moves from $-\infty$ to $+\infty$, and wind a results from ind a by adding $1/(2\pi)$ times the increment of the argument of a when x "jumps" from $+\infty$ to $-\infty$.

Let a be an invertible bounded continuous function on $\mathbf{R}$, $a \in GBC(\mathbf{R})$. Any continuous function $b : \mathbf{R} \to \mathbf{R}$ such that $a = |a|e^{ib}$ is called a *continuous argument* of a and is denoted by $\arg a$. Obviously, there are countably many continuous arguments of a, but any two of them differ only by an integral multiple of 2π. If $a \in GC(\overline{\mathbf{R}})$ (and, in particular, if $a \in GC(\dot{\mathbf{R}})$) and if $\arg a$ is any continuous argument of a, then the limits $(\arg a)(\pm\infty)$ exist and are finite. We define the *Cauchy index* of a by

$$\operatorname{ind} a := \frac{1}{2\pi}\Big((\arg a)(+\infty) - (\arg a)(-\infty)\Big). \tag{2.6}$$

Clearly, $\operatorname{ind} a$ is a real number and is independent of the particular choice of $\arg a$. Now suppose $a \in GC(\dot{\mathbf{R}})$. Then $\operatorname{ind} a$ is an integer. As x moves from $-\infty$ to $+\infty$, the point $a(x)$ traces out a closed continuous oriented curve in $\mathbf{C} \setminus \{0\}$. The number of times this curve surrounds the origin counter-clockwise is referred to as the *winding number* $\operatorname{wind} a$ of a. Obviously, $\operatorname{wind} a = \operatorname{ind} a$.

Theorem 2.15. *If $a \in C(\dot{\mathbf{R}}) \setminus \{0\}$ has a zero on $\dot{\mathbf{R}}$, then $W(a)$ is not normally solvable. If $a \in GC(\dot{\mathbf{R}})$, then $W(a)$ is Fredholm and*

$$\operatorname{Ind} W(a) = -\operatorname{ind} a = -\operatorname{wind} a.$$

Proof. From Corollary 2.8 we deduce that $W(a)$ cannot be normally solvable if $a \neq 0$ has a zero on $\dot{\mathbf{R}}$.

So suppose $a \in GC(\dot{\mathbf{R}})$. Since a is locally equivalent to the constant function $a_{x_0}(x) := a(x_0)$ at $x_0 \in \dot{\mathbf{R}}$ and since $W(a_{x_0}) = a(x_0)I$ is invertible, Theorem 2.14 implies that $W(a)$ is Fredholm. Another proof of the Fredholmness of $W(a)$ follows from Corollary 2.12, which shows that $W(a^{-1})$ is a regularizer of $W(a)$.

We are left with the index formula. Let $\operatorname{wind} a = n$. Then a is in the connected component of $GC(\dot{\mathbf{R}})$ which contains the function φ^n, where $\varphi(x) = (x-i)/(x+i)$. Since all functions in $GC(\dot{\mathbf{R}})$ generate Fredholm Wiener-Hopf operators, we infer from Theorem 2.1 that $\operatorname{Ind} W(a) = \operatorname{Ind} W(\varphi^n)$. By Corollary 2.12 and Theorems 2.2 and 2.3, we have $\operatorname{Ind} W(\varphi^n) = n \operatorname{Ind} W(\varphi)$. Thus, it remains to show that $\operatorname{Ind} W(\varphi) = -1$. Since $\operatorname{Ind} W(\varphi) = -\operatorname{Ind} W(\overline{\varphi})$ due to Theorem 2.4, we must show that $\operatorname{Ind} W(\overline{\varphi}) = 1$. The operator $W(\overline{\varphi})$ acts on $L^2(\mathbf{R}_+)$ by the formula

$$\big(W(\overline{\varphi})f\big)(t) = f(t) - 2\int_t^\infty e^{t-s} f(s)\, ds, \quad t \in \mathbf{R}_+,$$

because

$$\overline{\varphi}(x) = \frac{x+i}{x-i} = 1 - 2\int_{-\infty}^0 e^t e^{ixt}\, dt.$$

This easily implies that $\operatorname{Ker} W(\overline{\varphi}) = \{ce^{-t} : c \in \mathbf{C}\}$, whence $\operatorname{Ind} W(\overline{\varphi}) = 1$ by virtue of Corollary 2.6(b). □

Combining Corollary 2.6(c) and Theorem 2.15, we arrive at the conclusion that if $a \in C(\dot{\mathbf{R}})$, then $W(a)$ is invertible if and only if $a \in GC(\dot{\mathbf{R}})$ and $\operatorname{wind} a = 0$. In the language of spectra we have

$$\mathrm{sp_{ess}}\, W(a) = a(\dot{\mathbf{R}}),$$
$$\mathrm{sp}\, W(a) = a(\dot{\mathbf{R}}) \cup \Big\{\lambda \in \mathbf{C} \setminus a(\dot{\mathbf{R}}) : \mathrm{wind}\,(a-\lambda) \neq 0\Big\}.$$

Figure 2.1 shows an example.

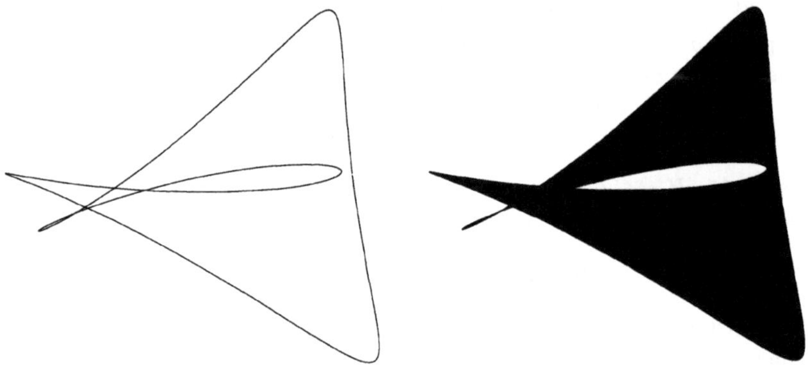

Figure 2.1: The essential range of a continuous function a (left) and the spectrum of $W(a)$ (right).

Let $\mathrm{alg}\, W(C)$ be the smallest closed subalgebra of $\mathcal{B}(L^2(\mathbf{R}_+))$ that contains the set $\{W(a) : a \in C(\dot{\mathbf{R}})\}$. One can show that $\mathrm{alg}\, W(C)$ contains all compact operators,

$$\mathcal{K}\big(L^2(\mathbf{R}_+)\big) \subset \mathrm{alg}\, W(C). \tag{2.7}$$

For $A \in \mathcal{B}(L^2(\mathbf{R}_+))$, put

$$A^\pi := A + \mathcal{K}\big(L^2(\mathbf{R}_+)\big), \tag{2.8}$$

and in case $A = W(a)$, let $W^\pi(a) := (W(a))^\pi$.

Theorem 2.16. *Every element of the quotient algebra*

$$\mathrm{alg}\, W^\pi(C) := \mathrm{alg}\, W(C)/\mathcal{K}\big(L^2(\mathbf{R}_+)\big)$$

is of the form $W^\pi(c)$ with some (uniquely determined) $c \in C(\dot{\mathbf{R}})$. The algebra $\mathrm{alg}\, W^\pi(C)$ is a commutative unital C^*-algebra whose maximal ideal space may be identified with $\dot{\mathbf{R}}$, and the Gelfand map is of the form

$$\Gamma : \mathrm{alg}\, W^\pi(C) \to C(\dot{\mathbf{R}}), \quad \big(\Gamma W^\pi(c)\big)(x_0) = c(x_0).$$

2.5 $C + H^\infty$ Symbols

Put $\mathbf{C}_+ := \{z \in \mathbf{C} : \operatorname{Im} z > 0\}$, $\mathbf{C}_- := \{z \in \mathbf{C} : \operatorname{Im} z < 0\}$, and denote by $H^\infty(\mathbf{C}_\pm)$ the set of all bounded and analytic functions in $\mathbf{C}_\pm$. *Fatou's theorem* says that functions in $H^\infty(\mathbf{C}_\pm)$ have non-tangential limits on $\mathbf{R}$ almost everywhere, and we denote by $H^\infty_\pm$ the set of all $a \in L^\infty(\mathbf{R})$ that are non-tangential limits of functions in $H^\infty(\mathbf{C}_\pm)$. It is well known that $H^\infty_\pm$ are closed subalgebras of $L^\infty(\mathbf{R})$.

The set of all functions f which are analytic in $\mathbf{C}_\pm$ and satisfy

$$\sup_{\pm y > 0} \int_{\mathbf{R}} |f(x+iy)|^p \, dy < \infty$$

is denoted by $H^p(\mathbf{C}_\pm)$. It is again a *theorem by Fatou* which tells us that functions in $H^p(\mathbf{C}_\pm)$ $(0 < p < \infty)$ have non-tangential limits almost everywhere on $\mathbf{R}$. We write $H^p_\pm$ for the set of the corresponding boundary functions on $\mathbf{R}$. If $1 \le p < \infty$, then $H^p_\pm$ is a closed subspace of $L^p(\mathbf{R})$.

The *Paley-Wiener theorem* states that

$$H^2_+ = FL^2(\mathbf{R}_+), \quad H^2_- = FL^2(\mathbf{R}_-), \tag{2.9}$$

where $L^2(\mathbf{R}_\pm)$ are regarded as subspaces of $L^2(\mathbf{R})$ via the identification $L^2(\mathbf{R}_\pm) = \chi_\pm L^2(\mathbf{R})$. The orthogonal projections of $L^2(\mathbf{R})$ onto $H^2_\pm$ are given by

$$P := F\chi_+F^{-1} : L^2(\mathbf{R}) \to H^2_+, \quad \text{and} \quad Q := F\chi_-F^{-1} : L^2(\mathbf{R}) \to H^2_-. \tag{2.10}$$

The operator P is also referred to as the *Riesz projection*. Taking into account that $F^{-1} = (2\pi)^{-1}FJ = (2\pi)^{-1}JF$ where $(Jf)(x) := f(-x)$, we obtain that

$$P := F\chi_+F^{-1} = FJ\chi_-JF^{-1} = F^{-1}\chi_-F, \tag{2.11}$$
$$Q := F\chi_-F^{-1} = FJ\chi_+JF^{-1} = F^{-1}\chi_+F, \tag{2.12}$$

whence

$$P = W^0(\chi_-), \quad Q = W^0(\chi_+). \tag{2.13}$$

Since $S = W^0(-\operatorname{sgn} x)$ (Example 1.17), we infer from (2.13) that

$$P = \frac{1}{2}(I + S), \quad Q = \frac{1}{2}(I - S). \tag{2.14}$$

The equalities (2.14) are referred to as the *Sokhotski-Plemelj formulas*.

Multiplication by functions in $H^\infty_\pm$ leaves $H^2_\pm$ invariant. Therefore,

$$a \in H^\infty_+ \Longrightarrow H(\tilde{a}) = J\chi_-W^0(a)\chi_+ = J\chi_-F^{-1}aF\chi_+ = 0, \tag{2.15}$$
$$a \in H^\infty_- \Longrightarrow H(a) = \chi_+W^0(a)\chi_-J = \chi_+F^{-1}aF\chi_-J = 0. \tag{2.16}$$

This simple observation is again of great importance throughout the theory of Wiener-Hopf operators.

Proposition 2.17. *If $a_\pm \in H^\infty_\pm$ and $b \in L^\infty(\mathbf{R})$, then*

$$W(a_- b a_+) = W(a_-)W(b)W(a_+).$$

Proof. Combine (2.4), (2.15), (2.16). □

Proposition 2.17 implies that the maps

$$W : H^\infty_+ \to \mathcal{B}\big(L^2(\mathbf{R}_+)\big), \quad W : H^\infty_- \to \mathcal{B}\big(L^2(\mathbf{R}_+)\big)$$

are multiplicative and thus Banach algebra homomorphisms. In particular, if $a = a_- a_+$ with $a_\pm \in GH^\infty_\pm$, then $W(a)$ is invertible and the inverse is $W(a_+^{-1})W(a_-^{-1})$. This is the foundation of so-called Wiener-Hopf factorization.

Sarason discovered that the smallest closed subalgebra of $L^\infty(\mathbf{R})$ that contains both $C(\dot{\mathbf{R}})$ and $H^\infty_\pm$ coincides with the sum of $C(\dot{\mathbf{R}})$ and $H^\infty_\pm$:

$$\operatorname{alg}\big(C(\dot{\mathbf{R}}), H^\infty_\pm\big) = C(\dot{\mathbf{R}}) + H^\infty_\pm.$$

The relevancy of $C(\dot{\mathbf{R}}) + H^\infty_\pm$ in Wiener-Hopf theory stems from the following result.

Theorem 2.18 (Hartman). *Let $a \in L^\infty(\mathbf{R})$. Then*

$$H(a) \text{ is compact} \Longleftrightarrow a \in C(\dot{\mathbf{R}}) + H^\infty_-,$$
$$H(\tilde{a}) \text{ is compact} \Longleftrightarrow a \in C(\dot{\mathbf{R}}) + H^\infty_+.$$

We remark that the implications "$\Longleftarrow$" are immediate consequences of (2.15), (2.16) and Proposition 2.11. The reverse implications are less trivial.

We now define the *winding number* wind a for $a \in G(C(\dot{\mathbf{R}}) + H^\infty_+)$. For a in $L^\infty(\mathbf{R})$, the *harmonic extension* into $\mathbf{C}_+$ is defined by

$$\hat{a}(x+iy) := \frac{1}{\pi}\int_{\mathbf{R}} \frac{y}{(x-\xi)^2+y^2} a(\xi)\,d\xi, \quad x+iy \in \mathbf{C}_+. \tag{2.17}$$

The function $\hat{a}$ is harmonic in $\mathbf{C}_+$, and its non-tangential limits on $\mathbf{R}$ coincide with the values of a on $\mathbf{R}$ almost everywhere. For $r \in (0,1)$ let $L_r \subset \mathbf{C}_+$ denote the counter-clockwise oriented circle with the center at i/r and the radius $1/r - r$. Let $\varrho_r : \dot{\mathbf{R}} \to L_r$ be any homeomorphism which preserves the orientation of L_r when $\dot{\mathbf{R}}$ is traced out from $-\infty$ to $+\infty$. If $a \in L^\infty(\mathbf{R})$, then $\hat{a} \circ \varrho_r$ belongs to $C(\dot{\mathbf{R}})$. Douglas showed that a function $a \in C(\dot{\mathbf{R}}) + H^\infty_+$ is invertible in $C(\dot{\mathbf{R}}) + H^\infty_+$ if and only if there is an $r_0 \in (0,1)$ such that $\hat{a} \circ \varrho_r \in GC(\dot{\mathbf{R}})$ for all $r \in (0, r_0)$. In that case the winding number $\operatorname{wind}(\hat{a} \circ \varrho_r)$ is constant for $r \in (0, r_0)$ and is denoted by wind a. It is easily seen that for $a \in GC(\dot{\mathbf{R}})$ the winding number defined here coincides with the winding number introduced in the beginning of this section.

Theorem 2.19 (Douglas). *Let $a \in C(\dot{\mathbf{R}}) + H^\infty_+$ and suppose a is not identically zero.*

(a) *If $a \notin GL^\infty(\mathbf{R})$, then $W(a)$ is not normally solvable.*

(b) *If $a \in GL^\infty(\mathbf{R})$ but $a \notin G(C(\dot{\mathbf{R}}) + H^\infty_+)$, then $W(a)$ is properly n-normal.*

(c) *If $a \in G(C(\dot{\mathbf{R}}) + H^\infty_+)$, then $W(a)$ is Fredholm, $W(a^{-1})$ is a regularizer of $W(a)$, and* $\operatorname{Ind} W(a) = -\operatorname{wind} a$.

Proof outline. For the index formula, we refer to the literature (see, e.g., [46] or [66]). The other assertions can be proved as follows.

Part (a) is immediate from Corollary 2.8. If $a \in G(C(\dot{\mathbf{R}}) + H^\infty_+)$, then (2.4) and (the trivial part of) Theorem 2.18 imply that $W(a^{-1})$ is a regularizer of $W(a)$. Hence, $W(a)$ is Fredholm.

Conversely, suppose $W(a)$ is Fredholm. Let R be a regularizer of $W(a)$. From (2.5) we get

$$0 = H(\tilde{a}^{-1}\tilde{a}) = W(\tilde{a}^{-1})H(\tilde{a}) + H(\tilde{a}^{-1})W(a), \tag{2.18}$$

and $H(\tilde{a})$ is compact by (the trivial part of) Theorem 2.18. On multiplying (2.18) by R from the right we obtain

$$0 = W(\tilde{a}^{-1})H(\tilde{a})R + H(\tilde{a}^{-1}) + H(\tilde{a}^{-1})\big(W(a)R - I\big),$$

which implies that $H(\tilde{a}^{-1})$ is compact. Consequently, $a^{-1} \in C(\dot{\mathbf{R}}) + H^\infty_+$ due to (the nontrivial part of) Theorem 2.18.

We have shown that $W(a)$ is Fredholm if and only if $a \in G(C(\dot{\mathbf{R}}) + H^\infty_+)$. If a is not invertible in $C(\dot{\mathbf{R}}) + H^\infty_+$ but $a^{-1} \in L^\infty(\mathbf{R})$, then

$$W(a^{-1})W(a) - I = H(a^{-1})H(\tilde{a})$$

is compact together with $H(\tilde{a})$ (trivial part of Theorem 2.18) and hence $W(a^{-1})$ is left-Fredholm. As $W(a)$ is not Fredholm, it follows that $W(a)$ is properly n-normal. □

2.6 *PC* Symbols

Given a function $a \in PC$, we denote by $a^\#(\dot{\mathbf{R}})$ the closed continuous curve that results from the essential range of a by filling in the line segment

$$[a(x_0 - 0),\, a(x_0 + 0)]$$

between the endpoints $a(x_0 - 0)$, $a(x_0 + 0)$ of each jump. The orientation of $\mathbf{R}$ from $-\infty$ to $+\infty$ induces an orientation of the curve $a^\#(\dot{\mathbf{R}})$ in a natural manner.

Thus, the segment $[a(x_0-0),\, a(x_0+0)]$ is oriented from $a(x_0-0)$ to $a(x_0+0)$. Also recall that $a(\infty-0) := a(+\infty)$, $a(\infty+0) := a(-\infty)$. If $0 \notin a^{\#}(\dot{\mathbf{R}})$, we denote by wind a the number of counter-clockwise circuits of the curve $a^{\#}(\dot{\mathbf{R}})$ around the origin and we call this number the *winding number* of a.

Theorem 2.20. *Let* $a \in PC \setminus \{0\}$.
(a) *If* $0 \in a^{\#}(\dot{\mathbf{R}})$, *then* $W(a)$ *is not normally solvable.*
(b) *If* $0 \notin a^{\#}(\dot{\mathbf{R}})$, *then* $W(a)$ *is Fredholm and*

$$\operatorname{Ind} W(a) = -\operatorname{wind} a.$$

Proof. Suppose $0 \notin a^{\#}(\dot{\mathbf{R}})$. At $x_0 \in \mathbf{R}$, the function a is locally equivalent to the function

$$a_{x_0}(x) := a(x_0-0)\chi_{(-\infty,x_0)}(x) + a(x_0+0)\chi_{(x_0,+\infty)}(x),$$

and at $x_0 = \infty$, the function a is locally equivalent to

$$a_\infty(x) := a(-\infty)\chi_{(-\infty,0)}(x) + a(+\infty)\chi_{(0,+\infty)}(x).$$

The operators $W(a_{x_0})$ are invertible (and thus Fredholm) for each $x_0 \in \dot{\mathbf{R}}$. This can be deduced from Theorem 2.7 as well as from the representation

$$W(a_{x_0}) = \frac{a(x_0-0)+a(x_0+0)}{2} I + \varepsilon \frac{a(x_0+0)-a(x_0-0)}{2} W\big(-\operatorname{sgn}(x-\alpha)\big),$$

where $\alpha = x_0$, $\varepsilon = -1$ for $x_0 \in \mathbf{R}$ and $\alpha = 0$, $\varepsilon = 1$ for $x_0 = \infty$, in conjunction with Example 1.24. From Theorem 2.14 we therefore obtain that $W(a)$ is Fredholm.

We now prove the index formula. First assume that the set Λ_a of the discontinuities of a is at most finite. For $\mu \in [0,1]$, we change a to a function a_μ such that $\mathcal{R}(a_\mu)$ is the union of $\mathcal{R}(a)$ and the segments

$$\Big[a(x_0-0),\, a(x_0-0) + \mu\big(a(x_0+0) - a(x_0-0)\big)\Big] \quad (x_0 \in \dot{\mathbf{R}}).$$

Clearly, this can be managed so that $a_\mu \in PC$, a_μ has no other discontinuities than a, $a_0 = a$, the map $[0,1] \to PC$, $\mu \mapsto a_\mu$ is continuous, and

$$a_\mu(x_0-0) = a(x_0-0) + \mu\big(a(x_0+0) - a(x_0-0)\big), \quad a_\mu(x_0+0) = a(x_0+0).$$

Since $0 \notin a_\mu^{\#}(\dot{\mathbf{R}})$, we have a continuous map $[0,1] \to \Phi\big(L^2(\mathbf{R}_+)\big)$, $\mu \mapsto W(a_\mu)$, and Theorem 2.1 implies that

$$\operatorname{Ind} W(a) = \operatorname{Ind} W(a_0) = \operatorname{Ind} W(a_1) = -\operatorname{wind} a_1 = -\operatorname{wind} a,$$

the equality $\operatorname{Ind} W(a_1) = -\operatorname{wind} a_1$ resulting from Theorem 2.15. Standard approximation and stability arguments now give the index formula for arbitrary functions $a \in PC$ subject to the condition $0 \notin a^{\#}(\dot{\mathbf{R}})$.

Finally, assume $0 \in a^{\#}(\dot{\mathbf{R}})$ but $W(a)$ is normally solvable and thus (Corollary 2.6(a)) semi-Fredholm. Approximating a by piecewise smooth functions and then slightly perturbing the piecewise smooth functions we can obtain two functions $b, c \in PC$ such that $0 \notin b^{\#}(\dot{\mathbf{R}})$, $0 \notin c^{\#}(\dot{\mathbf{R}})$, $|\operatorname{wind} b - \operatorname{wind} c| = 1$, and $\|b-a\|_\infty$, $\|c-a\|_\infty$ are as small as desired. However, this is impossible, since, by Theorem 2.1, the index of a semi-Fredholm operator is stable under small perturbations. This contradiction completes the proof. □

Clearly, Theorem 2.20 and Corollary 2.6 yield descriptions of the spectrum and the essential spectrum of Wiener-Hopf operators with *PC* symbols; see Figure 2.2.

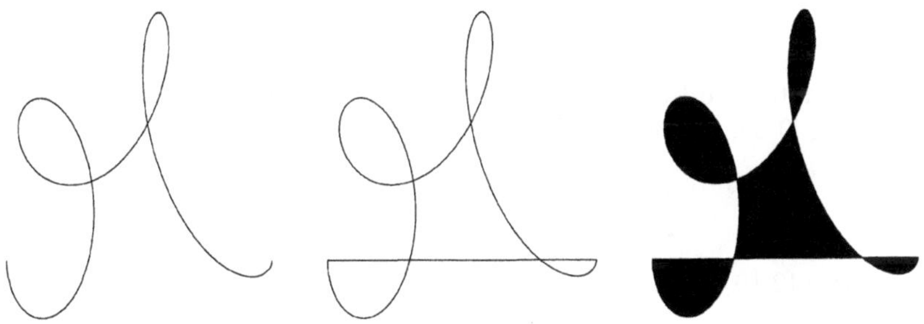

Figure 2.2: The essential range of a piecewise continuous function a with a single jump (left), the set $\operatorname{sp}_{\text{ess}} W(a) = a^{\#}(\dot{\mathbf{R}})$ (middle), and the spectrum $\operatorname{sp} W(a)$ (right).

Denote by $\operatorname{alg} W(PC)$ the smallest closed subalgebra of $\mathcal{B}(L^2(\mathbf{R}_+))$ which contains the set $\{W(a) : a \in PC\}$. Recall (2.7) and (2.8).

Theorem 2.21 (Gohberg-Krupnik). *The quotient algebra*

$$\operatorname{alg} W^\pi(PC) := \operatorname{alg} W(PC)/\mathcal{K}(L^2(\mathbf{R}_+))$$

is a commutative unital C^*-algebra whose maximal ideal space may be identified with the cylinder $\dot{\mathbf{R}} \times [0,1]$ with an exotic topology. The Gelfand map

$$\Gamma : \operatorname{alg} W^\pi(PC) \to C(\dot{\mathbf{R}} \times [0,1])$$

is for $a \in PC$ and $(x,\mu) \in \dot{\mathbf{R}} \times [0,1]$ given by

$$(\Gamma W^\pi(a))(x,\mu) = a(x-0)(1-\mu) + a(x+0)\mu.$$

The *exotic topology* on $\dot{\mathbf{R}} \times [0,1]$ mentioned in Theorem 2.21 can be described as follows. A neighborhood base of (x, μ) with $0 < \mu < 1$ is formed by the sets $\{x\} \times (\mu - \varepsilon, \mu + \varepsilon)$ $(\varepsilon > 0)$, a neighborhood base of $(x, 1)$ is given by the sets $(\{x\} \times (1-\varepsilon, 1]) \cup ((x, x+\delta) \times [0,1])$ $(\varepsilon, \delta > 0)$, and a neighborhood base of $(x, 0)$ is the sets $(\{x\} \times [0, \varepsilon)) \cup ((x - \delta, x) \times [0,1])$.

We conclude this section with some remarks on the Cauchy index and the winding number of piecewise continuous functions. We defined ind a for $a \in GC(\overline{\mathbf{R}})$ in Section 2.4, and wind a for $a \in GPC$ with $0 \notin a^{\#}(\dot{\mathbf{R}})$ was introduced before Theorem 2.20. A little plane geometry shows that if $a \in GC(\overline{\mathbf{R}})$ and $0 \notin a^{\#}(\dot{\mathbf{R}})$, then

$$\text{wind } a = \text{ind } a - \frac{1}{2} + \left\{ \frac{1}{2} + \frac{1}{2\pi} \arg \frac{a(-\infty)}{a(+\infty)} \right\} \tag{2.19}$$

where $\{\lambda\}$ denotes the fractional part of the real number λ: for $\lambda = k + \mu$ with $k \in \mathbf{Z}$ and $\mu \in [0,1)$, we put $\{\lambda\} := \mu$. Notice that in view of the identity

$$\frac{1}{2} - \left\{ \frac{1}{2} - x \right\} = -\frac{1}{2} + \left\{ \frac{1}{2} + x \right\}$$

we can rewrite (2.19) in the form

$$\text{wind } a = \text{ind } a + \frac{1}{2} - \left\{ \frac{1}{2} - \frac{1}{2\pi} \arg \frac{a(-\infty)}{a(+\infty)} \right\}. \tag{2.20}$$

In case $\arg(a(-\infty)/a(+\infty))$ is taken in $(-\pi, \pi)$, formulas (2.19) and (2.20) assume the form

$$\text{wind } a = \text{ind } a + \frac{1}{2\pi} \arg \frac{a(-\infty)}{a(+\infty)}. \tag{2.21}$$

Now suppose $a \in GPC$ has at most finitely many jumps. Denote by $\Lambda_a \subset \dot{\mathbf{R}}$ the set of points at which a is discontinuous. Given a connected component l of $\dot{\mathbf{R}} \setminus \Lambda_a$, we let $\text{ind}_l\, a$ stand for the increment of $(1/(2\pi)) \arg a$ on l, where $\arg a$ is any continuous argument of a on l. If $0 \notin a^{\#}(\dot{\mathbf{R}})$, the *Cauchy index* of a is

$$\begin{aligned} \text{ind } a &:= \sum_l \text{ind}_l\, a + \sum_{x \in \Lambda_a \setminus \{\infty\}} \left(-\frac{1}{2} + \left\{ \frac{1}{2} + \frac{1}{2\pi} \arg \frac{a(x+0)}{a(x-0)} \right\} \right) \\ &= \sum_l \text{ind}_l\, a + \sum_{x \in \Lambda_a \setminus \{\infty\}} \left(\frac{1}{2} - \left\{ \frac{1}{2} - \frac{1}{2\pi} \arg \frac{a(x+0)}{a(x-0)} \right\} \right) \\ &= \sum_l \text{ind}_l\, a + \sum_{x \in \Lambda_a \setminus \{\infty\}} \frac{1}{2\pi} \arg \frac{a(x+0)}{a(x-0)}, \end{aligned} \tag{2.22}$$

the latter formula being valid if and only if the argument is taken in $(-\pi, \pi)$. The geometric meaning of the three right-hand sides of (2.22) is as follows: if z moves along the naturally oriented curve $a^{\#}(\dot{\mathbf{R}})$ from $a(-\infty)$ to $a(+\infty)$, then the number (2.22) is just the increment of $(1/(2\pi)) \arg z$.

If $a \in GPC$ has countably many jumps and $0 \notin a^{\#}(\dot{\mathbf{R}})$, we can approximate a uniformly by functions $a_n \in GPC$ with finitely many jumps such that $0 \notin a_n^{\#}(\dot{\mathbf{R}})$ and $a_n(\pm\infty) = a(\pm\infty)$. One can then show that ind a_n is constant for sufficiently large n and one puts

$$\text{ind } a := \lim_{n\to\infty} \text{ind } a_n. \tag{2.23}$$

It is readily seen that with the definitions (2.22) and (2.23) the equalities (2.19) to (2.21) hold.

Finally, we should mention that if $a \in GPC$ and $0 \notin a^{\#}(\dot{\mathbf{R}})$, then wind a can also be given in terms of the harmonic extension (2.17) of a. Namely, there is an $r_0 \in (0,1)$ such that $\hat{a} \circ \varrho_r \in GC(\dot{\mathbf{R}})$ for all $r \in (0, r_0)$ (recall Section 2.4 for notation). For $r \in (0, r_0)$, the integer wind $(\hat{a} \circ \varrho_r)$ is independent of r and can be shown to coincide with wind a.

2.7 Mean Value and Bohr-Fourier Spectrum

We first establish two simple properties of almost periodic functions.

Proposition 2.22. *Let $A \subset (0,\infty)$ be an unbounded set and let $\{I_\alpha\}_{\alpha\in A} = \{(x_\alpha, y_\alpha)\}_{\alpha\in A}$ be a family of intervals $I_\alpha \subset \mathbf{R}$ such that $|I_\alpha| = y_\alpha - x_\alpha \to \infty$ as $\alpha \to \infty$. If $a \in AP$, then the limit*

$$M(a) := \lim_{\alpha\to\infty} \frac{1}{|I_\alpha|} \int_{I_\alpha} a(x)\,dx$$

exists, is finite, and is independent of the particular choice of the family $\{I_\alpha\}$.

Proof. Let first $p(x) = \sum_j r_j e^{i\lambda_j x}$ be an almost periodic polynomial and suppose $\lambda_0 = 0$. Because

$$\frac{1}{|I_\alpha|} \int_{I_\alpha} dx = 1, \quad \frac{1}{|I_\alpha|} \int_{I_\alpha} e^{i\lambda x}\,dx = \frac{e^{i\lambda y_\alpha} - e^{i\lambda x_\alpha}}{i\lambda(y_\alpha - x_\alpha)} \to 0 \quad (\lambda \neq 0),$$

it follows that $M(p)$ exists and equals r_0. Obviously, $|M(p)| \le \|p\|_\infty$. This shows that $M : AP^0 \to \mathbf{C}$ is a bounded linear operator. As AP^0 is dense in AP, the operator M extends to a bounded linear operator $M : AP \to \mathbf{C}$. This easily implies all assertions of the proposition. □

Proposition 2.23. *If $a \in AP$ then the set*

$$\Omega(a) := \left\{\lambda \in \mathbf{R} : M(ae_{-\lambda}) \neq 0\right\}$$

is at most countable.

Proof. If $p = \sum_{j=1}^{N} r_j e_{\lambda_j}$ is an almost periodic polynomial, then

$$M(pe_{-\lambda}) = \begin{cases} r_j & \text{if} \quad \lambda = \lambda_j, \\ 0 & \text{if} \quad \lambda \notin \{\lambda_1, \dots, \lambda_N\}, \end{cases}$$

and hence $\Omega(p) = \{r_1, \dots, r_N\}$ is finite. To prove that $\Omega(a)$ is at most countable for every $a \in AP$, it suffices to show that

$$\Omega_n(a) := \Big\{\lambda \in \mathbf{R} : \big|M(ae_{-\lambda})\big| > 1/n\Big\}$$

is at most finite for each $n \in \mathbf{N}$. Fix n and choose an almost periodic polynomial p such that $\|a - p\|_\infty < 1/(2n)$. Since

$$\big|M(ae_{-\lambda}) - M(pe_{-\lambda})\big| \le \|a - p\|_\infty < 1/(2n),$$

it results that $|M(pe_{-\lambda})| > 1/(2n)$ for every $\lambda \in \Omega_n(a)$. But as $\Omega(p)$ is finite, there is only a finite number of λ's such that $|M(pe_{-\lambda}| > 1/(2n)$. This proves that $\Omega_n(a)$ is at most finite. □

Definition 2.24. Let $a \in AP$. The number $M(a)$ given by Proposition 2.22 is called the *Bohr mean value* or simply the *mean value* of a. The set $\Omega(a)$ of Proposition 2.23 is referred to as the *Bohr-Fourier spectrum* of a. Figure 2.3 shows examples of AP functions and their mean value. □

Figure 2.3: Three almost periodic functions and their mean value (dot).

2.8 Bohr's Theorem and Mean Motion

Any continuous argument of an invertible continuous periodic function is the sum of a linear function and a periodic function. The following theorems reveals that any argument of an invertible *almost* periodic function can be represented as the sum of a linear function and an *almost* periodic function.

Theorem 2.25 (Bohr). *If $a \in GAP$, then there exist a real number $\kappa(a)$ and a function $b \in AP$ such that*

$$a(x) = e^{i\kappa(a)x} e^{b(x)} \quad \text{for all } x \in \mathbf{R}. \tag{2.24}$$

Proof. Let $\delta := \inf_{x \in \mathbf{R}} |a(x)|$ and choose an almost periodic polynomial

$$p(x) = \sum_{j=1}^{n} r_j e^{i\lambda_j x} \tag{2.25}$$

such that $\|a - p\|_\infty < \delta/2$. Then $p \in GAP$ and $\|1 - a/p\|_\infty < 1$. Hence, by (1.1), $a/p = e^c$ with some $c \in AP$. We are therefore left to prove the theorem in the case where $a = p$ is the almost periodic polynomial (2.25).

Let $\beta_1, \dots, \beta_m$ be as in Lemma 1.14. Then

$$p(x) = \sum_{j=1}^{n} r_j e^{i(k_1^{(j)}\beta_1 + \dots + k_m^{(j)}\beta_m)x}$$

with $k_1^{(j)}, \dots, k_m^{(j)} \in \mathbf{Z}$. Define $q : \mathbf{R}^m \to \mathbf{C}$ by

$$q(x_1, \dots, x_m) := \sum_{j=1}^{n} r_j e^{i(k_1^{(j)} x_1 + \dots + k_m^{(j)} x_m)}.$$

By Corollary 1.13, each point $(e^{ix_1}, \dots, e^{ix_m}) \in \mathbf{T}^m$ belongs to the closure of the set

$$\left\{ (e^{i\beta_1 x}, \dots, e^{i\beta_m x}) \in \mathbf{T}^m : x \in \mathbf{R} \right\}.$$

Consequently, $q(\mathbf{R}^m)$ is a subset of the closure of $p(\mathbf{R})$, which implies that $|q| \geq \delta/2$ on $\mathbf{R}^m$.

Let $\varphi : \mathbf{R}^m \to \mathbf{R}$ be any continuous function such that

$$q(x_1, \dots, x_m) = |q(x_1, \dots, x_m)| e^{i\varphi(x_1, \dots, x_m)}.$$

Since q and $|q|$ are 2π-periodic in each variable, there exist integers $l_1, \dots, l_m$ such that

$$\varphi(x_1, \dots, x_{j-1}, x_j + 2\pi, x_{j+1}, \dots, x_m) = \varphi(x_1, \dots, x_m) + 2\pi l_j$$

for all $(x_1, \dots, x_m) \in \mathbf{R}^m$. Letting

$$\psi(x_1, \dots, x_m) := \varphi(x_1, \dots, x_m) - (l_1 x_1 + \dots + l_m x_m)$$

we therefore obtain $\psi(x_1 + 2\pi, \ldots, x_m + 2\pi) = \psi(x_1, \ldots, x_m)$ for all $(x_1, \ldots, x_m)$ in $\mathbf{R}^m$. Thus,

$$q(x_1, \ldots, x_m) = e^{\theta(x_1, \ldots, x_m)} e^{i(l_1 x_1 + \ldots + l_m x_m)}$$

with $\theta = \log|q| + i\psi$, and

$$\theta(x_1 + 2\pi, \ldots, x_m + 2\pi) = \theta(x_1, \ldots, x_m) \tag{2.26}$$

for all $(x_1, \ldots, x_m) \in \mathbf{R}^m$. For $x \in \mathbf{R}$, we now get

$$p(x) = q(\beta_1 x, \ldots, \beta_m x) = e^{\theta(\beta_1 x, \ldots, \beta_m x)} e^{i(l_1 \beta_1 + \ldots + l_m \beta_m) x}.$$

In summary, we arrive at the representation (2.24) with

$$\kappa(a) = l_1 \beta_1 + \ldots + l_m \beta_m, \quad b(x) = c(x) + \theta(\beta_1 x, \ldots, \beta_m x),$$

and it remains to show that $x \mapsto \theta(\beta_1 x, \ldots, \beta_m x)$ is a function in AP. Since $\theta : \mathbf{R}^m \to \mathbf{C}$ is continuous and satisfies (2.26), we can approximate θ by trigonometric polynomial $s : \mathbf{R}^m \to \mathbf{C}$ (e.g., by the Fejér-Cesàro means). But if $s : \mathbf{R}^m \to \mathbf{C}$ is a trigonometric polynomial, then $x \mapsto s(\beta_1 x, \ldots, \beta_m x)$ is an almost periodic polynomial. This proves that $x \mapsto \theta(\beta_1 x, \ldots, \beta_m x)$ is an AP function. □

Definition 2.26. For $a \in GAP$, the number $\kappa(a) \in \mathbf{R}$ whose existence is guaranteed by Theorem 2.25 is obviously uniquely determined. It is called the *mean motion* of a. □

By Bohr's theorem, any continuous argument $\arg a$ of $a \in GAP$ is of the form $(\arg a)(x) = \kappa(a)x + b(x)$ where $\kappa(a)$ is the mean motion of a and b is a real-valued function in AP. Since b is necessarily bounded, we learn that the continuous arguments of GAP functions are essentially linear. Figures 2.4 and 2.5 exhibit two concrete examples. The following result shows a way to compute the mean motion of a function in GAP.

Proposition 2.27. *Let $A \subset (0, \infty)$ be an unbounded set and let $\{(x_\alpha, y_\alpha)\}_{\alpha \in A}$ be a family of intervals whose lengths $y_\alpha - x_\alpha$ approach ∞ as $\alpha \to \infty$. If $a \in GAP$ and $\arg a$ is any continuous argument of a, then*

$$\kappa(a) = \lim_{\alpha \to \infty} \frac{(\arg a)(y_\alpha) - (\arg a)(x_\alpha)}{y_\alpha - x_\alpha}.$$

Proof. By Theorem 2.25, $(\arg a)(x) = \kappa(a)x + b(x)$ with a real-valued function $b \in AP \subset L^\infty(\mathbf{R})$. Hence,

$$\lim_{\alpha \to \infty} \frac{(\arg a)(y_\alpha) - (\arg a)(x_\alpha)}{y_\alpha - x_\alpha} = \lim_{\alpha \to \infty} \frac{\kappa(a)(y_\alpha - x_\alpha) + b(y_\alpha) - b(x_\alpha)}{y_\alpha - x_\alpha},$$

and as $|b(y_\alpha) - b(x_\alpha)| \le 2\,\|b\|_\infty$, the limit is $\kappa(a)$. □

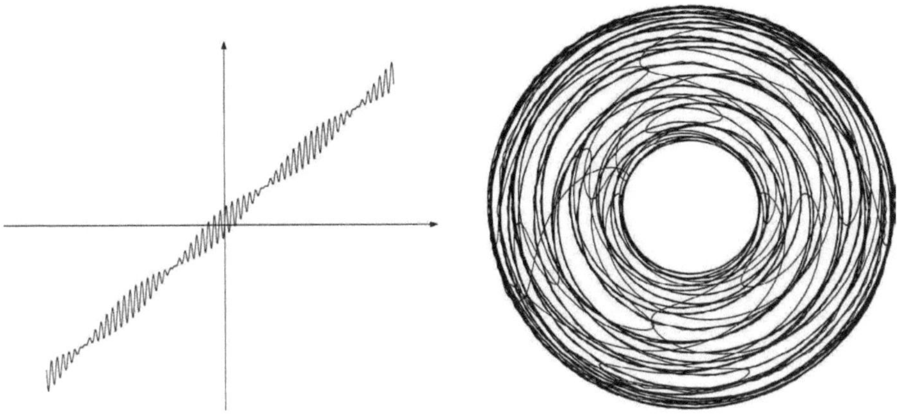

Figure 2.4: The function $\varphi(x) = x + 3\sin 3x + 3\sin 3.2x$ for x between -60 and 60 (left) and the function $(2+\sin 2x)e^{i\varphi(x)}$ for the same values of x (right).

Canonical choices of the families (x_α, y_α) occurring in Propositions 2.22 and 2.27 are $(-\alpha,\alpha)$, $(0,\alpha)$, $(-\alpha,0)$. One frequently thinks of α as the endpoint of a time interval and therefore writes T instead of α. Thus, Proposition 2.22 implies that the three limits

$$\lim_{T\to\infty}\frac{1}{2T}\int_{-T}^{T} a(x)\,dx, \quad \lim_{T\to\infty}\frac{1}{T}\int_{0}^{T} a(x)\,dx, \quad \lim_{T\to\infty}\frac{1}{T}\int_{-T}^{0} a(x)\,dx$$

are all equal to $M(a)$, while Proposition 2.27 shows that the three limits

$$\lim_{T\to\infty}\frac{(\arg a)(T)-(\arg a)(-T)}{2T}, \quad \lim_{T\to\infty}\frac{(\arg a)(T)-(\arg a)(0)}{T},$$
$$\lim_{T\to\infty}\frac{(\arg a)(0)-(\arg a)(-T)}{T}$$

are all equal to $\kappa(a)$.

2.9 *AP* Symbols

We are now in a position to consider Wiener-Hopf operators with symbols in AP.

Theorem 2.28 (Gohberg, Feldman, Coburn, Douglas). *Let $a \in AP$ and suppose a is not identically zero.*

(a) *If $a \notin GAP$, then $W(a)$ is not normally solvable.*

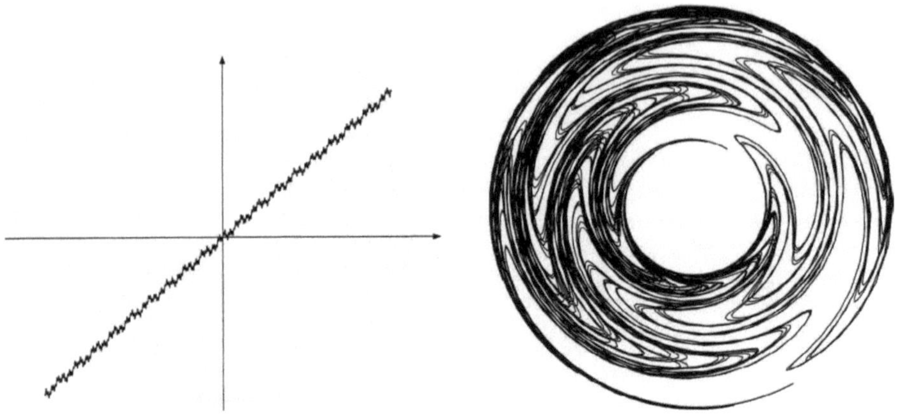

Figure 2.5: The function $\varphi(x) = x + \sin \pi x + \sin 10x$ for x between -60 and 60 (left) and the function $(2 + \sin 2x)e^{i\varphi(x)}$ for the same values of x (right).

(b) *If $a \in GAP$ and $\kappa(a) > 0$, then $W(a)$ is properly n-normal and left-invertible.*

(c) *If $a \in GAP$ and $\kappa(a) < 0$, then $W(a)$ is properly d-normal and right-invertible.*

(d) *If $a \in GAP$ and $\kappa(a) = 0$, then $W(a)$ is invertible.*

Proof. Part (a) follows from Corollary 2.8. So assume $a \in GAP$ and write $a(x) = e^{i\kappa(a)x}e^{b(x)}$ as in Theorem 2.25. Let $p = \sum_j r_j e_{\lambda_j}$ be an almost periodic polynomial such that $b = c + p$ and

$$\|e^c - 1\|_\infty \leq e^{\|c\|_\infty} - 1 < 1. \tag{2.27}$$

Write $p = p_- + p_+$ where $\Omega(p_-) \subset (-\infty, 0)$ and $\Omega(p_+) \subset [0, \infty)$ (in other words: p_- contains the terms of p in which $\lambda_j < 0$ and p_+ contains the terms of p in which $\lambda_j \geq 0$). We then have

$$a(x) = e^{p_-(x)}e^{c(x)}e^{i\kappa(a)x}e^{p_+(x)}.$$

Clearly, $e^{p_-} \in GH^\infty_-$ and $e^{p_+} \in GH^\infty_+$. Thus, by Proposition 2.17,

$$W(a(x)) = W(e^{p_-(x)})W(e^{c(x)}e^{i\kappa(a)x})W(e^{p_+(x)}),$$

and since the operators $W(e^{p_\pm})$ are invertible, the inverses being $W(e^{-p_\pm})$, it follows that $W(a)$ is strongly Φ-equivalent to

$$W(e^{c(x)}e^{i\kappa(a)x}). \tag{2.28}$$

If $\kappa(a) > 0$, then $e^{i\kappa(a)x}$ is a function in H^∞_+ and hence

$$W\big(e^{c(x)}e^{i\kappa(a)x}\big) = W\big(e^{c(x)}\big)W\big(e^{i\kappa(a)x}\big)$$

by Proposition 2.17. The operator $W(e^c)$ is invertible due to (2.27), and from Example 1.23 we obtain that $W(e^{i\kappa(a)x})$ is injective with an infinite-dimensional cokernel. Therefore $W(e^{i\kappa(a)x})$ and thus also $W(a)$ are properly n-normal. From Corollary 2.6(c) we deduce that $W(a)$ is left-invertible. This completes the proof of part (b). Part (c) can be proved analogously or can be derived from part (b) by passage to adjoints. Finally, if $\kappa(a) = 0$, then the operator (2.28) is invertible because of (2.27). □

Figure 2.6 illustrates Theorem 2.28.

Figure 2.6: Let $p(t) = -10it - 4it^2 - 5it^{-5}$. The functions $a(x) = p(e^{i\pi \sin x})$ and $b(x) = p(e^{ix})$ are both 2π-periodic and have the same range as p. This range is shown in the left picture. However, the spectrum of $W(a)$ is the set on the left, while the spectrum of $W(b)$ is the set on the right.

Using the results we have so far established, it is easy to explore the semi-Fredholm properties of Wiener-Hopf operators with symbols in $AP + C_0(\dot{\mathbf{R}})$. Let $a \in AP + C_0(\dot{\mathbf{R}})$ and write $a = b + f$ with $b \in AP$ and $f \in C_0(\dot{\mathbf{R}})$. By Theorem 2.7, $W(a)$ cannot be semi-Fredholm if $a \notin GL^\infty(\mathbf{R})$. So assume $a \in GL^\infty(\mathbf{R})$. Then $a \in G(AP + C_0(\dot{\mathbf{R}}))$ due to Proposition 1.4(a), and Theorem 1.16 implies that $b \in GAP$. Hence

$$a = b(1 + b^{-1}f)$$

with $b \in GAP$ and $1 + b^{-1}f \in GC(\dot{\mathbf{R}})$. Since

$$W(a) = W(b)W(1 + b^{-1}f) + K_1 = W(1 + b^{-1}f)W(b) + K_2$$

with compact operators K_1, K_2 (Corollary 2.12), it follows from Theorems 2.2, 2.15, 2.28, and Corollary 2.6(c) that

$$\begin{aligned} &W(a) \ \text{ is properly } n\text{-normal and left-invertible if } \ \kappa(b) > 0, \\ &W(a) \ \text{ is properly } d\text{-normal and right-invertible if } \ \kappa(b) < 0, \\ &W(a) \ \text{ is Fredholm and } \ \operatorname{Ind} W(a) = -\operatorname{wind}(1 + b^{-1}f) \ \text{ if } \ \kappa(b) = 0. \end{aligned}$$

One can reformulate this result in terms of only a (and not in terms of b and $1 + b^{-1}f$). We renounce doing this, since $AP + C_0(\dot{\mathbf{R}}) \subset SAP$ and SAP symbols are considered in detail in the next two chapters. We merely want to draw the reader's attention to the fact that $W(AP)$ does not contain Fredholm operators of nonzero index but that $W(AP + C_0(\dot{\mathbf{R}}))$ contains such operators.

Notes

The books [46], [47], [66], [70], [92], [99], [165] all contain the basics of the theory of scalar Wiener-Hopf operators. Proofs of the results quoted in Section 2.1 are in [99], for example. Theorem 2.5 and its Corollary 2.6 are due to Coburn [58] and Simonenko [207]; for operators with continuous symbols, such results were already established by Gohberg [88]. Theorem 2.7 unites the results by several mathematicians. Hartman and Wintner [106] showed that $a \in GL^\infty(\mathbf{R})$ if $W(a)$ is invertible, while Simonenko [207] proved that $a \in GL^\infty(\mathbf{R})$ if $W(a)$ is merely known to be semi-Fredholm. The inclusion $\operatorname{sp} W(a) \subset \operatorname{conv} \mathcal{R}(a)$ was first observed by Brown and Halmos [50]. The connectedness of $\operatorname{sp} W(a)$ and $\operatorname{sp}_{\text{ess}} W(a)$ was proved by Widom [227] and Douglas [66], respectively.

Standard texts on Hankel operators are the books [172] and [175]. By virtue of the role Hankel operators play in connection with Toeplitz and Wiener-Hopf operators, the basic results on Hankel operators are also treated in the books quoted in the beginning of the previous paragraph. Results like Proposition 2.10 have been used for many decades; in the form we have cited it, Proposition 2.10 appeared in Widom's paper [228] for the first time. Proposition 2.11 is a classical result. Theorem 2.14 has its origin in Simonenko's works [205], [206]. For more on this theorem and on so-called local principles we refer to [46], [66], [70], [99], [102], [165].

The results of Section 2.4 are Gohberg's [87]. Proofs of (2.7) and Theorem 2.16 are also in [46] and [66].

Proposition 2.17 is the basic idea of Wiener-Hopf factorization ([94], [143], [231]) and one of the most important tools for tackling Wiener-Hopf operators. Theorem 2.18 is due to Hartman [105], and this theorem clearly implies that $C(\dot{\mathbf{R}}) + H^\infty_-$ and $C(\dot{\mathbf{R}}) + H^\infty_+$ are closed subalgebras of $L^\infty(\mathbf{R})$. Sarason [195] was the first to make this explicit and thus to open the door to a new world in the analysis of Wiener-Hopf operators (and plenty of other topics). Theorem 2.19 was established by Douglas [65], [66].

The beautiful Theorem 2.20 has grown out of the work by Widom [224], Simonenko [203], Devinatz [63], Shamir [200], and Gohberg [89]. Theorem 2.21 is Gohberg and Krupnik's [95]. Full proofs of the theorems of Section 2.6 are also in [46], [47], [66], [70], [92], [99], [165].

The results of Sections 2.7 and 2.8 are standard in the theory of almost periodic functions and can be found in [60], [153], [154], for example. Theorem 2.28 was established independently by Coburn and Douglas [59] and Gohberg and Feldman [90].

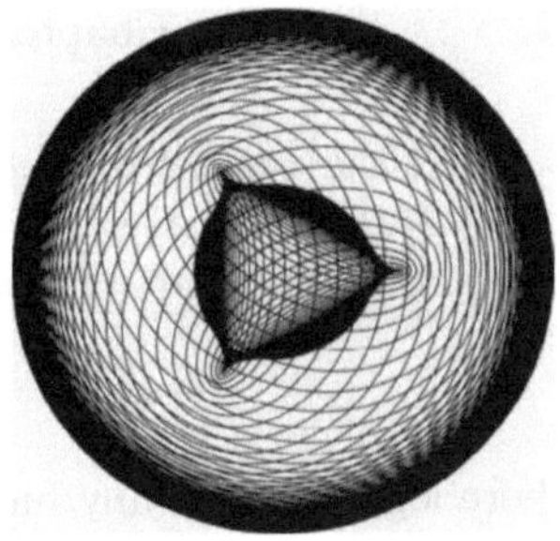

Chapter 3

Scalar Wiener-Hopf Operators with SAP Symbols

Semi-almost periodic (SAP) symbols constitute one of the simplest classes of symbols with discontinuities beyond jumps, and as shown in Chapter 1, SAP symbols emerge naturally in several settings. Armed with the results on $C+H^\infty$, PC, and AP symbols established in Chapter 2, we here present Sarason's theory of semi-Fredholmness for scalar Wiener-Hopf operators with SAP symbols. A key result in the analysis of the matter is Sarason's lemma, which says that the product of an almost periodic function in H^∞ with mean value zero and a function in $C(\overline{\mathbf{R}})$ is a function in $C+H^\infty$. Our approach is different from Sarason's, which can be characterized as an elegant combination of localization techniques and C^*-algebra arguments. We rather follow Duduchava and Saginashvili, who gave a proof of Sarason's theorem that can be carried over to the L^p case without undue difficulty.

3.1 The Geometric Mean

When considering Wiener-Hopf operators with symbols in SAP, we need one more number associated with functions in GAP.

Definition 3.1. Let $a \in GAP$. By Theorem 2.25,

$$a(x) = e^{i\kappa(a)x} e^{\psi(x)} \tag{3.1}$$

with $\psi \in AP$. The number

$$\mathbf{d}(a) := e^{M(\psi)}$$

is called the *geometric mean value* of the function a. Note that if $\psi_1, \psi_2 \in AP$ satisfy (3.1), then $\psi_1 - \psi_2 = 2k\pi i$ with $k \in \mathbf{Z}$ and hence $M(\psi_1) - M(\psi_2) = 2k\pi i$. This shows that $\mathbf{d}(a)$ is well-defined. If $\kappa(a) = 0$, we may also write

$$\mathbf{d}(a) = e^{M(\log a)},$$

where $\log a \in AP$ is any function for which $a = e^{\log a}$ (recall (1.1)). □

Example 3.2. (a) Let $a(x) = e^{i\mu \sin x}$, where μ is a real number. We have

$$\begin{aligned}
\mathbf{d}(a) &= \exp\left(i\mu \frac{1}{2\pi}\int_0^{2\pi} \sin x\, dx\right) = 1,\\
M(a) &= \frac{1}{2\pi}\int_0^{2\pi} e^{i\mu \sin x}\, dx = \frac{1}{2\pi}\int_0^{2\pi} \cos(\mu \sin x)\, dx\\
&= 1 + \frac{1}{2\pi}\sum_{n=1}^{\infty} \frac{(-1)^n \mu^{2n}}{(2n)!}\int_0^{2\pi} \sin^{2n} x\, dx\\
&= 1 + \sum_{n=1}^{\infty} \frac{(-1)^n}{(n!)^2}\left(\frac{\mu}{2}\right)^{2n}. \qquad (3.2)
\end{aligned}$$

(b) Now fix $\mu > 0$ and $\varepsilon \in (0, 1/4]$. Let ψ be the 1-periodic function that increases linearly from 0 to μ on $[0, \varepsilon]$, is constant equal to μ on $[\varepsilon, 1/2-\varepsilon]$, decreases linearly from μ to $-\mu$ on $[1/2-\varepsilon, 1/2+\varepsilon]$, is constant equal to $-\mu$ on $[1/2+\varepsilon, 1-\varepsilon]$, and increases linearly from $-\mu$ to 0 on $[1-\varepsilon, 1]$. Put $a(x) = e^{i\psi(x)}$. Easy computations give $\mathbf{d}(a) = 1$ and

$$M(a) = (1 - 4\varepsilon)\cos\mu + 4\varepsilon\frac{\sin\mu}{\mu}. \qquad (3.3)$$

Figures 3.1 and 3.2 illustrate the examples. Notice the different behavior of the geometric mean $\mathbf{d}(a)$ and the arithmetic mean $M(a)$. In particular, observe that $M(a)$ need not be zero if μ is an integral multiple of 2π. □

The question whether an operator $W(a)$ with $a \in GSAP$ is semi-Fredholm or Fredholm can be decided by computing the four numbers $\kappa(a_l)$, $\kappa(a_r)$, $\mathbf{d}(a_l)$, $\mathbf{d}(a_r)$ where a_l, a_r are the almost periodic representatives of a given by Theorem 1.21. Notice that this theorem implies that a_l, $a_r \in GAP$ whenever $a \in GSAP$, so that $\kappa(a_l)$, $\kappa(a_r)$, $\mathbf{d}(a_l)$, $\mathbf{d}(a_r)$ are well-defined in case $a \in GSAP$. The following result enables us to compute the numbers in terms of solely a.

Proposition 3.3. *Let $A \subset (0, \infty)$ be an unbounded subset and let $\{(x_\alpha, y_\alpha)\}_{\alpha\in A}$ be a family of intervals such that*

$$x_\alpha \geq 0 \text{ and } y_\alpha - x_\alpha \to +\infty \text{ as } \alpha \to \infty.$$

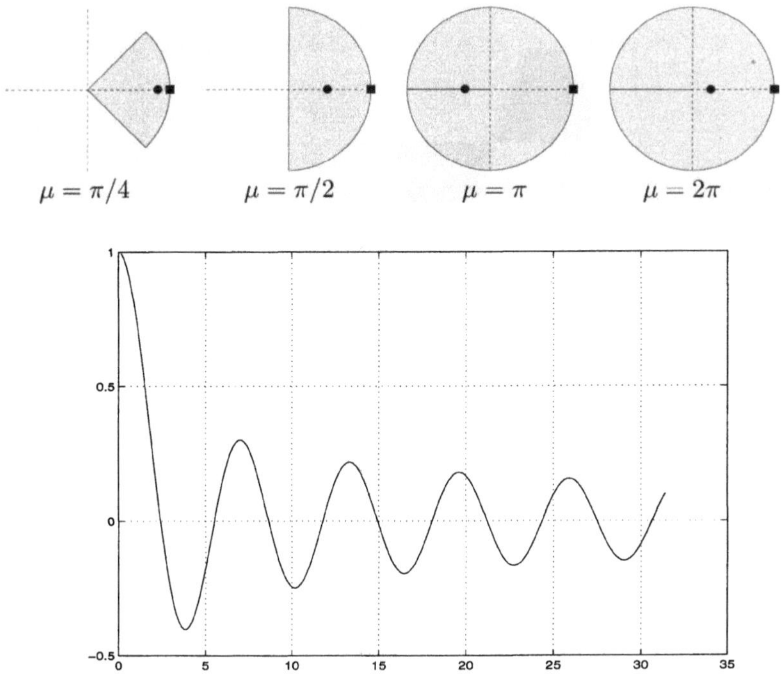

Figure 3.1: Let a be as in Example 3.2(a). In the top row we see $\mathbf{d}(a) = 1$ (small square) and $M(a)$ (dot) for four values of μ. The bottom plot shows function (3.2) for $\mu \in [0, 10\pi]$.

Suppose $a \in GSAP$ and $\arg a$ *is any continuous argument of a. Then*

$$\kappa(a_r) = \lim_{\alpha\to\infty} \frac{(\arg a)(y_\alpha) - (\arg a)(x_\alpha)}{y_\alpha - x_\alpha}, \tag{3.4}$$

$$\kappa(a_l) = \lim_{\alpha\to\infty} \frac{(\arg a)(-x_\alpha) - (\arg a)(-y_\alpha)}{y_\alpha - x_\alpha}. \tag{3.5}$$

Further, with $\psi(x) := \log|a(x)| + i(\arg a)(x)$, we have

$$\mathbf{d}(a_r) = \exp \lim_{\alpha\to\infty} \frac{1}{y_\alpha - x_\alpha} \int_{x_\alpha}^{y_\alpha} (\psi(x) - i\kappa(a_r)x)\, dx, \tag{3.6}$$

$$\mathbf{d}(a_l) = \exp \lim_{\alpha\to\infty} \frac{1}{y_\alpha - x_\alpha} \int_{-y_\alpha}^{-x_\alpha} (\psi(x) - i\kappa(a_l)x)\, dx. \tag{3.7}$$

Proof. Let $a_r(x) = e^{i\kappa(a_r)x} e^{\psi_r(x)}$ with $\psi_r \in AP$. By Theorem 1.21,

$$(\arg a)(x) = \kappa(a_r)x + \operatorname{Im}\psi_r(x) + 2k\pi + o(1) \quad \text{as} \quad x \to +\infty$$

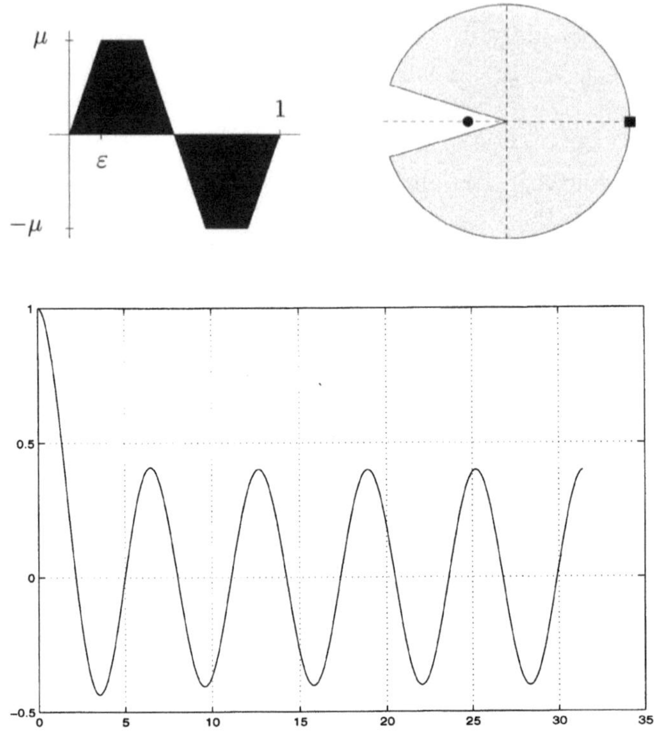

Figure 3.2: The upper left picture shows the function $\psi(x)$ of Example 3.2(b). The upper right plot depicts the point $\mathbf{d}(a)$ (small square) and the point $M(a)$ (dot) for $\varepsilon = 0.15$ and $\mu = 9\pi/10$. In the bottom picture we plotted the function (3.3) with $\varepsilon = 0.15$ for $\mu \in [0, 10\pi]$.

with $k \in \mathbf{Z}$. This gives (3.4). By definition, $\mathbf{d}(a_r) = \exp M(\psi_r)$. From Proposition 2.22 we know that

$$M(\psi_r) = \lim_{\alpha\to\infty} \frac{1}{y_\alpha - x_\alpha} \int_{x_\alpha}^{y_\alpha} \psi_r(x)\,dx.$$

Because

$$\begin{aligned} \psi_r(x) &= \operatorname{Re}\psi_r(x) + i\operatorname{Im}\psi_r(x) \\ &= \log|a(x)| + i\Big((\arg a)(x) - \kappa(a_r)x - 2k\pi - o(1)\Big) \end{aligned}$$

as $x \to +\infty$, we arrive at (3.6). The formulas (3.5) and (3.7) can be shown analogously. □

3.2 Sarason's Lemma

Recall that $e_\lambda(x) := e^{i\lambda x}$. Put

$$AP^+ := \mathrm{alg}_{L^\infty(\mathbf{R})}\{e_\lambda : \lambda \geq 0\}, \quad AP^- := \mathrm{alg}_{L^\infty(\mathbf{R})}\{e_\lambda : \lambda \leq 0\}. \tag{3.8}$$

Obviously, $AP^\pm$ are closed subalgebras of $AP \cap H^\infty_\pm$. Moreover, it is also clear that $AP^- \cap AP^+ = \mathbf{C}$ (since $H^\infty_- \cap H^\infty_+ = \mathbf{C}$) and that

$$\Omega(a) \subset [0, \infty) \ \text{ for } \ a \in AP^+, \quad \Omega(a) \subset (-\infty, 0] \ \text{ for } \ a \in AP^-.$$

Lemma 3.4. *We have $GAP^\pm = G_0 AP^\pm$, i.e., $a \in GAP^\pm$ if and only if there exists a $\psi \in AP^\pm$ such that $a = e^\psi$.*

Proof. The "if" part is trivial. To prove the "only if" part, suppose $a \in GAP^+$. Then $a \in GH^\infty_+$, and hence $W(a)$ is invertible by Proposition 2.17. From Theorem 2.28 we therefore deduce that $\kappa(a) = 0$. Hence $a = e^\psi$ with $\psi \in AP$ due to Bohr's theorem. Let p_n be almost periodic polynomials such that $\|\psi - p_n\|_\infty \to 0$. Then $\|e^{p_n} - e^\psi\|_\infty \to 0$ and hence, $e^{p_n} \in GAP^+$ for all sufficiently large n together with $e^\psi \in GAP^+$. Write $p_n = p_n^- + p_n^+$ with $p_n^\pm \in H^\infty_\pm$. Since $e^{p_n} = e^{p_n^-} e^{p_n^+}$ and $e^{p_n^+} \in GAP^+$, it follows that $e^{p_n^-} \in GAP^+$ for all n large enough. In particular, $e^{p_n^-} \in H^\infty_+$, whence $e^{p_n^-} \in H^\infty_+ \cap H^\infty_- = \mathbf{C}$ and thus $p_n^- \in \mathbf{C}$ for all sufficiently large n. Consequently, $p_n \in H^\infty_+$ for all n large enough, which implies that $\psi = \lim p_n$ is in AP^+. The proof is analogous for GAP^-. ☐

The key to Wiener-Hopf operators with SAP symbols is the following result by Sarason: if $h \in H^\infty_\pm \cong H^\infty(\mathbf{C}_\pm)$ and $h(iy) \to 0$ as $y \to \pm\infty$, then $hv \in C(\dot{\mathbf{R}}) + H^\infty_\pm$ for every function $v \in C(\overline{\mathbf{R}})$ (here "$\cong$" means that we identify $H^\infty_\pm$ and $H^\infty(\mathbf{C}_\pm)$). In fact, the following baby version of this result is sufficient for our purposes.

Lemma 3.5 (Sarason). *If $h \in AP^\pm$ and $M(h) = 0$, then $hv \in C(\dot{\mathbf{R}}) + H^\infty_\pm$ for every $v \in C(\overline{\mathbf{R}})$.*

Proof. Let us consider the case where $h \in AP^+$. Since $C(\dot{\mathbf{R}}) + H^\infty_+$ is a closed subalgebra of $L^\infty(\mathbf{R})$ and AP^+ is defined by (3.8), we can assume that $h(x) = e^{i\alpha x}$ with $\alpha > 0$.

The function

$$\varphi(z) := \pi \frac{\sin z}{z} \frac{1}{\pi^2 - z^2}$$

is analytic in $\mathbf{C}$, and there is a constant $M < \infty$ such that

$$|\varphi(z)| \leq M \frac{e^{|\mathrm{Im}\, z|}}{1 + |z|^3} \quad \text{for all } \ z \in \mathbf{C}.$$

Put

$$\Phi(z) := \int_0^z \varphi(\zeta)\,d\zeta. \tag{3.9}$$

Then Φ is also analytic in $\mathbf{C}$. If $\operatorname{Im} z \geq 0$, then taking the integral (3.9) along the line segment $[0, z]$ gives

$$|e^{iz}\Phi(z)| \leq e^{-\operatorname{Im} z} \int_0^z M \frac{e^{\operatorname{Im} z}}{1+|\zeta|^3}\,|d\zeta| \leq N < \infty,$$

which shows that $e^{iz}\Phi(z)$ is a function in $H^\infty(\mathbf{C}_+)$. Let $\sigma = \Phi|\mathbf{R}$ be the restriction of Φ to $\mathbf{R}$. We have

$$\begin{aligned}
\sigma(+\infty) &= \int_0^\infty \varphi(x)\,dx = \frac{1}{2}\int_{-\infty}^\infty \varphi(x)\,dx = \frac{\pi}{2}\int_{-\infty}^\infty \frac{\sin x}{x}\frac{dx}{\pi^2 - x^2} \\
&= \frac{\pi}{2}\int_{-\infty}^\infty \sin x \, \frac{1}{2\pi^2}\left(\frac{2}{x} - \frac{1}{x-\pi} - \frac{1}{x+\pi}\right) dx \\
&= \frac{1}{4\pi}\int_{-\infty}^\infty \left(\frac{2\sin x}{x} + \frac{\sin(x-\pi)}{x-\pi} + \frac{\sin(x+\pi)}{x+\pi}\right) dx \\
&= \frac{1}{\pi}\int_{-\infty}^\infty \frac{\sin x}{x} = 1
\end{aligned}$$

and $\sigma(-\infty) = -\sigma(+\infty) = -1$. Consequently, every function $v \in C(\overline{\mathbf{R}})$ can be written in the form

$$v(x) = \frac{v(-\infty) + v(+\infty)}{2} + \frac{v(+\infty) - v(-\infty)}{2}\sigma(\alpha x) + v_0(x) \tag{3.10}$$

where v_0 is some function in $C_0(\dot{\mathbf{R}})$. Since $e^{i\alpha x}v_0(x)$ is a function in $C_0(\dot{\mathbf{R}})$ and $e^{i\alpha x}\sigma(\alpha x)$ has been shown to be a function in H_+^∞, it follows that $e^{i\alpha x}v(x)$ is a function in $C(\dot{\mathbf{R}}) + H_+^\infty$. □

3.3 Canonical Representatives

In this section we consider some simple SAP symbols.

In Section 1.7, we defined the notion of strong Φ-equivalence. We now need a weaker equivalence relation: two operators $A \in \mathcal{B}(X)$ and $B \in \mathcal{B}(X)$ are called *weakly Φ-equivalent* if either none of them is normally solvable or if both A and B are properly n-normal, or both A and B are properly d-normal, or both A and B are Fredholm. An intermediate equivalence relation is the following: two operators $A \in \mathcal{B}(X)$ and $B \in \mathcal{B}(X)$ are said to be *Φ-equivalent* if either none of them is normally solvable or both A and B are semi-Fredholm with the same index.

Lemma 3.6. *Let $\alpha, \beta \in \mathbf{C}$, $\lambda, \mu \in \mathbf{R}$, and let*

$$b(x) := \alpha\big(1-u(x)\big)e^{i\lambda x} + \beta u(x)e^{i\mu x} + b_0(x)$$

with $b_0 \in C_0(\dot{\mathbf{R}})$. Put

$$c(x) := \alpha\chi_-(x)e^{i\lambda x} + \beta\chi_+(x)e^{i\mu x}, \tag{3.11}$$

where $\chi_- := \chi_{(-\infty,0)}$ and $\chi_+ := \chi_{(0,\infty)}$. If $b \in GSAP$ and $0 \notin [\alpha,\beta]$, then $W(b)$ and $W(c)$ are weakly Φ-equivalent.

Proof. Clearly, the function $p := b^{-1}(c-b)$ belongs to PC_0 and is continuous on $\dot{\mathbf{R}} \setminus \{0\}$. Because $c = b(1+p)$, we deduce from Theorem 2.13 that

$$W(c) = W(b)W(1+p) + K \tag{3.12}$$

with some compact operator K. Since $0 \notin [\alpha, \beta]$, the function $1+p = b^{-1}c$ is continuous and nonzero on $\dot{\mathbf{R}} \setminus \{0\}$ and we have

$$0 \notin \big[1+p(0-0),\, 1+p(0+0)\big] = \big(1/b(0)\big)[\alpha,\beta].$$

Hence, $W(1+p)$ is Fredholm due to Theorem 2.20. This in conjunction with (3.12) and Theorems 2.2 and 2.3 shows that $W(c)$ and $W(b)$ are weakly Φ-equivalent. □

The ranges $b(\mathbf{R})$ of the simple canonical representatives $b(x)$ of Lemma 3.6 are astonishingly beautiful. For 23 different choices of the parameters λ and μ in

$$b(x) = (1-u(x))e^{i\lambda x} + 3u(x)e^{i\mu x}, \quad u(x) = \frac{1}{2} + \frac{1}{\pi}\arctan\frac{x}{10}, \tag{3.13}$$

the range of $b([-300, 300])$ is plotted on the first pages of the 23 chapters of this book. In all these pictures λ and μ are nonzero. Figure 3.3 shows the result for $\lambda = 0$ or $\mu = 0$.

Lemma 3.7. *Let $\alpha, \beta \in \mathbf{C} \setminus \{0\}$, $\lambda, \mu \in \mathbf{R}$, and suppose $\max\{\lambda,\mu\} > 0$. If c is given by (3.11), then $\operatorname{Ker} W(c) = \{0\}$ and the algebraic complement of $\operatorname{Im} W(c)$ in $L^2(\mathbf{R}_+)$ is infinite-dimensional.*

Proof. For the sake of definiteness, assume that $\mu > 0$. Let $f_+, g_+ \in L^2(\mathbf{R}_+)$ and suppose $W(c)f_+ = g_+$. This is equivalent to the equality $\chi_+ F^{-1} c F f_+ = g_+$ and thus to

$$\alpha\chi_+F^{-1}e_\lambda FF^{-1}\chi_-Ff_+ + \beta\chi_+F^{-1}e_\mu FF^{-1}\chi_+Ff_+ = g_+.$$

By virtue of Example 1.10 and (2.13), the latter equality can be written in the form

$$\alpha\chi_+U_\lambda Pf_+ + \beta\chi_+U_\mu Qf_+ = g_+. \tag{3.14}$$

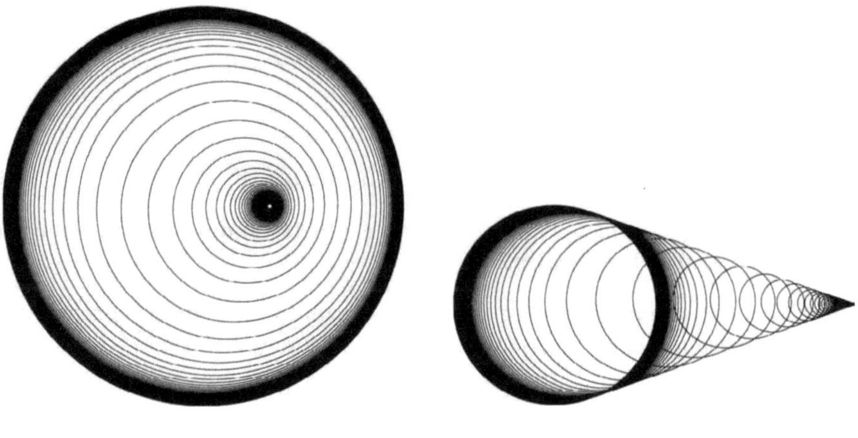

Figure 3.3: The function (3.13) for $\lambda = 0, \mu = 2$ (left) and $\lambda = 2, \mu = 0$ (right).

As $Q = I - P$, (3.14) holds if and only if

$$\alpha\chi_+ U_\lambda P f_+ - \beta\chi_+ U_\mu P f_+ + \beta\chi_+ U_\mu f_+ = g_+. \tag{3.15}$$

If $0 < x < \mu$, then $\beta\chi_+(x)(U_\mu f_+)(x) = \beta f_+(x - \mu) = 0$, and hence (3.15) implies that

$$\alpha(Pf_+)(x - \lambda) - \beta(Pf_+)(x - \mu) = g_+(x) \ \text{ for } x \in (0, \mu). \tag{3.16}$$

Now suppose $f_+ \in \operatorname{Ker} W(c)$, i.e., let $g_+ = 0$. Since $Pf_+ \in H^2_+$ and H^2_+ is translation invariant, the left-hand side of (3.16) is a function in H^2_+. As this function vanishes on $(0, \mu)$, it must vanish almost everywhere (F. and M. Riesz theorem). Thus $\alpha U_\lambda P f_+ = \beta U_\mu P f_+$. Inserting this into (3.14) (with $g_+ = 0$) we obtain

$$0 = \beta\chi_+ U_\mu P f_+ + \beta\chi_+ U_\mu Q f_+ = \beta\chi_+ U_\mu f_+,$$

and it follows that $f_+(x - \mu) = 0$ for all $x > 0$, which gives $f_+ = 0$. This proves that $\operatorname{Ker} W(c) = \{0\}$.

Now put $g_+ := \chi_{(0,\varepsilon)}$ where $\varepsilon \in (0, \mu)$. The left-hand side of (3.16) is again in H^2_+, and therefore $\chi_{(0,\varepsilon)}$ must be the restriction to $(0, \mu)$ of a function in H^2_+. Since $0 < \varepsilon < \mu$, this contradicts the Riesz brothers' theorem. Hence $\chi_{(0,\varepsilon)} \notin \operatorname{Im} W(c)$. As the functions $\chi_{(0,\varepsilon)}$ ($\varepsilon \in (0, \mu)$) are linearly independent, it results that the algebraic complement of $W(c)$ is of infinite dimension. □

Lemma 3.8. *Let $\alpha, \beta \in \mathbf{C} \setminus \{0\}$, $b_0 \in C_0(\dot{\mathbf{R}})$, and suppose that either*

$$b(x) = \alpha\big(1 - u(x)\big) + \beta u(x) e^{i\mu x} + b_0(x), \quad \mu > 0 \tag{3.17}$$

or

$$b(x) = \alpha\big(1 - u(x)\big)e^{i\lambda x} + \beta u(x) + b_0(x), \quad \lambda > 0. \tag{3.18}$$

If $b \in GSAP$ then $W(b)$ is properly n-normal.

Proof. For the sake of definiteness, let b be given by (3.17). Put

$$s_1 := (1/\alpha)(1-u) + (4/\beta)ue_{-\mu}$$

and choose $s_0 \in C_0(\dot{\mathbf{R}})$ so that $s := s_1 + s_0 \in GSAP$. Lemma 3.5 implies that

$$ue_\mu \in C(\dot{\mathbf{R}}) + H^\infty_+, \; ue_{-\mu} \in C(\dot{\mathbf{R}}) + H^\infty_-.$$

Hence, by Corollary 2.12 and Proposition 2.17,

$$\begin{aligned} W(s)W(b) &= \big(W(1-u)\big)^2 + W\Big(\frac{\beta}{\alpha}(1-u)ue_\mu\Big) \\ &+ W\Big(\frac{4\alpha}{\beta}u(1-u)e_{-\mu}\Big) + W(4u^2) + W(s_1b_0) + W(s_0b) + K \end{aligned}$$

with some compact operator K. Obviously,

$$\delta_0 := \frac{\beta}{\alpha}(1-u)ue_\mu + \frac{4\alpha}{\beta}u(1-u)e_{-\mu} + s_1b_0 + s_0b \in C_0(\dot{\mathbf{R}}),$$

and we have

$$W(s)W(b) = \big(W(1-u)\big)^2 + W(4u^2 + \delta_0) + K. \tag{3.19}$$

The operator $A := (W(1-u))^2 + W(4u^2+\delta_0)$ belongs to $\operatorname{alg} W(PC)$ and we can apply Theorem 2.21 to this operator. We have

$$(\Gamma A^\pi)(x,t) = \Big(\big(\Gamma W^\pi(1-u)\big)(x,t)\Big)^2 + \big(\Gamma W^\pi(4u^2+\delta_0)\big)(x,t).$$

For $x \in \mathbf{R}$, this is

$$\big(1-u(x)\big)^2 + 4u^2(x) + \delta_0(x) = s(x)b(x) \neq 0,$$

while for $x = \infty$ we get

$$\begin{aligned} &\Big(\big(1-u(+\infty)\big)(1-t) + \big(1-u(-\infty)\big)t\Big)^2 + 4u^2(+\infty)(1-t) + 4u^2(-\infty)t \\ &= t^2 + 4(1-t) = (t-2)^2 \neq 0 \quad \big(t \in [0,1]\big). \end{aligned}$$

Thus, Theorem 2.21 implies that A is Fredholm.

From (3.19) we now deduce that $W(b)$ is left-Fredholm. Assume $W(b)$ is even Fredholm. Since Fredholmness is stable under small perturbations (Theorem 2.1), we can suppose that $0 \notin [\alpha, \beta]$. Define c by (3.11) with $\lambda = 0$. From Lemma 3.6 we infer that $W(c)$ is also Fredholm, which contradicts Lemma 3.7. Thus, $W(b)$ cannot be Fredholm. This shows that $W(b)$ is properly n-normal. □

3.4 Sarason's Theorem

The following theorem settles the semi-Fredholm theory of Wiener-Hopf operators with SAP symbols. Figure 3.4 shows an example.

Theorem 3.9 (Sarason). *Let $a \in SAP \setminus \{0\}$.*

(a) *If $a \notin GSAP$, then $W(a)$ is not normally solvable.*

(b) *If $a \in GSAP$ and $\kappa(a_l) \cdot \kappa(a_r) < 0$, then $W(a)$ is not normally solvable.*

(c) *If $a \in GSAP$, $\kappa(a_l) \cdot \kappa(a_r) \geq 0$, and $\kappa(a_l) + \kappa(a_r) > 0$, then $W(a)$ is properly n-normal and left-invertible.*

(d) *If $a \in GSAP$, $\kappa(a_l) \cdot \kappa(a_r) \geq 0$, and $\kappa(a_l) + \kappa(a_r) < 0$, then $W(a)$ is properly d-normal and right-invertible.*

(e) *If $a \in GSAP$, $\kappa(a_l) = \kappa(a_r) = 0$, and $0 \in [\mathbf{d}(a_l), \mathbf{d}(a_r)]$, then $W(a)$ is not normally solvable.*

(f) *If $a \in GSAP$, $\kappa(a_l) = \kappa(a_r) = 0$, and $0 \notin [\mathbf{d}(a_l), \mathbf{d}(a_r)]$, then $W(a)$ is Fredholm.*

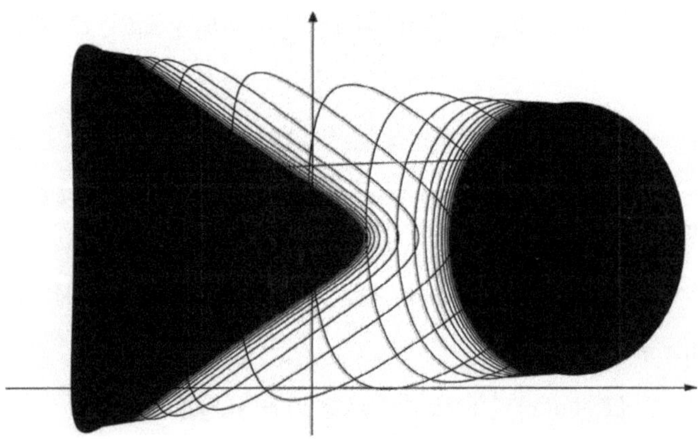

Figure 3.4: The picture shows the range of a function $a \in SAP$, the points at which $\kappa(a_l) \neq 0$ or $\kappa(a_r) \neq 0$, and the line segment $[\mathbf{d}(a_l), \mathbf{d}(a_r)]$. As this set of "forbidden" points does not contain the origin, the operator $W(a)$ is Fredholm. Notice that the set seen in the picture is *not* the essential spectrum of $W(a)$ (see Chapter 4).

The rest of this section is devoted to the proof of this theorem. Our proof is based on ideas by Saginashvili and Duduchava. This proof is less elegant than Sarason's original proof, but it is still simple and has the advantage that it can be carried over to operators on H^p and L^p spaces.

Lemma 3.10. *If $a, b \in L^\infty(\mathbf{R}) \setminus \{0\}$ and if there are functions $f_\pm \in G(C(\dot{\mathbf{R}}) + H^\infty_\pm)$ such that $a = f_- b f_+$, then $W(a)$ and $W(b)$ are weakly Φ-equivalent.*

Proof. By (2.4) and the trivial part of Theorem 2.18,

$$W(a) = W(f_-)W(b)W(f_+) + K$$

with some compact operator K. Since $W(f_\pm)$ are Fredholm due to Theorem 2.19(c), the assertion follows from Theorems 2.2, 2.3 and Corollary 2.6(a). □

Lemma 3.11. *Let $a, b \in GSAP$ and suppose the almost periodic representatives a_l, b_l, a_r, b_r are connected by*

$$a_l = \varphi_l^- b_l \varphi_l^+ \;\; \textit{with} \;\; \varphi_l^\pm \in GAP^\pm \;\; \textit{and} \;\; \mathbf{d}(\varphi_l^\pm) = 1,$$
$$a_r = \varphi_r^- b_r \varphi_r^+ \;\; \textit{with} \;\; \varphi_r^\pm \in GAP^\pm \;\; \textit{and} \;\; \mathbf{d}(\varphi_r^\pm) = 1.$$

Then there exist $f_\pm \in G(C(\dot{\mathbf{R}}) + H^\infty_\pm) \cap GSAP$ such that $a = f_- b f_+$.

Proof. By Lemma 3.4, there are $\psi_l^\pm \in AP^\pm$, $\psi_r^\pm \in AP^\pm$ such that

$$\varphi_l^\pm = e^{\psi_l^\pm}, \quad \varphi_r = e^{\psi_r^\pm}, \quad M(\psi_l^\pm) = M(\psi_r^\pm) = 0. \tag{3.20}$$

Let u be as in Theorem 1.21 and put

$$g_\pm := e^{(1-u)\psi_l^\pm + u\psi_r^\pm}. \tag{3.21}$$

Clearly, $g_\pm \in GSAP$. We can approximate ψ_l^+ and ψ_r^+ by almost periodic polynomials in AP^+ whose mean values are zero. If p is such an almost periodic polynomial, then $(1-u)p$ and up belong to $C(\dot{\mathbf{R}}) + H^\infty_+$ by virtue of Lemma 3.5. Hence $g_+ \in G(C(\dot{\mathbf{R}}) + H^\infty_+)$. Analogously, $g_- \in G(C(\dot{\mathbf{R}}) + H^\infty_-)$. By Theorem 1.21 and (3.20), (3.21),

$$g_\pm = (1-u)\varphi_l^\pm + u\varphi_r^\pm + \varphi_0^\pm$$

with $\varphi_0^\pm \in C_0(\dot{\mathbf{R}})$. Taking into account that $(1-u)^2 - (1-u)$, $u^2 - u$, $(1-u)u$ are functions in $C_0(\dot{\mathbf{R}})$, we therefore obtain

$$g_- b g_+ = (1-u)\varphi_l^- b_l \varphi_l^+ + u\varphi_r^- b_r \varphi_r^+ + \varphi_0$$

with $\varphi_0 \in C_0(\dot{\mathbf{R}})$. Thus, $g_- b g_+ = a + c_0$ with $c_0 \in C_0(\dot{\mathbf{R}})$. Since

$$g_- b g_+ = a(1 + a^{-1}c_0) \in GSAP \quad \text{and} \quad a \in GSAP,$$

it follows that $1 + a^{-1}c_0 \in GC(\dot{\mathbf{R}})$. Letting

$$f_- := g_-, \quad f_+ := g_+(1 + a^{-1}c_0)^{-1},$$

we arrive at the assertion. □

Proof of Theorem 3.9. Part (a) is immediate from Corollary 2.8. So assume

$$a = (1-u)a_l + ua_r + a_0 \in GSAP.$$

By Theorems 1.21 and 2.25 and by Definition 3.1,

$$a_l(x) = e^{i\kappa(a_l)x}\mathbf{d}(a_l)e^{i\psi_l(x)}, \quad a_r(x) = e^{i\kappa(a_r)x}\mathbf{d}(a_r)e^{i\psi_r(x)} \tag{3.22}$$

with $\mathbf{d}(a_l)\mathbf{d}(a_r) \neq 0$, $\psi_l, \psi_r \in AP$, and $M(\psi_l) = M(\psi_r) = 0$. Put

$$b(x) := \big(1-u(x)\big)\mathbf{d}(a_l)e^{i\kappa(a_l)x} + u(x)\mathbf{d}(a_r)e^{i\kappa(a_r)x} + b_0(x)$$

where $b_0 \in C_0(\dot{\mathbf{R}})$ is chosen so that $b \in GSAP$. We claim that $W(a)$ and $W(b)$ are weakly Φ-equivalent.

Suppose $W(a)$ is properly n-normal. Then there is an $\varepsilon > 0$ such that $W(v)$ is properly n-normal whenever $\|a - v\|_\infty < \varepsilon$ (Theorem 2.1). Choose almost periodic polynomials $p_l^\pm, p_r^\pm \in AP^\pm$ such that $M(p_l^\pm) = M(p_r^\pm) = 0$,

$$\|\psi_l - p_l^- - p_l^+\|_\infty < \delta, \quad \|\psi_r - p_r^- - p_r^+\|_\infty < \delta$$

and put

$$\begin{aligned} v(x) \;&:=\; \big(1-u(x)\big)\mathbf{d}(a_l)e^{p_l^-(x)}e^{i\kappa(a_l)x}e^{p_l^+(x)} \\ &+\; u(x)\mathbf{d}(a_r)e^{p_r^-(x)}e^{i\kappa(a_r)x}e^{p_r^+(x)} + a_0(x). \end{aligned}$$

Clearly, if $\delta > 0$ is sufficiently small, then $\|a - v\|_\infty < \varepsilon$, hence $W(v)$ is properly n-normal, and Theorem 2.7 yields that v is invertible in SAP together with a. Lemma 3.11 implies that there are $f_\pm \in G(C(\dot{\mathbf{R}}) + H^\infty_\pm)$ such that $v = f_- b f_+$, and from Lemma 3.10 we infer that $W(b)$ is properly n-normal. Analogously one can show that $W(b)$ is properly d-normal (resp. Fredholm) if $W(a)$ is properly d-normal (resp. Fredholm).

Conversely, assume $W(b)$ is properly n-normal. Then there exists an $\varepsilon > 0$ such that $W(v)$ is properly n-normal if only

$$\begin{aligned} v(x) \;&:=\; \big(1-u(x)\big)\mathbf{d}(a_l)e^{i\kappa(a_l)x}e^{c_l(x)} \\ &+\; u(x)\mathbf{d}(a_r)e^{i\kappa(a_r)x}e^{c_r(x)} + b_0(x) \end{aligned}$$

with $c_l, c_r \in AP$, $\|c_l\|_\infty < \varepsilon$, $\|c_r\|_\infty < \varepsilon$, and $b_0 \in C_0(\dot{\mathbf{R}})$. Using Lemmas 3.11 and 3.10 we obtain that $W(w)$ is properly n-normal whenever

$$\begin{aligned} w(x) \;&=\; \big(1-u(x)\big)\mathbf{d}(a_l)e^{p_l^-(x)}e^{i\kappa(a_l)x}e^{c_l(x)}e^{p_l^+(x)} \\ &+\; u(x)\mathbf{d}(a_r)e^{p_r^-(x)}e^{i\kappa(a_r)x}e^{c_r(x)}e^{p_r^+(x)} + w_0(x) \end{aligned}$$

where $c_l, c_r \in AP$, $\|c_l\|_\infty < \varepsilon$, $\|c_r\|_\infty < \varepsilon$, $p_l^\pm \in AP^\pm$, $p_r^\pm \in AP^\pm$ are almost periodic polynomials such that $M(p_l^\pm) = M(p_r^\pm) = 0$, and $w_0 \in C_0(\dot{\mathbf{R}})$ is chosen so that $w \in GSAP$. But given $a \in GSAP$ with almost periodic representatives (3.22), we can clearly put $w_0 = a_0$ and find $p_l^\pm \in AP^\pm$, $p_r^\pm \in AP^\pm$ such that $M(p_l^\pm) = M(p_r^\pm) = 0$,

$$\psi_l = c_l + p_l^- + p_l^+, \quad \psi_r = c_r + p_r^- + p_r^+,$$

$c_l, c_r \in AP$, $\|c_l\|_\infty < \varepsilon$, $\|c_r\|_\infty < \varepsilon$. Hence, $W(a)$ is properly n-normal. It can be shown analogously that if $W(b)$ is properly d-normal (resp. Fredholm), then $W(a)$ is properly d-normal (resp. Fredholm).

Finally, since a nonzero Wiener-Hopf operator is either not normally solvable or semi-Fredholm (Corollary 2.6(a)), we arrive at the conclusion that $W(a)$ is normally solvable if and only if so is $W(b)$. At this point the proof of the weak Φ-equivalence of $W(a)$ and $W(b)$ is complete.

Suppose now that $\kappa(a_l) = \kappa(a_r) = 0$. Then $b \in GC(\overline{\mathbf{R}})$ and Theorem 2.20 shows that $W(b)$ is Fredholm if $0 \notin [\mathbf{d}(a_l), \mathbf{d}(a_r)]$ and that $W(b)$ is not normally solvable if $0 \in [\mathbf{d}(a_l), \mathbf{d}(a_r)]$. This gives parts (e) and (f) of the theorem.

If $\kappa(a_l) > 0$ and $\kappa(a_r) > 0$, then $b \in C(\dot{\mathbf{R}}) + H_+^\infty$ due to Lemma 3.5. From Theorem 2.19 we infer that $W(b)$ is Fredholm or properly n-normal. Since Fredholmness is preserved under small perturbations, we can without loss of generality assume that $0 \notin [\mathbf{d}(a_l), \mathbf{d}(a_r)]$. Thus, we can employ Lemmas 3.6 and 3.7 to deduce that $W(b)$ cannot be Fredholm. Therefore $W(b)$ is properly n-normal. If $\kappa(a_l) = 0$ and $\kappa(a_r) > 0$ or if $\kappa(a_l) > 0$ and $\kappa(a_r) = 0$, we deduce from Lemma 3.8 that $W(b)$ is properly n-normal. This and Corollary 2.6(c) complete the proof of part (c) of the theorem. Part (d) follows from part (c) and Theorem 2.4 after passage to adjoints.

Finally, suppose $\kappa(a_l)\kappa(a_r) < 0$. For the sake of definiteness, let $\kappa(a_l) < 0$ and $\kappa(a_r) > 0$. Contrary to what we want, let us assume that $W(b)$ is normally solvable and thus semi-Fredholm. Then small perturbations of $W(b)$ are also semi-Fredholm (Theorem 2.1), which allows us to assume that $0 \notin [\mathbf{d}(a_l), \mathbf{d}(a_r)]$. From Lemmas 3.6 and 3.7 we obtain that $W(b)$ and $(W(b))^* = W(\overline{b})$ are properly n-normal, which contradicts Theorem 2.4. Thus, $W(b)$ cannot be normally solvable, and the proof of part (b) is complete. □

3.5 Index Formula

Let $GSAP_{0,0}$ be the set of all functions $a \in GSAP$ for which $\kappa(a_l) = \kappa(a_r) = 0$. Theorem 3.9 tells us that if $a \in SAP$, then $W(a)$ is Fredholm if and only if $a \in GSAP_{0,0}$ and $0 \notin [\mathbf{d}(a_l), \mathbf{d}(a_r)]$. In this section we establish an index formula for Fredholm Wiener-Hopf operators with symbols in SAP.

Lemma 3.12. *Let $A \subset (0, \infty)$ be an unbounded set and let*

$$\{I_\alpha\}_{\alpha \in A} = \{(x_\alpha, y_\alpha)\}_{\alpha \in A}$$

be a family of intervals such that $x_\alpha \geq 0$ *and* $|I_\alpha| = y_\alpha - x_\alpha \to \infty$ *as* $\alpha \to \infty$. *If* $a \in GSAP_{0,0}$ *and* $\arg a$ *is any continuous argument of* a, *then the limit*

$$\frac{1}{2\pi} \lim_{\alpha\to\infty} \frac{1}{|I_\alpha|} \int_{I_\alpha} \big((\arg a)(x) - (\arg a)(-x)\big)\, dx \tag{3.23}$$

exists, is finite, and is independent of the particular choices of $\{(x_\alpha, y_\alpha)\}_{\alpha\in A}$ *and* $\arg a$.

Proof. There are continuous arguments $\arg a_r \in AP$ and $\arg a_l \in AP$ of the almost periodic representatives a_r and a_l of a such that

$$\begin{aligned}(\arg a)(x) &= (\arg a_r)(x) + o(1) \text{ as } x \to +\infty,\\ (\arg a)(-x) &= (\arg a_l)(-x) + o(1) \text{ as } x \to +\infty.\end{aligned}$$

From Proposition 2.22 we therefore obtain that the limit (3.23) is

$$(1/(2\pi))\big(M(\arg a_r) - M(\arg a_l)\big). \tag{3.24}$$

□

Definition 3.13. For $a \in GSAP_{0,0}$, we denote the limit (3.23) by ind a and call it the *Cauchy index* of a. It is easily seen that if $a \in GC(\overline{\mathbf{R}})$, then (3.23) equals (2.6), so that this definition is consistent with that of Section 2.4.

If $a \in GSAP_{0,0}$ (and thus $\mathbf{d}(a_l)\mathbf{d}(a_r) \neq 0$) and if $0 \notin (\mathbf{d}(a_l), \mathbf{d}(a_r))$, we define the *winding number* wind a by

$$\begin{aligned}\text{wind } a &:= \text{ind } a - \frac{1}{2} + \left\{\frac{1}{2} + \frac{1}{2\pi}\arg\frac{\mathbf{d}(a_l)}{\mathbf{d}(a_r)}\right\}\\ &= \text{ind } a + \frac{1}{2} - \left\{\frac{1}{2} - \frac{1}{2\pi}\arg\frac{\mathbf{d}(a_l)}{\mathbf{d}(a_r)}\right\}\\ &= \text{ind } a + \frac{1}{2\pi}\arg\frac{\mathbf{d}(a_l)}{\mathbf{d}(a_r)},\end{aligned} \tag{3.25}$$

the latter equality being valid if and only if $\arg(\mathbf{d}(a_l)/\mathbf{d}(a_r))$ is taken from $(-\pi, \pi)$ (recall (2.19) to (2.21)). We claim that wind a is an integer. Indeed, we have

$$a_l(x) = \mathbf{d}(a_l)e^{\psi_l(x)}, \quad a_r(x) = \mathbf{d}(a_r)e^{\psi_r(x)} \tag{3.26}$$

with $\psi_l, \psi_r \in AP$ and $M(\psi_l) = M(\psi_r) = 0$. From the proof of Lemma 3.12 we know that, for appropriate choices of $\arg a_r$ and $\arg a_l$, the Cauchy index ind a equals (3.24). Let $\arg \mathbf{d}(a_l)$ and $\arg \mathbf{d}(a_r)$ be any arguments of $\mathbf{d}(a_l)$ and $\mathbf{d}(a_r)$. Then

$$\arg a_l = \arg \mathbf{d}(a_l) + \operatorname{Im}\psi_l + 2\pi k_l \quad (k_l \in \mathbf{Z})$$

and hence, $M(\arg a_l) = \arg \mathbf{d}(a_l) + 2\pi k_l$. Analogously, $M(\arg a_r) = \arg \mathbf{d}(a_r) + 2\pi k_r$ with some $k_r \in \mathbf{Z}$. Thus,

$$\begin{aligned} \text{ind } a &= \frac{1}{2\pi}\big(\arg \mathbf{d}(a_r) - \arg \mathbf{d}(a_l)\big) + k_r - k_l \\ &= -\frac{1}{2\pi}\arg\frac{\mathbf{d}(a_l)}{\mathbf{d}(a_r)} + k \quad (k \in \mathbf{Z}) \end{aligned}$$

where $\arg(\mathbf{d}(a_l)/\mathbf{d}(a_r)) \in (-\pi, \pi)$. This gives wind $a = k$ and proves our claim. □

Theorem 3.14 (Sarason). *If $a \in GSAP_{0,0}$ and $0 \notin [\mathbf{d}(a_l), \mathbf{d}(a_r)]$ then* $\operatorname{Ind} W(a) = -\operatorname{wind} a$.

Proof. We have $a = (1-u)a_l + ua_r + f_0$ where $f_0 \in C_0(\dot{\mathbf{R}})$, a_l, a_r are of the form (3.26), $\psi_l, \psi_r \in AP$, and $M(\psi_l) = M(\psi_r) = 0$. Without loss of generality assume that $u \in C(\overline{\mathbf{R}})$ is real-valued. Consider

$$\begin{aligned} b &:= ae^{-(1-u)\psi_l - u\psi_r} \\ &= \big((1-u)\mathbf{d}(a_l)e^{\psi_l} + u\mathbf{d}(a_r)e^{\psi_r} + f_0\big)e^{-(1-u)\psi_l - u\psi_r}. \end{aligned}$$

Obviously, $b(-\infty) = \mathbf{d}(a_l)$, $b(+\infty) = \mathbf{d}(a_r)$, and hence b belongs to $GC(\overline{\mathbf{R}})$ and the origin is not in $b^{\#}(\dot{\mathbf{R}})$. For $\mu \in [0,1]$, put

$$a_\mu := be^{\mu((1-u)\psi_l + u\psi_r)}.$$

Clearly, $a_1 = a$ and $a_0 = b$. For the almost periodic representatives $(a_\mu)_l$ and $(a_\mu)_r$ we get

$$(a_\mu)_l = \mathbf{d}(a_l)e^{\mu\psi_l}, \quad (a_\mu)_r = \mathbf{d}(a_r)e^{\mu\psi_r}.$$

This shows that $a_\mu \in GSAP_{0,0}$ and $0 \notin [\mathbf{d}((a_\mu)_l), \mathbf{d}((a_\mu)_r)]$. Consequently, $W(a_\mu)$ is Fredholm for all $\mu \in [0,1]$ by virtue of Theorem 3.9(f). The map

$$[0,1] \to \mathcal{B}\big(L^2(\mathbf{R}_+)\big), \ \mu \mapsto W(a_\mu)$$

is easily seen to be continuous. Theorem 2.1 therefore implies that

$$\operatorname{Ind} W(a) = \operatorname{Ind} W(a_0) = \operatorname{Ind} W(b), \tag{3.27}$$

and from Theorem 2.20 and formula (2.19) we know that

$$\operatorname{Ind} W(b) = -\text{ ind } b + \frac{1}{2} - \left\{\frac{1}{2} + \frac{1}{2\pi}\arg\frac{b(-\infty)}{b(+\infty)}\right\}. \tag{3.28}$$

By the definition of the Cauchy index,

$$\begin{aligned}
\operatorname{ind} b &= \operatorname{ind} a - \operatorname{ind}\left(e^{(1-u)\psi_l + u\psi_r}\right) \\
&= \operatorname{ind} a - \frac{1}{2\pi}\lim_{T\to+\infty}\frac{1}{T}\int_T^{2T}\Big[(1-u(x))\operatorname{Im}\psi_l(x) + u(x)\operatorname{Im}\psi_r(x) \\
&\quad -(1-u(-x))\operatorname{Im}\psi_l(-x) - u(-x)\operatorname{Im}\psi_r(-x)\Big]\,dx \\
&= \operatorname{ind} a - \frac{1}{2\pi}\lim_{T\to+\infty}\frac{1}{T}\int_T^{2T}\Big[u(x)\operatorname{Im}\psi_r(x) - (1-u(-x))\operatorname{Im}\psi_l(-x)\Big]\,dx \\
&= \operatorname{ind} a - \frac{1}{2\pi}\lim_{T\to+\infty}\frac{1}{T}\int_T^{2T}\left[\operatorname{Im}\psi_r(x) - \operatorname{Im}\psi_l(-x)\right]dx \\
&= \operatorname{ind} a - \frac{1}{2\pi}\left[M(\operatorname{Im}\psi_r) - M(\operatorname{Im}\psi_l)\right] = \operatorname{ind} a.
\end{aligned}$$

Taking into account that $b(-\infty) = \mathbf{d}(a_l)$, $b(+\infty) = \mathbf{d}(a_r)$, we obtain from (3.27) and (3.28) that

$$\operatorname{Ind} W(a) = -\operatorname{ind} a + \frac{1}{2} - \left\{\frac{1}{2} + \frac{1}{2\pi}\arg\frac{\mathbf{d}(a_l)}{\mathbf{d}(a_r)}\right\} =: -\operatorname{wind} a. \qquad \square$$

3.6 $SAP + PC_0$ Symbols

Combining the results of Sections 2.9, 3.4, 3.5, it is easy to describe the semi-Fredholm properties of Wiener-Hopf operators with symbols in the C^*-algebra

$$SAP + PC_0 = \operatorname{alg}(AP, PC).$$

Proposition 3.15. (a) *If $a \in SAP + PC_0$, then there are uniquely determined functions $b_l, b_r \in AP$ and $a_0 \in PC_0$ such that*

$$a = (1-u)b_l + ub_r + a_0,$$

where u is as in Theorem 1.21.

(b) *If $a \in G(SAP + PC_0)$, then there exist functions $b \in GSAP$ and $c \in GPC$ such that $c(-\infty) = c(+\infty) = 1$, $a = bc$, and*

$$W(a) = W(b)W(c) + K_1 = W(c)W(b) + K_2 \tag{3.29}$$

with compact operators K_1, K_2. The almost periodic representatives of b are the functions b_l, b_r of part (a).

Proof. Part (a) is immediate from Theorem 1.21. To prove part (b), suppose a is in $G(SAP + PC_0)$ and put $f := (1-u)b_l + ub_r$. Then $a = f + a_0$. There is an

$M \in (0, \infty)$ such that $|f(x)|$ is bounded away from zero for $|x| > M$, and therefore we can find a function $f_0 \in C_0(\dot{\mathbf{R}})$ such that $b := f + f_0 \in GSAP$. We have

$$a = b + a_0 - f_0 = b\big(1 + b^{-1}(a_0 - f_0)\big) =: bc,$$

and it is clear that $c = b^{-1}a \in GPC$ and $c(-\infty) = c(+\infty) = 1$. Since b is continuous on $\mathbf{R}$ and c is continuous at ∞, we deduce from Theorem 2.13 that (3.29) holds with compact operators K_1, K_2. □

Given $a \in G(SAP + PC_0)$, we write a as in Proposition 3.15 and define

$$\kappa_l(a) = \kappa(b_l), \quad \kappa_r(a) = \kappa(b_r), \quad \mathbf{d}_l(a) = \mathbf{d}(b_l), \quad \mathbf{d}_r(a) = \mathbf{d}(b_r).$$

If, in addition, $b \in GSAP_{0,0}$, $0 \notin [\mathbf{d}_l(a), \mathbf{d}_r(a)]$, and $0 \notin c^{\#}(\dot{\mathbf{R}})$, we put

$$\text{wind } a := \text{wind } b + \text{wind } c. \tag{3.30}$$

We remark that there are definitions of $\kappa_l(a)$, $\kappa_r(a)$, $\mathbf{d}_l(a)$, $\mathbf{d}_r(a)$, wind a in terms of only a (and not in terms of b and c), but we will not embark on this matter.

Theorem 3.16. *Let $a \in SAP + PC_0$ and suppose a is not identically zero.*
(a) *If $a \in G(SAP + PC_0)$, $\kappa_l(a) = \kappa_r(a) = 0$, and*

$$0 \notin \big[\mathbf{d}_l(a), \mathbf{d}_r(a)\big] \cup \bigcup_{x \in \mathbf{R}} \big[a(x-0), a(x+0)\big],$$

then $W(a)$ is Fredholm and $\operatorname{Ind} W(a) = -\,\text{wind } a$.
(b) *If $a \in G(SAP + PC_0)$, $\kappa_l(a) \cdot \kappa_r(a) \geq 0$, $\kappa_l(a) + \kappa_r(a) > 0$, and*

$$0 \notin \bigcup_{x \in \mathbf{R}} \big[a(x-0), a(x+0)\big],$$

then $W(a)$ is properly n-normal and left-invertible.
(c) *If $a \in G(SAP + PC_0)$, $\kappa_l(a) \cdot \kappa_r(a) \geq 0$, $\kappa_l(a) + \kappa_r(a) < 0$, and*

$$0 \notin \bigcup_{x \in \mathbf{R}} \big[a(x-0), a(x+0)\big],$$

then $W(a)$ is properly d-normal and right-invertible.
(d) *In all other cases the operator $W(a)$ is not normally solvable.*

Proof. If $a \notin G(SAP + PC_0)$, we see from Corollary 2.8 that $W(a)$ cannot be normally solvable. So assume that $a \in G(SAP + PC_0)$. We then write $a = bc$ as in Proposition 3.15. Taking into account (3.29) we see that $W(a)$ is Fredholm if and only if both $W(b)$ and $W(c)$ are Fredholm, which, by Theorems 2.20 and 3.9, happens if and only if the conditions of part (a) are satisfied. From (3.29) and Theorems 2.2, 2.3, 2.20, 3.14 we also infer that

$$\begin{aligned}\operatorname{Ind} W(a) &= \operatorname{Ind} W(b) + \operatorname{Ind} W(c)\\ &= -\operatorname{wind} b - \operatorname{wind} c = -\operatorname{wind} a.\end{aligned}$$

Since $W(c)$ is Fredholm under the conditions of parts (b) and (c) (Theorem 2.20), we infer from Theorem 2.3 that $W(a)$ is properly n-normal (resp. properly d-normal) if and only if so is $W(b)$. We therefore obtain part (b) (resp. part(c)) from Theorem 3.9 and Corollary 2.6(c).

To complete the proof, we use the following well known fact: if $A, B \in \mathcal{B}(X)$ and AB is n-normal (resp. d-normal), then B is n-normal (resp. A is d-normal). This, Theorem 2.2, and (3.29) show that $W(a)$ is n-normal (resp. d-normal) if and only if so are both $W(b)$ and $W(c)$, and hence we get part (d) for $a \in G(SAP + PC_0)$ from Corollary 2.6(a) and Theorems 2.20 and 3.9. □

Notes

The geometric mean value is an important characteristic: it plays the same role for oscillating functions as one-sided limits play for piecewise continuous functions. In the Fredholm theory of Toeplitz operators with semi almost periodic symbols, it was first used by Sarason [197].

Lemma 3.5 was proved by Sarason in [196]. It is a key ingredient in the consideration of semi-almost periodic symbols, because it allows us to reduce the SAP case to the (previously known) piecewise continuous case. The main results of Sections 3.4 and 3.5 are also Sarason's [197]. The proofs given in Sections 3.3 to 3.5 follow Duduchava and Saginashvili [72], [192].

Section 3.6 is based on the results of Sections 9.26 and 9.27 of [46].

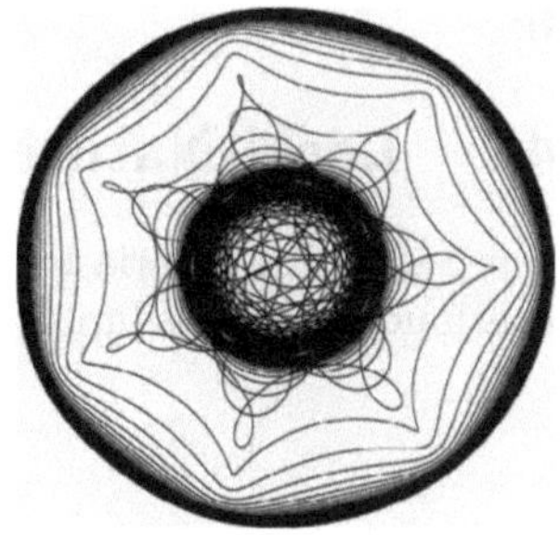

Chapter 4

Some Phenomena Caused by SAP Symbols

It is not only phenomena, it is in fact some surprises (optimistic view) and all of the evil with Wiener-Hopf operators (pessimistic view) that begins with SAP symbols.

First, while the Fredholm criteria for operators with $C + H^\infty$, PC, or AP symbols can be almost at once rephrased into descriptions of the essential spectra of such operators, this is no longer the case for SAP symbols. As will be shown here, the essential spectra of Wiener-Hopf operators with SAP symbols can assume a variety of unexpectedly beautiful shapes. Secondly, if a sequence a_n of functions in $C + H^\infty$, PC, or AP converges uniformly to some function a, then the spectra $\operatorname{sp} W(a_n)$ converge in the Hausdorff metric to the spectrum $\operatorname{sp} W(a)$. It has been suspected that this is true for general functions $a_n \in L^\infty(\mathbf{R})$. We will show that this suspicion is wrong, and it will be symbols in SAP that will produce counterexamples. Finally, it is well known that amplitude modulation does not affect Fredholmness, index, and invertibility of Wiener-Hopf operators with L^∞ symbols. It is also well known that the Fredholm properties of operators with PC symbols are stable under frequency modulation. We show that frequency modulation can nevertheless destroy Fredholmness of Wiener-Hopf operators with AP symbols. The proof of this fact is based on constructing an AP symbol that after frequency modulation becomes a SAP symbol and on subsequently invoking Sarason's Fredholm criterion.

4.1 Local Nature of the Essential Spectrum

The Fredholm criteria for Wiener-Hopf operators with continuous and piecewise continuous symbols immediately delivered the essential spectra of these operators:

$$\mathrm{sp}_{\mathrm{ess}} W(a) = a(\dot{\mathbf{R}}) \text{ for } a \in C(\dot{\mathbf{R}}),$$
$$\mathrm{sp}_{\mathrm{ess}} W(a) = a^{\#}(\dot{\mathbf{R}}) \text{ for } a \in PC.$$

For almost periodic symbols, $a \in AP$, Theorem 2.28 gives

$$\mathrm{sp}_{\mathrm{ess}} W(a) = a(\mathbf{R}) \cup \{\lambda \notin a(\mathbf{R}) : a - \lambda \notin GAP \text{ or } \kappa(a-\lambda) \neq 0\}, \tag{4.1}$$

which provides us with a still acceptable description of the essential spectrum. Things are more involved for symbols in SAP. In that case Theorem 3.9 enables us to decide whether $W(a - \lambda)$ is Fredholm for a given point λ in the plane, but the problem of characterizing the set of *all* λ for which $W(a - \lambda)$ is not Fredholm is nevertheless intricate. Figure 4.1 shows $\mathrm{sp}_{\mathrm{ess}} W(a)$ for several $a \in SAP$.

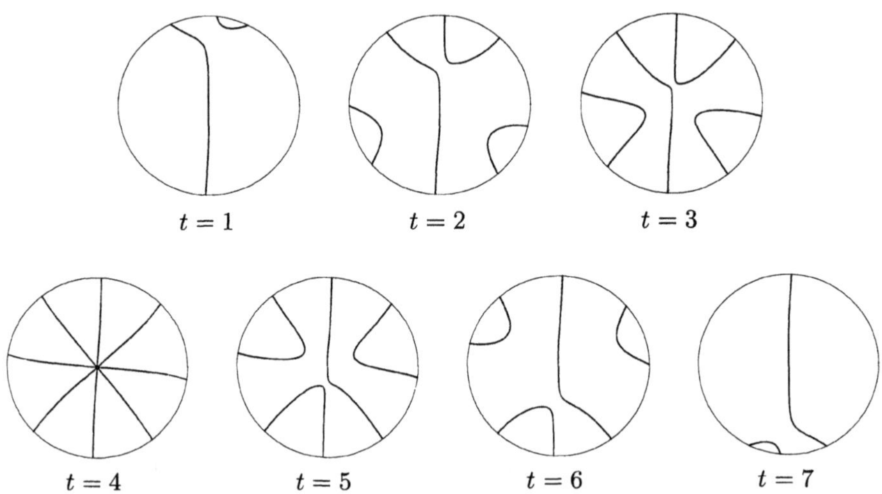

Figure 4.1: Snapshots of $\mathrm{sp}_{\mathrm{ess}} W(a_t)$ for a unimodular symbol $a_t \in SAP$ depending continuously on a parameter t.

For $a \in SAP$, we divide $\mathrm{sp}_{\mathrm{ess}} W(a)$ into two disjoint parts. We define

$$\mathrm{sp}_{AP}\, W(a) := a(\mathbf{R}) \cup \mathrm{sp}_{\mathrm{ess}} W(a_l) \cup \mathrm{sp}_{\mathrm{ess}} W(a_r), \tag{4.2}$$

where a_l and a_r are the almost periodic representatives of a. The set $\mathrm{sp}_{AP}\, W(a)$ can be identified using (4.1) and (4.2). The interesting part of $\mathrm{sp}_{\mathrm{ess}} W(a)$ is the set

$$\mathrm{sp}_{PC}\, W(a) := \mathrm{sp}_{\mathrm{ess}} W(a) \setminus \mathrm{sp}_{AP}\, W(a).$$

If λ is in the closure of $a(\mathbf{R})$ but not in $a(\mathbf{R})$ itself, then $a_l - \lambda$ or $a_r - \lambda$ is not invertible in AP. From (4.1) and (4.2) we therefore see that $\mathrm{sp}_{AP} W(a)$ is compact. Hence $\mathbf{C} \setminus \mathrm{sp}_{AP} W(a)$ is always open. Theorem 3.9 shows that

$$\mathrm{sp}_{PC} W(a) = \{\lambda \in \mathbf{C} \setminus \mathrm{sp}_{AP} W(a) : 0 \in (\mathbf{d}(a_l - \lambda), \mathbf{d}(a_r - \lambda))\},$$

which can also be written in the form

$$\mathrm{sp}_{PC} W(a) = \left\{ \lambda \in \mathbf{C} \setminus \mathrm{sp}_{AP} W(a) \colon \frac{\mathbf{d}(a_r - \lambda)}{\mathbf{d}(a_l - \lambda)} \in (-\infty, 0) \right\}. \tag{4.3}$$

We call a set $X \subset \mathbf{C}$ an *analytic star* if there are a point $\lambda_0 \in \mathbf{C}$, an open neighborhood $U \subset \mathbf{C}$ of λ_0, and a nonconstant analytic function $f : U \to \mathbf{C}$ such that

$$f(\lambda_0) = 0 \quad \text{and} \quad X = \{\lambda \in U : f(\lambda) \in \mathbf{R}\}.$$

If $f'(\lambda_0) \neq 0$, then $X \cap V$ is an analytic arc through λ_0 whenever $V \subset \mathbf{C}$ is a sufficiently small open neighborhood of λ_0. If

$$f'(\lambda_0) = \ldots = f^{(k-1)}(\lambda_0) = 0, \quad f^{(k)}(\lambda_0) \neq 0,$$

then, for every sufficiently small open neighborhood $V \subset \mathbf{C}$ of the point λ_0, the set $(X \cap V) \setminus \{\lambda_0\}$ is composed of $2k$ pairwise disjoint analytic arcs whose closures all contain the point λ_0. We call λ_0 the *center* of the star and refer to the number of arcs in $(X \cap V) \setminus \{\lambda_0\}$ as the *valency* of the star.

Theorem 4.1 (Sarason). *Let a be a function in SAP and let Ω be a connected component of the (open) set $\mathbf{C} \setminus \mathrm{sp}_{AP} W(a)$. We then have exactly one of the following three possibilities:*

(i) $\Omega \cap \mathrm{sp}_{PC} W(a) = \emptyset$;

(ii) $\Omega \subset \mathrm{sp}_{PC} W(a)$;

(iii) $\Omega \cap \mathrm{sp}_{PC} W(a) \neq \emptyset$ *and each point* $\lambda_0 \in \Omega \cap \mathrm{sp}_{PC} W(a)$ *has an open neighborhood* $U \subset \Omega$ *such that* $U \cap \mathrm{sp}_{PC} W(a)$ *is an analytic star.*

Proof. Fix a connected component Ω of $\mathbf{C} \setminus \mathrm{sp}_{AP} W(a)$. Let $\lambda_0 \in \Omega$. If $\lambda \in \Omega$ and $|\lambda - \lambda_0|$ is sufficiently small, then

$$\begin{aligned} \mathbf{d}(a_r - \lambda) &= \mathbf{d}(a_r - \lambda_0 - (\lambda - \lambda_0)) \\ &= \mathbf{d}(a_r - \lambda_0) \exp\left(M(\log(1 - (\lambda - \lambda_0)(a_r - \lambda_0)^{-1})\right) \\ &= \mathbf{d}(a_r - \lambda_0) \exp\left(- \sum_{k=1}^{\infty} \frac{(\lambda - \lambda_0)^k}{k} M\left((a_r - \lambda_0)^{-k}\right) \right) \end{aligned} \tag{4.4}$$

and, analogously,

$$\mathbf{d}(a_l - \lambda) = \mathbf{d}(a_l - \lambda_0) \exp\left(- \sum_{k=1}^{\infty} \frac{(\lambda - \lambda_0)^k}{k} M\left((a_l - \lambda_0)^{-k}\right) \right) \tag{4.5}$$

This shows that

$$f(\lambda) := \frac{\mathbf{d}(a_r - \lambda)}{\mathbf{d}(a_l - \lambda)}$$

is analytic in Ω and that

$$f(\lambda) = \frac{\mathbf{d}(a_r - \lambda_0)}{\mathbf{d}(a_l - \lambda_0)} g(\lambda) \tag{4.6}$$

with

$$g(\lambda) := \exp\left(-\sum_{k=1}^{\infty} \frac{(\lambda - \lambda_0)^k}{k} \left(M\left((a_r - \lambda_0)^{-k}\right) - M\left((a_l - \lambda_0)^{-k}\right)\right)\right). \tag{4.7}$$

Suppose (i) is not satisfied. Then there is a point $\lambda_0 \in \Omega \cap \operatorname{sp}_{PC} W(a)$. Pick a point $\lambda \in \Omega$ in a sufficiently small open neighborhood $U \subset \Omega$ of λ_0. By (4.3), $\lambda \in \operatorname{sp}_{PC} W(a)$ if and only if

$$f(\lambda) \in (-\infty, 0). \tag{4.8}$$

If

$$M\left((a_l - \lambda_0)^{-k}\right) = M\left((a_r - \lambda_0)^{-k}\right) \text{ for all } k \in \mathbf{N}, \tag{4.9}$$

then (4.6) and (4.7) imply that f is constant in U. As f is analytic in Ω, it results that $f(\lambda) = f(\lambda_0) \in (-\infty, 0)$ for all $\lambda \in \Omega$. Thus, we have (ii). So assume (4.9) does not hold. Let $n \in \mathbf{N}$ be the smallest number for which

$$\delta_n := M\left((a_r - \lambda_0)^{-n}\right) - M\left((a_l - \lambda_0)^{-n}\right) \neq 0. \tag{4.10}$$

Then, by (4.6) and (4.7),

$$f(\lambda) = f(\lambda_0) - \frac{f(\lambda_0)\delta_n}{n}(\lambda - \lambda_0)^n + O\left((\lambda - \lambda_0)^{n+1}\right). \tag{4.11}$$

Because $f(\lambda_0) \in (-\infty, 0)$, the set of all λ given by (4.8) and (4.11) is an analytic star of the valency $2n$ and hence this set is as described by (iii). □

We remark that $\operatorname{sp}_{PC} W(a)$ may be empty; this happens, for instance, if a is in $C(\dot{\mathbf{R}})$, that is, if a_l and a_r are constant. Here is an immediate consequence of Theorem 4.1.

Corollary 4.2. *Let $a \in SAP$, suppose $\operatorname{sp}_{PC} W(a) \neq \emptyset$, and let $\lambda_0 \in \operatorname{sp}_{PC} W(a)$. Then there exists an open neighborhood $U \subset \mathbf{C}$ of λ_0 such that $\operatorname{sp}_{PC} W(a) \cap U$ is either all of U or an analytic star.* □

In the next two sections we show that Corollary 4.2 is sharp and hence provides us with a complete description of the local geometry of the essential spectra of Wiener-Hopf operators with SAP symbols.

4.2 Prescribed Stars in the Essential Spectrum

In this section we prove the following two theorems.

Theorem 4.3. *Given any analytic star X, there exists a function $a \in SAP$ such that $\mathrm{sp}_{PC}\, W(a) \cap U = X \cap U$ for some open set $U \subset \mathbf{C}$ containing the center of the star X.*

Theorem 4.4. *Given any distinct points $\lambda_1, \ldots, \lambda_n$ in the complex plane and any even natural numbers $2k_1, \ldots, 2k_n$, there exist $a \in SAP$ such that $\lambda_1, \ldots, \lambda_n$ are all located in a single component Ω of $\mathbf{C} \setminus \mathrm{sp}_{AP}\, W(a)$ and such that, for each j, the set $\mathrm{sp}_{PC}\, W(a)$ is an analytic star with the center λ_j and the valency $2k_j$ in some neighborhood of λ_j.*

Throughout this section we assume that a is a unimodular function in SAP, that is, $|a(x)| = 1$ for all $x \in \mathbf{R}$. From Theorem 1.21 we infer that then $|a_l(x)| = |a_r(x)| = 1$ for all $x \in \mathbf{R}$. Furthermore, we assume that $\kappa(a_l) = \kappa(a_r) = 0$. This implies that $\kappa(a_l - \lambda) = \kappa(a_r - \lambda) = 0$ for all $\lambda \in \mathbf{C} \setminus \mathbf{T}$. For $\lambda \in \mathbf{C} \setminus \mathbf{T}$, we can write $a_l(x) - \lambda = e^{ib_{l,\lambda}(x)}$ and $a_r(x) - \lambda = e^{ib_{r,\lambda}(x)}$, where $b_{l,\lambda}, b_{r,\lambda}$ are functions in AP. It follows that $\mathbf{d}(a_l - \lambda) = e^{iM(b_{l,\lambda})}$ and $\mathbf{d}(a_r - \lambda) = e^{iM(b_{r,\lambda})}$. From Theorems 2.28 and 2.7 we therefore deduce that $\mathrm{sp}_{AP}\, W(a) \subset \mathbf{T}$ and $\mathrm{sp}_{PC}\, W(a) \subset \mathbf{D}$. In Section 4.1 we showed that $\mathbf{d}(a_r - \lambda)/\mathbf{d}(a_l - \lambda) = -e^{h(\lambda)}$, where $h(\lambda)$ is analytic for $|\lambda| < 1$. Because $\mathbf{d}(a_r)/\mathbf{d}(a_l) = e^{i(M(b_{r,0}) - M(b_{l,0}))}$ and the functions $b_{l,0}$ and $b_{r,0}$ can be chosen to be real-valued, we see that $h(0) \in i\mathbf{R}$. Consequently, we arrive at the following result.

Proposition 4.5. *If $a \in SAP$ is unimodular and $\kappa(a_l) = \kappa(a_r) = 0$, then*

$$\mathrm{sp}_{PC}\, W(a) = \{\lambda \in \mathbf{D} : h(\lambda) \in \mathbf{R} + 2\pi i \mathbf{Z}\}$$

with some function h which is analytic in $\mathbf{D}$ and satisfies $h(0) \in i\mathbf{R}$. □

The construction of symbols whose spectra have prescribed local shapes is based on the following lemma.

Lemma 4.6. *Let $R > 1$. If $h(\lambda)$ is analytic for $|\lambda| < R$, $h(0) \in i\mathbf{R}$, and*

$$\max_{|t|=1} |h'(t)| < 1/2,$$

then there is a unimodular function $a \in SAP$ such that $\kappa(a_l - \lambda) = \kappa(a_r - \lambda) = 0$ and

$$\frac{\mathbf{d}(a_r - \lambda)}{\mathbf{d}(a_l - \lambda)} = -e^{h(\lambda)}$$

for $|\lambda| < 1$.

Proof. Let $h(\lambda) = i\mu + h_1\lambda + h_2\lambda^2 + \ldots$ $(|\lambda| < R)$. For $t \in \mathbf{T}$, put

$$p(t) = 1 - \sum_{m=1}^{\infty} \left(m h_m t^m + m\overline{h_m} t^{-m} \right);$$

note that the series converges absolutely, because $h(\lambda)$ was supposed to be analytic in $|\lambda| < R$ for some $R > 1$. Since

$$\left| \sum_{m=1}^{\infty} m h_m t^m \right| = |t h'(t)| < 1/2 \text{ for } t \in \mathbf{T},$$

we see that $p(t) > 0$ for $t \in \mathbf{T}$. Choose a real number δ and define a function c by

$$c(y) = -\pi + \frac{1}{2} \int_{-\delta}^{y} p(e^{i\xi})\, d\xi.$$

Obviously,

$$c(-\delta) = -\pi, \quad c(2\pi - \delta) = -\pi + \frac{1}{2} \int_{-\delta}^{2\pi-\delta} p(e^{i\xi}) d\xi = 0,$$

and $c'(y) = p(e^{iy})/2 > 0$ for $y \in [-\delta, 2\pi - \delta]$. Thus, $c : [-\delta, 2\pi - \delta] \to [-\pi, 0]$ is strictly monotonous. Let $b_r : [-\pi, 0] \to [-\delta, 2\pi - \delta]$ be the inverse function and extend b_r to a 2π-periodic and even function on all of $\mathbf{R}$. For $|\lambda| < 1$, we have

$$\frac{1}{2\pi} \int_{-\pi}^{\pi} \log\left(e^{ib_r(x)} - \lambda\right) dx = \frac{i}{2\pi} \int_{-\pi}^{\pi} b_r(x) dx + \frac{1}{2\pi} \int_{-\pi}^{\pi} \log\left(1 - \lambda e^{-ib_r(x)}\right) dx \tag{4.12}$$

and

$$\frac{1}{2\pi} \int_{-\pi}^{\pi} \log\left(1 - \lambda e^{-ib_r(x)}\right) dx = -\frac{1}{2\pi} \sum_{m=1}^{\infty} \frac{\lambda^m}{m} \int_{-\pi}^{\pi} e^{-imb_r(x)} dx$$

$$= -\frac{1}{\pi} \sum_{m=1}^{\infty} \frac{\lambda^m}{m} \int_{-\pi}^{0} e^{-imb_r(x)} dx = -\frac{1}{\pi} \sum_{m=1}^{\infty} \frac{\lambda^m}{m} \int_{-\delta}^{2\pi-\delta} e^{-imy} c'(y) dy$$

$$= -\frac{1}{2\pi} \sum_{m=1}^{\infty} \frac{\lambda^m}{m} \int_{-\delta}^{2\pi-\delta} e^{-imy} p(e^{iy}) dy = -\frac{1}{2\pi} \sum_{m=1}^{\infty} \frac{\lambda^m}{m} \int_{0}^{2\pi} e^{-imy} p(e^{iy}) dy$$

$$= -\frac{1}{2\pi} \sum_{m=1}^{\infty} \frac{\lambda^m}{m} (-2\pi m h_m) = \sum_{m=1}^{\infty} h_m \lambda^m.$$

Hence, (4.12) equals

$$\frac{i}{2\pi} \int_{-\pi}^{\pi} b_r(x) dx + \sum_{m=1}^{\infty} h_m \lambda^m. \tag{4.13}$$

On appropriately choosing $\delta \in \mathbf{R}$, we can achieve that $(2\pi)^{-1} \int_{-\pi}^{\pi} b_r(x)dx$ assumes any prescribed value $\mu \in \mathbf{R}$. From (4.12) and (4.13) we therefore obtain that

$$\frac{1}{2\pi}\int_{-\pi}^{\pi} \log\left(e^{ib_r(x)} - \lambda\right) dx = i\mu + \sum_{m=1}^{\infty} h_m \lambda^m. \tag{4.14}$$

Repeating the above construction in the special case where $h_1 = h_2 = \ldots = 0$, we get a function b_l such that

$$\frac{1}{2\pi}\int_{-\pi}^{\pi} \log\left(e^{ib_l(x)} - \lambda\right) dx = \frac{i}{2\pi}\int_{-\pi}^{\pi} b_l(x)dx = -i\pi. \tag{4.15}$$

Now define $a \in SAP$ by $a(x) = \exp\Big((1-u(x))ib_l(x) + u(x)ib_r(x)\Big)$, where u in $C(\overline{\mathbf{R}})$ is any function for which $u(-\infty) = 0$ and $u(+\infty) = 1$. Clearly, the almost periodic representatives of a are $a_l(x) = e^{ib_l(x)}$ and $a_r(x) = e^{ib_r(x)}$, and we have $\kappa(a_l) = \kappa(a_r) = 0$. The function a_l is 2π-periodic. This implies that

$$\mathbf{d}(a_l - \lambda) = \exp \lim_{T\to\infty} \frac{1}{2T}\int_{-T}^{T} \log(a_l(x) - \lambda)dx = \exp \frac{1}{2\pi}\int_{-\pi}^{\pi} \log(a_l(x) - \lambda)dx.$$

An analogous formula is true for $\mathbf{d}(a_r - \lambda)$. Hence, from (4.14) and (4.15) we infer that

$$\frac{\mathbf{d}(a_r - \lambda)}{\mathbf{d}(a_l - \lambda)} = e^{i(\mu+\pi)+h_1\lambda+h_2\lambda^2+\ldots} = -e^{h(\lambda)}. \qquad \square$$

The following result in conjunction with Proposition 4.5 yields a fairly good idea of the variety of sets that can be assumed by $\mathrm{sp}_{PC}\, W(a)$. See also Figure 4.2.

Proposition 4.7. *Let $R > 1$ and suppose that $h(\lambda)$ is analytic for $|\lambda| < R$. Then there is a function $a \in SAP$ such that*

$$\mathrm{sp}_{PC}\, W(a) = \{\lambda \in \mathbf{D} : h(\lambda) \in \mathbf{R}\}.$$

Proof. Let $h(0) = \nu + i\mu$ with $\nu, \mu \in \mathbf{R}$ and put $g(\lambda) = h(\lambda) - \nu$. Then $g(0) = i\mu$ lies on $i\mathbf{R}$. If $M > 0$ is sufficiently large, then

$$\max_{|t|=1} \frac{|g'(t)|}{M} < \frac{1}{2}, \quad \max_{|\lambda|\le 1} \frac{|g(\lambda)|}{M} < 2\pi.$$

Thus, Lemma 4.6 ensures the existence of a unimodular function $a \in SAP$ such that

$$\frac{d(a_r - \lambda)}{d(a_l - \lambda)} = -e^{g(\lambda)/M} \text{ for } |\lambda| < 1.$$

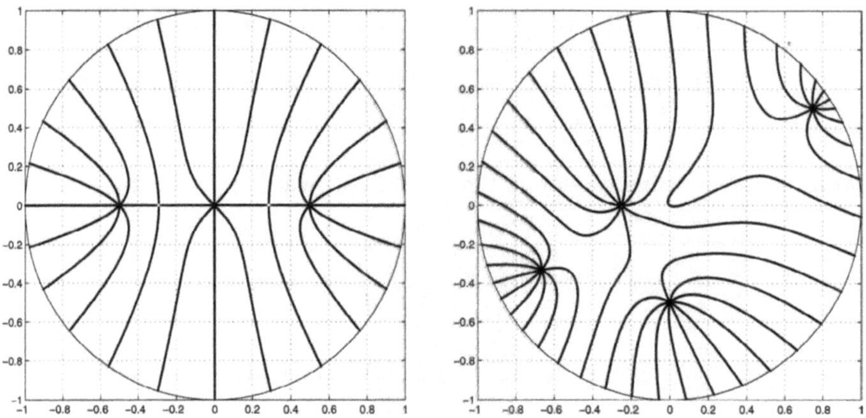

Figure 4.2: The set $\{\lambda \in \mathbf{D} : h(\lambda) \in \mathbf{R}\}$ for $h(\lambda) = (\lambda - 1/2)^4\lambda^4(\lambda + 1/2)^4$ (left) and $h(\lambda) = \lambda(\lambda + 1/4)^5(\lambda + i/2)^5(\lambda - 3/4 - i/2)^5(\lambda + 2/3 + i/3)^5$ (right).

By Proposition 4.5, it results that $\mathrm{sp}_{PC}\, W(a) = \{\lambda \in \mathbf{D} : g(\lambda)/M \in \mathbf{R} + 2\pi i\mathbf{Z}\}$. Since $|g(\lambda)|/M < 2\pi$ for $|\lambda| < 1$, we see that actually

$$\mathrm{sp}_{PC}\, W(a) = \{\lambda \in \mathbf{D} : g(\lambda)/M \in \mathbf{R}\} = \{\lambda \in \mathbf{D} : h(\lambda) \in \mathbf{R}\}. \qquad \square$$

Proof of Theorem 4.3. Let X have its center at the origin and suppose $X = \{\lambda \in V : f(\lambda) \in \mathbf{R}\}$, where f is analytic in V and $f(0) = 0$. Choose $N > 0$ so large that $f(\lambda/N)$ is analytic for $|\lambda| < R$ with some $R > 1$. From Proposition 4.7 we deduce that there is an a in SAP such that $\mathrm{sp}_{PC}\, W(a) = \{\lambda \in \mathbf{D} : f(\lambda/N) \in \mathbf{R}\}$. Hence, if $U \subset \mathbf{C}$ is a sufficiently small neighborhood of the origin, then $\mathrm{sp}_{PC}\, W(a) \cap U = NX \cap U$. Since

$$\mathrm{sp}_{PC}\, W(\alpha + \beta a) = \mathrm{sp}_{PC}\, (\alpha I + \beta W(a)) = \alpha + \beta\, \mathrm{sp}_{PC}\, W(a)$$

for all $\alpha, \beta \in \mathbf{C}$, we obtain that

$$\mathrm{sp}_{PC}\, W(\lambda_0 + a/N) \cap (\lambda_0 + U) = (\lambda_0 + X) \cap (\lambda_0 + U). \qquad \square$$

Proof of Theorem 4.4. If $N > 0$ is large enough, then $\lambda_j/N \in \mathbf{D}$ for all j. Put $h(\lambda) = (\lambda - \lambda_1/N)^{k_1} \dots (\lambda - \lambda_n/N)^{k_n}$. Proposition 4.7 tells us that there exist $a \in SAP$ such that

$$\mathrm{sp}_{PC}\, W(a) = \{\lambda \in \mathbf{D} : h(\lambda) \in \mathbf{R}\}. \tag{4.16}$$

Clearly, $\mathrm{sp}_{PC}\, W(a)$ is an analytic star with the center λ_j/N and the valency $2k_j$ in a neighborhood of λ_j/N. This implies that $\mathrm{sp}_{PC}\, W(Na) = N\mathrm{sp}_{PC}\, W(a)$ is an analytic star centered at λ_j and of the valency $2k_j$ in some neighborhood of λ_j. $\square$

4.3 Essential Spectra of Prescribed Global Shapes

A *Jordan curve* is a subset of the plane that is homeomorphic to the unit circle. The purpose of this section is to prove the following two theorems.

Theorem 4.8. *Given any open bounded subset G of the complex plane whose boundary ∂G is a Jordan curve, there exist $a \in SAP$ such that*

$$\mathrm{sp}_{PC}\, W(a) = G.$$

Theorem 4.9. *Let $G \subset \mathbf{C}$ be an open bounded set whose boundary is a Jordan curve and let f be a conformal map of $\mathbf{D}$ onto G. Then there exist $a \in SAP$ such that*

$$\mathrm{sp}_{PC}\, W(a) = f\big((0,1)\big).$$

Let Γ be a Jordan curve in the complex plane and let D_Γ^+ and D_Γ^- denote the bounded and the unbounded component of $\mathbf{C} \setminus \Gamma$, respectively. We suppose that $0 \in D_\Gamma^+$. Let $f_\pm$ be any conformal maps of $\mathbf{D}$ onto $D_\Gamma^\pm$ such that

$$f_+(0) = 0, \quad f_-(z) = \frac{c_{-1}}{z} + c_0 + c_1 z + \dots \quad (|z| < 1),$$

and denote by $F_\pm : D_\Gamma^\pm \to \mathbf{D}$ the inverse maps of $f_\pm$. Our assumptions that Γ be a Jordan curve guarantees that $f_\pm$ extend to homeomorphisms of $\mathbf{T}$ onto Γ. Define $\varphi_\pm : \mathbf{T} \to \Gamma$ by

$$\varphi_\pm(e^{ix}) = \begin{cases} f_\pm(e^{-2ix}) & \text{for } x \in (-\pi, 0], \\ f_\pm(e^{2ix}) & \text{for } x \in [0, \pi]. \end{cases} \tag{4.17}$$

Cut the complex plane along any curves $\gamma_\pm \subset D_\Gamma^-$ joining $f_\pm(1)$ and the point at infinity and fix any continuous branches of $\arg(f_\pm(1) - \lambda)$ for $\lambda \in \mathbf{C} \setminus \gamma_\pm$. Then, for each $\lambda \in \mathbf{C} \setminus (\Gamma \cup \gamma_\pm)$, we can choose $\arg(f_\pm(e^{iy}) - \lambda)$ and $\arg(\varphi_\pm(e^{ix}) - \lambda)$ so that the functions

$$\begin{aligned} (0, 2\pi] \to \mathbf{R}, &\quad y \mapsto \arg(f_\pm(e^{iy}) - \lambda) \\ (-\pi, \pi] \to \mathbf{R}, &\quad x \mapsto \arg(\varphi_\pm(e^{ix}) - \lambda) \end{aligned}$$

are continuous. Finally, for $\lambda \in \mathbf{C} \setminus (\Gamma \cup \gamma_\pm)$, put

$$\begin{aligned} \log(f_\pm(e^{iy}) - \lambda) &= \log|f_\pm(e^{iy}) - \lambda| + i \arg(f_\pm(e^{iy}) - \lambda), \\ \log(\varphi_\pm(e^{ix}) - \lambda) &= \log|\varphi_\pm(e^{ix}) - \lambda| + i \arg(\varphi_\pm(e^{ix}) - \lambda), \end{aligned}$$

and consider

$$\delta_\pm(\lambda) := \frac{1}{2\pi} \int_{-\pi}^{\pi} \log(\varphi_\pm(e^{ix}) - \lambda)\,dx.$$

Obviously,

$$\begin{aligned}\delta_\pm(\lambda) &= \frac{1}{2\pi}\int_{-\pi}^{0}\log(f_\pm(e^{-2ix})-\lambda)dx+\frac{1}{2\pi}\int_0^{\pi}\log(f_\pm(e^{2ix})-\lambda)dx\\ &= \frac{1}{\pi}\int_0^{\pi}\log(f_\pm(e^{2ix})-\lambda)dx=\frac{1}{2\pi}\int_0^{2\pi}\log(f_\pm(e^{iy})-\lambda)dy\\ &= \frac{1}{2\pi i}\int_{\mathbf{T}}\log(f_\pm(t)-\lambda)\frac{dt}{t}. \qquad (4.18)\end{aligned}$$

Lemma 4.10. *We have* $e^{\delta_+(0)}=-f'_+(0)$ *and* $e^{\delta_-(0)}=-c_{-1}$.

Proof. From (4.18) we see that

$$\delta_+(0)=\frac{1}{2\pi i}\int_{\mathbf{T}}\log f_+(t)\frac{dt}{t}. \qquad (4.19)$$

Let $\log z$ be the branch of the logarithm which is analytic in $\mathbf{C}\setminus[0,\infty)$ and assumes the value $i\pi$ at the point $z=-1$. By the construction of $\log f_+(t)$, the function

$$\log\frac{f_+(t)}{t}:=\log f_+(t)-\log t$$

is continuous on $\mathbf{T}\setminus\{1\}$ and has equal one-sided limits at $t=1$. Hence, this function can be continued to a function $\log(f_+(z)/z)$ that is analytic for all $z\in\mathbf{D}$. From (4.14) we therefore get

$$\begin{aligned}\delta_+(0) &= \frac{1}{2\pi i}\int_{\mathbf{T}}\log\frac{f_+(t)}{t}\frac{dt}{t}+\frac{1}{2\pi i}\int_{\mathbf{T}}\log t\frac{dt}{t}\\ &= \log\left(\frac{f_+(z)}{z}\right)\bigg|_{z=0}+\frac{1}{2\pi}\int_0^{2\pi}ix\,dx=\log f'_+(0)+i\pi,\end{aligned}$$

whence $e^{\delta_+(0)}=-f'_+(0)$. Analogously one can show that

$$\begin{aligned}\delta_-(0) &= \frac{1}{2\pi i}\int_{\mathbf{T}}\log f_-(t)\frac{dt}{t}\\ &= \frac{1}{2\pi i}\int_{\mathbf{T}}\log\Big(f_-(t)t\Big)\frac{dt}{t}-\frac{1}{2\pi i}\int_{\mathbf{T}}\log t\frac{dt}{t}\\ &= \log\Big(f_-(z)z\Big)\Big|_{z=0}-\frac{1}{2\pi}\int_0^{2\pi}ix\,dx=\log c_{-1}-i\pi,\end{aligned}$$

which gives $e^{\delta_-(0)}=-c_{-1}$. □

Lemma 4.11. *We have*

$$e^{\delta_+(\lambda)}=\begin{cases}-\frac{\lambda}{F_+(\lambda)} & \text{for } \lambda\in D_\Gamma^+\setminus\{0\},\\ -\lambda & \text{for } \lambda\in D_\Gamma^-\end{cases}$$

and

$$e^{\delta_-(\lambda)} = \begin{cases} -c_{-1} \text{ for } \lambda \in D_\Gamma^+, \\ -\frac{c_{-1}}{F_-(\lambda)} \text{ for } \lambda \in D_\Gamma^-. \end{cases}$$

Proof. From (4.18) we infer that

$$\delta_+'(\lambda) = -\frac{1}{2\pi i}\int_{\mathbf{T}} \frac{dt}{(f_+(t)-\lambda)t}. \tag{4.20}$$

If $\lambda \in D_\Gamma^+ \setminus \{0\}$, then the residue theorem shows that the right-hand side of (4.20) is

$$-\left(\frac{1}{f_+'(F_+(\lambda))\,F_+(\lambda)} - \frac{1}{\lambda}\right) = \frac{1}{\lambda} - \frac{F_+'(\lambda)}{F_+(\lambda)}.$$

It follows that

$$\left(e^{\delta_+(\lambda)}\frac{F_+(\lambda)}{\lambda}\right)' = e^{\delta_+(\lambda)}\left(\delta_+'(\lambda)\frac{F_+(\lambda)}{\lambda} + \frac{F_+'(\lambda)\lambda - F_+(\lambda)}{\lambda^2}\right)$$
$$= e^{\delta_+(\lambda)}\left(\left(\frac{1}{\lambda} - \frac{F_+'(\lambda)}{F_+(\lambda)}\right)\frac{F_+(\lambda)}{\lambda} + \frac{F_+'(\lambda)\lambda - F_+(\lambda)}{\lambda^2}\right) = 0.$$

Thus, $e^{\delta_+(\lambda)} = c\lambda/F_+(\lambda)$ for $\lambda \in D_\Gamma^+ \setminus \{0\}$ with some constant c. As

$$\lim_{\lambda\to 0} e^{\delta_+(\lambda)} = e^{\delta_+(0)}, \quad \lim_{\lambda\to 0}\frac{\lambda}{F_+(\lambda)} = \frac{1}{F_+'(0)} = f_+'(0),$$

we deduce from Lemma 4.10 that $c = -1$.

Now let $\lambda \in D_\Gamma^-$. There is a cut γ_+ such that $\lambda \in D_\Gamma^- \setminus \gamma_+$. From (4.20) and the residue theorem we obtain that $\delta_+'(\lambda) = 1/\lambda$, whence

$$\left(\frac{e^{\delta_+(\lambda)}}{\lambda}\right)' = e^{\delta_+(\lambda)}\left(\delta_+'(\lambda)\frac{1}{\lambda} - \frac{1}{\lambda^2}\right) = 0$$

and thus $e^{\delta_+(\lambda)} = c\lambda$ for $\lambda \in D_\Gamma^- \setminus \gamma_+$ with some constant c. By (4.18),

$$\begin{aligned}\delta_+(\lambda) &= \frac{1}{2\pi i}\int_{\mathbf{T}}\left(\log(-\lambda) + \log\left(1 - \frac{f_+(t)}{\lambda}\right)\right)\frac{dt}{t} \\ &= \log(-\lambda) + O(1/\lambda) \text{ as } \lambda \to \infty.\end{aligned}$$

Consequently, $e^{\delta_+(\lambda)} = -\lambda + O(1)$ as $\lambda \to \infty$. This gives $c = -1$.

Again by (4.18),

$$\delta_-'(\lambda) = -\frac{1}{2\pi i}\int_{\mathbf{T}} \frac{dt}{(f_-(t)-\lambda)t}. \tag{4.21}$$

If $\lambda \in D_\Gamma^+$, then $f_-(z) - \lambda \neq 0$ for $|z| \le 1$, and hence (4.21) shows that $\delta'_-(\lambda) = 0$. Thus, $e^{\delta_-(\lambda)}$ is some constant c for $\lambda \in D_\Gamma^+$. By Lemma 4.10, $c = -c_{-1}$. Now suppose $\lambda \in D_\Gamma^-$. We can find a cut γ_- such that $\lambda \in D_\Gamma^- \setminus \gamma_-$. Then (4.21) in conjunction with the residue theorem gives

$$\delta'_-(\lambda) = -\frac{1}{f'_-(F_-(\lambda))\, F_-(\lambda)} = -\frac{F'_-(\lambda)}{F_-(\lambda)}.$$

Hence

$$\left(e^{\delta_-(\lambda)} F_-(\lambda)\right)' = e^{\delta_-(\lambda)}\left(\delta'_-(\lambda) F_-(\lambda) + F'_-(\lambda)\right) = 0,$$

and it results that $e^{\delta_-(\lambda)} = c/F_-(\lambda)$ for $\lambda \in D_\Gamma^- \setminus \gamma_-$ with some constant c. From (4.18) we obtain as above that $e^{\delta_-(\lambda)} = -\lambda + O(1)$ as $\lambda \to \infty$, and since $F_-(\lambda) = c_{-1}/\lambda + O(1)$ as $\lambda \to \infty$, we conclude that $c = -c_{-1}$. □

Proof of Theorem 4.8. Without loss of generality assume that $0 \in G$. Let Γ be the boundary of G and let $D_\Gamma^\pm$, f_-, F_- be as above. Clearly, $D_\Gamma^+ = G$. Define $a \in SAP$ by

$$a(x) = (1 - u(x))a_l(x) + u(x)a_r(x) + a_0(x),$$

where $u \in C(\overline{\mathbf{R}})$, $u(-\infty) = 0$, $u(+\infty) = 1$, $a_0 \in C(\dot{\mathbf{R}})$, $a_0(\infty) = 0$,

$$\begin{aligned} a_l(x) &= \begin{cases} f_-(e^{-2ix}) & \text{for } x \in (-\pi, 0] + 2\pi\mathbf{Z}, \\ f_-(e^{2ix}) & \text{for } x \in [0, \pi] + 2\pi\mathbf{Z}, \end{cases} \\ a_r(x) &= \begin{cases} f_-(-e^{-2ix}) & \text{for } x \in (-\pi, 0] + 2\pi\mathbf{Z}, \\ f_-(-e^{2ix}) & \text{for } x \in [0, \pi] + 2\pi\mathbf{Z}. \end{cases} \end{aligned}$$

The function a_0 is chosen so that $a(x) \in \Gamma$ for all $x \in \mathbf{R}$. This in conjunction with Theorem 2.28 implies that $\mathrm{sp}_{AP}\, W(a) = \Gamma$. Given $\lambda \in \mathbf{C} \setminus \Gamma$, we have

$$\mathbf{d}(a_l - \lambda) = e^{\delta_-(\lambda)},$$

and Lemma 4.11 shows that

$$\mathbf{d}(a_l - \lambda) = \begin{cases} -c_{-1} & \text{for } \lambda \in G, \\ -\frac{c_{-1}}{F_-(\lambda)} & \text{for } \lambda \in \mathbf{C} \setminus (G \cup \Gamma). \end{cases}$$

To obtain $\mathbf{d}(a_r - \lambda)$, we apply Lemma 4.11 to the situation where $f_-(z)$ and $F_-(\lambda)$ are replaced by $f_-(-z)$ and $-F_-(\lambda)$, respectively. This gives

$$\mathbf{d}(a_r - \lambda) = e^{\delta_-(\lambda)} = \begin{cases} -(-c_{-1}) & \text{for } \lambda \in G, \\ -\frac{-c_{-1}}{-F_-(\lambda)} & \text{for } \lambda \in \mathbf{C} \setminus (G \cup \Gamma). \end{cases}$$

In summary,

$$\frac{\mathbf{d}(a_r - \lambda)}{\mathbf{d}(a_l - \lambda)} = \begin{cases} -1 & \text{for } \lambda \in G, \\ 1 & \text{for } \lambda \in \mathbf{C} \setminus (G \cup \Gamma), \end{cases}$$

and consequently, by (4.3), $\mathrm{sp}_{PC}\, W(a) = G$. □

Proof of Theorem 4.9. We can without loss of generality assume that $0 \in G$ and that $f(0) = 0$. Let Γ be the boundary of G and let $D_\Gamma^+ = G$, $D_\Gamma^- = \mathbf{C} \setminus (\Gamma \cup G)$ be as above. The map $f_+(z) := f(-z)$ is a conformal map of $\mathbf{D}$ onto G and $f_+((-1,0)) = f((0,1))$. Define $a \in SAP$ by $a(x) = u(x)\varphi_+(x) + a_0(x)$, where $u \in C(\overline{\mathbf{R}})$, $u(-\infty) = 0$, $u(+\infty) = 1$, $a_0 \in C(\dot{\mathbf{R}})$, $a_0(\infty) = 0$, and φ_+ is given by (4.17). Obviously, $a_l(x) = 0$ and $a_r(x) = \varphi_+(x)$. We can choose a_0 so that $a(\mathbf{R}) = \Gamma \cup \mathcal{L}$ where $\mathcal{L} \subset G$ is some curve starting at the origin and terminating on Γ such that $\mathcal{L} \cap f((0,1)) = \emptyset$. In this way we get $\mathrm{sp}_{AP}\, W(a) = \Gamma \cup \mathcal{L}$ due to Theorem 2.28. We have $\mathbf{d}(a_l - \lambda) = -\lambda$ for all $\lambda \in \mathbf{C} \setminus \Gamma$. For $\lambda \in \mathbf{C} \setminus (\Gamma \cup \mathcal{L})$, we obtain from Lemma 4.11 that

$$\mathbf{d}(a_r - \lambda) = e^{\delta_+(\lambda)} = \begin{cases} -\frac{\lambda}{F_+(\lambda)} & \text{for } \lambda \in G \setminus \mathcal{L}, \\ -\lambda & \text{for } \lambda \in \mathbf{C} \setminus (G \cup \Gamma), \end{cases}$$

and hence

$$\frac{\mathbf{d}(a_r - \lambda)}{\mathbf{d}(a_l - \lambda)} = \begin{cases} \frac{1}{F_+(\lambda)} & \text{for } \lambda \in G \setminus \mathcal{L}, \\ 1 & \text{for } \lambda \in \mathbf{C} \setminus (G \cup \Gamma). \end{cases}$$

Thus, by (4.3),

$$\begin{aligned} \mathrm{sp}_{PC}\, W(a) &= \{\lambda \in G \setminus \mathcal{L} \colon 1/F_+(\lambda) \in (-\infty, 0)\} \\ &= \{\lambda \in G \setminus \mathcal{L} \colon F_+(\lambda) \in (-1, 0)\} = f_+((-1,0)). \end{aligned}$$ □

A subset $\mathcal{S}$ of the plane is called a *simple curve* if there exists an injective and continuous map $\varphi : [0,1] \to \mathbf{C}$ such that $\mathcal{S} = \varphi((0,1))$. The points $\varphi(0)$ and $\varphi(1)$ are referred to as the endpoints of $\mathcal{S}$. The next result is a consequence of Theorem 4.9, and it shows that $\mathrm{sp}_{PC}\, W(a)$ may mimic any prescribed simple curve in the plane.

Corollary 4.12. *Given any simple curve $\mathcal{S} \subset \mathbf{C}$ with the endpoints A and B and any $\delta > 0$, there exist $a \in SAP$ such that $\mathrm{sp}_{PC}\, W(a)$ is a simple curve that lies entirely in the δ-neighborhood of $\mathcal{S}$ and that has one endpoint in the δ-neighborhood of A and the other endpoint in the δ-neighborhood of B.*

Proof. Pick $\varepsilon \in (0, \delta)$. Let $\overline{\mathcal{S}} = \mathcal{S} \cup \{A, B\}$ and $V_\varepsilon = \{\lambda \in \mathbf{C} : \mathrm{dist}\,(\lambda, \mathcal{S}) < \varepsilon\}$. For each point $\lambda \in \overline{\mathcal{S}}$, we have $U_\varepsilon(\lambda) := \{z \in \mathbf{C} : |z - \lambda| < \varepsilon\} \subset V_\varepsilon$. As $\overline{\mathcal{S}}$ is compact, there exist finitely many $\lambda_1, \dots, \lambda_n \in \overline{\mathcal{S}}$ such that $\overline{\mathcal{S}} \subset G \subset V_\varepsilon$ with $G := \cup_{j=1}^n U_\varepsilon(\lambda_j)$. Obviously, the boundary of G is a Jordan curve. Let $\varphi : \mathbf{D} \to G$ be a conformal map for which $\varphi(0) = A$. There is a point $t_0 \in \mathbf{T}$ such that $\varphi(t_0) \in \partial U_\varepsilon(B) \subset U_\delta(B)$. Letting $f(z) := \varphi(t_0^{-1} z)$, we get a conformal map $f : \mathbf{D} \to G$ satisfying $f(0) = A$ and $f(1) \in U_\delta(B)$. By Theorem 4.9 there are $a \in SAP$ such that $\mathrm{sp}_{PC}\, W(a) = f((0,1))$. Clearly, these a have the required properties. □

4.4 Separated Almost Periodic Representatives

In this section we describe the essential spectrum of $W(a)$ in the case where a is in SAP and the essential ranges of the almost periodic representatives of a can be separated by a straight line.

Theorem 4.13. *Let $a \in SAP$ have the almost periodic representatives a_l and a_r. Then every straight line in the plane which separates $\mathcal{R}(a_l)$ and $\mathcal{R}(a_r)$ contains at most one point of* $\mathrm{sp}_{PC} W(a)$.

Proof. Let $\Pi = \{\alpha + t\beta : t \in (-\infty, +\infty)\}$ be a straight line separating $\mathcal{R}(a_l)$ and $\mathcal{R}(a_r)$. We give Π the orientation induced by the movement of t from $-\infty$ to $+\infty$. Assume that $\mathcal{R}(a_l)$ lies in the open half-plane on the left of Π and that $\mathcal{R}(a_r)$ is contained in the open half-plane on the right of Π.

Clearly, we can choose arguments so that the two functions

$$\mathbf{R} \times \mathbf{R} \to \mathbf{R}, \quad (x,t) \mapsto \arg(a_l(x) - \alpha - t\beta),$$
$$\mathbf{R} \times \mathbf{R} \to \mathbf{R}, \quad (x,t) \mapsto \arg(a_r(x) - \alpha - t\beta),$$

are continuous. Since $\mathcal{R}(a_l)$ lies to the left of Π, we have, for each $x \in \mathbf{R}$,

$$t_1 < t_2 \implies \arg(a_l(x) - \alpha - t_1\beta) < \arg(a_l(x) - \alpha - t_2\beta).$$

Moreover, $\arg(a_l(x) - \alpha - t_2\beta) - \arg(a_l(x) - \alpha - t_1\beta)$ is the angle at which the segment $\{\alpha + t\beta : t_1 \le t \le t_2\}$ is seen at the point $a_l(x)$, and because $\mathcal{R}(a_l)$ is compact, there is an $\varepsilon(t_1, t_2) > 0$ such that

$$\arg(a_l(x) - \alpha - t_2\beta) - \arg(a_l(x) - \alpha - t_1\beta) \ge \varepsilon(t_1, t_2)$$

for all $x \in \mathbf{R}$. Hence

$$\begin{aligned} &M\Big(\arg(a_l(x) - \alpha - t_2\beta)\Big) - M\Big(\arg(a_l(x) - \alpha - t_1\beta)\Big) \\ &\quad = \lim_{T\to\infty} \frac{1}{2T}\int_{-T}^{T} \Big(\arg(a_l(x) - \alpha - t_2\beta) - \arg(a_l(x) - \alpha - t_1\beta)\Big)dx \\ &\quad \ge \lim_{T\to\infty} \frac{1}{2T}\int_{-T}^{T} \varepsilon(t_1,t_2)dx = \varepsilon(t_1,t_2) > 0. \end{aligned}$$

This shows that $L(t) := M(\arg(a_l - \alpha - t\beta))$ is a strictly monotonically increasing continuous function. Analogously one can prove that $R(t) := M(\arg(a_r - \alpha - t\beta))$ is a strictly monotonically decreasing continuous function.

Now let $\lambda = \alpha + t\beta \in \Pi$. We have

$$\mathbf{d}(a_r - \lambda)/\mathbf{d}(a_l - \lambda) \in (-\infty, 0) \tag{4.22}$$

if and only if $\arg \mathbf{d}(a_r - \lambda) - \arg \mathbf{d}(a_l - \lambda) \in \pi + 2\pi\mathbf{Z}$, and because

$$\begin{aligned}\arg \mathbf{d}(a_r - \lambda) &= M(\arg(a_r - \lambda)) = R(t),\\ \arg \mathbf{d}(a_l - \lambda) &= M(\arg(a_l - \lambda)) = L(t),\end{aligned}$$

condition (4.22) is satisfied if and only if

$$R(t) - L(t) \in \pi + 2\pi\mathbf{Z}. \tag{4.23}$$

A moment's thought reveals that there is a $k \in \mathbf{Z}$ such that

$$R(-\infty) - L(-\infty) = 2k\pi, \quad R(+\infty) - L(+\infty) = 2k\pi - 2\pi.$$

As $R(t)-L(t)$ is a strictly monotonically decreasing continuous function, it follows that (4.23) has exactly one solution $t = t_0 \in (-\infty, +\infty)$. Thus, there is exactly one point $\lambda = \lambda_0 \in \Pi$ which satisfies (4.22). As

$$\mathrm{sp}_{PC}\, W(a) \cap \Pi = \left\{\lambda \in \Pi \setminus a(\mathbf{R}) : \frac{\mathbf{d}(a_r - \lambda)}{\mathbf{d}(a_l - \lambda)} \in (-\infty, 0)\right\},$$

we see that $\mathrm{sp}_{PC}\, T(a) \cap \Pi$ is the empty set if $\lambda_0 \in a(\mathbf{R})$ and that $\mathrm{sp}_{PC}\, W(a) \cap \Pi = \{\lambda_0\}$ if $\lambda_0 \notin a(\mathbf{R})$. □

The previous theorem can be restated in slightly sharper form. Let $\mathrm{conv}E$ denote the convex hull of a set $E \in \mathbf{C}$. We call a simple curve $\mathcal{S} \subset \mathbf{C}$ a *simple analytic curve* if every point $\lambda_0 \in \mathcal{S}$ has an open neighborhood $U \subset \mathbf{C}$ such that $\mathcal{S} \cap U$ is an analytic arc (= analytic star of the valency 2).

Corollary 4.14. *Let $a \in SAP$, let a_l and a_r be the almost periodic representatives of a, and suppose* $\mathrm{conv}\mathcal{R}(a_l) \cap \mathrm{conv}\mathcal{R}(a_r) = \emptyset$. *Put*

$$\Sigma = \mathbf{C} \setminus \Big(\mathrm{conv}\mathcal{R}(a_l) \cup \mathrm{conv}\mathcal{R}(a_r)\Big).$$

Then the intersection of $\mathrm{sp}_{\mathrm{ess}} W(a)$ *and* Σ *is the union of* $a(\mathbf{R}) \cap \Sigma$ *and a simple analytic curve* $\mathcal{S}$ *with one endpoint on* $\mathrm{conv}\mathcal{R}(a_l)$ *and the other endpoint on* $\mathrm{conv}\mathcal{R}(a_r)$. *Moreover, every straight line which separates $\mathcal{R}(a_l)$ and $\mathcal{R}(a_r)$ contains exactly one point of $\mathcal{S}$.*

Proof. The set $\mathrm{sp}_{\mathrm{ess}} W(a) \cap \Sigma$ is the union of $a(\mathbf{R}) \cap \Sigma$ and the set

$$\mathcal{S} := \left\{\lambda \in \Sigma : \frac{\mathbf{d}(a_r - \lambda)}{\mathbf{d}(a_l - \lambda)} \in (-\infty, 0)\right\}. \tag{4.24}$$

The argument of the proof of Theorem 4.1 shows that there are exactly three possibilities: (i) $\mathcal{S}$ is empty; (ii) $\mathcal{S}$ is all of Σ; (iii) $\mathcal{S}$ is an analytic star in some neighborhood of each of its points. When proving Theorem 4.13, we observed that $\mathcal{S} \cap \Pi$ is a singleton whenever Π is a straight line which separates $\mathcal{R}(a_l)$ and $\mathcal{R}(a_r)$.

This implies that we actually have case (iii) and that $\mathcal{S}$ is a simple analytic curve whose endpoints are located on $\text{conv}\mathcal{R}(a_l)$ and $\text{conv}\mathcal{R}(a_r)$. □

Theorems 4.13 and Corollary 4.14 are in particular applicable if $\mathcal{R}(a_l)$ and $\mathcal{R}(a_r)$ are circles whose convex hulls have no common points. The following is another nice special case. Figure 4.3 illustrates things.

Corollary 4.15. *Let $a \in SAP$ and suppose $\mathcal{R}(a_l)$ and $\mathcal{R}(a_r)$ are line segments that form two opposite sides AB and CD of a convex quadrangle $ABCD$. Then $\text{sp}_{\text{ess}}T(a)$ is the union of $a(\mathbf{R})$, $\mathcal{R}(a_l)$, $\mathcal{R}(a_r)$, and a simple analytic curve $\mathcal{S}$ with one endpoint on $\mathcal{R}(a_l)$ and the other endpoint on $\mathcal{R}(a_r)$. The curve $\mathcal{S}$ is contained in the quadrangle $ABCD$, and each line segment having one endpoint on BC and one endpoint of DA intersects $\mathcal{S}$ exactly once.* □

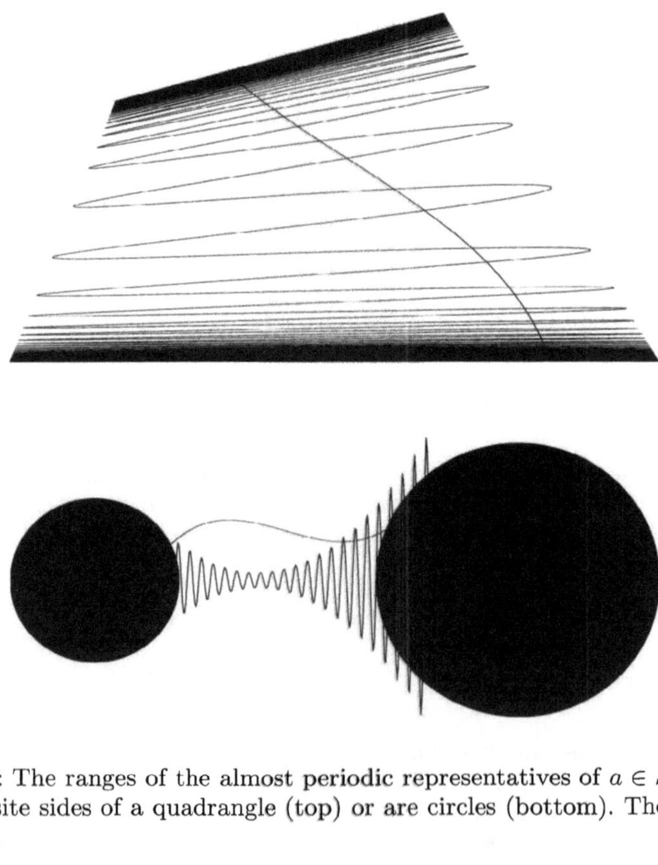

Figure 4.3: The ranges of the almost periodic representatives of $a \in SAP$ constitute opposite sides of a quadrangle (top) or are circles (bottom). The plots show $\text{sp}_{\text{ess}}W(a)$.

4.5 The Spectrum is Discontinuous

Let $\{A_n\}_{n=1}^\infty$ be a sequence of operators $A_n \in \mathcal{B}(H)$ on some Hilbert space H. We denote by $\liminf \operatorname{sp} A_n$ the set of all $\lambda \in \mathbf{C}$ for which there are $\lambda_n \in \operatorname{sp} A_n$ such that $\lambda_n \to \lambda$, and we let $\limsup \operatorname{sp} A_n$ stand for the set of all $\lambda \in \mathbf{C}$ for which there exist $n_1 < n_2 < \dots$ and $\lambda_{n_k} \in \operatorname{sp} A_{n_k}$ such that $\lambda_{n_k} \to \lambda$. Clearly,

$$\liminf_{n\to\infty} \operatorname{sp} A_n \subset \limsup_{n\to\infty} \operatorname{sp} A_n. \tag{4.25}$$

Now suppose the operators $A_n \in \mathcal{B}(H)$ converge uniformly to $A \in \mathcal{B}(H)$. Then, obviously,

$$\limsup_{n\to\infty} \operatorname{sp} A_n \subset \operatorname{sp} A, \tag{4.26}$$

and it is well known that in general neither in (4.25) nor in (4.26) equality holds. Here are two simple examples.

Example 4.16. The two eigenvalues of the matrix $B(a,b) := \begin{pmatrix} a & b \\ b & -a \end{pmatrix}$ are $\pm\sqrt{a^2+b^2}$. Choose any sequence $\{a_n\}_{n=1}^\infty$ of numbers $a_n \in (0,1)$ and define b_n in $(0,1)$ by $a_n^2 + b_n^2 = 1$. Put

$$A = \operatorname{diag}(B(a_1,b_1), B(a_2,b_2), B(a_3,b_3), \dots).$$

Clearly, $\operatorname{sp} A = \operatorname{sp}_{\mathrm{ess}} A = \{-1,1\}$. For the even and odd sections A_{2m} and A_{2m-1} we have $\operatorname{sp} A_{2m} = \{-1,1\}$ and $\operatorname{sp} A_{2m-1} = \{-1, a_m, 1\}$, whence $\liminf \operatorname{sp} A_n = \{-1,1\}$, whereas $\limsup \operatorname{sp} A_n$ is the union of $\{-1,1\}$ and the set of all partial limits of the sequence $\{a_n\}$. □

Example 4.17. Let $A = (A_{jk})_{j,k\in\mathbf{Z}}$ be an infinite matrix that generates a bounded operator on $l^2(\mathbf{Z})$. Given a number $\omega \in \mathbf{C}$, we denote by $E_j\omega E_k$ the infinite matrix whose j,k entry is ω and all other entries of which are zero. It is easily seen that

$$\operatorname{sp}(A + E_j\omega E_k) = \operatorname{sp} A \cup \left\{\lambda \notin \operatorname{sp} A : 1 + [(A-\lambda I)^{-1}]_{kj}\omega = 0\right\}.$$

Now let $U : l^2(\mathbf{Z}) \to l^2(\mathbf{Z})$ be the forward shift: $(Ux)_j = x_{j-1}$. Then $\operatorname{sp} U = \mathbf{T}$, and $[(U-\lambda I)^{-1}]_{21} = 1$ for $|\lambda| < 1$ and $[(U-\lambda I)^{-1}]_{21} = 0$ for $|\lambda| > 1$. Consequently, $1 + [(U-\lambda I)^{-1}]_{21}\omega$ is $1+\omega$ for $|\lambda| < 1$ and is 1 for $|\lambda| > 1$, which implies that

$$\operatorname{sp}(U + E_1\omega E_2) = \begin{cases} \mathbf{T} & \text{for} \quad \omega \neq -1, \\ \overline{\mathbf{D}} & \text{for} \quad \omega = -1, \end{cases}$$

and shows that the spectrum is discontinuous on the set of all bounded linear operators. □

In [74], Farenick and Lee raised the question as to whether or not one has equalities in (4.25) and (4.26) provided the operators are Toeplitz operators. According to

[115], Farenick and Lee conjectured that if $a_n \in L^\infty(\mathbf{R})$ and $\|a_n - a\|_\infty \to 0$ for some $a \in L^\infty(\mathbf{R})$, then

$$\liminf_{n\to\infty} \operatorname{sp} W(a_n) = \operatorname{sp} W(a). \tag{4.27}$$

In [74] and [115], equality (4.27) was verified for several classes of symbols, e.g., for symbols a_n in $C(\dot{\mathbf{R}}) + H^\infty_\pm$, in PQC (the piecewise quasicontinuous functions), or in $AP + C(\dot{\mathbf{R}})$.

We prove that (4.27) is nevertheless false in general. Our counterexample is constructed with symbols in SAP. Here is our result.

Theorem 4.18. *There exist a_n and a in SAP such that*

$$\|a_n - a\|_\infty \to 0, \quad \operatorname{sp} W(a_n) = \operatorname{sp}_{\text{ess}} W(a_n) = \mathbf{T}, \quad \operatorname{sp} W(a) = \operatorname{sp}_{\text{ess}} W(a) = \overline{\mathbf{D}}.$$

In particular, if a_n and a are as in this theorem, then

$$\liminf_{n\to\infty} \operatorname{sp} W(a_n) = \limsup_{n\to\infty} \operatorname{sp} W(a_n) \neq \operatorname{sp} W(a).$$

Also notice that when choosing $b_n = a_n$ for odd n and $b_n = a$ for even n, then we get a uniformly convergent sequence $\{b_n\}$ such that

$$\liminf_{n\to\infty} \operatorname{sp} W(b_n) \neq \limsup_{n\to\infty} \operatorname{sp} W(b_n).$$

To prove Theorem 4.18, let $\beta : \mathbf{R} \to [0, 2\pi]$ be the 2π-periodic function which increases linearly from 0 to 2π on $[-\pi, 0]$ and decreases linearly from 2π to 0 on $[0, \pi]$. The mean values $M(e^{ik\beta})$ are equal to zero for all $k \in \mathbf{N}$:

$$\begin{aligned} M(e^{ik\beta}) &= \frac{1}{2\pi}\int_{-\pi}^{\pi} e^{ik\beta(x)}dx = \frac{1}{\pi}\int_0^{\pi} e^{ik\beta(x)}dx \\ &= \frac{1}{\pi}\int_0^{\pi} e^{ik2(\pi-x)}dx = \frac{e^{2k\pi i}-1}{2k\pi i} = 0. \end{aligned} \tag{4.28}$$

Let further φ be any unimodular function in $C(\overline{\mathbf{R}})$, pick $\mu \in [0,1]$, and define $f_\mu \in SAP$ by

$$f_\mu(x) = e^{-i\mu\beta(x)}\varphi(x).$$

Lemma 4.19. *If the origin is not contained in the convex hull of the essential range of the function φ, then $W(f_\mu)$ is invertible for all $\mu \in [0,1]$.*

Proof. We first show that $W(f_\mu)$ is Fredholm for all $\mu \in [0,1]$. Since $(f_\mu)_l = e^{-i\mu\beta}\varphi(-\infty)$ and $(f_\mu)_r = e^{-i\mu\beta}\varphi(+\infty)$, we deduce from Theorem 3.9 that $0 \notin$

$\mathrm{sp}_{\mathrm{ess}} W(f_\mu)$ if and only if $0 \notin [\varphi(-\infty), \varphi(+\infty)]$, which is certainly satisfied if 0 is not in the convex hull of the essential range of φ.

Thus, $W(f_\mu)$ is Fredholm for all $\mu \in [0,1]$. As the map $[0,1] \to L^\infty(\mathbf{R})$, $\mu \mapsto f_\mu$ is continuous, it follows that $\operatorname{Ind} W(f_\mu) = \operatorname{Ind} W(f_0) = \operatorname{Ind} W(\varphi)$. Theorems 2.5 and 2.7 now imply the invertibility of $W(f_\mu)$. □

Lemma 4.20. *We have*

$$\mathrm{sp}_{\mathrm{ess}} W(f_1) = \begin{cases} \mathbf{T} & \text{if } \varphi(+\infty)/\varphi(-\infty) \neq -1, \\ \overline{\mathbf{D}} & \text{if } \varphi(+\infty)/\varphi(-\infty) = -1. \end{cases}$$

Proof. Obviously, $(f_1)_l = e^{-i\beta}\varphi(-\infty)$, $(f_1)_r = e^{-i\beta}\varphi(+\infty)$. From Theorem 2.7 we see that $\mathbf{T} \subset \mathrm{sp}_{\mathrm{ess}} W(f_1) \subset \overline{\mathbf{D}}$. So fix $\lambda \in \mathbf{D}$. Applying Theorem 3.9 to $f_1 - \lambda$, we obtain that $\lambda \in \mathrm{sp}_{\mathrm{ess}} W(f_1)$ if and only if $0 \in (\mathbf{d}((f_1)_l - \lambda), \mathbf{d}((f_1)_r - \lambda))$, which in turn holds if and only if

$$\mathbf{d}((f_1)_r - \lambda)/\mathbf{d}((f_1)_l - \lambda) \in (-\infty, 0). \tag{4.29}$$

Clearly,

$$\begin{aligned} \mathbf{d}((f_1)_r - \lambda) &= \mathbf{d}(e^{-i\beta}\varphi(+\infty) - \lambda) = \mathbf{d}\left(e^{-i\beta}\varphi(+\infty)\left(1 - \frac{\lambda}{\varphi(+\infty)}e^{i\beta}\right)\right) \\ &= e^{-iM(\beta)}\varphi(+\infty)\exp\left(M\left(\log\left(1 - \frac{\lambda}{\varphi(+\infty)}e^{i\beta}\right)\right)\right) \\ &= e^{-iM(\beta)}\varphi(+\infty)\exp\left(-\sum_{k=1}^{\infty}\frac{\lambda^k}{k\varphi(+\infty)^k}M(e^{ik\beta})\right) \\ &= e^{-iM(\beta)}\varphi(+\infty), \end{aligned}$$

the last equality resulting from (4.28). Analogously, $\mathbf{d}((f_1)_l - \lambda) = e^{-iM(\beta)}\varphi(-\infty)$. Thus, (4.29) is equivalent to the requirement $\varphi(+\infty)/\varphi(-\infty) \in (-\infty, 0)$. As $|\varphi(+\infty)| = |\varphi(-\infty)| = 1$, we arrive at the assertion. □

Proof of Theorem 4.18. Let $a_n = e^{-i\beta}\varphi_n$ and $a = e^{-i\beta}\varphi$ with

$$\varphi_n(x) = \exp\left(i\left(1 - \frac{1}{n}\right)\arctan x\right), \quad \varphi(x) = \exp(i \arctan x). \tag{4.30}$$

It is clear that $\|a_n - a\|_\infty \to 0$. Since

$$\varphi_n(+\infty)/\varphi_n(-\infty) = e^{i(1-1/n)\pi}, \quad \varphi(+\infty)/\varphi(-\infty) = -1,$$

we obtain from Lemma 4.20 that $\mathrm{sp}_{\mathrm{ess}} W(a_n) = \mathbf{T}$ and $\mathrm{sp}_{\mathrm{ess}} W(a) = \overline{\mathbf{D}}$. Since $\mathrm{sp}_{\mathrm{ess}} W(a) \subset \mathrm{sp}\, W(a) \subset \overline{\mathbf{D}}$ (the last inclusion being a consequence of Theorem 2.7), we arrive at the conclusion that $\mathrm{sp}\, W(a) = \overline{\mathbf{D}}$. To show that $\mathrm{sp}\, W(a_n) = \mathbf{T}$, we compute the index of $W(a_n - \lambda)$ for $\lambda \in \mathbf{D}$. Since $\mathrm{Ind} W(a_n - \lambda)$ is constant on $\mathbf{D}$ and $\mathrm{Ind} W(a_n) = 0$ by virtue of Lemma 4.19, it follows that $W(a_n - \lambda)$ has index zero and is therefore invertible for all $\lambda \in \mathbf{D}$ (Corollary 2.6(c)). Thus, $\mathrm{sp}\, W(a_n) = \mathbf{T}$. □

One can even produce Toeplitz operators for which both the inclusions (4.25) and (4.26) are strict.

Theorem 4.21. *There exist c_n and c in $L^\infty(\mathbf{R})$ such that $\|c_n - c\|_\infty \to 0$ and*

$$\begin{aligned} \mathrm{sp}\, T(c) &= \mathrm{sp}_{\mathrm{ess}} T(c) = \overline{\mathbf{D}} \cup (2 + \overline{\mathbf{D}}), \\ \mathrm{sp}\, T(c_n) &= \mathrm{sp}_{\mathrm{ess}} T(c_n) = \begin{cases} \mathbf{T} \cup (2 + \mathbf{T}) & \text{if } n \text{ is odd}, \\ \mathbf{T} \cup (2 + \overline{\mathbf{D}}) & \text{if } n \text{ is even}. \end{cases} \end{aligned}$$

In particular,

$$\liminf_{n \to \infty} \mathrm{sp}\, T(c_n), \quad \limsup_{n \to \infty} \mathrm{sp}\, T(c_n), \quad \mathrm{sp}\, T(c)$$

are three different sets.

A full proof of this theorem is in [31]. This proof shows that Theorem 4.21 is true with functions c_n and c that are continuous on $\mathbf{R} \setminus \{0\}$ and have discontinuities of "semi-almost periodic type" at 0 and ∞.

4.6 Amplitude Modulation Preserves Fredholmness

Let $a \in GL^\infty(\mathbf{R})$. We say that $b \in GL^\infty(\mathbf{R})$ results from a by amplitude modulation if $b(x) = \mu(x)a(x)$ for $x \in \mathbf{R}$ with some positive function $\mu \in GL^\infty(\mathbf{R})$. Figure 4.4 shows some examples.

Theorem 4.22. *Suppose b results from a by amplitude modulation. Then $W(b)$ and $W(a)$ are strongly Φ-equivalent.*

Since $b = \mu a$ with a positive function $\mu \in GL^\infty(\mathbf{R})$ if and only if $b/|b| = a/|a|$, Theorem 4.22 is equivalent to the following theorem.

Theorem 4.23. *If $a \in GL^\infty(\mathbf{R})$, then the operators $W(a)$ and $W(a/|a|)$ are strongly Φ-equivalent.*

The proof of the two theorems can be based on the fact that every positive function $\mu \in GL^\infty(\mathbf{R})$ can be written in the form $\mu = |h|^2$ with $h \in GH^\infty_+$. We then have $W(b) = W(\overline{h})W(a)W(h)$ by Proposition 2.17, and since $W(h)$ and $W(\overline{h}) = W^*(h)$ are invertible, the desired result follows.

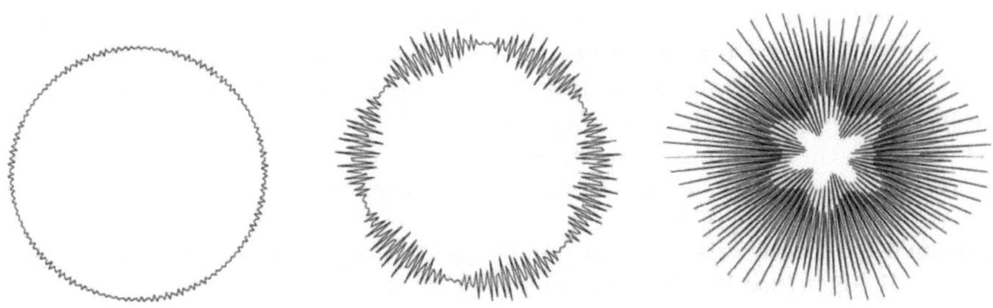

Figure 4.4: The range of the function $b(x) = (10 + \alpha \sin 40\pi x + \beta \sin 42\pi x)e^{ix}$ for $\alpha = 0.2, \beta = 0.1$ (left), $\alpha = 1, \beta = 1$ (middle), and $\alpha = 7, \beta = 1$ (right). All three functions result from $a(x) = e^{ix}$ by amplitude modulation and hence each $W(b)$ is strongly Φ-equivalent to the shift operator $W(a)$.

For symbols in SAP, Theorem 4.23 can also be deduced from Theorem 3.9. Indeed, let

$$a_l(x) = e^{i\kappa_l x} e^{\psi_l(x) + i\varphi_l(x)}, \quad a_r(x) = e^{i\kappa_r x} e^{\psi_r(x) + i\varphi_r(x)}$$

with real-valued functions $\psi_l, \psi_r, \varphi_l, \varphi_r \in AP$ (Theorem 2.25). Then

$$(a/|a|)_l(x) = e^{i\kappa_l x} e^{i\varphi_l(x)}, \quad (a/|a|)_r(x) = e^{i\kappa_r x} e^{i\varphi_r(x)},$$

whence

$$\begin{aligned} &\kappa(a_l) = \kappa((a/|a|)_l) = \kappa_l, \quad \kappa(a_r) = \kappa((a/|a|)_r) = \kappa_r \\ &\mathbf{d}(a_l) = \exp(M(\psi_l) + iM(\varphi_l)), \quad \mathbf{d}((a/|a|)_l) = \exp(iM(\varphi_l)), \\ &\mathbf{d}(a_r) = \exp(M(\psi_r) + iM(\varphi_r)), \quad \mathbf{d}((a/|a|)_r) = \exp(iM(\varphi_r)), \end{aligned}$$

and thus

$$\frac{\mathbf{d}(a_r)}{\mathbf{d}(a_l)} = \frac{e^{M(\psi_r)}}{e^{M(\psi_l)}} \frac{\mathbf{d}((a(|a|)_r))}{\mathbf{d}((a/|a|_l))}.$$

Theorem 4.23 can now be derived from Theorems 3.9 and 3.14.

4.7 Frequency Modulation Can Destroy Fredholmness

In this section we study the influence of frequency modulation on the Fredholm properties of Wiener-Hopf operators. This means that we replace the function $a(x) = |a(x)|e^{i\varphi(x)}$ by the function

$$b(x) := a(\alpha(x)) = |a(\alpha(x))|e^{i\varphi(\alpha(x))} \quad (x \in \mathbf{R})$$

where $\alpha : \mathbf{R} \to \mathbf{R}$ is an orientation preserving homeomorphism. The Fredholm criterion for Wiener-Hopf operators with symbols in $C(\overline{\mathbf{R}})$ immediately implies that if $a \in C(\overline{\mathbf{R}})$, then $W(a)$ and $W(b)$ are either both Fredholm (of the same index) or neither of them is Fredholm. The same is, of course, also true for symbols in $PC(\dot{\mathbf{R}})$. As the following theorem shows, things change when passing to almost periodic symbols. Figure 4.5 illustrates the following proof.

Theorem 4.24. *There exist $a \in GAP$ and orientation preserving homeomorphisms $\alpha : \mathbf{R} \to \mathbf{R}$ such that $W(a)$ is Fredholm on H^2 but the operator $W(b)$ generated by the function $b(x) = a(\alpha(x))$ is not Fredholm on H^2.*

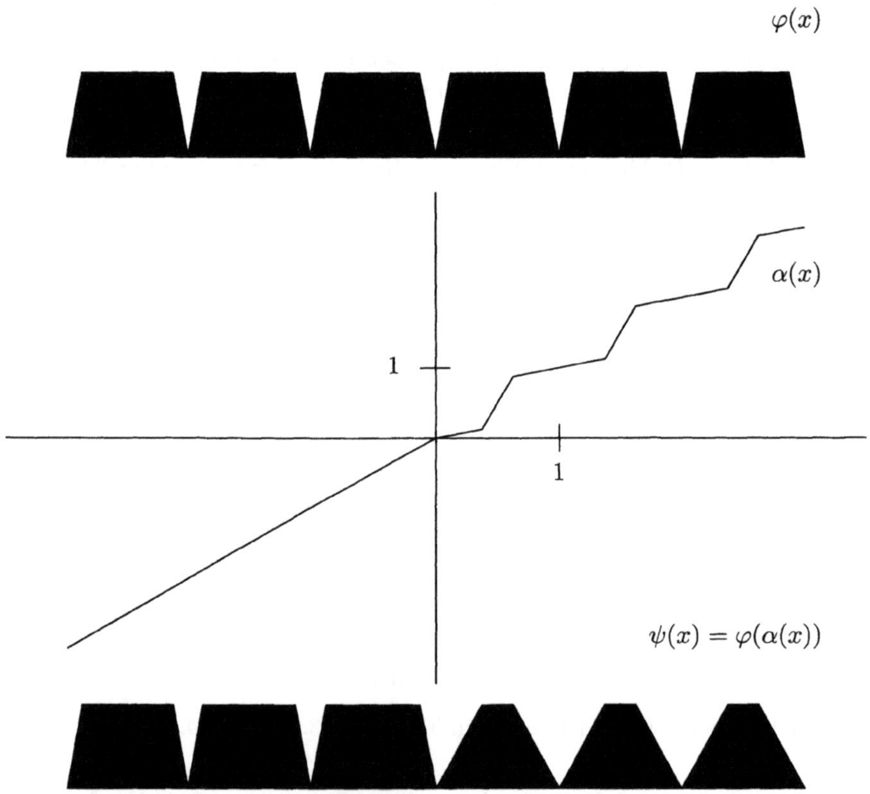

Figure 4.5: Frequency modulation transforms an almost periodic function into a semi-almost periodic function. The functions $\varphi(x)$, $\alpha(x)$, $\psi(x)$ are as in the proof of Theorem 4.24. Note the different scales on the x-axes and y-axes; the bases of the trapezoids have length 1 and their heights are 4π.

Proof. Put $a(x) = e^{i\varphi(x)}$ where φ is 1-periodic and

$$\varphi(x) = \begin{cases} 32\pi x & \text{for} \quad 0 \le x \le 1/8, \\ 4\pi & \text{for} \quad 1/8 \le x \le 7/8, \\ 32\pi(1-x) & \text{for} \quad 7/8 \le x \le 1. \end{cases}$$

Obviously, $a \in GAP$ and $\kappa(a) = 0$. Hence, $W(a)$ is Fredholm due to Theorem 2.28. Let $\alpha : \mathbf{R} \to \mathbf{R}$ be the following homeomorphism: $\alpha(x) = x$ for $x \le 0$ and

$$\alpha(n+x) = n + \begin{cases} x/3 & \text{for} \quad 0 \le x \le 3/8, \\ 3x - 1 & \text{for} \quad 3/8 \le x \le 5/8, \\ x/3 + 2/3 & \text{for} \quad 5/8 \le x \le 1 \end{cases}$$

for $n \in \{0, 1, 2, \dots\}$ and $x \in [0, 1)$. Now consider $b(x) = a(\alpha(x))$. We then have $a(x) = e^{\psi(x)}$ where $\psi(x) = \varphi(x)$ for $x \le 0$, while ψ is 1-periodic for $x \ge 0$ and

$$\psi(x) = \begin{cases} 32\pi x/3 & \text{for} \quad 0 \le x \le 3/8, \\ 4\pi & \text{for} \quad 3/8 \le x \le 5/8, \\ 32\pi(1-x)/3 & \text{for} \quad 5/8 \le x \le 1. \end{cases}$$

Consequently, $b \in GSAP$ and $b_l = e^{i\varphi}$, $b_r = e^{i\psi}$. It follows that $b_l \in GAP$, $b_r \in GAP$, $\kappa(b_l) = 0$, $\kappa(b_r) = 0$, and a simple computation gives

$$\mathbf{d}(b_l) = \exp\left(i \int_{-1}^{0} \varphi(x)dx\right) = \exp(7\pi i/2),$$

$$\mathbf{d}(b_r) = \exp\left(i \int_{0}^{1} \psi(x)dx\right) = \exp(5\pi i/2).$$

Since $0 \in [\mathbf{d}(b_l), \mathbf{d}(b_r)]$, we deduce from Theorem 3.9 that the operator $W(a)$ is not Fredholm. □

In fact, the above proof actually yields $a \in AP$ and α such that $W(a)$ is invertible but $W(a \circ \alpha)$ is not even normally solvable (see Theorem 3.9).

Notes

That the passage from the invertibility criterion to the description of the spectrum is non-trivial for Toeplitz operators with SAP symbols was already realized by Sarason [197]. He established in particular Theorem 4.1. However, this phenomenon had not been fully appreciated (at least, by the authors) until it was shown up in connection with the question of the continuity of the spectrum of Toeplitz operators. This question, raised in [74], was answered negatively in [31], and Section 4.5 is based on the latter paper. A more involved treatment of the spectra of Toeplitz operators with SAP symbols was subsequently undertaken in [33]. Sections 4.2 to 4.4 follow [33].

The result of Section 4.6 is well known and goes back to Widom [225] and Devinatz [63]. Frequency modulations of AP and SAP symbols were studied in [32]. On top of the observation that such modulations are capable of destroying Fredholmness (Section 4.7), the paper [32] contains a number of positive results, that is, conditions on the modulation under which the Fredholm properties are preserved.

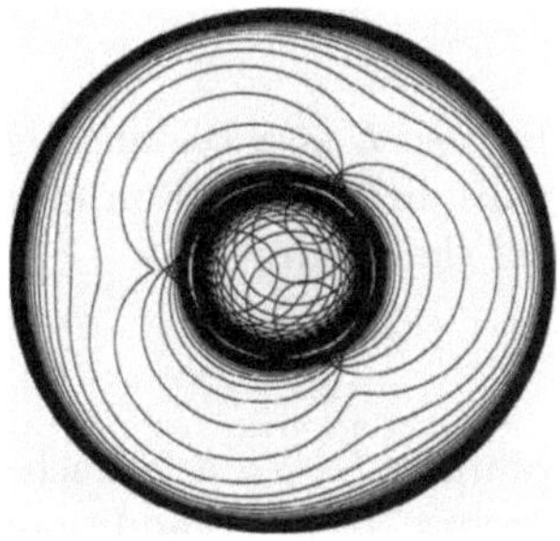

Chapter 5

Introduction to Matrix Wiener-Hopf Operators

We now leave the vivid world of scalar Wiener-Hopf operators and enter the rugged realm of operators with matrix-valued symbols. In this chapter we cite a few well known results on operators with matrix symbols in $C + H^\infty$ and PC. Invoking a nice theorem on operator determinants whose entries commute modulo compact operators, we can easily develop the Fredholm theory of operators with $C + H^\infty$ symbols on the basis of its scalar counterpart. For PC symbols, it is an appropriate triangularization result that reduces the Fredholm theory of the matrix operators to the scalar case. However, having climbed the hills connected with matrix-valued $C + H^\infty$ and PC symbols, we will be confronted with a serious wall of rocks: the Fredholm theory of Wiener-Hopf operators with AP symbols. It turns out that for operators with AP symbols Fredholmness is equivalent to invertibility. Thus, the search for a Fredholm criterion is at the same time the search for an invertibility criterion, and it is well known that invertibility of matrix operators is a delicate matter...

5.1 General Remarks and Normal Solvability

Given a set E, we denote by E_N and $E_{N\times N}$ the columns of length N and the $N \times N$ matrices with entries in E, respectively. If E is a Banach or Hilbert space, then so also is E_N with natural algebraic operations and the norm

$$\|(f_1, \dots , f_N)^T\| := \left(\|f_1\|^2 + \dots + \|f_N\|^2\right)^{1/2}.$$

In case E is a Banach algebra and $a \in E_{N\times N}$, we denote by aI the operator of multiplication by a on E_N. On providing $E_{N\times N}$ with natural algebraic operations and the norm

$$\|a\|_{E_{N\times N}} := \|aI\|_{\mathcal{B}(E_N)},$$

we make $E_{N\times N}$ a Banach algebra. If E is a C^*-algebra, then $E_{N\times N}$ is also a C^*-algebra. To avoid additional parantheses or brackets, we make use of the notation

$$L^\infty_{N\times N}(\mathbf{R}) := \left[L^\infty(\mathbf{R})\right]_{N\times N}, \quad L^2_N(\mathbf{R}) := \left[L^2(\mathbf{R})\right]_N$$

and of similar modifications. Also notice that $GA_{N\times N}$ always means $G(A_{N\times N})$ and not $(GA)_{N\times N}$ (although for $N = 1$ both are the same). Finally, instead of $\mathbf{C}_N$ and $\mathbf{C}_{N\times N}$ we will employ the standard notations $\mathbf{C}^N$ and $\mathbf{C}^{N\times N}$.

We now consider Wiener-Hopf operators $W(a)$ with matrix symbols a in the algebra $L^\infty_{N\times N}(\mathbf{R})$ on the vector-valued space $L^2_N(\mathbf{R}_+)$.

Some of the results of Chapter 2 extend to the matrix case, the others are pure scalar case results and have no matrix analogues. Proposition 2.10, i.e., the identities

$$W(ab) = W(a)W(b) + H(a)H(\tilde{b}), \tag{5.1}$$
$$H(ab) = W(a)H(b) + H(a)W(\tilde{b}), \tag{5.2}$$

also hold in the matrix case. Furthermore, we have

$$H(c) \text{ is compact} \iff c \in [C(\dot{\mathbf{R}}) + H^\infty_-]_{N\times N}, \tag{5.3}$$
$$H(\tilde{c}) \text{ is compact} \iff c \in [C(\dot{\mathbf{R}}) + H^\infty_+]_{N\times N} \tag{5.4}$$

(recall Theorem 2.18). Theorem 2.14 is also true in the matrix case, that is, if $a \in L^\infty_{N\times N}(\mathbf{R})$ is at each $x_0 \in \dot{\mathbf{R}}$ locally equivalent to an $a_{x_0} \in L^\infty_{N\times N}(\mathbf{R})$ for which $W(a_{x_0})$ is (left/right) Fredholm, then $W(a)$ itself is (left/right) Fredholm. The following theorem extends the last assertion of Theorem 2.7 to matrix Wiener-Hopf operators.

Theorem 5.1 (Simonenko). *If $a \in L^\infty_{N\times N}(\mathbf{R})$ and $W(a)$ is semi-Fredholm then $a \in GL^\infty_{N\times N}(\mathbf{R})$.*

One of the big problems in the matrix case is the failure of the theorem by Coburn and Simonenko (Theorem 2.5) and its Corollary 2.6 for matrix symbols. The question whether a matrix Wiener-Hopf operator is (left/right) invertible cannot simply be answered by computing the index – it requires determining the so-called right partial indices of the symbol, which will be introduced later.

While the zero operator is the only normally solvable scalar Wiener-Hopf operator which is not semi-Fredholm, there is a large gap between normal solvability and semi-Fredholmness for matrix symbols. Note that if, for example,

$$a = \begin{pmatrix} b & 0 \\ 0 & c \end{pmatrix}, \quad W(b) \in \Phi_n\big(L^2(\mathbf{R}_+)\big), \quad W(c) \in \Phi_d\big(L^2(\mathbf{R}_+)\big),$$

then $W(a)$ is normally solvable whereas $n(W(a)) = d(W(a)) = \infty$, so that $W(a)$ is not semi-Fredholm. The question of deciding whether a matrix Wiener-Hopf operator is normally solvable is difficult and concerned with phenomena that are not present in the scalar case.

The following theorem provides us with two useful criteria for normal solvability of general bounded Hilbert space operators.

Theorem 5.2. *Let H be a Hilbert space and $A \in \mathcal{B}(H)$. Then the following are equivalent:*

(i) *A is normally solvable;*

(ii) *A^* is normally solvable;*

(iii) *there is a number $d > 0$ such that* $\operatorname{sp} A^*A \subset \{0\} \cup [d^2, \infty)$.

Condition (iii) says that the origin is at most an isolated point of the spectrum of A^*A (= the set of the squares of the singular values of A). Obviously, in (iii) we may replace A^*A by AA^*. The adjoint of the Wiener-Hopf operator $W(a)$ is the operator $W(a^*)$ where

$$a^*(x) = \Big[\big(a_{jk}(x)\big)_{j,k=1}^N\Big]^* := \Big(\overline{a_{kj}(x)}\Big)_{j,k=1}^N.$$

Here is a necessary condition for the normal solvability of Wiener-Hopf operators.

Theorem 5.3. *Let $a \in L^\infty_{N\times N}(\mathbf{R})$ and let U be a connected subset of $\dot{\mathbf{R}}$ such that a is continuous on U. If $W(a)$ is normally solvable on $L^2_N(\mathbf{R}_+)$, then*

$$\text{either } \inf_{x\in U} |\det a(x)| > 0 \text{ or } \det a(x) = 0 \text{ for all } x \in U.$$

Proof outline. If $W(a)$ is normally solvable then, by Theorem 5.2,

$$\operatorname{sp}_{\mathrm{ess}} W(a^*)W(a) \subset \operatorname{sp} W(a^*)W(a) \subset \{0\} \cup [d^2, \infty)$$

for some $d > 0$. Taking into account that $a|U$ is continuous and using a local principle, one can show that

$$\bigcup_{x\in U} \operatorname{sp}\big(a^*(x)a(x)\big) \subset \operatorname{sp}_{\mathrm{ess}} W(a^*)W(a),$$

where $\operatorname{sp}\big(a^*(a)a(x)\big)$ stands for the spectrum (= set of eigenvalues) of $a^*(x)a(x)$ as an element of $\mathbf{C}^{N\times N}$. It follows that

$$\operatorname{sp}\big(a^*(x)a(x)\big) \subset \{0\} \cup [d^2, \infty) \text{ for all } x \in U. \tag{5.5}$$

By the Wielandt-Hoffman theorem (see, e.g., [112, Theorem 6.3.5]), the eigenvalues of $a^*(x)a(x)$ depend continuously on $x \in U$. Hence, if $0 \in \text{sp}\,(a^*(x)a(x))$ for some $x \in U$, then, by (5.5), 0 must be an eigenvalue of $a^*(x)a(x)$ for all $x \in U$, which shows that $\det a(x) = 0$ for all $x \in U$. On the other hand, if $0 \notin \text{sp}\,(a^*(x)a(x))$ for all $x \in U$, then $\text{sp}\,(a^*(x)a(x)) \subset [d^2, \infty)$ for all $x \in U$ and consequently,

$$|\det a(x)|^2 = \det\left(a^*(x)a(x)\right) \geq d^{2N} \quad \text{for all } x \in U. \qquad \square$$

We remark that Theorem 5.3 does not hold in case a is discontinuous on U.

Corollary 5.4. *If $a \in L^\infty_{N\times N}(\mathbf{R})$ is continuous on $\mathbf{R}$ and $W(a)$ is normally solvable, then*

$$\textit{either } a \in GL^\infty_{N\times N}(\mathbf{R}) \textit{ or } \det a(x) = 0 \textit{ for all } x \in \mathbf{R}. \qquad \square$$

The previous corollary is in particular applicable to matrix symbols in $SAP_{N\times N}$. For $a \in SAP_{N\times N}$ with identically vanishing determinant, $W(a)$ may be normally solvable or not. For example, if

$$a(x) = \begin{pmatrix} 1 & 1 \\ 1 & 1 \end{pmatrix} \quad \text{for all } x \in \mathbf{R},$$

then $W(a)$ is surjective and thus normally solvable. On the other hand, if b is in $AP \setminus \{0\}$ and has a zero on $\mathbf{R}$, then $W(b)$ is not normally solvable (Theorem 2.28) and hence, if

$$a(x) = \begin{pmatrix} b(x) & b(x) \\ b(x) & b(x) \end{pmatrix},$$

then $W(a)$ is clearly not normally solvable, too. One can also take $b, c \in C(\dot{\mathbf{R}}) \setminus \{0\}$ such that $b(x)c(x) = 0$ for all $x \in \mathbf{R}$ and put

$$a(x) = \begin{pmatrix} b(x) & 0 \\ 0 & c(x) \end{pmatrix}$$

to obtain an operator $W(a)$ which is not normally solvable although $\det a$ vanishes identically.

5.2 Matrix-Valued $C + H^\infty$ Symbols

Wiener-Hopf operators with symbols in $[C + H^\infty_\pm]_{N\times N}$ can be most easily tackled with the help of the following theorem.

Let X be a Banach space and let $A \in \mathcal{B}(X_N)$ be given by the operator matrix $(A_{jk})_{j,k=1}^N$. If the operators A_{jk} commute pairwise modulo compact operators, then $\det A \in \mathcal{B}(X)$ is well-defined modulo compact operators. By Theorem 2.2, the semi-Fredholm properties of $\det A$ are independent of the manner the determinant is computed.

Theorem 5.5. (a) *If the operator entries of $A \in \mathcal{B}(X_N)$ commute pairwise modulo compact operators, then A is Fredholm (properly n-normal, resp. properly d-normal) if and only if* $\det A$ *has the corresponding property.*

(b) *If the operator entries of $A \in \mathcal{B}(X_N)$ commute modulo finite rank operators and A is semi-Fredholm, then*

$$\operatorname{Ind} A = \operatorname{Ind} \det A.$$

Here are the matrix analogues of Theorems 2.15 and 2.19.

Theorem 5.6 (Gohberg and Krein). *Let $a \in C_{N\times N}(\dot{\mathbf{R}})$*

(a) *If $a \notin GC_{N\times N}(\dot{\mathbf{R}})$, then $W(a)$ is not semi-Fredholm.*

(b) *If $a \in GC_{N\times N}(\dot{\mathbf{R}})$, then $W(a)$ is Fredholm and*

$$\operatorname{Ind} W(a) = -\operatorname{wind} \det a.$$

Theorem 5.7 (Douglas). *Let $a \in [C(\dot{\mathbf{R}}) + H^\infty_+]_{N\times N}$.*

(a) *If $a \notin GL^\infty_{N\times N}(\mathbf{R})$, then $W(a)$ is not semi-Fredholm.*

(b) *If $a \in GL^\infty_{N\times N}(\mathbf{R})$ but $a \notin G[C(\dot{\mathbf{R}}) + H^\infty_+]_{N\times N}$, then $W(a)$ is properly n-normal and $W(a^{-1})$ is a left regularizer of $W(a)$.*

(c) *If $a \in G[C(\dot{\mathbf{R}}) + H^\infty_+]_{N\times N}$, then $W(a)$ is Fredholm, $W(a^{-1})$ is a regularizer of $W(a)$, and*

$$\operatorname{Ind} W(a) = -\operatorname{wind} \det a. \tag{5.6}$$

We remark that the invertibility of a in $L^\infty_{N\times N}(\mathbf{R})$, $C_{N\times N}(\dot{\mathbf{R}})$, $[C(\dot{\mathbf{R}})+H^\infty_+]_{N\times N}$ is equivalent to the invertibility of $\det a$ in $L^\infty(\mathbf{R})$, $C(\dot{\mathbf{R}})$, $C(\dot{\mathbf{R}}) + H^\infty_+$, respectively. In (5.6), the winding number of $\det a$ is understood as in Section 2.5.

Proof outline. Except for the index formula, the above two theorems can be deduced from Theorem 5.5(a) and Theorems 2.15 and 2.19. Of course, parts (a) also follow from Theorem 5.1, and part (c) can be derived from (5.1) and (5.3). Furthermore, the equivalence

$$W(a) \text{ is Fredholm} \iff a \in G\big[C(\dot{\mathbf{R}}) + H^\infty_+]_{N\times N}$$

can be shown as in the proof of Theorem 2.19.

As for the index formula, notice first that $H(b)$ and $H(\tilde{b})$ have finite rank whenever b is a finite sum of the form

$$b(x) = \sum_{k=-m}^{m} b_k \Big(\frac{x-i}{x+i}\Big)^k, \quad x \in \mathbf{R}$$

(recall the proof of Theorem 2.15). Denote the collection of all such finite sums by $\mathcal{P}$. Obviously, $\mathcal{P}$ is dense in $C(\dot{\mathbf{R}})$. Now let $a = c + h$ with $c \in C_{N\times N}(\dot{\mathbf{R}})$ and $h \in [H^\infty_+]_{N\times N}$. Choose $b_n \in \mathcal{P}_{N\times N}$ so that $\|c - b_n\|_\infty \to 0$. If $W(a)$ is Fredholm, then $W(b_n + h)$ is Fredholm for all sufficiently large n and

$$\operatorname{Ind} W(a) = \lim_{n\to\infty} \operatorname{Ind} W(b_n + h), \ \operatorname{Ind} W(\det a) = \lim_{n\to\infty} \operatorname{Ind} W\big(\det(b_n + h)\big).$$

Since $\operatorname{Ind} W(b_n + h) = \operatorname{Ind} W(\det(b_n + h))$ by virtue of Theorem 5.5(b), we arrive at (5.6). □

Let $\operatorname{alg} W(C_{N\times N})$ stand for the smallest closed subalgebra of $\mathcal{B}(L^2_N(\mathbf{R}_+))$ which contains the set $\{W(a) : a \in C_{N\times N}(\dot{\mathbf{R}})\}$, and for $A \in \operatorname{alg} W(C_{N\times N})$, denote by A^π the coset $A + \mathcal{K}(L^2_N(\mathbf{R}_+))$ in the Calkin algebra. One can show that $\mathcal{K}(L^2_N(\mathbf{R}_+))$ is a subset of $\operatorname{alg} W(C_{N\times N})$.

Theorem 5.8. *The map Γ given for $a \in C_{N\times N}(\dot{\mathbf{R}})$ by*

$$\big(\Gamma W^\pi(a)\big)(x) := a(x), \quad x \in \dot{\mathbf{R}},$$

extends to a C^*-algebra isomorphism of

$$\operatorname{alg} W^\pi(C_{N\times N}) := \operatorname{alg} W(C_{N\times N})/\mathcal{K}\big(L^2_N(\mathbf{R}_+)\big)$$

onto $C_{N\times N}(\dot{\mathbf{R}})$. An operator K is compact if and only if $(\Gamma K^\pi)(x) = 0$ for all $x \in \dot{\mathbf{R}}$. An operator $A \in \operatorname{alg} W(C_{N\times N})$ is Fredholm if and only if $\det(\Gamma A^\pi)(x) \neq 0$ for all $x \in \dot{\mathbf{R}}$.

5.3 Matrix-Valued PC Symbols

For $a \in PC_{N\times N}$, we define the function $a_2 : \dot{\mathbf{R}} \times [0,1] \to \mathbf{C}^{N\times N}$ by

$$a_2(x,\mu) := (1-\mu)a(x-0) + \mu a(x+0), \quad (x,\mu) \in \dot{\mathbf{R}} \times [0,1].$$

Recall that $a(\infty - 0) := a(+\infty)$ and $a(\infty + 0) := a(-\infty)$. The subscript 2 in a_2 indicates that we are considering operators on L^2. The function $\det a_2$ is given by

$$\det a_2(x,\mu) = \det\big((1-\mu)a(x-0) + \mu a(x+0)\big), \quad (x,\mu) \in \dot{\mathbf{R}} \times [0,1]$$

and maps $\dot{\mathbf{R}} \times [0,1]$ into $\mathbf{C}$. We emphasize that in general $\det a_2 \neq (\det a)_2$. For $x \in \dot{\mathbf{R}}$, the set

$$\big\{\det a_2(x,\mu) : \mu \in [0,1]\big\} \tag{5.7}$$

is a continuous curve joining $\det a(x-0)$ to $\det a(x+0)$; of course, if $\det a(x-0) = \det a(x+0)$, then (5.7) is a singleton. In the case where $\det a_2(x,\mu) \neq 0$ for all $(x,\mu) \in \dot{\mathbf{R}} \times [0,1]$, we denote by $\operatorname{wind}(\det a_2)$ the number of the counter-clockwise circuits around the origin of the naturally oriented closed and continuous curve which results from $\mathcal{R}(\det a)$ by filling in the curves (5.7).

Theorem 5.9. *Let $a \in PC_{N\times N}$.*

(a) *If $\det a_2(x_0,\mu_0) = 0$ for some $(x_0,\mu_0) \in \dot{\mathbf{R}} \times [0,1]$, then $W(a)$ is not semi-Fredholm.*

(b) *If $\det a_2(x,\mu) \neq 0$ for all $(x,\mu) \in \dot{\mathbf{R}} \times [0,1]$, then $W(a)$ is Fredholm and*

$$\operatorname{Ind} W(a) = -\operatorname{wind}(\det a_2).$$

Proof outline. If $a \in GPC_{N\times N}$ has only finitely many jumps, one can write $a = b\varphi c$ with $b, c \in GC_{N\times N}(\dot{\mathbf{R}})$ and an upper-triangular function $\varphi \in GPC_{N\times N}$ (see, e.g., [57, p. 171]). Let $\varphi_1, \dots, \varphi_N$ be the diagonal entries of $W(\varphi)$. Since $W(a) - W(b)W(\varphi)W(c)$ is compact, the operator $W(a)$ is Fredholm if so are all the operators $W(\varphi_j)$, in which case

$$\operatorname{Ind} W(a) = \operatorname{Ind} W(b) + \sum_{j=1}^{N} \operatorname{Ind} W(\varphi_j) + \operatorname{Ind} W(c).$$

Theorem 2.20 therefore gives part (b) for symbols with at most finitely many jumps. For symbols with countably many jumps, the Fredholm condition of part (b) now follows from the matrix version of Theorem 2.14, after which the index formula can be proved by approximation. Part (a) follows from part (b) by an index perturbation argument (recall the proof of Theorem 2.20).

We remark that the Fredholm criterion contained in Theorem 5.9 can also be obtained by combining Theorems 5.5(a) and 2.21. □

The purpose of what follows is to define the *Cauchy index* $\operatorname{ind}(\det a_2)$ and to reformulate Theorem 5.9(b) in other terms.

Suppose $\det a_2(x,\mu) \neq 0$ for all $(x,\mu) \in \dot{\mathbf{R}} \times [0,1]$. Then $a(x-0)$ and $a(x+0)$ are invertible for every $x \in \dot{\mathbf{R}}$. Clearly,

$$\begin{aligned} &\det\big[(1-\mu)a(x-0) + \mu a(x+0)\big] \neq 0 \quad \forall \mu \in (0,1) \\ \Longleftrightarrow\ &\det\big[-\xi I + a^{-1}(x-0)a(x+0)\big] \neq 0 \quad \forall \xi \in (-\infty,0) \\ \Longleftrightarrow\ &\frac{1}{2} - \frac{1}{2\pi}\arg \xi_k(x) \notin \mathbf{Z} \quad \forall k \in \{1,\dots,N\} \end{aligned}$$

where $\xi_1(x), \dots, \xi_k(x)$ are the eigenvalues of $a^{-1}(x-0)a(x+0)$.

Assume the set $\Lambda_a := \{x \in \mathbf{R} : a(x-0) \neq a(x+0)\}$ is finite. For a connected component l of $\dot{\mathbf{R}} \setminus \Lambda_a$, we denote by $\operatorname{ind}_l(\det a)$ the increment of any continuous argument of $\det a$ on l. Given a continuous function $f : [0,1] \to \mathbf{C} \setminus \{0\}$, we let $\{\arg f(\mu)\}_{\mu\in[0,1]}$ stand for the increment of any continuous argument of $f(\mu)$ as μ changes from 0 to 1. We define

$$\operatorname{ind}(\det a_2) := \sum_l \operatorname{ind}_l(\det a) + \sum_{x\in\Lambda_a} \big\{ \arg\, \det a_2(x,\mu)\big\}_{\mu\in[0,1]}.$$

Fix $x \in \Lambda_a$ and choose $C_x \in G\mathbf{C}^{N\times N}$ so that

$$J_x := C_x a^{-1}(x-0)a(x+0)C_x$$

is the Jordan form of $a^{-1}(x-0)a(x+0)$. We have

$$\begin{aligned}
&\big\{\arg\det a_2(x,\mu)\big\}_{\mu\in[0,1]}\\
&= \Big\{\arg\det\big[(1-\mu)I + \mu a^{-1}(x-0)a(x+0)\big]\Big\}_{\mu\in[0,1]}\\
&= \Big\{\arg\det\big[(1-\mu)I + \mu J_x\big]\Big\}_{\mu\in[0,1]}\\
&= \Big\{\arg\prod_{k=1}^N\big[(1-\mu)I + \mu\xi_k(x)\big]\Big\}_{\mu\in[0,1]}\\
&= \sum_{k=1}^N\Big\{\arg\big[(1-\mu)I + \mu\xi_k(x)\big]\Big\}_{\mu\in[0,1]}\\
&= \sum_{k=1}^N\left(\frac{1}{2} - \left\{\frac{1}{2} - \frac{1}{2\pi}\arg\xi_k(x)\right\}\right)
\end{aligned}$$

(recall Section 2.6 for the last equality). Thus,

$$\operatorname{ind}(\det a_2) = \sum_l \operatorname{ind}_l(\det a) + \sum_{x\in\Lambda_a}\sum_{k=1}^N\left(\frac{1}{2} - \left\{\frac{1}{2} - \frac{1}{2\pi}\arg\xi_k(x)\right\}\right). \tag{5.8}$$

Analogously,

$$\operatorname{ind}(\det a_2) = \sum_l \operatorname{ind}_l(\det a) + \sum_{x\in\Lambda_a}\sum_{k=1}^N\left(-\frac{1}{2} + \left\{\frac{1}{2} - \frac{1}{2\pi}\arg\eta_k(x)\right\}\right) \tag{5.9}$$

where $\eta_1(x),\dots,\eta_k(x)$ are the eigenvalues of $a^{-1}(x+0)a(x-0)$ (again recall Section 2.6). Taking into account the possible jump at infinity, we see that the *winding number* of $\det a_2$ introduced before Theorem 5.9 can be given by

$$\begin{aligned}
\operatorname{wind}(\det a_2) &= \operatorname{ind}(\det a_2) + \sum_{k=1}^N\left(\frac{1}{2} - \left\{\frac{1}{2} - \frac{1}{2\pi}\arg\xi_k(\infty)\right\}\right) &(5.10)\\
&= \operatorname{ind}(\det a_2) + \sum_{k=1}^N\left(-\frac{1}{2} + \left\{\frac{1}{2} - \frac{1}{2\pi}\arg\eta_k(\infty)\right\}\right). &(5.11)
\end{aligned}$$

In summary, we arrive at the following.

Theorem 5.10. *Let* $a \in GPC_{N\times N}$ *and put*

$$\begin{aligned}\operatorname{sp}\big(a^{-1}(x-0)a(x+0)\big) &= \{\xi_1(x),\dots,\xi_N(x)\},\\ \operatorname{sp}\big(a^{-1}(x+0)a(x-0)\big) &= \{\eta_1(x),\dots,\eta_N(x)\}.\end{aligned}$$

Then for $W(a)$ *to be Fredholm it is necessary and sufficient that one of the following four equivalent conditions is satisfied:*

(i) $\operatorname{sp}\big(a^{-1}(x-0)a(x+0)\big)\cap(-\infty,0]=\emptyset$ *for all* $x\in\dot{\mathbf{R}}$;

(ii) $\operatorname{sp}\big(a^{-1}(x+0)a(x-0)\big)\cap(-\infty,0]=\emptyset$ *for all* $x\in\dot{\mathbf{R}}$;

(iii) $\frac{1}{2}-\frac{1}{2\pi}\arg\xi_k(x)\notin\mathbf{Z}$ *for all* $x\in\dot{\mathbf{R}}$ *and all* $k\in\{1,\dots,N\}$;

(iv) $\frac{1}{2}-\frac{1}{2\pi}\arg\eta_k(x)\notin\mathbf{Z}$ *for all* $x\in\dot{\mathbf{R}}$ *and all* $k\in\{1,\dots,N\}$.

If $W(a)$ *is Fredholm and* a *has at most finitely many jumps then*

$$\operatorname{Ind}W(a) = -\operatorname{wind}(\det a_2)$$

where $\operatorname{wind}(\det a_2)$ *is given by* (5.8) *to* (5.11). *Choosing the arguments in* $(-\pi,\pi)$, *we also have*

$$\begin{aligned}\operatorname{Ind}W(a) &= -\operatorname{ind}(\det a_2)-\frac{1}{2\pi}\sum_{k=1}^{N}\arg\xi_k(\infty)\\ &= -\operatorname{ind}(\det a_2)+\frac{1}{2\pi}\sum_{k=1}^{N}\arg\eta_k(\infty).\end{aligned}$$

□

Example 5.11. Let $u\in C(\overline{\mathbf{R}})$, $0\le u\le 1$, $u(-\infty)=0$, $u(+\infty)=1$, and consider the matrix function

$$b(x)=\big(1-u(x)\big)\begin{pmatrix} i & 0\\ 0 & i\end{pmatrix}+u(x)\begin{pmatrix}1 & 0\\ 0 & 1\end{pmatrix},\quad x\in\mathbf{R}.$$

We have $\det b(x)=((1-u(x))i+u(x))^2\ne 0$ for all $x\in\dot{\mathbf{R}}$, and the eigenvalues of $b^{-1}(+\infty)b(-\infty)$ are i,i. Hence, by Theorem 5.10, $W(b)$ is Fredholm and

$$\begin{aligned}\operatorname{Ind}W(b) &= -\operatorname{ind}(\det b)-\frac{1}{2\pi}(\arg i+\arg i)\quad\big(\arg i\in(-\pi,\pi)\big)\\ &= -\frac{1}{2\pi}2\Big(-\frac{\pi}{2}\Big)-\frac{1}{2\pi}\Big(\frac{\pi}{2}+\frac{\pi}{2}\Big)=0.\end{aligned}$$

Since $\det b(\pm\infty)=\pm 1$, the operator $W(\det b)$ is not Fredholm (and not even normally solvable). Moral: there are diagonal $b\in[C(\overline{\mathbf{R}})]_{2\times 2}$ such that $W(b)$ is Fredholm but $W(\det b)$ is not Fredholm. See Figure 5.1. In particular, the difference

$W(\det b) - \det W(b)$ is not compact. Notice that here $\det W(b)$ is well-defined up to a compact additive term, because the operator $W(a)W(c) - W(c)W(a)$ is compact on $L^2(\mathbf{R}_+)$ for all $a, c \in PC$ (see, e.g., Theorem 5.12 below). □

$\mathrm{sp}_{\mathrm{ess}} W(b)$ $\qquad$ $\mathrm{sp}_{\mathrm{ess}} W(\det b)$ $\qquad$ $\mathrm{sp}\, W(b)$ $\qquad$ $\mathrm{sp}\, W(\det b)$

Figure 5.1: Spectra of the operators of Example 5.11.

Let $\mathrm{alg}\, W(PC_{N\times N})$ be the smallest closed subalgebra of $\mathcal{B}(L^2_N(\mathbf{R}_+))$ which contains the set $\{W(a) : a \in PC_{N\times N}\}$, and put $A^\pi := A + \mathcal{K}(L^2_N(\mathbf{R}_+))$.

Theorem 5.12. *The map* Γ *given for* $a \in PC_{N\times N}$ *by*

$$(\Gamma W^\pi(a))(x,\mu) := (1-\mu)a(x-0) + \mu a(x+0), \quad (x,\mu) \in \dot{\mathbf{R}} \times [0,1]$$

extends to a C^**-algebra isomorphism of*

$$\mathrm{alg}\, W^\pi(PC_{N\times N}) := \mathrm{alg}\, W(PC_{N\times N})/\mathcal{K}(L^2_N(\mathbf{R}_+))$$

onto $C_{N\times N}(\dot{\mathbf{R}} \times [0,1])$, *where* $\dot{\mathbf{R}} \times [0,1]$ *has the Gelfand topology as the maximal ideal space of* $\mathrm{alg}\, W^\pi(PC)$. *An operator* $K \in \mathrm{alg}\, W(PC_{N\times N})$ *is compact if and only if* $(\Gamma K^\pi)(x,\mu) = 0$ *for all* $(x,\mu) \in \dot{\mathbf{R}} \times [0,1]$. *An operator* $A \in \mathrm{alg}\, W(PC_{N\times N})$ *is Fredholm if and only if*

$$\det(\Gamma A^\pi)(x,\mu) \neq 0 \ \text{ for all } \ (x,\mu) \in \dot{\mathbf{R}} \times [0,1].$$

5.4 The Gohberg-Krein Theorem

So far we have established Fredholm criteria and index formulas for scalar and matrix Wiener-Hopf operators with symbols from several classes. In the scalar case, a Fredholm criterion and an index formula provide us with an invertibility criterion (Corollary 2.6). The problem of finding the inverse operator leads to the task of constructing a so-called Wiener-Hopf factorization of the symbol.

In Example 1.9, we encountered symbols of the form $c + Fk$ with $c \in \mathbf{C}$ and $k \in L^1(\mathbf{R})$. The set of all these functions is denoted by W or $W(\mathbf{R})$ and referred to as the *Wiener algebra of* $\mathbf{R}$:

$$W := \mathbf{C} + FL^1(\mathbf{R}) = \{c + Fk : c \in \mathbf{C}, k \in L^1(\mathbf{R})\}.$$

It is well known (and easily seen) that W is a unital commutative Banach algebra under pointwise operations and the norm

$$\|c + Fk\|_W := |c| + \|k\|_{L^1(\mathbf{R})}.$$

Gelfand showed that the maximal ideal space of W can be identified with $\dot{\mathbf{R}}$ and that the Gelfand transform is given by

$$\Gamma : W \to C(\dot{\mathbf{R}}), \ (\Gamma(c + Fk))(x) = c + (Fk)(x).$$

Theorem 1.6 and formula (1.1) therefore imply the following *theorem by Wiener*: if $a \in W$ and $a(x) \neq 0$ for all $x \in \dot{\mathbf{R}}$, then $a^{-1} \in W$, and if, in addition, $\operatorname{wind} a = 0$, then a has a logarithm in W, that is, there exists a function $b \in W$ such that $a = e^b$.

We define W^+ and W^- by

$$W^\pm := \mathbf{C} + FL^1(\mathbf{R}_\pm) = \{c + Fk\colon c \in \mathbf{C}, k \in L^1(\mathbf{R}), k(x) = 0 \ \text{for} \ \pm x < 0\}.$$

The sets W^+ and W^- are closed unital subalgebras of W. The Paley-Wiener theorem tells us that $W^\pm = W \cap H^\infty_\pm$ and one can show that $GW^\pm = W \cap GH^\infty_\pm$. Given $a \in W \cap GL^\infty(\mathbf{R}) = GW$, let $\varkappa = \operatorname{wind} a$. By Wiener's theorem, we can write

$$a(x) = \left(\frac{x-i}{x+i}\right)^\varkappa e^{b(x)}$$

with $b = c + Fg \in W$. Put $g_\pm = \chi_\pm g$, where $\chi_\pm$ is the characteristic function of $\mathbf{R}_\pm$. We obtain that $b = b_- + b_+$ with $b_- = Fg_- \in W^-$ and $b_+ = c + Fg_+ \in W^+$, whence

$$a(x) = a_-(x) \left(\frac{x-i}{x+i}\right)^\varkappa a_+(x) \tag{5.12}$$

with $a_\pm = e^{b_\pm} \in GW^\pm$. Factorization (5.12) is called a *Wiener-Hopf factorization of a in W*. Let $d(x)$ denote the middle factor in (5.12). From Proposition 2.17 we deduce that $W(a) = W(a_-)W(d)W(a_+)$ and that $W(a_+^{-1})W(d^{-1})W(a_-^{-1})$ is a one-sided inverse and a regularizer of $W(a)$, which becomes the inverse of $W(a)$ in case $\varkappa = 0$ and thus $d = 1$.

Definition 5.13. Let $a \in GW_{N\times N} = W_{N\times N} \cap GL^\infty_{N\times N}(\mathbf{R})$. A *right Wiener-Hopf factorization of a in W* is a representation

$$a = a_- d a_+, \tag{5.13}$$

where $a_\pm$ are in $GW^\pm_{N\times N} = W^\pm_{N\times N} \cap G[H^\infty_\pm]_{N\times N}$ and

$$d(x) = \operatorname{diag}\left[\left(\frac{x-i}{x+i}\right)^{\varkappa_1}, \ldots, \left(\frac{x-i}{x+i}\right)^{\varkappa_N}\right]$$

with $\varkappa_1, \ldots, \varkappa_n \in \mathbf{Z}$. □

Let $a \in GW_{N\times N}$ have a Wiener-Hopf factorization (5.13). Then $W(a)$ is equal to $W(a_-)W(d)W(a_+)$, and since $W(a_\pm)$ have the inverses $W(a_\pm^{-1})$ and

$$n\left(W\left(\left(\frac{x-i}{x+i}\right)^{\varkappa}\right)\right) = \max(-\varkappa, 0), \quad d\left(W\left(\left(\frac{x-i}{x+i}\right)^{\varkappa}\right)\right) = \max(+\varkappa, 0)$$

(recall the proof of Theorem 2.15), it follows that

$$n(W(a)) = \sum_{\varkappa_j<0} |\varkappa_j|, \quad d(W(a)) = \sum_{\varkappa_j>0} \varkappa_j.$$

In particular, $W(a)$ is invertible if and only if $\varkappa_j = 0$ for all j, in which case $W^{-1}(a) = W(a_+^{-1})W(a_-^{-1})$.

We know from (5.12) that every scalar-valued function $a \in GW$ possesses a Wiener-Hopf factorization in W. That the same is true for matrix-valued functions is the contents of the following theorem.

Theorem 5.14 (Gohberg and Krein). *Every matrix function in $GW_{N\times N}$ has a right Wiener-Hopf factorization in W.*

This theorem ensures the *existence* of a right Wiener-Hopf factorization in W of every matrix function in $GW_{N\times N}$. The *construction* of this factorization and *the problem of at least determining the numbers $\varkappa_j$* are in general very difficult and no universal algorithms are known for $N > 1$.

5.5 Outlook

Following the line of Chapter 2, our next task would be to establish a Fredholm criterion for Wiener-Hopf operators with symbols in $AP_{N\times N}$. Unfortunately, this is an extremely difficult problem.

Let APW be the "Wiener version" of AP, that is, let APW denote the set of all functions a which can be written in the form

$$a(x) = \sum_j a_j e^{i\lambda_j x} \ (x \in \mathbf{R}), \quad \sum_j |a_j| < \infty,$$

where $\lambda_j \in \mathbf{R}$ and $a_j \neq 0$ for at most countably many j. Clearly, $APW \subset AP$. We will prove the following.

Theorem 5.15. *If $a \in APW_{N\times N}$, then $W(a)$ is Fredholm if and only if $W(a)$ is invertible.*

Thus, a Fredholm criterion for Wiener-Hopf operators with Wienerian AP symbols (which are much simpler than general AP symbols) is at the same time an invertibility criterion, and we know from the preceding section that invertibility of matrix Wiener-Hopf operators is a delicate matter.

A right APW factorization of a matrix function $a \in APW_{N\times N}$ is a representation

$$a = a_- d a_+ \tag{5.14}$$

with $a_\pm \in G[APW \cap H^\infty_\pm]_{N\times N}$ and

$$d(x) = \operatorname{diag}\left(e^{i\mu_1 x}, \dots, e^{i\mu_N x}\right), \tag{5.15}$$

where $\mu_j \in \mathbf{R}$ for all j. The factorization (5.14), (5.15) is called canonical if $\mu_j = 0$ for all j. The following result will be proved later.

Theorem 5.16. *If $a \in APW_{N\times N}$, then $W(a)$ is invertible if and only if a has a canonical right APW factorization.*

The Gohberg-Krein theorem tells us that every matrix function in $W_{N\times N}$ has a right Wiener-Hopf factorization in W and leaves us with the problem of finding the numbers $\varkappa_1, \dots, \varkappa_N$. As will be shown later, things are dramatically more complicated for matrix functions in $APW_{N\times N}$ because of the following result.

Theorem 5.17. *Not every matrix function in $APW_{N\times N}$ has a right APW factorization.*

Notes

Theorem 5.1 is proved in [207]. Theorem 5.2 is a standard fact from the theory of Hilbert space operators, following from the polar decomposition of A with subsequent use of spectral decomposition of its modulus $(A^*A)^{1/2}$ (see also [47, Section 4.7]). We have not found Theorem 5.3 and Corollary 5.4 in the literarure.

Theorem 5.5(a) was established by Krupnik (see [146]) for Fredholm operators, and by Köhler and Silbermann [141] for semi-Fredholm operators; Part (b) of that theorem is due to Markus and Feldman [157].

For several subclasses of continuous matrix functions and in different disguises (Toeplitz operators, systems of singular integral equations, matrix-valued boundary value problems), particular cases of Theorems 5.6 and 5.8 were studied by many authors; see the monographs [57], [80], [155], [161] for more on this topic. For symbols in the Wiener algebra W, Theorem 5.6 was explicitly stated and proved in Gohberg and Krein's paper [94]. Gohberg and Krein were also the first to understand the Banach algebraic background of the Fredholm theory of Wiener-Hopf operators with continuous symbols. Thus, it seems us to be justified to attribute both Theorems 5.6 and 5.8 to Gohberg and Krein. From the modern point of view, the Fredholm criterion of Theorem 5.6 is an immediate consequence of the local principle, according to which the general case can be reduced to the case of constant matrix symbols. Theorem 5.7 was established in [65].

The case of piecewise continuous matrix functions was disposed of in its full generality by Gohberg and Krupnik [97], [98]; see also the expositions in [57], [155].

Thees approaches use factorization techniques. An alternative approach, based on reduction of the problem to integral operators with homogeneous kernels via a local principle and on consecutive application of the Mellin transform, is in [183] and [209].

The description of the maximal ideal space of W cited in Section 5.4 can be found in [85], [164], and the proof of Wiener's theorem based on it was one of the first demonstrations of the usefulness of the abstract Banach algebras approach. The original proof of this theorem is in [230]. Theorem 5.14 is from [94].

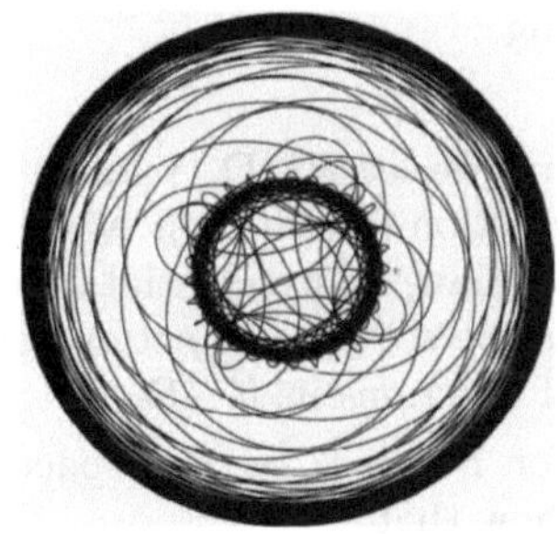

Chapter 6

Factorizations of Matrix Functions

We now begin the trek to a criterion for the Fredholmness or, equivalently, the invertibility of Wiener-Hopf operators with matrix-valued AP symbols. In one or another way, all invertibility criteria for matrix Wiener-Hopf operators are connected with certain factorizations of their symbols. In this chapter we introduce the notions of Wiener-Hopf factorization and almost periodic factorization. The first of these two types of factorizations has been used in Wiener-Hopf theory for many decades, while almost periodic factorization is less known. Both factorizations are special cases of factorizations of matrix functions into a product a_-ua_+ in which $a_\pm$ are invertible analytic matrix functions and u is a unitary matrix function. We therefore conclude this chapter by collecting some results on the Fredholmness and invertibility of Wiener-Hopf operators with unitary-valued symbols.

6.1 Hardy Spaces

In Section 2.4, we defined the *Hardy spaces* $H^p(\mathbf{C}_\pm)$ and $H^p_\pm := H^p_\pm(\mathbf{R})$. In what follows we will need the analogues of these spaces for the disk and the circle. Put

$$\mathbf{D}_+ := \{z \in \mathbf{C} : |z| < 1\}, \quad \mathbf{D}_- := \{z \in \mathbf{C} : |z| > 1\} \cup \{\infty\}.$$

We denote by $H^\infty(\mathbf{D}_\pm)$ the set of all bounded analytic functions in $\mathbf{D}_\pm$. For $1 \le p < \infty$, we let $H^p(\mathbf{D}_+)$ be the set of all functions φ which are analytic in $\mathbf{D}_+$ and satisfy

$$\sup_{r\in(0,1)} \int_0^{2\pi} |\varphi(re^{i\theta})|^p \, d\theta < \infty,$$

and we let $H^p(\mathbf{D}_-)$ stand for the functions $\varphi(z)$ $(z \in \mathbf{D}_-)$ for which $\varphi(1/z)$ is a function in $H^p(\mathbf{D}_+)$. Again by a *theorem of Fatou*, functions in $H^p(\mathbf{D}_\pm)$ $(1 \le p \le \infty)$ have nontangential limits almost everywhere on $\mathbf{T} = \partial\mathbf{D}_\pm$. We define $H^p_\pm(\mathbf{T})$ as the set of all functions on $\mathbf{T}$ which can be represented as nontangential limits of functions in $H^p(\mathbf{D}_\pm)$.

For $1 \le p \le \infty$, the spaces $H^p_\pm(\mathbf{T})$ are closed subspaces of $L^p(\mathbf{T})$ and one can show that

$$H^p_\pm(\mathbf{T}) = \{\varphi \in L^p(\mathbf{T}) : \varphi_n = 0 \ \text{for} \ \pm n < 0\} \tag{6.1}$$

where

$$\varphi_n := \frac{1}{2\pi}\int_0^{2\pi} \varphi(e^{i\theta})e^{-in\theta}\, d\theta \quad (n \in \mathbf{Z})$$

is the nth Fourier coefficient of φ. Also notice that $H^\infty_\pm(\mathbf{T})$ are closed subalgebras of $L^\infty(\mathbf{T})$. The spaces $H^p(\mathbf{D}_\pm)$ and $H^p_\pm(\mathbf{T})$ are called *Hardy spaces* of the disk and the circle, respectively.

The map $\omega : z \mapsto \frac{z-i}{z+i}$ maps $\mathbf{C}_\pm$ conformally onto $\mathbf{D}_\pm$ and $\mathbf{R}$ onto $\mathbf{T} \setminus \{1\}$. Obviously,

$$f \in H^\infty_\pm \iff f = \varphi \circ \omega \ \text{with} \ \varphi \in H^\infty_\pm(\mathbf{T}).$$

Things are a little more complicated for $1 \le p < \infty$. For $\sigma \in \mathbf{R}$, denote by $(x \pm i)^\sigma$ any branch of the function which is analytic in $\mathbf{C}_\pm$ and let $H^p_\pm(\sigma)$ or, in other notation, $(x \pm i)^{-\sigma} H^p_\pm$ be the set of all functions f on $\mathbf{R}$ for which $f(x)(x \pm i)^\sigma$ is a function in $H^p_\pm$. It can be shown that

$$f \in H^p_\pm(-2/p) \iff f = \varphi \circ \omega \ \text{with} \ \varphi \in H^p_\pm(\mathbf{T}). \tag{6.2}$$

In particular,

$$\{\varphi \circ \omega : \varphi \in H^2_\pm(\mathbf{T})\} = H^2_\pm(-1) = (x \pm i)H^2_\pm. \tag{6.3}$$

We know from Section 1.5 that the Cauchy singular integral operator S on $L^2(\mathbf{R})$ can be written as $S = W^0(-\operatorname{sgn} x)$ and that

$$P := \frac{1}{2}(I + S) \ \text{and} \ Q := \frac{1}{2}(I - S)$$

are the orthogonal projections of $L^2(\mathbf{R})$ onto H^2_+ and H^2_-, respectively. Also recall formulas (2.11) and (2.12).

Let S_0 be the *Cauchy singular integral operator* on $L^2(\mathbf{T})$:

$$(S_0\varphi)(t) := \frac{1}{\pi i}\lim_{\varepsilon\to 0}\int_{|\tau - t|\ge\varepsilon} \frac{\varphi(\tau)}{\tau - t}\, d\tau, \quad t \in \mathbf{T}.$$

As above, define

$$P_0 := \frac{1}{2}(I + S_0), \quad Q_0 := \frac{1}{2}(I - S_0).$$

One can show that P_0 is the orthogonal projection of $L^2(\mathbf{T})$ onto $H^2_+(\mathbf{T})$ and that Q_0 is the orthogonal projection of $L^2(\mathbf{T})$ onto

$$\dot{H}^2_-(\mathbf{T}) := \{\varphi \in H^2_-(\mathbf{T}) : \varphi_0 = 0\}.$$

Notice that $H^2_+(\mathbf{T}) \cap H^2_-(\mathbf{T}) = \mathbf{C}$, while $L^2(\mathbf{T})$ is the orthogonal (and thus direct) sum of $H^2_+(\mathbf{T})$ and $\dot{H}^2_-(\mathbf{T})$.

The map B given by

$$(B\varphi)(x) = \frac{\sqrt{2}}{x+i}\varphi\Big(\frac{x-i}{x+i}\Big), \quad x \in \mathbf{R},$$

is an isometric isomorphism of $L^2(\mathbf{T})$ onto $L^2(\mathbf{R})$, of $H^2_+(\mathbf{T})$ onto H^2_+, and of $\dot{H}^2_-(\mathbf{T})$ onto H^2_-. We have

$$(B^{-1}f)(t) = \frac{i\sqrt{2}}{1-t}f\Big(i\frac{1+t}{1-t}\Big), \quad t \in \mathbf{T} \setminus \{1\},$$

and a straightforward computation shows that

$$S = BS_0B^{-1}, \quad P = BP_0B^{-1}, \quad Q = BQ_0B^{-1}. \tag{6.4}$$

For $a \in L^\infty(\mathbf{R})$, put

$$(B_0a)(t) := a\Big(i\frac{1+t}{1-t}\Big), \quad t \in \mathbf{T} \setminus \{1\}. \tag{6.5}$$

Obviously, B_0 is an isometric isomorphism of $L^\infty(\mathbf{R})$ onto $L^\infty(\mathbf{T})$ and of $H^\infty_\pm$ onto $H^\infty_\pm(\mathbf{T})$. Clearly,

$$(B_0^{-1}b)(x) = b\Big(\frac{x-i}{x+i}\Big), \quad x \in \mathbf{R}.$$

It can be verified directly that

$$B^{-1}aB = (B_0a)I, \tag{6.6}$$

that is, B transforms multiplication by a on $L^2(\mathbf{R})$ to multiplication by B_0a on $L^2(\mathbf{T})$.

Since $L^\infty(\mathbf{T}) \subset L^2(\mathbf{T})$, the function S_0b is a well-defined function in $L^2(\mathbf{T})$ for every $b \in L^\infty(\mathbf{T})$ (note that S_0 actually maps $L^\infty(\mathbf{T})$ into $BMO(\mathbf{T})$). Hence, for

$a \in L^\infty(\mathbf{R})$, we can consider the function $S_0(B_0 a) \in L^2(\mathbf{T})$. Again by straightforwardly computing one gets

$$\big(S_0(B_0 a)\big)\Big(\frac{x-i}{x+i}\Big) = (B_0^{-1} S_0 B_0 a)(x) = \frac{x+i}{\pi i} \int_{\mathbf{R}} \frac{1}{y+i} \frac{a(y)}{y-x}\, dy, \quad x \in \mathbf{R},$$

or equivalently,

$$\big(P_0(B_0 a)\big)\Big(\frac{x-i}{x+i}\Big) = (x+i) P\Big(\frac{1}{x+i} a(x)\Big), \quad x \in \mathbf{R}. \tag{6.7}$$

Clearly, the right-hand side of (6.7) has to be interpreted as $(x+i)(Pb)(x)$ with $b(x) = (x+i)^{-1} a(x)$.

6.2 Wiener-Hopf Factorization

Definition 6.1. A matrix function $a \in GL^\infty_{N\times N}(\mathbf{R})$ is said to admit a *right Wiener-Hopf factorization* or simply a *right WH factorization* if it can be represented in the form

$$a(x) = a_-(x) d(x) a_+(x) \ \text{ for almost all } \ x \in \mathbf{R}$$

where

$$\text{(i)} \quad d(x) = \operatorname{diag}\left[\Big(\frac{x-i}{x+i}\Big)^{\varkappa_1}, \dots, \Big(\frac{x-i}{x+i}\Big)^{\varkappa_N}\right] \ \text{ with } \ \varkappa_j \in \mathbf{Z}, \tag{6.8}$$

$$\text{(ii)} \quad a_-(x)/(x-i) \in \big[H^2_-\big]_{N\times N}, \ a_-^{-1}(x)/(x-i) \in \big[H^2_-\big]_{N\times N}, \tag{6.9}$$

$$a_+(x)/(x+i) \in \big[H^2_+\big]_{N\times N}, \ a_+(x)/(x+i) \in \big[H^2_+\big]_{N\times N}, \tag{6.10}$$

(iii) the operator $a_- S a_-^{-1} I$ is bounded on $L^2_N(\mathbf{R})$.

A right WH factorization in which $d(x)$ is the $N \times N$ identity matrix I is referred to as a *canonical right WH factorization.* □

A few comments on this definition are in order.

Remark 6.2. The above definition has an analogue for (and is more popular in the context of) the unit circle. A right WH factorization of a matrix function $b \in GL^\infty_{N\times N}(\mathbf{T})$ is a representation

$$b(t) = b_-(t) v(t) b_+(t) \ \text{ for almost all } \ t \in \mathbf{T} \tag{6.11}$$

where

$$\text{(i)} \quad v(t) = \operatorname{diag}(t^{\varkappa_1}, \ldots, t^{\varkappa_N}) \text{ with } \varkappa_j \in \mathbf{Z}, \tag{6.12}$$

$$\text{(ii)} \quad b_- \in [H^2_-(\mathbf{T})]_{N\times N}, \; b_-^{-1} \in [H^2_-(\mathbf{T})]_{N\times N}, \tag{6.13}$$

$$b_+ \in [H^2_+(\mathbf{T})]_{N\times N}, \; b_+^{-1} \in [H^2_+(\mathbf{T})]_{N\times N}, \tag{6.14}$$

(iii) the operator $b_- S_0 b_-^{-1} I$ is bounded on $L^2_N(\mathbf{T})$.

Replacing t by $(x-i)/(x+i)$ in (6.11), we get $a = a_- d a_+$ with

$$a_\pm(x) = b_\pm\Big(\frac{x-i}{x+i}\Big), \quad d(x) = \operatorname{diag}\left[\Big(\frac{x-i}{x+i}\Big)^{\varkappa_j}\right]_{j=1}^N,$$

and (6.3) tells us that this replacement transforms (6.13), (6.14) into (6.9), (6.10), while (6.4) and (6.6) imply that the boundedness of $b_- S_0 b_-^{-1} I$ is equivalent to the boundedness of $a_- S a_-^{-1} I$. In other words, $a \in GL^\infty_{N\times N}(\mathbf{R})$ has a right WH factorization on $\mathbf{R}$ if and only if $B_0 a \in GL^\infty_{N\times N}(\mathbf{T})$ has a right WH factorization on $\mathbf{T}$. □

Remark 6.3. If f is continuous and has compact support, then $a_-^{-1} f \in L^2_N(\mathbf{R})$ by virtue of (6.9). Hence, $S(a_-^{-1} f)$ is a well-defined function in $L^2_N(\mathbf{R})$. Condition (iii) of Definition 6.1 has to be interpreted as follows: the function $a_- S(a_-^{-1} f)$ belongs to $L^2_N(\mathbf{R})$ and

$$\|a_- S(a_-^{-1} f)\|_2 \le M \|f\|_2 \tag{6.15}$$

for all continuous f with compact support, where $M < \infty$ is a constant independent of f. Since the continuous functions with compact support are dense in $L^2_N(\mathbf{R})$, (6.15) implies that $a_- S a_-^{-1} I$ extends to a bounded operator on $L^2_N(\mathbf{R})$. Also note that in Definition 6.1 one can replace the operator $a_- S a_-^{-1} I$ in condition (iii) by any of the operators $a_+^{-1} S a_+ I$, $a_+^{-1} d^{-1} S a_-^{-1} I$, $a_+^{-1} S d^{-1} a_-^{-1} I$, $a_- S d a_+ I$, $a_- d S a_+ I$. This follows from the fact that the two operators $(dS - Sd) a_+ I$ and $(d^{-1} S - S d^{-1}) a_-^{-1} I$ are bounded from $L^2_N(\mathbf{R})$ into $L^\infty_N(\mathbf{R})$. □

If $\mu_1, \ldots, \mu_N$ is a permutation of $\varkappa_1, \ldots, \varkappa_N$ then

$$\operatorname{diag}\left[\Big(\frac{x-i}{x+i}\Big)^{\mu_j}\right] = \pi^{-1} \operatorname{diag}\left[\Big(\frac{x-i}{x+i}\Big)^{\varkappa_j}\right] \pi$$

with an appropriate (constant) permutation matrix π. Thus, if a right WH factorization of a exists, then there exists also a right WH factorization such that

$$\varkappa_1 \ge \varkappa_2 \ge \ldots \ge \varkappa_N \tag{6.16}$$

in (6.8). The following theorem concerns the uniqueness of right WH factorization.

Theorem 6.4. *Let $a \in GL^\infty_{N\times N}(\mathbf{R})$ and let $a = a_- d a_+$ be a right WH factorization such that* (6.16) *is satisfied. Then all possible right WH factorizations of a are exhausted by $a = b_- f b_+$ where*

$$b_- = a_- d h^{-1} d^{-1} \pi, \quad f = \pi^{-1} d \pi, \quad b_+ = \pi^{-1} h a_+,$$

π is a permutation matrix and $h = (h_{jk})_{j,k=1}^N$ is a matrix function in $GL^\infty_{N\times N}(\mathbf{R})$ whose jk entry h_{jk} is a polynomial of degree at most $\varkappa_k - \varkappa_j$ in $(x-i)/(x+i)$ (with the convention that a polynomial of negative degree is the zero).

Note that if h is a matrix function as in Theorem 6.4, then automatically $h \in G[H^\infty_+]_{N\times N}$ and $dh^{-1}d^{-1} \in G[H^\infty_-]_{N\times N}$.

Theorem 6.4 implies in particular that the integers $\varkappa_1, \dots, \varkappa_N$ in (6.8) are uniquely defined up to their arrangement and that they are unique provided (6.16) is required. They are called the *right partial indices* of the matrix function a. Of course, a matrix function which does not admit a right WH factorization does not have right partial indices.

The interest in Wiener-Hopf factorization is motivated by the following theorem on matrix Wiener-Hopf operators, which, at the same time, may also be interpreted as a criterion for the existence of a right WH factorization.

Theorem 6.5 (Simonenko). *Let $a \in L^\infty_{N\times N}(\mathbf{R})$. The operator $W(a)$ is Fredholm on $L^2_N(\mathbf{R}_+)$ if and only if $a \in GL^\infty_{N\times N}(\mathbf{R})$ and a admits a right WH factorization. In that case*

$$n(W(a)) = \sum_{\varkappa_j<0} |\varkappa_j|, \quad d(W(a)) = \sum_{\varkappa_j>0} \varkappa_j,$$

and hence

$$\operatorname{Ind} W(a) = -(\varkappa_1 + \dots + \varkappa_N), \tag{6.17}$$

where $\varkappa_1, \dots, \varkappa_N$ are the right partial indices of a. In particular, $W(a)$ is invertible if and only if a admits a canonical right WH factorization.

We remark that if $a = a_- a_+$ is a canonical right WH factorization, then, at least formally, $W(a) = W(a_-)W(a_+)$. Since $a_- S a_-^{-1} I$ is bounded if and only if so is $a_- P a_-^{-1} I$, it is not difficult to show that

$$\begin{aligned} W(a_+^{-1})W(a_-^{-1}) &= F^{-1} P a_+^{-1} P a_-^{-1} P F | \operatorname{Im} \chi_+ \\ &= F^{-1} a_+^{-1} P a_-^{-1} F | \operatorname{Im} \chi_+ \\ &= F^{-1} a^{-1} a_- P a_-^{-1} F | \operatorname{Im} \chi_+ \end{aligned}$$

is the inverse of $W(a)$ (also recall (2.11)).

By virtue of (6.17), the sum $\varkappa_1 + \dots + \varkappa_N$ of the right partial indices is referred to as the *right total index* of a.

One says that a matrix function $a \in GL^{\infty}_{N\times N}(\mathbf{R})$ admits a *left WH factorization* if

$$a(x) = a_+(x)\, d(x)\, a_-(x) \text{ for almost all } x \in \mathbf{R}$$

where

$$d(x) = \operatorname{diag}\left[\left(\frac{x-i}{x+i}\right)^{\nu_1}, \dots, \left(\frac{x-i}{x+i}\right)^{\nu_N}\right] \text{ with } \nu_j \in \mathbf{Z},$$

(6.9), (6.10) are satisfied, and $a_-^{-1} S a_- I$ is a bounded operator on $L^2_N(\mathbf{R})$. A left WH factorization with $d(x) = I$ is called a *canonical left WH factorization.*

For $a \in GL^{\infty}_{N\times N}(\mathbf{R})$, define $\tilde{a} \in GL^{\infty}_{N\times N}(\mathbf{R})$ by $\tilde{a}(x) := a(-x)$. It is readily seen that the following are equivalent:

(i) $a = a_+ d a_-$ is a left WH factorization of a;

(ii) $a^{-1} = a_-^{-1} d^{-1} a_+^{-1}$ is a right WH factorization of a^{-1};

(iii) $\tilde{a} = \tilde{a}_+ \tilde{d} \tilde{a}_-$ is a right WH factorization of $\tilde{a}$.

Thus, applying Theorem 6.4 to a^{-1} (or $\tilde{a}$), we get a uniqueness result for left WH factorization. It follows in particular that the so-called *left partial indices* $\nu_1, \dots, \nu_N$ are uniquely determined up to their arrangement. The sum $\nu_1 + \dots + \nu_N$ is referred to as the *left total index* of a.

Example 6.6. If $a \in GL^{\infty}_{N\times N}(\mathbf{R})$ admits both a left and a right WH factorization, then the collections of its left and right partial indices may be different. Indeed, let

$$a(x) = \begin{pmatrix} t & 1 \\ 0 & t^{-1} \end{pmatrix}, \quad t := \frac{x-i}{x+i}.$$

Then

$$a(x) = \begin{pmatrix} 1 & 0 \\ 0 & 1 \end{pmatrix}\begin{pmatrix} t & 0 \\ 0 & t^{-1} \end{pmatrix}\begin{pmatrix} 1 & t^{-1} \\ 0 & 1 \end{pmatrix}, \quad a(x) = \begin{pmatrix} 1 & 0 \\ t^{-1} & -1 \end{pmatrix}\begin{pmatrix} 1 & 0 \\ 0 & 1 \end{pmatrix}\begin{pmatrix} t & 1 \\ 1 & 0 \end{pmatrix}$$

are left and right WH factorizations, respectively. Hence, the left partial indices are $1, -1$ while the right partial indices are $0, 0$. □

Example 6.7. There are matrix functions $a \in GL^{\infty}_{N\times N}(\mathbf{R})$ which admit a right WH factorization but do not have a left WH factorization. An example is the matrix function

$$a(x) = \begin{pmatrix} e^{ix} & 1 \\ 0 & e^{-ix} \end{pmatrix},$$

i.e., $a(x) = \begin{pmatrix} t & 1 \\ 0 & t^{-1} \end{pmatrix}$ with $t = e^{ix}$. We have $a = b_- b_+$ with

$$b_-(x) = \begin{pmatrix} 1 & 0 \\ e^{-ix} & -1 \end{pmatrix}, \quad b_+(x) = \begin{pmatrix} e^{ix} & 1 \\ 1 & 0 \end{pmatrix},$$

and since $W(b_\pm)$ are invertible, the inverses being $W(b_\mp^{-1})$, the operator $W(a)$ is also invertible. From Theorem 6.5 we deduce that a has a canonical right WH factorization. On the other hand, $\tilde{a} = c_- d c_+$ with

$$c_-(x) = \begin{pmatrix} 1 & 0 \\ 0 & 1 \end{pmatrix}, \quad d(x) = \begin{pmatrix} e^{-ix} & 0 \\ 0 & e^{ix} \end{pmatrix}, \quad c_+(x) = \begin{pmatrix} 1 & e^{ix} \\ 0 & 1 \end{pmatrix},$$

and the operators $W(c_\pm)$ have the inverses $W(c_\mp^{-1})$. If $W(\tilde{a})$ were Fredholm, then so also were $W(d)$. But the latter operator obviously has infinite-dimensional kernel and cokernel (recall Example 1.10). This implies that $W(\tilde{a})$ is not Fredholm. Consequently, by Theorem 6.5, the matrix function $\tilde{a}$ has no right WH factorization, which showes that a does not possess a left WH factorization. □

More on the connection between left and right WH factorization will be said in Chapter 11.

6.3 Almost Periodic Factorization

In this section we introduce a type of factorization of matrix functions a in $GAP_{N\times N}$ which mimics both the factorization of scalar functions $a \in GAP$ ensured by Bohr's theorem (Theorem 2.25) and Wiener-Hopf factorization (Definition 6.1).

Definition 6.8. A matrix function $a \in GAP_{N\times N}$ is said to admit a *right AP factorization* if it can be represented in the form

$$a(x) = a_-(x)\, d(x)\, a_+(x) \text{ for all } x \in \mathbf{R} \tag{6.18}$$

with

$$a_- \in GAP^-_{N\times N}, \quad a_+ \in GAP^+_{N\times N} \tag{6.19}$$

and

$$d(x) = \operatorname{diag}\big(e^{i\lambda_1 x}, \dots, e^{i\lambda_N x}\big), \quad \lambda_j \in \mathbf{R}. \tag{6.20}$$

Recall that $AP^\pm$ is defined by (3.8). Later we will show that $AP^\pm = AP \cap H^\infty_\pm$. A right AP factorization with $d(x) = I$ is referred to as a *canonical right AP factorization.* When replacing (6.18) by

$$a(x) = a_+(x)\, d(x)\, a_-(x) \text{ for all } x \in \mathbf{R},$$

we speak of *left AP* and *canonical left AP* factorizations. □

Clearly, if a admits a right AP factorization, then $W(a)$ and $W(d)$ are strongly Φ-equivalent (Proposition 2.17). Also notice that Theorem 2.28 shows that if a in

$GAP_{N\times N}$ admits a right AP factorization and

$$\lambda_j \geq 0 \text{ for all } j, \ \lambda_1 + \ldots + \lambda_N > 0$$
$$(\text{resp. } \lambda_j \leq 0 \text{ for all } j, \ \lambda_1 + \ldots + \lambda_N < 0),$$

then $W(a)$ is properly n-normal and left-invertible (resp. properly d-normal and right invertible), while the existence of λ_j, λ_k with $\lambda_j < 0 < \lambda_k$ implies that $W(a)$ cannot be semi-Fredholm.

We now recall the definition of the class APW given in Section 5.5.

Definition 6.9. By virtue of Proposition 2.23, the Bohr-Fourier spectrum $\Omega(a) := \{\lambda \in \mathbf{R} : M(ae_{-\lambda}) \neq 0\}$ is at most countable for every $a \in AP$. Thus, we can associate the series

$$\sum_j a_j e^{i\lambda_j x}, \quad \lambda_j \in \Omega(a), \quad a_j = M\big(a(x)e^{-i\lambda_j x}\big)$$

with every function $a(x)$ in AP. We denote by APW the functions in AP for which this series is absolutely convergent. Equivalently, $a \in APW$ if and only if a can be represented in the form

$$a(x) = \sum_j a_j e^{i\lambda_j x}, \quad \lambda_j \in \mathbf{R}, \quad \|a\|_W := \sum_j |a_j| < \infty.$$

It is easily seen that APW is a Banach algebra with pointwise algebraic operations and the norm $\|\cdot\|_W$. Let APW^+ and APW^- stand for the set of all functions $a \in APW$ for which $\Omega(a) \subset [0,\infty)$ and $\Omega(a) \subset (-\infty, 0]$, respectively. Obviously, $APW^\pm$ are closed subalgebras of APW. □

Definition 6.10. A right/left AP or canonical AP factorization of a matrix function $a \in GAP_{N\times N}$ is called a *right/left APW or canonical APW factorization* if instead of (6.19) the stronger condition

$$a_- \in GAPW^-, \ a_+ \in GAPW^+ \tag{6.21}$$

is satisfied. It is clear that if $a \in GAP_{N\times N}$ admits a left or right APW factorization, then necessarily $a \in GAPW_{N\times N}$. □

6.4 Unitary Middle Factors

A matrix function $u \in L^\infty_{N\times N}(\mathbf{R})$ is said to be *unitary-valued* if $uu^* = u^*u = I$ almost everywhere on $\mathbf{R}$. Clearly, unitary-valued matrix functions belong to $GL^\infty_{N\times N}(\mathbf{R})$.

Definition 6.11. Let $u \in L^\infty_{N\times N}(\mathbf{R})$ be unitary-valued. We say that a matrix function $a \in GL^\infty_{N\times N}(\mathbf{R})$ admits a *right $WH(u)$ factorization* if

$$a(x) = a_-(x)\, u(x)\, a_+(x) \ \text{ for almost all } x \in \mathbf{R} \tag{6.22}$$

and conditions (ii), (iii) of Definition 6.1 are satisfied. Given a subset A of $L^\infty(\mathbf{R})$, we say that $a \in GL^\infty_{N\times N}(\mathbf{R})$ has a *right $WH(u)$ factorization in A* if a can be represented in the form (6.22) with

$$a_-^{\pm 1} \in [A \cap H^\infty_-]_{N\times N}, \quad a_+^{\pm 1} \in [A \cap H^\infty_+]_{N\times N}.$$

Clearly, a right $WH(u)$ factorization in $A \subset L^\infty(\mathbf{R})$ is automatically a right $WH(u)$ factorization. □

Thus, a matrix function has a right WH factorization if and only if it possesses a right $WH(u)$ factorization for some u of the form

$$u(x) = \operatorname{diag}\left[\left(\frac{x-i}{x+i}\right)^{\varkappa_1}, \dots, \left(\frac{x-i}{x+i}\right)^{\varkappa_N}\right] \quad \text{with } \varkappa_j \in \mathbf{Z}.$$

The matrix functions with a canonical right WH factorization are just the matrix functions having a right $WH(I)$ factorization.

A right $WH(u)$ factorization in AP or APW will be referred to as a right $AP(u)$ and right $APW(u)$ factorization. Thus, $a \in GL^\infty_{N\times N}$ has a right $AP(u)$ factorization if and only if it can be written in the form (6.22) with $a_\pm \in GAP^\pm_{N\times N}$.

Lemma 6.12. *If $a \in GL^\infty_{N\times N}(\mathbf{R})$ admits a right $WH(u)$ factorization in $L^\infty(\mathbf{R})$, then $W(a)$ and $W(u)$ are strongly Φ-equivalent.*

Proof. Immediate from Proposition 2.17. □

The following theorem shows that every $a \in GL^\infty_{N\times N}(\mathbf{R})$ admits a right $WH(u)$ factorization in $L^\infty(\mathbf{R})$ *for some* unitary valued u (even with $a_- = I$).

Theorem 6.13 (Pousson and Rabindranathan). *If $a \in GL^\infty_{N\times N}(\mathbf{R})$, then there exist a unitary-valued u and an $h \in G[H^\infty_+]_{N\times N}$ such that $a = uh$ almost everywhere on $\mathbf{R}$.*

We remark that in the scalar case ($N = 1$) Theorem 6.13 amounts to the well-known fact that if $a \in GL^\infty(\mathbf{R})$, then there is a function $h \in GH^\infty_+$ such that $|a| = |h|$ almost everywhere on $\mathbf{R}$. We used this fact already in Section 4.6 in order to show that in the scalar case $W(a)$ and $W(a/|a|)$ are strongly Φ-equivalent.

A function $u \in H^\infty_+$ is called an *inner function* if $|u| = 1$ almost everywhere on $\mathbf{R}$. In other words, the inner functions are the unitary-valued functions in H^∞_+. The functions in GH^∞_+ are (special) *outer functions.* If $h \in H^\infty_+ \cap GL^\infty(\mathbf{R})$ then, by what was said in the preceding paragraph, there is an outer function h_e such that $|h| = |h_e|$, i.e., $h = h_i h_e$ with $|h_i| = 1$ almost everywhere. Since obviously $h_i = hh_e^{-1} \in H^\infty_+$, the function h_i is inner. Consequently, every function in $H^\infty_+ \cap GL^\infty(\mathbf{R})$ is the product of an inner and an outer function (the *inner-outer factorization theorem* states that this is true for every function in H^∞_+).

Wiener-Hopf operators with unitary-valued symbols can be tackled by having recourse to the following theorem, which is the quantitative matrix version of the Nehari and Hartman theorems. Given a subset E of $L^\infty_{N\times N}(\mathbf{R})$ and a function $a \in GL^\infty_{N\times N}(\mathbf{R})$, we put

$$\operatorname{dist}(a,E) := \inf_{f\in E} \|a-f\|_{L^\infty_{N\times N}(\mathbf{R})},$$

where the norm on $L^\infty_{N\times N}(\mathbf{R})$ is the one introduced in the beginning of Section 5.1. The *essential norm* $\|A\|_{\mathrm{ess}}$ of an operator $A \in \mathcal{B}(X)$ is the norm of the coset $A+\mathcal{K}(X)$ in the quotient algebra $\mathcal{B}(X)/\mathcal{K}(X)$, i.e., the distance of A to the compact operators:

$$\|A\|_{\mathrm{ess}} := \operatorname{dist}\big(A,\mathcal{K}(X)\big) := \inf_{K\in\mathcal{K}(X)} \|A+K\|.$$

Theorem 6.14. *If $a \in L^\infty_{N\times N}(\mathbf{R})$, then for the Hankel operator $H(a)$ on $L^2_N(\mathbf{R}_+)$ we have*

$$\begin{aligned}\|H(a)\| &= \operatorname{dist}\big(a,[H^\infty_-]_{N\times N}\big),\\ \|H(a)\|_{\mathrm{ess}} &= \operatorname{dist}\Big(a,\big[C(\dot{\mathbf{R}})+H^\infty_-\big]_{N\times N}\Big).\end{aligned}$$

Using this theorem, it is not too difficult to establish the following criteria.

Theorem 6.15 (Douglas and Sarason). *If $u \in L^\infty_{N\times N}(\mathbf{R})$ is unitary valued, then*

(a) $W(u)$ *is invertible* $\Longleftrightarrow \operatorname{dist}\big(u,G[H^\infty_+]_{N\times N}\big) < 1$
$\Longleftrightarrow \operatorname{dist}\big(u,G[H^\infty_-]_{N\times N}\big) < 1,$

(b) $W(u)$ *is left-invertible* $\Longleftrightarrow \operatorname{dist}\big(u,[H^\infty_+]_{N\times N}\big) < 1,$

(b') $W(u)$ *is right-invertible* $\Longleftrightarrow \operatorname{dist}\big(u,[H^\infty_-]_{N\times N}\big) < 1,$

(c) $W(u)$ *is Fredholm* $\Longleftrightarrow \operatorname{dist}\Big(u,G\big[C(\dot{\mathbf{R}})+H^\infty_+\big]_{N\times N}\Big) < 1$
$\Longleftrightarrow \operatorname{dist}\Big(u,G\big[C(\dot{\mathbf{R}})+H^\infty_-\big]_{N\times N}\Big) < 1,$

(d) $W(u)$ *is n-normal* $\Longleftrightarrow \operatorname{dist}\Big(u,\big[C(\dot{\mathbf{R}})+H^\infty_+\big]_{N\times N}\Big) < 1,$

(d') $W(u)$ *is d-normal* $\Longleftrightarrow \operatorname{dist}\Big(u,\big[C(\dot{\mathbf{R}})+H^\infty_-\big]_{N\times N}\Big) < 1.$

Proof. Assertion (b') follows from Proposition 2.10 and Theorem 6.14, and (b) results from (b') by passage to adjoints.

To prove the "+" version of (a), notice first that if $W(u)$ is invertible, then, by (b), there is an $h \in [H^\infty_+]_{N\times N}$ such that $\|u-h\|_\infty < 1$. Since

$$\|I-W^*(u)W(h)\| = \|W(I-u^*h)\| = \|I-u^*h\|_\infty = \|u-h\|_\infty < 1, \tag{6.23}$$

the operator $W^*(u)W(h)$ is invertible. This implies that $W(h)$ is also invertible. Hence, $h \in GL^\infty_{N\times N}(\mathbf{R})$. As the matrix equation $hf = I$ has a unique solution $f \in [H^2_+]_{N\times N}$, we deduce that

$$h^{-1} \in L^\infty_{N\times N}(\mathbf{R}) \cap [H^2_+]_{N\times N} = [H^\infty_+]_{N\times N}.$$

Thus, $h \in G[H^\infty_+]_{N\times N}$. Conversely, if there exists an $h \in G[H^\infty_+]_{N\times N}$ such that $\|u-h\|_\infty < 1$, then (6.23) again gives the invertibility of $W^*(u)W(h)$. As $W(h)$ is invertible, so also must be $W(u)$. This completes the proof of the "+" version of (a). The remaining part of (a) now follows by passing to adjoints.

Replacing in the above reasoning $\|\cdot\|$ by $\|\cdot\|_{\text{ess}}$, we get (c), (d), (d'). Recall that n-normality (resp. d-normality) of Hilbert space operators is equivalent to left-Fredholmness (resp. right-Fredholmness). □

In the scalar case we can apply Theorem 6.15 to $W(a/|a|)$ in order to obtain semi-Fredholm criteria for $W(a)$. The matrix case is more complicated.

A matrix function $s \in L^\infty_{N\times N}(\mathbf{R})$ is called *sectorial* if there exist a real number $\varepsilon > 0$ and two (constant) matrices $b, c \in G\mathbf{C}^{N\times N}$ such that

$$\operatorname{Re}\big(bs(x)cz, z\big) \geq \varepsilon\|z\|^2$$

for almost all $x \in \mathbf{R}$ and all $z \in \mathbf{C}^N$. In the scalar case, this means that the essential range of s is contained in some open half-plane whose boundary passes through the origin. Equivalently, $a \in L^\infty(\mathbf{R})$ is sectorial if and only if $a/|a| = e^{i(c+v)}$ with $c \in \mathbf{R}$ and a real-valued function v such that $\|v\|_\infty < \pi/2$.

Lemma 6.16. *If E is a subset of $L^\infty_{N\times N}(\mathbf{R})$, $u \in L^\infty_{N\times N}(\mathbf{R})$ is unitary-valued, and* $\operatorname{dist}(u, E) < 1$, *then there exist a matrix function $b \in E$ and a sectorial matrix function $s \in GL^\infty_{N\times N}(\mathbf{R})$ such that $u = sb$.*

A proof of this lemma is, for example, in [46, p. 104].

Using Theorems 6.13, 6.15, and Lemma 6.16, one can show the following result.

Theorem 6.17. *Let $\mathcal{S}$ stand for the sectorial functions. If $a \in GL^\infty_{N\times N}(\mathbf{R})$, then*

$$\begin{aligned}
W(a) \text{ is invertible} &\Longleftrightarrow a = sh,\ s \in \mathcal{S},\ h \in G[H^\infty_+]_{N\times N}\\
&\Longleftrightarrow a = hs,\ s \in \mathcal{S},\ h \in G[H^\infty_-]_{N\times N},\\
W(a) \text{ is left-invertible} &\Longleftrightarrow a = sh,\ s \in \mathcal{S},\ h \in [H^\infty_+]_{N\times N},\\
W(a) \text{ is right-invertible} &\Longleftrightarrow a = hs,\ s \in \mathcal{S},\ h \in [H^\infty_-]_{N\times N},\\
W(a) \text{ is Fredholm} &\Longleftrightarrow a = sh,\ s \in \mathcal{S},\ h \in G\big[C(\dot{\mathbf{R}}) + H^\infty_+\big]_{N\times N}\\
&\Longleftrightarrow a = hs,\ s \in \mathcal{S},\ h \in G\big[C(\dot{\mathbf{R}}) + H^\infty_-\big]_{N\times N},\\
W(a) \text{ is } n\text{-normal} &\Longleftrightarrow a = sh,\ s \in \mathcal{S},\ h \in \big[C(\dot{\mathbf{R}}) + H^\infty_+\big]_{N\times N},\\
W(a) \text{ is } d\text{-normal} &\Longleftrightarrow a = hs,\ s \in \mathcal{S},\ h \in \big[C(\dot{\mathbf{R}}) + H^\infty_-\big]_{N\times N}.
\end{aligned}$$

Notes

Detailed expositions of Hardy space theory can be found in [84], [73], [110], [142]. Factorization of matrix functions in the context of boundary value problems for analytic vector functions and systems of singular integral equations has its roots in the work of Plemelj and Gakhov; see [80], [161]. Such factorizations come in different flavors, depending on the spaces of functions involved [57], [155]. Definition 6.1 is due to Simonenko [207], [208]. The name Wiener-Hopf factorization is justified by the close relationship between this factorization and the method proposed by Wiener and Hopf for solving convolution type equations in 1931. In the form cited here, Theorem 6.5 was established by Simonenko [207]. The famous Example 6.6 was constructed by Gohberg and Krein [94]. Almost periodic factorization of matrix functions was introduced in [128]; see also [132]. Definition 6.11, that is, the concept of factorizations with a unitary but not necessarily diagonal middle factor u was proposed (with some additional conditions on u) in [136].

Theorem 6.13 was obtained independently in Pousson [174] and Rabindranathan [178]. In the scalar case, the norm of Hankel operators was computed by Nehari [163]; the formula for the essential norm follows from there and the result of Hartman [105]; see [175]. Matrix-valued (and even operator-valued) generalizations of Nehari's theorem were proved by Adamyan, Arov, Krein [1] and Page [168].

In the scalar setting, Theorem 6.15 is Douglas and Sarason's [67]. In the matrix and operator cases, statement (a) of Theorem 6.15 was established by Rabindranathan [178]. The other statements of Theorem 6.15 can be found in [46], [57], [155]. The notion of sectorial matrix functions was introduced into the factorization business by Simonenko [204]. Theorem 6.17 is taken from [46], [57].

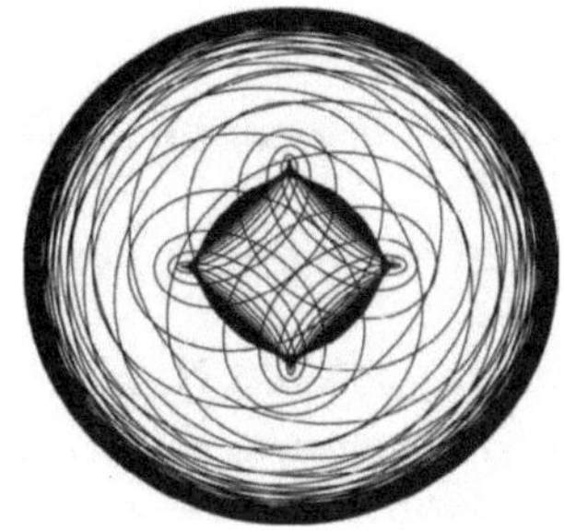

Chapter 7

Bohr Compactification

In this chapter we record some results from harmonic analysis on locally compact Abelian groups. These results will be needed in the following chapters. In particular, we need the fact that an almost periodic function with Bohr-Fourier spectrum in some set E can be uniformly approximated by almost periodic polynomials whose Bohr-Fourier spectrum is supported in E. This result will be proved by making use of the Bochner-Fejér operators, which are the almost periodic analogues of the Fejér-Cesàro means we know from approximation by trigonometric polynomials. Furthermore, the abstract context allows us to interpret Wiener-Hopf and almost periodic factorization as $AP(G)$ factorizations corresponding to the groups $G = \mathbf{T}$ and $G = \mathbf{R}$, respectively. Finally, we introduce the Besicovitch space B^2. When studying Wiener-Hopf operators with symbols in $W_{N\times N}$, it is often useful to think of these operators as acting not only on $L^2_N(\mathbf{R})$ but also on W_N. The role of W_N is played by B^2_N when considering Wiener-Hopf operators with symbols in $AP_{N\times N}$.

7.1 Some Commutative Harmonic Analysis

In this section we review a few facts from abstract commutative harmonic analysis we will need in the following.

Let G be a locally compact Abelian group and write G additively. The dual group G^* is the locally compact Abelian group of all continuous characters of G, that is, of all continuous functions $\chi : G \to \mathbf{T}$ such that $\chi(g_1 + g_2) = \chi(g_1)\chi(g_2)$ for all $g_1, g_2 \in G$. We write G^* also additive. In particular, the inverse of $\chi \in G^*$ is denoted by $-\chi$. We equip G^* with the usual topology: a neighborhood base of

$\chi_0 \in G^*$ is formed by the sets

$$U_{g_1,\dots,g_n;\varepsilon}(\chi_0) := \Big\{\chi \in G^* : |\chi(g_j) - \chi_0(g_j)| < \varepsilon \ \text{ for all } \ j\Big\},$$

where $\varepsilon > 0$, $n \in \mathbf{N}$, $g_1, \dots, g_n \in G$. For $g \in G$ and $\chi \in G^*$, we define $g(\chi) := \chi(g)$. A *trigonometric polynomial* is a function $p : G \to \mathbf{C}$ that can be represented in the form

$$p(g) = \sum_j p_j \chi_j(g), \ \ p_j \in \mathbf{C}, \ \chi_j \in G^*,$$

the sum being finite. The set $AP(G)$ of all *almost periodic functions on* G is defined as the closure of the set of the trigonometric polynomials in $BC(G)$, that is, in the bounded continuous functions on G.

Example 7.1. Let $G = \mathbf{R}$. Then

$$\mathbf{R} \to G^*, \ \lambda \mapsto \chi_\lambda \ \text{ where } \ \chi_\lambda(x) := e^{i\lambda x},$$

is a bijection between $\mathbf{R}$ and G^*. We identify G^* and $\mathbf{R}$. A trigonometric polynomial $p : G = \mathbf{R} \to \mathbf{C}$ is a finite sum of the form

$$p(x) = \sum_j p_j \chi_{\lambda_j}(x), \ p_j \in \mathbf{C}, \ \lambda_j \in \mathbf{R},$$

and hence $p(x) = \sum_j p_j e^{i\lambda_j x}$ is just what we call an almost periodic polynomial. Consequently, $AP(G) = AP(\mathbf{R})$ is the set of almost periodic functions given by Definition 1.11. □

Example 7.2. Suppose $G = \mathbf{T}$. Now

$$\mathbf{Z} \to G^*, \ n \mapsto \chi_n \ \text{ where } \ \chi_n(t) = t^n$$

effects a bijection between $\mathbf{Z}$ and G^*. We identify G^* with $\mathbf{Z}$. A trigonometric polynomial $p : G = \mathbf{T} \to \mathbf{C}$ is a finite sum

$$p(t) = \sum_j p_j \chi_{n_j}(t), \ p_j \in \mathbf{C}, \ n_j \in \mathbf{Z},$$

and hence a function of the form $p(t) = \sum_j p_j t^{n_j}$. These are the usual trigonometric polynomials. Thus, $AP(G) = AP(\mathbf{T})$ coincides with $C(\mathbf{T})$. □

Example 7.3. Finally, let $G = \mathbf{Z}$. In this case

$$\mathbf{T} \to G^*, \ t \mapsto \chi_t \ \text{ where } \ \chi_t(n) := t^n$$

is a bijection of $\mathbf{T}$ onto G^*. By identifying G^* with $\mathbf{T}$, we see that the trigonometric polynomials $p : G = \mathbf{Z} \to \mathbf{C}$ are the finite sums

$$p(n) = \sum_j p_j \chi_{t_j}(n), \; p_j \in \mathbf{C}, \; t_j \in \mathbf{T},$$

i.e., the finite sums $p(n) = \sum_j p_j t_j^n$. The closure of all such functions in $BC(\mathbf{Z}) = l^\infty(\mathbf{Z})$ is $AP(G) = AP(\mathbf{Z})$. □

Let G_d^* denote the group G^* with the discrete topology. The dual group $(G_d^*)^*$ of G_d^* is denoted by G_B and called the *Bohr compactification* of G. Since, by Pontryagin's duality theorem, G is the set of all continuous characters of G^* and since G_B is the set of all – not necessarily continuous – characters of G^*, it is clear that $G \subset G_B$. One can show that G is dense in G_B.

Theorem 7.4. *We have $AP(G) = C(G_B)$ in the following sense: every function in $AP(G)$ admits a (unique) extension to a function in $C(G_B)$ and conversely, the restriction of every function in $C(G_B)$ to G is a function in $AP(G)$.*

By virtue of Theorem 7.4 we can (and will) identify $AP(G)$ and $C(G_B)$. Notice that Theorem 7.4 is a special case of Theorem 1.7(a): $AP(G)$ is a commutative unital C^*-algebra whose maximal ideal space is homeomorphic to G_B.

Let μ be the *normalized Haar measure* on G_B. Thus, μ is a positive regular Borel measure such that $\mu(U) = \mu(U+x)$ for every Borel set $U \subset G_B$ and every $x \in G_B$ (invariance) and $\mu(G_B) = 1$ (normalization). Since G_B is a compact Abelian group, such a measure exists and is unique.

For $a \in C(G_B)$, the number

$$M(a) := \int_{G_B} a(x)\, d\mu(x)$$

is referred to as the *Bohr mean value* of a. One can show that

$$\begin{aligned} M(a) &= \lim_{T\to\infty} \frac{1}{2T} \int_{-T}^{T} a(x)\, dx \;\text{ for } G = \mathbf{R}, \\ M(a) &= \int_0^{2\pi} a(e^{i\theta}) \frac{d\theta}{2\pi} \;\text{ for } G = \mathbf{T}, \\ M(a) &= \lim_{n\to\infty} \frac{1}{2n+1} \sum_{k=-n}^{n} a(k) \;\text{ for } G = \mathbf{Z} \end{aligned}$$

(recall Proposition 2.22 for the first formula). For $\chi \in G^*$, the *Fourier coefficient* $\hat{a}(\chi)$ of $a \in C(G_B)$ is defined by

$$\hat{a}(\chi) = \int_{G_B} a(x)\, \chi(-x)\, d\mu(x)$$

(note that we write G_B additive and that therefore $\chi(-x) = 1/\chi(x) = (-\chi)(x)$). The set

$$\Omega(a) := \{\chi \in G^* : \hat{a}(\chi) \neq 0\}$$

is called the *Bohr-Fourier spectrum* of $a \in C(G_B)$. For $G = \mathbf{R}$, this definition coincides with Definition 2.24.

7.2 The Bochner-Fejér Operators

In Section 3.2, we defined AP^+ and AP^- as the closures of the almost periodic polynomials with Bohr-Fourier spectrum in $[0,\infty)$ and $(-\infty,0]$, respectively. This implies that

$$AP^+ \subset \{a \in AP : \Omega(a) \subset [0,\infty)\}, \quad AP^- \subset \{a \in AP : \Omega(a) \subset (-\infty,0]\}. \quad (7.1)$$

The reverse inclusions are less trivial, and the purpose of this section is to show that in (7.1) actually equality holds.

Clearly, equality in (7.1) will follow once we know that every function $a \in AP$ can be uniformly approximated as closely as desired by almost periodic polynomials p_n such that $\Omega(p_n) \subset \Omega(a)$. For functions in $AP(\mathbf{T}) = C(\mathbf{T})$, this follows from the fact that every $a \in C(\mathbf{T})$ can be uniformly approximated by its Fejér-Cesàro means: $\|a - \sigma_n a\|_\infty \to 0$ where

$$\begin{aligned} (\sigma_n a)(e^{i\theta}) &:= \int_0^{2\pi} K_n(\theta - \varphi)a(e^{i\theta})\frac{d\varphi}{2\pi} = \sum_{k=-n}^{n}\left(1 - \frac{|k|}{n+1}\right)\hat{a}(k)e^{ik\theta}, \\ K_n(\theta) &:= \frac{1}{n+1}\left(\frac{\sin((n+1)\theta/2)}{\sin(\theta/2)}\right)^2. \end{aligned}$$

For functions in $AP = AP(\mathbf{R})$ or $AP(G)$ with general G, things are more complicated, but eventually everything can be managed.

Let $\mathcal{U}$ be the collection of all open neighborhoods $U \subset G_B$ of the zero and order $\mathcal{U}$ partially by inclusion. By Uryson's lemma, for every $U \in \mathcal{U}$ there is a $\varphi_U \in C(G_B)$ such that $\varphi_U \geq 0$, supp $\varphi_U \subset U$ and $M(\varphi_U) = 1$. Given $\varepsilon > 0$, choose a trigonometric polynomial $p_{U,\varepsilon} \in AP(G) = C(G_B)$ so that

$$\sup_{g \in G} |\varphi_U(g) - p_{U,\varepsilon}(g)| = \max_{x \in G_B} |\varphi_U(x) - p_{U,\varepsilon}(x)| < \varepsilon$$

and define the trigonometric polynomial $q_{U,\varepsilon}$ by

$$q_{U,\varepsilon} := \frac{p_{U,\varepsilon} + \varepsilon}{M(p_{U,\varepsilon}) + \varepsilon}.$$

On $\Gamma := \mathcal{U} \times (0,1)$, a partial order $\prec$ is given by defining $\gamma_1 = (U_1, \varepsilon_1) \prec \gamma_2 = (U_2, \varepsilon_2)$ if and only if $U_1 \supset U_2$ and $\varepsilon_1 \geq \varepsilon_2$. The net of trigonometric polynomials $\{q_\gamma\}$ just constructed has the following properties:

$$\begin{aligned} &q_\gamma \geq 0 \text{ on } G_B, \quad M(q_\gamma) = 1, \\ &\lim_\Gamma \Big(\sup_{x \in G_B \setminus V} q_\gamma(x) \Big) = 0 \text{ for every } V \in \mathcal{U}. \end{aligned} \tag{7.2}$$

For $a \in AP(G) = C(G_B)$ and $\gamma \in \Gamma$, put

$$(\sigma_\gamma a)(x) := \int_{G_B} q_\gamma(x - \xi) a(\xi)\, d\mu(\xi) = \int_{G_B} q_\gamma(\xi) a(x - \xi)\, d\mu(\xi).$$

In other terms, $\sigma_\gamma a$ is the Bohr mean value of the convolution $q_\gamma * a$. Obviously, $\sigma_\gamma \in \mathcal{B}(C(G_B))$ and $\|\sigma_\gamma\| \leq 1$. The operators σ_γ $(\gamma \in \Gamma)$ are called the *Bochner-Fejér operators* of G.

Theorem 7.5. *Let $a \in AP(G) = C(G_B)$. Then*
(a) *$\sigma_\gamma a$ is a trigonometric polynomial in $AP(G)$;*
(b) *$\Omega(\sigma_\gamma a) \subset \Omega(a)$;*
(c) *$\lim_\Gamma \sigma_\gamma a = a$ in the norm of $C(G_B)$.*

Proof. (a) We can write the trigonometric polynomial q_γ as a finite sum

$$q_\gamma(x) = \sum_j r_j^{(\gamma)} \chi_j(x), \quad r_j^{(\gamma)} \in \mathbf{C}, \quad \chi_j \in G^*.$$

This gives

$$\begin{aligned} (\sigma_\gamma a)(x) &= \sum_j r_j^{(\gamma)} \int_{G_B} \chi_j(x - \xi) a(\xi)\, d\mu(\xi) \\ &= \sum_j r_j^{(\gamma)} \chi_j(x) \int_{G_B} a(\xi) \chi_j(-\xi)\, d\mu(\xi) = \sum_j r_j^{(\gamma)} \hat{a}(\chi_j) \chi_j(x) \end{aligned} \tag{7.3}$$

and shows that $\sigma_\gamma a$ is a trigonometric polynomial.
(b) We have $\int_{G_B} \chi(\xi)\, d\mu(\xi) = 0$ whenever $\chi \in G^* \setminus \{0\}$: indeed, if $\chi \in G^* \setminus \{0\}$, then $\chi(x_0) \neq 1$ for some $x_0 \in G_B$, and since

$$\int_{G_B} \chi(\xi)\, d\mu(\xi) = \chi(x_0) \int_{G_B} \chi(\xi - x_0)\, d\mu(\xi) = \chi(x_0) \int_{G_B} \chi(\xi)\, d\mu(\xi),$$

it follows that $\int_{G_B} \chi(\xi)\, d\mu(\xi) = 0$. We therefore infer from (7.3) that $\Omega(\sigma_\gamma a)$ is the set $\{\chi_j : \hat{a}(\chi_j) \neq 0\} \subset \Omega(a)$.

(c) Let $V \in \mathcal{U}$ and $x \in G_B$. We have $|a(x) - (\sigma_\gamma a)(x)| \le I_1(x) + I_2(x)$ where

$$\begin{aligned} I_1(x) &= \int_V q_\gamma(\xi)|a(x) - a(x-\xi)|\, d\mu(\xi) \\ &\le \varepsilon \int_{G_B} q_\gamma(\xi)\, d\mu(\xi) = \varepsilon \end{aligned}$$

if $|a(x) - a(x-\xi)| < \varepsilon$, which holds for sufficiently small V due to the (uniform) continuity of a on G_B, and

$$\begin{aligned} I_2(x) &= \int_{G_B \setminus V} q_\gamma(\xi)|a(x) - a(x-\xi)|\, d\mu(\xi) \\ &\le 2\|a\|_\infty \sup_{\xi \in G_B \setminus V} q_\gamma(\xi), \end{aligned}$$

which goes to zero by virtue of (7.2). □

Given a nonempty subset E of $\mathbf{R}$, denote by AP_E the closure of all almost periodic polynomials $p = \sum_j r_j e_{\lambda_j}$ with $\lambda_j \in E$. In particular, $AP^+ = AP_{[0,\infty)}$ and $AP^- = AP_{(-\infty,0]}$.

Corollary 7.6. *We have* $AP_E = \{a \in AP : \Omega(a) \subset E\}$.

Proof. Immediate from Theorem 7.5. □

Corollary 7.7. *The equalities* $AP^\pm = AP \cap H^\infty_\pm$ *hold.*

Proof. Let us prove the "+" version. The inclusion $AP^+ \subset AP \cap H^\infty_+$ follows from (3.8). For $a \in L^\infty(\mathbf{R})$, the value $\hat{a}(iy)$ of the harmonic extension at $iy \in \mathbf{C}_+$ is given by

$$\hat{a}(iy) = \frac{1}{\pi} \int_{\mathbf{R}} \frac{y a(x)}{x^2 + y^2}\, dx. \tag{7.4}$$

If $\lambda \ge 0$, then e_λ is analytic and bounded in $\mathbf{C}_+$, whence $(e_\lambda)\hat{}(iy) = e^{-\lambda y}$. In the case $\lambda < 0$, the substitution $x \mapsto -x$ in the integral gives $(e_\lambda)\hat{}(iy) = e^{\lambda y}$. Hence $(e_\lambda)\hat{}(iy) = e^{-|\lambda| y}$ for all $\lambda \in \mathbf{R}$. As the map $a \mapsto \hat{a}$ is continuous in the L^∞ norm, it follows that

$$M(a) = \lim_{y \to +\infty} \hat{a}(iy) \quad \text{for every } a \in AP.$$

Thus, if $a \in AP \cap H^\infty_+$ and $\lambda > 0$, then

$$\begin{aligned} M(ae_\lambda) &= \lim_{y \to +\infty} (ae_\lambda)\hat{}(iy) = \lim_{y \to +\infty} \hat{a}(iy)(e_\lambda)\hat{}(iy) \\ &= \lim_{y \to +\infty} a(iy) e^{-|\lambda| y} = 0, \end{aligned}$$

which implies that $\Omega(a) \subset [0, \infty)$. Corollary 7.6 now shows that $a \in AP^+$. □

7.3 Another Look at AP Factorization

Let G be a locally compact Abelian group. In Section 7.1, we introduced the C^*-algebra $AP(G)$. Given a subset H of G^*, we let $AP_H(G)$ stand for the closure in $AP(G)$ of the trigonometric polynomials $p : G \to \mathbf{C}$ of the form

$$p(g) = \sum_j p_j \chi_j(g), \quad p_j \in \mathbf{C}, \quad \chi_j \in H,$$

and we denote by $APW_H(G)$ the set of all functions $a : G \to \mathbf{C}$ such that

$$a(g) = \sum_j a_j \chi_j(g), \quad a_j \in \mathbf{C}, \quad \chi_j \in H, \quad \sum_j |a_j| < \infty.$$

Let now $a \in G[AP(G)]_{N\times N}$ (of course, the first G stands for "group of invertible elements of") and let H, K be subsets of G^* such that $H \cup K = G^*$. An $AP(G)$ factorization with respect to (H, K) is a representation $a = a_H d a_K$ with

$$a_H^{\pm 1} \in \left[AP_H(G)\right]_{N\times N}, \quad a_K^{\pm 1} \in \left[AP_K(G)\right]_{N\times N}, \tag{7.5}$$

$$d(g) = \operatorname{diag}\left(\chi_1(g), \dots, \chi_N(g)\right), \quad \chi_j \in G^*. \tag{7.6}$$

If $a \in G[APW(G)]_{N\times N}$ and AP is replaced by APW in (7.5), we speak of an $APW(G)$ factorization with respect to (H, K).

Example 7.8. Let $G = \mathbf{T}$ and put $\mathbf{Z}_+ := \{n \in \mathbf{Z} : n \geq 0\}$, $\mathbf{Z}_- := \{n \in \mathbf{Z} : n \leq 0\}$. In that case the diagonal matrix (7.6) may be rewritten as

$$d(t) = \operatorname{diag}\left(t^{\varkappa_1}, \dots, t^{\varkappa_N}\right), \quad \varkappa_j \in \mathbf{Z}$$

where $t \in G = \mathbf{T}$ (Example 7.2). Thus, $AP(\mathbf{T})$ and $APW(\mathbf{T})$ factorization with respect to $(\mathbf{Z}_-, \mathbf{Z}_+)$ is equivalent to right WH factorization on $\mathbf{T}$ within $C(\mathbf{T})$ and $W(\mathbf{T})$, respectively. □

Example 7.9. Let $G = \mathbf{R}$ and set $\mathbf{R}_+ := [0, \infty)$, $\mathbf{R}_- := (-\infty, 0]$. Now (7.6) reads

$$d(x) = \operatorname{diag}\left(e^{i\lambda_1 x}, \dots, e^{i\lambda_N x}\right), \quad \lambda_j \in \mathbf{R},$$

with $x \in G = \mathbf{R}$ (Example 7.1). Therefore, $AP(\mathbf{R})$ and $APW(\mathbf{R})$ factorization with respect to $(\mathbf{R}_-, \mathbf{R}_+)$ is what we simply call right AP and right APW factorization, respectively. □

7.4 The Besicovitch Space

Recall that AP^0 stands for the set of all almost periodic polynomials on $\mathbf{R}$. The *Besicovitch space* B^2 is defined as the completion of AP^0 with respect to the norm

$$\|f\|_{B^2} := \left(\sum_\lambda |f_\lambda|^2\right)^{1/2} = \left(M(|f|^2)\right)^{1/2}$$

where $f = \sum_\lambda f_\lambda e_\lambda \in AP^0$.

In Section 7.1, we introduced the Bohr compactification $\mathbf{R}_B = (\mathbf{R}_d^*)^*$ of $\mathbf{R}$ and stated that AP may be identified with $C(\mathbf{R}_B)$. Also as in Section 7.1, let μ be the normalized Haar measure on $\mathbf{R}_B$. One can show that B^2 may be identified with $L^2(\mathbf{R}_B, d\mu)$. Thus, B^2 is a (nonseparable) Hilbert space, and the inner product in $B^2 = L^2(\mathbf{R}_B, d\mu)$ is given by

$$(f,g) = \int_{\mathbf{R}_B} f(\xi)\overline{g(\xi)}\, d\mu(\xi). \tag{7.7}$$

For $f, g \in AP$ we also have

$$(f,g) = \lim_{T\to\infty} \frac{1}{2T} \int_{-T}^{T} f(x)\overline{g(x)}\, dx.$$

In what follows we frequently think of the elements of the space B^2 as functions in $L^2(\mathbf{R}_B,\, d\mu)$. Since $\mu(\mathbf{R}_B) = 1$ is finite, we have $AP \subset B^2$. Moreover, AP is a dense subset of B^2.

The Cauchy-Schwarz inequality shows that the mean value

$$M(f) := \int_{\mathbf{R}_B} f(\xi)\, d\mu(\xi)$$

exists and is finite for every $f \in B^2$. The set

$$\Omega(f) := \{\lambda \in \mathbf{R} : M(fe_{-\lambda}) \neq 0\}$$

is called the Bohr-Fourier spectrum of f and can be shown to be at most countable. Taking into account (7.7), one can prove that

$$\|f\|_{B^2} = \sum_{\lambda\in\Omega(f)} |M(fe_{-\lambda})|^2 \quad \text{(Parseval's equality)}$$

for every $f \in B^2$.

Let $l^2(\mathbf{R})$ denote the collection of all functions $f : \mathbf{R} \to \mathbf{C}$ for which the set $\{\lambda \in \mathbf{C} : f(\lambda) \neq 0\}$ is at most countable and

$$\|f\|_{l^2(\mathbf{R})}^2 := \sum |f(\lambda)|^2 < \infty.$$

Further, $l^\infty(\mathbf{R})$ is defined as the set of all functions $f : \mathbf{R} \to \mathbf{C}$ such that

$$\|f\|_{l^\infty(\mathbf{R})} := \sup_{\lambda\in\mathbf{R}} |f(\lambda)| < \infty.$$

Note that $l^2(\mathbf{R})$ is a (nonseparable) Hilbert space with pointwise operations and the inner product

$$(f,g) := \sum_{\lambda\in\mathbf{R}} f(\lambda)\overline{g(\lambda)}$$

and that $l^\infty(\mathbf{R})$ is a C^*-algebra with pointwise operations and the norm $\|\cdot\|_{l^\infty(\mathbf{R})}$.

The map $F_B : l^2(\mathbf{R}) \to B^2$ which sends a function $f \in l^2(\mathbf{R})$ with a finite support to the function

$$(F_B f)(x) = \sum_{\lambda \in \mathbf{R}} f(\lambda) e^{i\lambda x}, \; x \in \mathbf{R},$$

can be extended by continuity to all of $l^2(\mathbf{R})$. It is referred to as the *Bohr-Fourier transform.* The operator F_B is an isometric isomorphism. The inverse Bohr-Fourier transform acts by the rule

$$F_B^{-1} : B^2 \to l^2(\mathbf{R}), \; (F_B^{-1} f)(\lambda) = M(f e_{-\lambda}), \quad \lambda \in \mathbf{R}.$$

Given $a \in l^\infty(\mathbf{R})$, we denote by aI the operator of multiplication by a on $l^2(\mathbf{R})$. As usual, we omit the I in case aI follows another operator. If $a \in l^\infty(\mathbf{R})$, then the operator

$$\psi(a) : B^2 \to B^2, \quad \psi(a) := F_B a F_B^{-1}$$

is bounded. For $h \in \mathbf{R}$, define $e_h \in l^\infty(\mathbf{R})$ by $e_h(\lambda) = e^{ih\lambda}$. Obviously, the operator

$$T_h := \psi(e_h) : B^2 \to B^2, \quad T_h := \psi(e_h) := F_B e_h F_B^{-1}$$

acts by the rule

$$(T_h f)(x) = f(x+h) \;\text{ for }\; f \in AP \;\text{ and }\; x \in \mathbf{R}, \tag{7.8}$$

and it can be extended to a continuous operator on the whole B^2. An operator $A \in \mathcal{B}(B^2)$ is said to be *translation invariant* if

$$A = T_h A T_h^{-1} \;\text{ for all }\; h \in \mathbf{R}.$$

Proposition 7.10. *An operator $A \in \mathcal{B}(B^2)$ is translation invariant if and only if $A = \psi(a)$ for some $a \in l^\infty(\mathbf{R})$.*

Proof. The "if" part is trivial, because

$$T_h \psi(a) T_h^{-1} = F_B e_h a e_{-h} F_B^{-1} = \psi(a).$$

To prove the "only if" part, let $A \in \mathcal{B}(B^2)$ be translation invariant. The operator $F_B^{-1} A F_B \in \mathcal{B}(l^2(\mathbf{R}))$ can be represented by a matrix $(a_{\lambda\mu})_{\lambda,\mu \in \mathbf{R}}$ with entries in $\mathbf{C}$, and we must show that this matrix is diagonal (the boundedness of A implies that $\sup |a_{\lambda\lambda}| < \infty$). The equality $A = T_h A T_h^{-1}$ is equivalent to the matrix equality

$$(a_{\lambda\mu}) = \operatorname{diag}(e^{ih\lambda})(a_{\lambda\mu}) \operatorname{diag}(e^{-ih\mu})$$

and thus to the equalities $a_{\lambda\mu} = e^{ih(\lambda-\mu)} a_{\lambda\mu}$. If $\lambda \neq \mu$, we can choose $h \in \mathbf{R}$ so that $e^{ih(\lambda-\mu)} \neq 1$, which implies that $a_{\lambda\mu} = 0$. □

Notes

All the facts from harmonic analysis on commutative locally compact groups we need here (including Pontryagin's duality theorem) can be found in [107], [108], [188], and in many other standard textbooks on the subject. More specific results related to almost periodic functions, including Bohr compactification, Bochner-Fejér operators, and Besicovitch spaces, are in [152], [153], [154], and in Chapter 1 of [171]. In particular, Corollary 7.6 for $E = \mathbf{R}_\pm$ and Corollary 7.7 follow from [152, Chap. VI, Theorem 1]. Proposition 7.10 is from [11]. The abstract view at AP factorization presented in Section 7.3 covers also the notion of the Wiener-Hopf factorization in the setting of discrete linearly ordered Abelian groups as considered in [2].

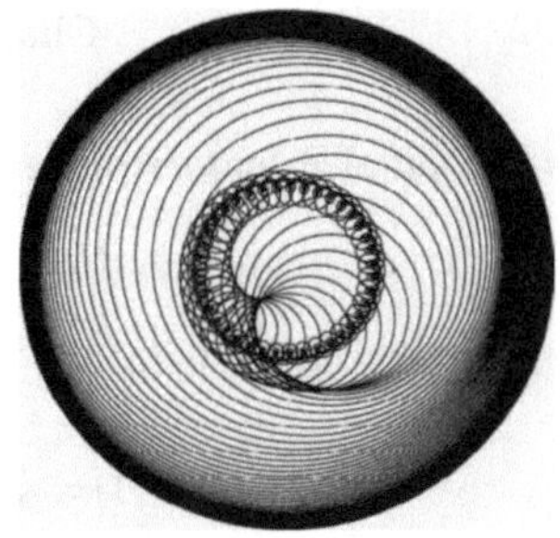

Chapter 8

Existence and Uniqueness of AP Factorization

We first prove that if a matrix function has an AP factorization, then this factorization is essentially unique. This allows us to define the almost periodic indices of an almost periodic matrix function and the geometric mean in the case where all almost periodic indices coincide. The question on the existence of an AP factorization is much more complicated. Satisfactory answers will be given for scalar-valued functions and for periodic matrix functions. We also construct an invertible almost periodic matrix polynomial that has no AP factorization. The existence of such matrix polynomials uncovers a big difference between WH and AP factorization. We finally study the question whether the almost periodic indices are stable under small perturbations, after which we can prove that the geometric mean depends continuously on the matrix function.

8.1 Uniqueness of AP Factorization

In this section we establish the analogue of Theorem 6.4 for AP factorization.

Theorem 8.2 will show that any two right AP factorizations $a = a_-da_+$ and $a = b_-fb_+$ are strongly related. The following theorem implies that relations even hold if $b_-^{\pm 1}$ and $b_+^{\pm 1}$ are not required to belong to $AP^-_{N\times N}$ and $AP^+_{N\times N}$ but to certain (much bigger) Hardy spaces. Recall that $H^p_\pm(-\sigma)$ stands for the functions c for which $c(x)/(x \pm i)^\sigma$ is a function in $H^p_\pm$ (Section 6.1).

Theorem 8.1. *Let $a \in GAP_{N\times N}$. Suppose*

$$a = a_-da_+, \quad d(x) = \operatorname{diag}(e^{i\lambda_j x}), \quad \lambda_1 \geq \ldots \geq \lambda_N$$

is a right AP factorization and

$$a = b_- f b_+, \quad f(x) = \operatorname{diag}(e^{i\mu_j x}), \quad \mu_1 \geq \ldots \geq \mu_N$$

is a factorization with

$$b_-^{\pm 1} \in \left[H_-^p(-\sigma)\right]_{N\times N}, \quad b_+^{\pm 1} \in \left[H_+^p(-\sigma)\right]_{N\times N}, \quad p \geq 1, \quad \sigma < 1 + \frac{1}{p}.$$

Then

$$b_- = a_- d h^{-1} d^{-1}, \quad f = d, \quad b_+ = h a_+,$$

where $h = (h_{jk})_{j,k=1}^N$ is a matrix function with nonzero and constant determinant whose entries are entire functions and

$$h_{jk}(z) = \begin{cases} 0 & \text{if} \quad \lambda_k < \lambda_j, \\ c_{jk} = \text{const} & \text{if} \quad \lambda_k = \lambda_j. \end{cases} \tag{8.1}$$

Example: if $\lambda_1 > \lambda_2 = \lambda_3 > \lambda_4 = \lambda_5$, we have

$$h(z) = \begin{pmatrix} c_{11} & 0 & 0 & 0 & 0 \\ h_{21}(z) & c_{22} & c_{23} & 0 & 0 \\ h_{31}(z) & c_{32} & c_{33} & 0 & 0 \\ h_{41}(z) & h_{42}(z) & h_{43}(z) & c_{44} & c_{45} \\ h_{51}(z) & h_{52}(z) & h_{53}(z) & c_{54} & c_{55} \end{pmatrix}.$$

Proof. The equality $a_- d a_+ = b_- f b_+$ implies that

$$b_-^{-1} a_- d = f b_+ a_+^{-1}, \tag{8.2}$$

and letting $h_- := b_-^{-1} a_-$, $h_+ := b_+ a_+^{-1}$, we can rewrite (8.2) in the form

$$h_{jk}^-(x) = e^{i(\mu_j - \lambda_k)x} h_{jk}^+(x), \tag{8.3}$$

where $h_{jk}^{\pm}$ is the jk entry of $h_{\pm}$. Since $AP^{\pm} \subset H_{\pm}^{\infty}$, we have

$$\begin{aligned} h_+ &= b_+ a_+^{-1} \in b_+ AP_{N\times N}^+ \subset \left[H_+^p(-\sigma)\right]_{N\times N}, && (8.4) \\ h_- &= b_-^{-1} a_- \in b_-^{-1} AP_{N\times N}^- \subset \left[H_-^p(-\sigma)\right]_{N\times N}. && (8.5) \end{aligned}$$

Suppose $\mu_j \geq \lambda_k$. We then deduce from (8.4) and (8.5) that the left-hand side of (8.3) belongs to $H_-^p(-\sigma)$ and the right-hand side of (8.3) is in $H_+^p(-\sigma)$. Taking into account that $\sigma < 1 + 1/p \leq 2$, we can use Hölder's inequality to see that $H_{\pm}^p(-\sigma) \subset H_{\pm}^1(-2)$. Consequently, after passing from $\mathbf{C}_{\pm}$ to $\mathbf{D}_{\pm}$ by means of the conformal map $\omega^{-1}(z) = i(1+z)/(1-z)$, (8.3) becomes

$$h_{jk}^-\left(i\frac{1+t}{1-t}\right) = \exp\left\{-(\mu_j - \lambda_k)\frac{1+t}{1-t}\right\} h_{jk}^+\left(i\frac{1+t}{1-t}\right), \tag{8.6}$$

and, by (6.2), the left-hand side of (8.6) belongs to $H^1_-(\mathbf{T})$, while the right-hand side of (8.6) is in $H^1_+(\mathbf{T})$. As $H^1_-(\mathbf{T}) \cap H^1_+(\mathbf{T}) = \mathbf{C}$, it follows that there is a constant c_{jk} such that

$$h^-_{jk}(x) = c_{jk}, \quad h^+_{jk}(x) = c_{jk}\, e^{i(\lambda_k - \mu_j)x}.$$

If $\mu_j > \lambda_k$, the function $e^{i(\lambda_k-\mu_j)x}$ does not lie in $H^p_+(-\sigma)$ and hence $c_{jk} = 0$. Thus, we have proved that

$$h^+_{jk}(x) = \begin{cases} 0 & \text{if} \quad \lambda_k < \mu_j, \\ c_{jk} = \text{const} & \text{if} \quad \lambda_k = \mu_j. \end{cases} \tag{8.7}$$

Now assume $\mu_j < \lambda_k$. The left-hand side of (8.3) extends to an analytic function in $\mathbf{C}_-$ and the right-hand side of (8.3) extends to an analytic function in $\mathbf{C}_+$. Hence, $h^+_{jk}(x)$ is the restriction to the real line of an entire function.

We now prove that $\mu_j = \lambda_j$ for all j. Assume first that $\mu_j > \lambda_j$ for some j. Then $\mu_l > \lambda_k$ for $l \le j \le k$ and from (8.7) we infer that $h^+_{lk}(x) = 0$ for $l \le j \le k$. This and the Laplace expansion theorem show that $\det h_+(x) = 0$ for all $x \in \mathbf{R}$, which is impossible because $h_+ = b_+a_+^{-1}$. If $\mu_j < \lambda_j$ for some j, we can repeat the above reasoning starting with $da_+b_+^{-1} = a_-^{-1}b_-f$ instead of (8.2) to obtain a contradiction. Thus, $\mu_j = \lambda_j$ for all j and therefore $f = d$.

Letting $h = h_+$, we see that h is an entire function which, by virtue of the equality $f = d$ and (8.7), satisfies (8.1). The block-triangular structure of h implies that $\det h(z)$ is a constant, and since $h = b_+a_+^{-1}$, this constant cannot be zero. Finally, (8.2) gives $h_- = dhd^{-1}$, whence $b_+ = ha_+$, $b_- = a_-h_-^{-1} = a_-dh^{-1}d^{-1}$. □

Theorem 8.2. *Let $a \in GAP_{N\times N}$ and suppose*

$$a = a_-da_+, \quad d(x) = \operatorname{diag}(e^{i\lambda_j x}), \quad \lambda_1 \ge \ldots \ge \lambda_N$$

is a right AP factorization. Then all possible right AP factorizations of a are exhausted by $a = b_-fb_+$ where

$$b_- = a_-dh^{-1}d^{-1}\pi, \quad f = \pi^{-1}d\pi, \quad b_+ = \pi^{-1}ha_+,$$

π is a permutation matrix and $h = (h_{jk})_{j,k=1}^N$ is a matrix function in $GAP_{N\times N}$ such that $\Omega(h_{jk}) \subset [0, \lambda_k - \lambda_j]$ (with the convention that $h_{jk} = 0$ if $\lambda_k < \lambda_j$).

Proof. Let h be as in the theorem. Then $h \in AP^+_{N\times N}$ due to Corollary 7.6. The block-triangular structure of h and the invertibility of h imply that $\det h$ is a nonzero constant. Consequently, $h^{-1} \in AP^+_{N\times N}$. The jk entry of $d(x)h(x)d^{-1}(x)$ is

$$e^{i(\lambda_j-\lambda_k)x}h_{jk}(x) \tag{8.8}$$

and hence $dhd^{-1} \in AP^-_{N\times N}$ by Corollary 7.6. As $\det(dhd^{-1}) = \det h$ is a nonzero constant, it follows that $dh^{-1}d^{-1} \in AP^-_{N\times N}$. Thus, for each h subject to the conditions of the theorem and for each permutation matrix π, $a = b_- f b_+$ is also a right AP factorization.

Conversely, suppose $a = b_- f b_+$, $f = \operatorname{diag}(e_{\nu_j})$ is a right AP factorization. Then there is a π such that

$$\pi f \pi^{-1} = \operatorname{diag}(e_{\nu_j}),\ \nu_1 \geq \ldots \geq \nu_N,$$

and we have $a = (b_-\pi^{-1})(\pi f \pi^{-1})(\pi b_+)$. Theorem 8.1 implies that $\pi f \pi^{-1} = d$, $b_-\pi^{-1} = a_- dh^{-1}d^{-1}$, $\pi b_+ = ha_+$ where h is an entire matrix function with nonzero constant determinant such that (8.1) holds. We are left to show that $\Omega(h_{jk}) \subset [0, \lambda_k - \lambda_j]$ whenever $\lambda_k > \lambda_j$. But this is obvious from the facts that $dhd^{-1} = \pi b_-^{-1} a_-$ belongs to $AP^-_{N\times N}$ and that the jk entry of dhd^{-1} is given by (8.8). □

8.2 *AP* Indices and Geometric Mean

Definition 8.3. Theorem 8.2 shows that if $a \in GAP_{N\times N}$ admits a right AP factorization $a = a_- d a_+$, $d = \operatorname{diag}(e_{\lambda_j})$, then the numbers $\lambda_1, \ldots, \lambda_N \in \mathbf{R}$ are up to their arrangement independent of the particular choice of the factorization. We call $\lambda_1, \ldots, \lambda_N$ the *right AP indices* of a. If a admits any right AP factorization then, by Theorem 8.2, there is a right AP factorization such that $\lambda_1 \geq \ldots \geq \lambda_N$; the vector of the monotonously decreasing right AP indices will be denoted by $\kappa(a)$, i.e.,

$$\kappa(a) := (\lambda_1, \ldots, \lambda_N), \quad \lambda_1 \geq \ldots \geq \lambda_N.$$

Analogously, if $a \in GAP_{N\times N}$ admits a left AP factorization $a = a_+ f a_-$, $f = \operatorname{diag}(e_{\mu_j})$, then the numbers $\mu_1, \ldots, \mu_N$ are uniquely determined provided we require that $\mu_1 \geq \ldots \geq \mu_N$. We call $\mu_1, \ldots, \mu_N$ the *left AP indices* of a and we put

$$\tilde{\kappa}(a) := (\mu_1, \ldots, \mu_N), \quad \mu_1 \geq \ldots \geq \mu_N.$$ □

If $a \in GAP_{N\times N}$ has both a right and a left AP factorization, then $\kappa(a)$ and $\tilde{\kappa}(a)$ need not coincide: from Example 6.6 we infer that if

$$a(x) = \begin{pmatrix} t & 1 \\ 0 & t^{-1} \end{pmatrix}, \quad t = e^{ix},$$

then $\kappa(a) = (0, 0)$ and $\tilde{\kappa}(a) = (1, -1)$. However, passage to determinants gives

$$\lambda_1 + \ldots + \lambda_N = \mu_1 + \ldots + \mu_N = \kappa(\det a),$$

where $\kappa(\det a)$ is the mean motion of $\det a$ (recall Definition 2.26). This shows that the sums of the left and right *AP* indices always coincide. It is also clear that the following are equivalent:

(i) a has a left/right *AP* factorization with the left/right *AP* indices $\nu_1, \ldots, \nu_N$;

(ii) a^*, the Hermitian adjoint of a, has a left/right *AP* factorization with the left/right *AP* indices $-\nu_1, \ldots, -\nu_N$;

(iii) a^T, the transposed of a, has a right/left *AP* factorization with the right/left *AP* indices $\nu_1, \ldots, \nu_N$;

(iv) a^{-1} has a right/left *AP* factorization whose right/left *AP* indices are the numbers $-\nu_1, \ldots, -\nu_N$;

(v) $\tilde{a}$, where $\tilde{a}(x) := a(-x)$, has a right/left *AP* factorization with the right/left *AP* indices $-\nu_1, \ldots, -\nu_N$.

Proposition 8.4. *If $a \in GAP_{N\times N}$ has a right AP factorization with equal right AP indices,*

$$a = a_- d a_+, \quad d(x) = \operatorname{diag}(e^{i\lambda x}, \ldots, e^{i\lambda x}),$$

then the matrix

$$\mathbf{d}(a) := M(a_-)M(a_+) \in \mathbf{C}^{N\times N}$$

is independent of the particular choice of the right AP factorization.

Proof. Let $a = b_- d b_+$ be another right *AP* factorization. Then, by Theorem 8.2, $b_- = a_- h^{-1}$ and $b_+ = h a_+$ where $h \in \mathbf{C}^{N\times N}$ and $\det h \neq 0$. Consequently,

$$M(b_-)M(b_+) = M(a_-)h^{-1}hM(a_+) = M(a_-)M(a_+). \qquad \square$$

Definition 8.5. The matrix $\mathbf{d}(a)$ $(= \mathbf{d}_r(a))$ given by Proposition 8.4 for every matrix function $a \in GAP_{N\times N}$ with a right *AP* factorization and with coinciding right *AP* indices is called the *geometric mean* of a or, more precisely, the *right* geometric mean of a. Analogously, if $a \in GAP_{N\times N}$ has a left *AP* factorization of the form

$$a = b_+ f b_-, \quad f(x) = \operatorname{diag}(e^{i\mu x}, \ldots, e^{i\mu x}),$$

we define the *left geometric mean* by

$$\mathbf{d}_l(a) := M(b_+)M(b_-).$$

In the scalar case ($N = 1$), we clearly have $\mathbf{d}_r(a) = \mathbf{d}_l(a)$. In Definition 3.1, we introduced $\mathbf{d}(a)$ for scalar functions $a \in GAP$ by $\mathbf{d}(a) = e^{M(\psi)}$ where $a(x) = e^{i\kappa(a)x}e^{\psi(x)}$ is the factorization ensured by Bohr's theorem. If $a \in GAP$ has a right *AP* factorization $a = a_- e_\lambda a_+$, then the present definition is consistent with

Definition 3.1. Indeed, it is clear that $\kappa(a) = \lambda$, and Lemma 3.4 implies that $a_- = e^{\psi_-}$, $a_+ = e^{\psi_+}$ with $\psi_\pm \in AP^\pm$. Therefore $\psi = \psi_- + \psi_+ + 2k\pi i$ with $k \in \mathbf{Z}$, whence

$$
\begin{aligned}
M(a_-)M(a_+) &= M(e^{\psi_-})M(e^{\psi_+}) = \Big(\lim_{y\to-\infty} e^{\psi_-(iy)}\Big)\Big(\lim_{y\to+\infty} e^{\psi_+(iy)}\Big) \\
&= e^{\lim_{y\to-\infty}\psi_-(iy)} e^{\lim_{y\to+\infty}\psi_+(iy)} = e^{M(\psi_-)}e^{M(\psi_+)} \\
&= e^{M(\psi_-+\psi_+)} = e^{M(\psi)}e^{-2k\pi i} = e^{M(\psi)}. \qquad \square
\end{aligned}
$$

8.3 Scalar Case

The problem of deciding whether a concrete function has a WH factorization is difficult. Let us first consider the circle case.

Example 8.6. The *Wiener algebra* $W(\mathbf{T})$ is the set of all functions $b \in L^\infty(\mathbf{T})$ with absolutely convergent Fourier series:

$$b(t) = \sum_{n\in\mathbf{Z}} b_n t^n \quad (t \in \mathbf{T}), \quad \|b\|_W := \sum_{n\in\mathbf{Z}} |b_n| < \infty.$$

It is well known that $W(\mathbf{T})$ is a Banach algebra with the norm $\|\cdot\|_W$. A theorem of Wiener implies that if $b \in W(\mathbf{T}) \cap GL^\infty(\mathbf{T})$, then $b \in GW(\mathbf{T})$ and $b(t) = t^\varkappa e^{c(t)}$ for all $t \in \mathbf{T}$ with certain $\varkappa \in \mathbf{Z}$ and $c \in W(\mathbf{T})$. We therefore have

$$c(t) = \sum_{n\in\mathbf{Z}} c_n t^n \quad (t \in \mathbf{T}), \quad \sum_{n\in\mathbf{Z}} |c_n| < \infty,$$

and letting

$$c_-(t) := (Q_0 c)(t) = \sum_{n\le -1} c_n t^n, \quad c_+(t) := (P_0 c)(t) = \sum_{n\ge 0} c_n t^n,$$

we obtain

$$b(t) = e^{c_-(t)} t^\varkappa e^{c_+(t)}, \quad t \in \mathbf{T}. \tag{8.9}$$

Since $e^{\pm c_-} \in H^\infty_-(\mathbf{T})$ and $e^{\pm c_+} \in H^\infty_+(\mathbf{T})$, it is clear that (8.9) is a right WH factorization of b. Thus, every function $b \in W(\mathbf{T}) \cap GL^\infty(\mathbf{T}) = GW(\mathbf{T})$ has a right WH factorization and this factorization can be constructed as above. $\square$

One of the differences between the circle and the line cases is that $L^\infty(\mathbf{T})$ is contained in $L^2(\mathbf{T})$, whereas $L^\infty(\mathbf{R})$ is not a subset of $L^2(\mathbf{R})$. Let P and Q be the projections on $L^2(\mathbf{R})$ given by (2.11) to (2.14) and let $L^2(-1)$ be the Hilbert space of all functions f on $\mathbf{R}$ for which $f(x)/(x+i)$ is a function in $L^2(\mathbf{R})$. The norm in $L^2(-1)$ is given by

$$\|f\| := \left(\int_{\mathbf{R}} \left| \frac{f(x)}{x+i} \right|^2 dx \right)^{1/2}.$$

Since P and Q are bounded (orthogonal) projections on $L^2(\mathbf{R})$, the operators

$$\mathbf{P} := (x+i)P(x+i)^{-1}I, \quad \mathbf{Q} := (x+i)Q(x+i)^{-1}I \tag{8.10}$$

are bounded (orthogonal) projections on $L^2(-1)$. Because $L^\infty(\mathbf{R}) \subset L^2(-1)$, we can in particular apply $\mathbf{P}$ and $\mathbf{Q}$ to functions in $L^\infty(\mathbf{R})$.

In Section 5.4 we introduced the Wiener algebra $W = \mathbf{C} + FL^1(\mathbf{R})$. The following lemma reveals the action of $\mathbf{P}$ and $\mathbf{Q}$ on the Wiener algebra.

For $k \in L^1(\mathbf{R})$, put $k_- := \chi_{(-\infty,0)}k$, $k_+ := \chi_{(0,\infty)}k$.

Lemma 8.7. *If $a = c + Fk \in W$, then*

$$\mathbf{P}a = c + \int_{-\infty}^0 e^s k(s)\, ds + Fk_+, \quad \mathbf{Q}a = -\int_{-\infty}^0 e^s k(s)\, ds + Fk_-.$$

In particular, $\mathbf{P}$ and $\mathbf{Q}$ are bounded operators on W.

Proof. Obviously,

$$\begin{aligned} P\Big(\frac{1}{x+i}(Fk)(x)\Big) &= P\Big(\frac{1}{x+i}(Fk_+)(x) + \frac{1}{x+i}(Fk_-)(x)\Big) \\ &= \frac{1}{x+i}(Fk_+)(x) + P\Big(\frac{1}{x+i}(Fk_-)(x)\Big). \end{aligned}$$

The function $1/(x+i)$ is the Fourier transform of the function $-i\chi_+(t)e^{-t}$ where $\chi_+ := \chi_{(0,\infty)}$. Hence,

$$\begin{aligned} &P\Big(\frac{1}{x+i}(Fk_-)(x)\Big) = F\chi_+\big(-i\chi_+(t)e^{-t} * k_-(t)\big) \\ &= F\Big(-i\chi_+(t)\int_{-\infty}^0 \chi_+(t-s)e^{-(t-s)}k_-(s)\, ds\Big) \\ &= F\Big(-i\chi_+(t)e^{-t}\int_{-\infty}^0 e^s k(s)\, ds\Big) = \frac{1}{x+i}\int_{-\infty}^0 e^s k(s)\, ds. \end{aligned}$$

This proves the formula for $\mathbf{P}a$. The formula for $\mathbf{Q}a$ can be verified analogously or can be derived from the equality $\mathbf{Q}a = a - \mathbf{P}a$. □

We remark that if in (8.10) we would replace $x+i$ by $x-i$, then Lemma 8.7 would go over into

$$\begin{aligned}(x-i)P\Big(\frac{1}{x-i}a(x)\Big) &= -\int_0^\infty e^{-s}k(s)\,ds + (Fk_+)(x),\\ (x-i)Q\Big(\frac{1}{x-i}a(x)\Big) &= c+\int_0^\infty e^{-s}k(s)\,ds + (Fk_-)(x).\end{aligned}$$

Using Lemma 8.7 it is not difficult to show that if $a \in W$ is given by

$$a(x) = \sum_{n\in\mathbf{Z}} a_n\Big(\frac{x-i}{x+i}\Big)^n \quad (x \in \mathbf{R}), \quad \sum_{n\in\mathbf{Z}} |a_n| < \infty,$$

then

$$(\mathbf{P}a)(x) = \sum_{n\geq 0} a_n\Big(\frac{x-i}{x+i}\Big)^n, \quad (\mathbf{Q}a)(x) = \sum_{n\leq -1} a_n\Big(\frac{x-i}{x+i}\Big)^n. \tag{8.11}$$

Indeed, we need only take into account that

$$\Big(\frac{x-i}{x+i}\Big)^n = 1-(Fl_n)(x), \quad n \in \mathbf{Z},$$

where

$$l_n(x) := \begin{cases} \sum_{j=1}^{|n|}\binom{j}{|n|}\frac{2^j}{(j-1)!}t^{j-1}e^t\chi_-(t), & n=-1,-2,\dots,\\ 0, & n=0,\\ \sum_{j=1}^{n}\binom{j}{n}\frac{2^j}{(j-1)!}(-t)^{j-1}e^{-t}\chi_+(t), & n=1,2,\dots\end{cases}$$

(see, e.g., [92]) and that

$$\int_{-\infty}^0 e^s l_n(s)ds = \begin{cases} 0, & n=0,1,\dots,\\ 1, & n=-1,-2,\dots.\end{cases}$$

Of course, the direct computation of $\mathbf{P}a$ and $\mathbf{Q}a$ on the basis of (8.10) is simpler.

Example 8.8. We know from Section 5.4 that every function $a \in W \cap GL^\infty(\mathbf{R})$ admits a right WH factorization in W. Given such a function, we can write

$$a(x) = \Big(\frac{x-i}{x+i}\Big)^{\varkappa} e^{b(x)}, \quad x \in \mathbf{R},$$

with $\varkappa \in \mathbf{Z}$ and $b \in W$, and taking into account Lemma 8.7 one easily sees that

$$a(x) = e^{(\mathbf{Q}b)(x)}\Big(\frac{x-i}{x+i}\Big)^{\varkappa} e^{(\mathbf{P}b)(x)} \tag{8.12}$$

is a right WH factorization in W. Notice that all factors on the right of (8.12) belong to GW. □

Things are more complicated if all we know is that the matrix function is continuous. Here is a well known scalar result.

Theorem 8.9. *Let* $a \in GC(\dot{\mathbf{R}})$ *and* wind $a = \varkappa$, *so that*

$$a(x) = \left(\frac{x-i}{x+i}\right)^{\varkappa} e^{c(x)}$$

with $c \in C(\dot{\mathbf{R}})$. *The function* a *admits a right* WH *factorization in* $C(\dot{\mathbf{R}})$ *if and only if* $\mathbf{P}c$ *is a function in* $C(\dot{\mathbf{R}})$. *In that case a right* WH *factorization in* $C(\dot{\mathbf{R}})$ *is given by*

$$a(x) = e^{(\mathbf{Q}c)(x)} \left(\frac{x-i}{x+i}\right)^{\varkappa} e^{(\mathbf{P}c)(x)}.$$

Example 8.10. There are functions $c \in C(\dot{\mathbf{R}})$ such that $\mathbf{P}c$ does not belong to $C(\dot{\mathbf{R}})$. For instance, the function

$$c(x) = \sum_{n=2}^{\infty} \frac{1}{n \log n} \left(\left(\frac{x-i}{x+i}\right)^{n} - \left(\frac{x+i}{x-i}\right)^{n} \right)$$

is in $C(\dot{\mathbf{R}})$, but the function

$$(\mathbf{P}c)(x) = \sum_{n=2}^{\infty} \frac{1}{n \log n} \left(\frac{x-i}{x+i}\right)^{n}$$

(recall (8.11)) has a discontinuity at infinity.

Here is another and very instructive example. If $v \in L^2(\mathbf{T})$ is a real-valued function, then the conjugate function $H_0 v$ is the real-valued function in $L^2(\mathbf{T})$ for which the harmonic extension of $v + iH_0 v$ is analytic in the open unit disk $\mathbf{D}$ and takes the value 0 at the origin. Let $P_0 = (I + S_0)/2$ be as in Section 6.1. It is well known that

$$P_0 v = \frac{1}{2}(v + iH_0 v) + \frac{1}{2} v_0$$

where v_0 is the zeroth Fourier coefficient of v. Now let $\Omega \subset \mathbf{C}$ be the domain

$$\Omega = \{x + iy \in \mathbf{C} : 0 < x < 1,\ 0 < y < 1/x\}$$

and let $f : \mathbf{D} \to \Omega$ be a conformal map such that $f(z) \to \infty$ as $z \to 1$. The map f extends to a continuous function $f : \mathbf{D} \cup (\mathbf{T} \setminus \{1\}) \to \mathbf{C}$. Put $v := \operatorname{Re} f$. Clearly, v is continuous on $\mathbf{T} \setminus \{1\}$ and $v(1-0) = v(1+0)$, that is, $v \in C(\mathbf{T})$. However, $\operatorname{Im} f$ is unbounded on $\mathbf{T} \setminus \{1\}$, and since $H_0 v$ differs from $\operatorname{Im} f$ merely by a constant, it follows that $H_0 v$ is also unbounded. Consequently, $P_0 v \notin L^\infty(\mathbf{T})$.

Now define $c \in C(\dot{\mathbf{R}})$ by $c(x) = v((x-i)/(x+i))$. Then $v = B_0 c$ (recall (6.5)), and from (6.7) we deduce that $\mathbf{P}c$ is unbounded. □

By Bohr's theorem, every function $a \in GAP$ can be written in the form

$$a(x) = e^{i\kappa(a)x} e^{b(x)}, \quad \kappa(a) \in \mathbf{R}, \quad b \in AP, \tag{8.13}$$

and hence the problem of finding a (right = left) AP or APW factorization of a is equivalent to the task of constructing an AP or APW factorization of e^b.

We know that the operators

$$\mathbf{P} : L^2(-1) \to H^2_+(-1),\ (\mathbf{P}f)(x) := (x+i)P\Big(\frac{1}{x+i} f(x)\Big),$$
$$\mathbf{Q} : L^2(-1) \to H^2_-(-1),\ (\mathbf{Q}f)(x) := (x+i)Q\Big(\frac{1}{x+i} f(x)\Big)$$

are the orthogonal projections. Obviously, $\mathbf{P}+\mathbf{Q} = I$. In particular, for $f \in L^\infty(\mathbf{R})$, both $\mathbf{P}f$ and $\mathbf{Q}f$ are well-defined functions in $H^2_+(-1)$ and $H^2_-(-1)$, respectively. Let $b \in APW$. Then $b = \sum b_j e_{\lambda_j}$ with $\sum |b_j| < \infty$. We claim that

$$\mathbf{P}e_\lambda = \begin{cases} e^\lambda & \text{if } \lambda < 0, \\ e_\lambda & \text{if } \lambda \ge 0, \end{cases} \quad \mathbf{Q}e_\lambda = \begin{cases} e_\lambda - e^\lambda & \text{if } \lambda < 0, \\ 0 & \text{if } \lambda \ge 0; \end{cases} \tag{8.14}$$

here, of course, e^λ is the number $\exp \lambda$, while e_λ stands for the function given by $e_\lambda(x) = e^{i\lambda x}$. To prove (8.14), notice first that if $\lambda \ge 0$, then the function $(x+i)^{-1} e_\lambda(x)$ belongs to H^2_+, whence

$$(\mathbf{P}e_\lambda)(x) = (x+i)P\big((x+i)^{-1} e_\lambda(x)\big) = e_\lambda(x).$$

If $\lambda < 0$, then $(x-i)^{-1} e_\lambda(x)$ is a function in H^2_-, so $P((x-i)^{-1} e_\lambda(x)) = 0$, and we get

$$\begin{aligned}
(\mathbf{P}e_\lambda)(x) &= (x+i)P\Big((x+i)^{-1} e_\lambda(x)\Big) - (x-i)P\Big((x-i)^{-1} e_\lambda(x)\Big) \\
&= \frac{1}{2}(x+i)S\Big((x+i)^{-1} e_\lambda(x)\Big) - \frac{1}{2}(x-i)S\Big((x-i)^{-1} e_\lambda(x)\Big) \\
&= \frac{1}{2\pi i}\int_{\mathbf{R}} \Big(\frac{x+i}{y+i} - \frac{x-i}{y-i}\Big)\frac{e^{i\lambda y}}{y-x}\,dy = \frac{1}{\pi}\int_{\mathbf{R}} \frac{e^{i\lambda y}}{y^2+1}\,dy = e^\lambda
\end{aligned}$$

(also recall the proof of Corollary 7.7). This proves the first part of (8.14). The other part of (8.14) follows from the identity $\mathbf{Q} = I - \mathbf{P}$. Since $\sum |b_j| < \infty$, we therefore obtain

$$(\mathbf{P}b)(x) = \sum_{\lambda_j \ge 0} b_j e_{\lambda_j}(x) + \sum_{\lambda_j < 0} b_j e^{\lambda_j}, \tag{8.15}$$

$$(\mathbf{Q}b)(x) = \sum_{\lambda_j < 0} b_j e_{\lambda_j}(x) - \sum_{\lambda_j < 0} b_j e^{\lambda_j}. \tag{8.16}$$

Although $\mathbf{P}$ and $\mathbf{Q}$ are projections of APW into APW^+ and APW^-, respectively, it is not convenient to work with them in APW. We rather employ the projections $\tilde{P}$ and $\tilde{Q}_0$ given by

$$\tilde{P}e_\lambda = \begin{cases} 0 & \text{if} \quad \lambda < 0, \\ e_\lambda & \text{if} \quad \lambda \geq 0, \end{cases} \qquad \tilde{Q}_0 e_\lambda = \begin{cases} e_\lambda & \text{if} \quad \lambda < 0, \\ 0 & \text{if} \quad \lambda \geq 0. \end{cases} \tag{8.17}$$

For $b = \sum b_j e_{\lambda_j} \in APW$ we then have

$$\tilde{P}b = \sum_{\lambda_j \geq 0} b_j e_{\lambda_j}, \quad \tilde{Q}_0 b = \sum_{\lambda_j < 0} b_j e_{\lambda_j}. \tag{8.18}$$

Thus, $\tilde{P}$ is the projection of APW onto APW^+ parallel to $APW_0^- := \{b \in APW^- : M(b) = 0\}$, and $\tilde{Q}_0$ is the projection of APW onto APW_0^- parallel to APW^+.

Theorem 8.11. *If $a \in GAPW$ is given by* (8.13), *then $b \in APW$ and*

$$a(x) = e^{(\mathbf{Q}b)(x)} e^{i\kappa(a)x} e^{(\mathbf{P}b)(x)} \tag{8.19}$$

is an APW factorization of a.

Proof. It suffices to prove that $b \in APW$. It is clear that $e^b \in APW$, and since $[0,1] \to GAPW$, $\mu \mapsto e^{\mu b}$ is continuous and is the constant function 1 at $\mu = 0$, we deduce from (1.1) that $b = c + 2k\pi i$ with some $c \in APW$ and $k \in \mathbf{Z}$. This shows that $b \in APW$. ☐

For $b = \sum b_j e_{\lambda_j} \in APW$ we see from (8.15), (8.16) and (8.18) that

$$\mathbf{P}b = \tilde{P}b + \sum_{\lambda_j < 0} b_j e^{\lambda_j}, \quad \mathbf{Q}b = \tilde{Q}_0 b - \sum_{\lambda_j < 0} b_j e^{\lambda_j}.$$

Hence, we can also replace (8.19) with

$$a(x) = e^{(\tilde{Q}_0 b)(x)} e^{i\kappa(a)x} e^{(\tilde{P}b)(x)}. \tag{8.20}$$

Theorem 8.12. *Let $a \in GAP$ be given by* (8.13). *The function a admits an AP factorization if and only if $\mathbf{P}b \in AP^+$ or, equivalently, $\mathbf{Q}b \in AP^-$. In that case* (8.19) *is an AP factorization of a.*

Proof. Suppose a admits an AP factorization. Then $e^b = a_- a_+$ with $a_\pm \in GAP^\pm$. From Lemma 3.4 we infer that $a_\pm = e^{b_\pm}$ with $b_\pm \in AP^\pm$. It follows that $b_- + b_+ = b + 2k\pi i$ with $k \in \mathbf{Z}$, and because $b = \mathbf{Q}b + \mathbf{P}b$, we get

$$b_- - \mathbf{Q}b = \mathbf{P}b - b_+ + 2k\pi i.$$

The left-hand side of this equality belongs to $H^2_-(-1)$, while its right-hand side is in $H^2_+(-1)$. Because of (6.3) and the equality $H^2_-(\mathbf{T}) \cap H^2_+(\mathbf{T}) = \mathbf{C}$, we obtain that there is a constant $\gamma \in \mathbf{C}$ such that $\mathbf{Q}b = b_- - \gamma$ and $\mathbf{P}b = b_+ + 2k\pi i + \gamma$. Since $b_\pm \in AP^\pm$, it results that $\mathbf{Q}b \in AP^-$ and $\mathbf{P}b \in AP^+$.

Conversely, if $\mathbf{P}b \in AP^+$ or $\mathbf{Q}b \in AP^-$, then (8.19) is obviously an AP factorization of a. □

Example 8.13. We show that there are $b \in AP$ such that $\mathbf{P}b \notin AP^+$, which implies that *not every function $a \in GAP$ admits an AP factorization.*

In Example 8.10, we constructed a function $v \in C(\mathbf{T})$ such that $P_0 v \notin L^\infty(\mathbf{T})$. Put $b(x) := v(e^{ix})$ for $x \in \mathbf{R}$. Since v can be approximated by trigonometric polynomials, it follows that $b \in AP$. Let v_n $(n \in \mathbf{Z})$ be the nth Fourier coefficient of v. Then $b(x) = \sum_{n\in\mathbf{Z}} v_n e^{inx}$, the series being convergent in $L^2(-1)$. Thus, by (8.14),

$$(\mathbf{P}b)(x) = \sum_{n\geq 0} v_n e^{inx} + \sum_{n<0} v_n e^n = (P_0 v)(e^{ix}) + \sum_{n<0} v_n e^n$$

for almost all $x \in \mathbf{R}$; note that the series $\sum_{n<0} v_n e^n$ is absolutely convergent because $\sup_{n\in\mathbf{Z}} |v_n| < \infty$. In summary, $b \in AP$ but $\mathbf{P}b \notin L^\infty(\mathbf{R})$. □

8.4 Periodic Matrix Functions

Continuous and periodic matrix functions are almost periodic. Indeed, let a be in $C_{N\times N}(\mathbf{R})$ and, for the sake of simplicity, suppose a is 2π-periodic, that is, $a(x + 2\pi) = a(x)$ for all $x \in \mathbf{R}$. Then

$$a(x) = c(e^{ix}), \quad x \in \mathbf{R}, \tag{8.21}$$

where $c \in C_{N\times N}(\mathbf{T})$. As c can be uniformly approximated by its Fejér-Cesàro means, we see that $a \in AP_{N\times N}$.

In Remark 6.2, we defined right WH factorization on the unit circle $\mathbf{T}$. Such a factorization is said to be a right WH factorization within $C(\mathbf{T})$ if all its factors belong to $C(\mathbf{T})$. Equivalently, a matrix function $a \in GC_{N\times N}(\mathbf{T})$ has a right WH factorization on $\mathbf{T}$ within $C(\mathbf{T})$ if and only if the function $B_0^{-1}c \in GC_{N\times N}(\dot{\mathbf{R}})$ has a right WH factorization in $C(\dot{\mathbf{R}})$; recall that $(B_0^{-1}c)(x) := c((x-i)/(x+i))$.

Theorem 8.14. *Let $c \in GC_{N\times N}(\mathbf{T})$ and let $a \in GAP_{N\times N}$ be given by (8.21). The matrix function a admits a right AP factorization if and only if c has a right WH factorization on $\mathbf{T}$ within $C(\mathbf{T})$:*

$$c(t) = c_-(t)\,\mathrm{diag}\,(t^{\varkappa_1}, \ldots, t^{\varkappa_N})c_+(t), \quad t \in \mathbf{T}. \tag{8.22}$$

In that case

$$a(x) = c_-(e^{ix})\,\mathrm{diag}\,(e^{i\varkappa_1 x}, \ldots, e^{i\varkappa_N x})c_+(e^{ix}), \quad x \in \mathbf{R}, \tag{8.23}$$

is a right AP factorization of a. In particular, the right AP indices of a are the integers $\varkappa_1, \dots, \varkappa_N$.

Proof. Suppose that (8.22) is a right WH factorization on $\mathbf{T}$ within $C(\mathbf{T})$. Then

$$c_-^{\pm 1} \in \left[C(\mathbf{T}) \cap H_-^\infty(\mathbf{T})\right]_{N\times N}, \quad c_+^{\pm 1} \in \left[C(\mathbf{T}) \cap H_+^\infty(\mathbf{T})\right]_{N\times N},$$

and hence the matrix functions $a_\pm$ defined by $a_\pm(x) = c_\pm(e^{ix})$ belong to $GAP^\pm$ (Corollary 7.7). This shows that (8.23) is a right AP factorization of a.

Conversely, let

$$a(x) = a_-(x) \operatorname{diag}\left(e^{i\lambda_1 x}, \dots, e^{i\lambda_N x}\right) a_+(x), \quad x \in \mathbf{R}, \tag{8.24}$$

be a right AP factorization of a. Put

$$b(x) := c\big((x-i)/(x+i)\big), \quad x \in \mathbf{R}.$$

Then $b \in GC_{N\times N}(\dot{\mathbf{R}})$. Since Propositions 2.10 and 2.11 remain literally true in the matrix case, we see that $W(b^{-1})$ is a regularizer of $W(b)$. Thus, $W(b)$ is Fredholm and Theorem 6.5 gives a right WH factorization

$$b(x) = b_-(x) \operatorname{diag}\left[\left(\frac{x-i}{x+i}\right)^{\varkappa_1}, \dots, \left(\frac{x-i}{x+i}\right)^{\varkappa_N}\right] b_+(x), \quad x \in \mathbf{R}.$$

Letting $x = i(1+t)/(1-t)$ in the latter equality, we get the right WH factorization (on $\mathbf{T}$)

$$c(t) = c_-(t) \operatorname{diag}\left(t^{\varkappa_1}, \dots, t^{\varkappa_N}\right) c_+(t), \quad t \in \mathbf{T} \tag{8.25}$$

with $c_\pm(t) = b_\pm(i(1+t)/(1-t))$.

Put $g_\pm(x) := c_\pm(e^{ix})$. We claim that $g_\pm(x)$ are functions in $(x \pm i)\,[H_\pm^2]_{N\times N}$. It suffices to prove this under the assumption that $N = 1$. We have $c_\pm \in H_\pm^2(\mathbf{T})$ by the definition of WH factorization on $\mathbf{T}$. If $y > 0$, then

$$\int_{\mathbf{R}} \left|\frac{g_+(x+iy)}{(x+iy+i)}\right|^2 dx = \int_{\mathbf{R}} \frac{|c_+(e^{ix}e^{-y})|^2}{x^2+(y+1)^2}\, dx = \sum_{n\in\mathbf{Z}} \int_{2n\pi}^{2(n+1)\pi} \frac{|c_+(e^{ix}e^{-y})|^2}{x^2+(y+1)^2}\, dx$$

$$\le 2\int_0^{2\pi} |c_+(e^{ix}e^{-y})|^2\, dx + 2\sum_{m\ge 1} \frac{1}{(2m\pi)^2} \int_0^{2\pi} |c_+(e^{ix}e^{-y})|^2\, dx$$

and since

$$\sup_{0<r<1} \int_0^{2\pi} |c_+(re^{ix})|^2\, dx < \infty$$

by the definition of $H_+^2(\mathbf{T}) \cong H^2(\mathbf{D}_+)$, it follows that $g_+(x)$ is a function in $(x+i)H_+^2$. Analogously one can show that $g_-(x)$ is a function in $(x-i)H_-^2$. This completes the proof of our claim.

From (8.25) we get

$$a(x) = g_-(x)\,\mathrm{diag}\,(e^{i\varkappa_1 x}, \dots, e^{i\varkappa_N x})g_+(x), \quad x \in \mathbf{R}. \tag{8.26}$$

Applying Theorem 8.1 (with $p = 2$ and $\sigma = 1$) to (8.24) and (8.26), we obtain that the indices $\lambda_1, \dots, \lambda_N$ and $\varkappa_1, \dots, \varkappa_N$ coincide and that

$$\begin{aligned} g_+(x) &= \pi^{-1}h(x)a_+(x),\\ g_-(x) &= a_-(x)\,\mathrm{diag}\,(e^{i\varkappa_j x})h^{-1}(x)\,\mathrm{diag}\,(e^{-i\varkappa_j x})\pi \end{aligned}$$

with some permutation matrix π and a certain entire matrix function h. As $a_\pm$ and h are continuous on $\mathbf{R}$, so also are $g_\pm$, and because $g_\pm(x) = c_\pm(e^{ix})$, it results that $c_\pm \in C_{N\times N}(\mathbf{T})$. □

Recall the definition of $W(\mathbf{T})$ (Example 8.6). It is easily seen that the 2π-periodic functions a in APW are just the functions (8.21) with $c \in W(\mathbf{T})$.

Theorem 8.15. *If $a \in GAPW_{N\times N}$ is given by (8.21) with $c \in W_{N\times N}(\mathbf{T})$, then a admits both a left and a right APW factorization with integral AP indices.*

Proof. The circle analogue of Theorem 5.14 is true: every matrix function in $GW_{N\times N}(\mathbf{T})$ admits a left and a right WH factorization on $\mathbf{T}$ within $W(\mathbf{T})$. The assertion is therefore an immediate consequence of the "if" portion of Theorem 8.14. □

We finally consider a special class of 2π-periodic matrix functions. Given a sequence $\{a_k\}_{k\in\mathbf{Z}}$ of matrices $a_k \in \mathbf{C}^{N\times N}$, we denote by $T_m(a)$ the block Toeplitz matrix

$$T_m(a) = \begin{pmatrix} a_0 & a_{-1} & \dots & a_{-(m-1)} \\ a_1 & a_0 & \dots & a_{-(m-2)} \\ \vdots & \vdots & \ddots & \vdots \\ a_{m-1} & a_{m-2} & \dots & a_0 \end{pmatrix}.$$

Theorem 8.16. *Let $a_k \in \mathbf{C}^{N\times N}$, $\sum \|a_k\| < \infty$, $m \in \mathbf{N}$. The matrix function*

$$a(x) := \begin{pmatrix} e^{-imx}I & 0 \\ \sum_{k\in\mathbf{Z}} a_k e^{ikx} & e^{imx}I \end{pmatrix} \in GAPW_{2N\times 2N}$$

always has a right APW factorization. It has a canonical right APW factorization if and only if $\det T_m(a) \neq 0$.

Proof. Put

$$c(t) := \begin{pmatrix} t^{-m}I & 0 \\ \sum_{k\in\mathbf{Z}} a_k t^k & t^m I \end{pmatrix}, \quad t \in \mathbf{T}.$$

By the $W(\mathbf{T})$ analogue of Theorem 5.14, c has a right WH factorization on $\mathbf{T}$ within $W(\mathbf{T})$,

$$c(t) = c_-(t)d(t)c_+(t), \quad t \in \mathbf{T}.$$

This implies that $a(x) = c(e^{ix})$ has a right APW factorization and that the right AP indices of a are the right partial indices of c. Thus, a has a canonical right APW factorization if and only if c has a canonical right WH factorization on the unit circle $\mathbf{T}$. By the circle analogue of Theorem 6.5, this is equivalent to the invertibility of the Toeplitz operator $T(c)$ given by the block Toeplitz matrix $(c_{j-k})_{j,k=0}^{\infty}$ on $l^2_{2N}(\mathbf{Z}_+)$, where $\{c_n\}_{n\in\mathbf{Z}}$ is the sequence of the Fourier coefficients of c. Finally, by the discrete analogue of Theorem 1.25, $T(c)$ is invertible if and only if so is $T_m(a)$. □

The above results can be easily extended to matrix functions $a \in C_{N\times N}(\mathbf{R})$ with an arbitrary period $T > 0$. Such functions admit the representation

$$a(x) = c(e^{2\pi ix/T}) \text{ with } c \in C_{N\times N}(\mathbf{T}). \tag{8.27}$$

For example, the period T version of Theorem 8.14 says that the matrix function (8.27) has a right AP factorization if and only if c possesses a right WH factorization on $\mathbf{T}$ within $C(\mathbf{T})$; in that case the right AP indices of a are $2\pi\varkappa_j/T$, where $\varkappa_j$ are the right partial indices of c.

8.5 An Invertible *APW* Polynomial Without *APW* Factorization

Theorem 5.14 tells us that every matrix function in $GW_{N\times N}$ admits both a right and a left WH factorization in W. Theorems 8.11 and 8.15 nourish the hope that an analogous result holds for matrix functions in $GAPW_{N\times N}$, i.e., that every $a \in GAPW_{N\times N}$ admits both a left and a right APW factorization. It is one of the big surprises of AP factorization that this is not true: even invertible almost periodic matrix **polynomials** need not possess an AP factorization (and thus, all the more, need not have an APW factorization). Here is a counterexample, which also proves Theorem 5.17.

Theorem 8.17. *The almost periodic matrix polynomial*

$$a(x) = \begin{pmatrix} e^{-i(1+\alpha)x} & 0 \\ e^{-i\alpha x} - 1 + e^{ix} & e^{i(1+\alpha)x} \end{pmatrix}, \quad \alpha = \frac{\sqrt{5}-1}{2}$$

is invertible but does not admit a right AP factorization.

Proof. Assume a has a right AP factorization

$$a(x) = a_-(x)\operatorname{diag}\left(e^{i\lambda_1 x}, e^{i\lambda_2 x}\right)a_+(x), \quad \lambda_1 \geq \lambda_2. \tag{8.28}$$

Put

$$b_-(x) := \begin{pmatrix} 1 - e^{-i\alpha x} & e^{-i(1+\alpha)x} \\ -e^{i(1-2\alpha)x} & 1 + e^{-i\alpha x} + e^{-2i\alpha x} \end{pmatrix},$$

$$b_+(x) := \begin{pmatrix} 1 + e^{ix} & -e^{i\alpha x} \\ -e^{i(1-\alpha)x} & 1 \end{pmatrix}.$$

Since $\det b_-(x) = \det b_+(x) = 1$, we have $b_\pm \in GAP^\pm_{2\times 2}$. A straightforward computation (along with the identity $\alpha^2 + \alpha - 1 = 0$) shows that

$$b_-(x)a(x)b_+(x) = a\big((1-\alpha)x\big). \tag{8.29}$$

From (8.28) and (8.29) we get two right AP factorizations of $a((1-\alpha)x)$:

$$\begin{aligned} a\big((1-\alpha)x\big) &= a_-\big((1-\alpha)x\big)\operatorname{diag}\big(e^{i(1-\alpha)\lambda_j x}\big)a_+\big((1-\alpha)x\big) \\ &= b_-(x)a_-(x)\operatorname{diag}\big(e^{i\lambda_j x}\big)a_+(x)b_+(x). \end{aligned}$$

Theorem 8.2 implies that either $(1-\alpha)\lambda_j = \lambda_j$ or $\lambda_j = (1-\alpha)\lambda_{3-j}$. Hence in both cases $\lambda_1 = \lambda_2 = 0$. Thus,

$$a\big((1-\alpha)x\big) = a_-\big((1-\alpha)x\big)a_+\big((1-\alpha)x\big) = b_-(x)a_-(x)a_+(x)b_+(x)$$

are two canonical right AP factorizations. Again having recourse to Theorem 8.2 we obtain

$$a_+\big((1-\alpha)x\big) = ha_+(x)b_+(x) \text{ with some } h \in G\mathbf{C}^{2\times 2}.$$

Passage to mean values gives

$$M\Big(a_+\big((1-\alpha)x\big)\Big) = M(h)M\big(a_+(x)\big)M\big(b_+(x)\big),$$

and since

$$\begin{aligned} &M\Big(a_+\big((1-\alpha)x\big)\Big) = M\big(a_+(x)\big)\ \Big(= \big[M(a_+^{-1})\big]^{-1}\Big) \in GC^{2\times 2}, \\ &M(h) = h,\ M\big(b_+(x)\big) = I, \end{aligned}$$

it follows that $h = I$. Consequently, $a_+((1-\alpha)x) = a_+(x)b_+(x)$ for all $x \in \mathbf{R}$, which at $x = 0$ results in the equality

$$a_+(0) = a_+(0)\begin{pmatrix} 2 & -1 \\ -1 & 1 \end{pmatrix}.$$

As $a_+(0) \in G\mathbf{C}^{2\times 2}$, this is impossible. □

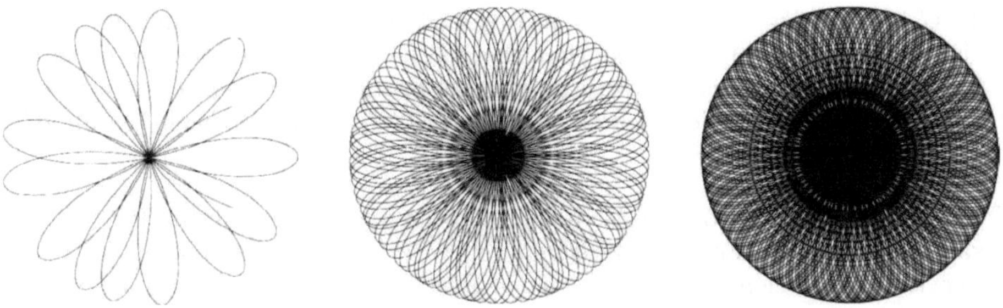

Figure 8.1: The image of $e^{-i\alpha x} - 1 + e^{ix}$ ($\alpha = (\sqrt{5}-1)/2$) for x in $[-30, 30]$ (left), $[-200, 200]$ (middle), and $[-600, 600]$ (right).

Figure 8.1 shows the trinomial in the lower-left corner of the matrix function of Theorem 8.17.

In Theorem 8.17, the left lower entry of $a(x)$ is a trinomial. We will later show that matrix functions of the form

$$a(x) = \begin{pmatrix} e^{-i\lambda x} & 0 \\ p(x) & e^{i\lambda x} \end{pmatrix}, \quad \lambda > 0,$$

always have a right APW factorization in case $p(x)$ is a monomial or a binomial.

8.6 Stability of AP Indices

We now study the behavior of the right AP indices under small perturbations of the matrix function. We begin with the scalar case.

Proposition 8.18. *The mean motion of invertible functions in AP is stable under small perturbations, i.e., for every $a \in GAP$ there is an $\varepsilon > 0$ such that if $b \in AP$ and $\|b - a\|_\infty < \varepsilon$, then $b \in GAP$ and $\kappa(b) = \kappa(a)$.*

Proof. We give an "operator theoretic" proof. If $a \in GAP$ then $a(x) = e^{i\kappa(a)x}e^{\psi(x)}$ with $\psi \in AP$ by Bohr's theorem. The operator $W(e^{\psi})$ is invertible by virtue of Theorem 2.28(d). Hence, if $\varepsilon > 0$ is sufficiently small and $\|b - a\|_\infty < \varepsilon$, then

$$\sup_{x \in \mathbf{R}} \left| e^{-i\kappa(a)x} b(x) - e^{\psi(x)} \right| = \|b - a\|_\infty < \varepsilon$$

and therefore the operator $W(e^{-i\kappa(a)x}b(x))$ (as well as its symbol) is also invertible. Theorem 2.28 implies that the mean motion of $e^{-i\kappa(a)x}b(x)$ is zero, which shows that $\kappa(b) = \kappa(a)$. □

Since $GAPW$ is dense in GAP and every function in $GAPW$ has an AP factorization (Theorem 8.11), it follows that the AP factorable functions are dense in GAP. However, the functions with no AP factorization are also dense in GAP. Indeed, let $a(x) = e^{i\kappa(a)x}e^{\psi(x)}$ be a function in GAP and suppose a has an AP factorization.

Then $\mathbf{P}\psi \in AP^+$ by Theorem 8.12. Let $\varphi \in AP$ be a function such that $\mathbf{P}\varphi \notin AP^+$ (Example 8.13), and for $\delta \in \mathbf{R}$, put $b(x) := e^{i\kappa(a)x}e^{\psi(x)+\delta\varphi(x)}$. Clearly, $\|a-b\|_\infty < \varepsilon$ whenever δ is sufficiently small, but since $\mathbf{P}(\psi + \delta\varphi) = \mathbf{P}\psi + \delta\mathbf{P}\varphi \notin AP^+$, the function b has no AP factorization.

Thus, when considering the behavior of AP indices under small perturbations, we have to restrict ourselves to perturbations within the set of AP factorable (matrix) functions.

Theorem 8.19. *If $a \in GAP$ has an AP factorization with the AP index λ, then there is an $\varepsilon > 0$ such that every AP factorable function $b \in GAP$ with $\|b-a\|_\infty < \varepsilon$ also has the AP index λ.*

Proof. This follows from Proposition 8.18 and the fact that the AP index of an AP factorable function is equal to its mean motion (Theorem 8.12). □

As will be shown below, in the matrix case small perturbations may change the AP indices.

We define an *elementary operation* on the points $(\lambda_1, \dots, \lambda_N)$ in $\mathbf{R}^N$ as follows: one picks two indices r, s such that $\lambda_r > \lambda_s$ and a number $\theta \in [0, \lambda_r - \lambda_s]$ and then one replaces λ_r by $\lambda_r - \theta$ and λ_s by $\lambda_s + \theta$, while the numbers λ_j with $j \notin \{r, s\}$ remain as they are; if $\lambda_1 = \dots = \lambda_N$, the only elementary operation applicable is the one that leaves $(\lambda_1, \dots, \lambda_N)$ unchanged. On defining $(\mu_j) \prec (\lambda_j)$ if there is a finite number of elementary operations which take (λ_j) to (μ_j), we obtain a partial order on $\mathbf{R}^N$. Obviously, for two points (μ_j) and (λ_j) to be comparable in this order it is necessary that $\sum \mu_j = \sum \lambda_j$. It is also clear that the minimal elements of this order are just the points $(\lambda, \lambda, \dots, \lambda)$.

Theorem 8.20. *Let $a \in GAP_{N\times N}$ admit a right AP factorization and denote by $\lambda_1, \dots, \lambda_N$ the right AP indices. Then for every $\varepsilon > 0$ and every $(\mu_1, \dots, \mu_N)$ in $\mathbf{R}^N$ satisfying $(\mu_j) \prec (\lambda_j)$ there exists a $b \in GAP_{N\times N}$ such that $\|b-a\|_\infty < \varepsilon$ and b has a right AP factorization with the right AP indices $\mu_1, \dots, \mu_N$.*

Proof. It suffices to prove the assertion in the case where (μ_j) results from (λ_j) by a single elementary operation $\lambda_r \mapsto \lambda_r - \theta$, $\lambda_s \mapsto \lambda_s + \theta$. Let $a = a_-da_+$, $d = \operatorname{diag}(e_{\lambda_j})$ be a right AP factorization of a. For $\eta > 0$, put

$$b_\eta := a_-(d + \eta K)a_+, \quad K(x) := e^{i(\lambda_s+\theta)x}E_{r,s},$$

where $E_{r,s} \in \mathbf{C}^{N\times N}$ is the matrix whose only nonzero entry is a 1 at the intersection of the rth row and the sth column. The identity

$$\begin{pmatrix} e^{i\lambda_r x} & \eta e^{i(\lambda_s+\theta)x} \\ 0 & e^{i\lambda_s x} \end{pmatrix}$$
$$= \begin{pmatrix} 0 & 1 \\ -1/\eta & (1/\eta)e^{-i\theta x} \end{pmatrix} \begin{pmatrix} e^{i(\lambda_r-\theta)x} & 0 \\ 0 & e^{i(\lambda_s+\theta)x} \end{pmatrix} \begin{pmatrix} 1 & 0 \\ e^{i(\lambda_r-\lambda_s-\theta)x} & \eta \end{pmatrix}$$

shows that $d+\eta K$ and thus also b_η has a right AP factorization with the right AP indices $\mu_1, \dots, \mu_N$. We have $\|b_\eta - a\|_\infty = \eta\|a_- K a_+\|_\infty$, and this is smaller than $\varepsilon > 0$ if only $\eta > 0$ is sufficiently small. $\square$

The converse of Theorem 8.20 is also true.

Theorem 8.21. *Let $a \in GAP_{N\times N}$ admit a right AP factorization with the right AP indices $\lambda_1, \dots, \lambda_N$. Then there exists an $\varepsilon > 0$ such that if $b \in GAP_{N\times N}$ has a right AP factorization with the right AP indices $\mu_1, \dots, \mu_N$ and $\|b-a\|_\infty < \varepsilon$, then $(\mu_j) \prec (\lambda_j)$.*

Combining Theorems 8.20 and 8.21 we arrive at the following stability criterion.

Corollary 8.22. *The right AP indices of a matrix function in $GAP_{N\times N}$ are stable under small perturbations (within the class of matrix functions having a right AP factorization) if and only if they are equal to one another.*

A proof of Theorem 8.21 is in [130]. This proof makes use of tools that are beyond the scope of this book. As we will not make use of Theorem 8.21 and Corollary 8.22 in what follows, we omit the proof.

We now prove the APW version of Corollary 8.22. To do this, we need the following result, which is of independent interest. Let $\tilde{P}$ and $\tilde{Q}_0$ be the projections on $APW_{N\times N}$ given by (8.18), and let B_1 stand for the open unit ball of $APW_{N\times N}$,

$$B_1 := \{c \in APW_{N\times N} : \|c\|_W < 1\}.$$

Proposition 8.23. *There exist continuous maps*

$$\Phi_\pm : I - B_1 \to GAPW^\pm_{N\times N}$$

such that $a = \Phi_-(a)\Phi_+(a)$ for all $a \in I - B_1$. In other words, if $a = I - c$ with $c \in APW_{N\times N}$ and $\|c\|_W < 1$, then a admits a canonical right APW factorization $a = a_- a_+$ and the factors $a_\pm$ can be chosen to depend continuously on a. Moreover, $\mathbf{d}(a) = M(a_-)M(a_+)$ also depends continuously on a and

$$\mathbf{d}(a) = \left(\sum_{j=0}^{\infty} M(L_c^j I)\right)^{-1} \tag{8.30}$$

where $L_c x := \tilde{P}(cx)$ for $x \in APW_{N\times N}$.

Proof. Since $\|c\|_W < 1$ and thus $\|L_c\| < 1$, the equation

$$(I - L_c)x = x - \tilde{P}(cx) = I \tag{8.31}$$

has a unique solution $x \in APW_{N\times N}$. As $\tilde{P} = I - \tilde{Q}_0$, this equation can be rewritten in the form $(I-c)x = I - \tilde{Q}_0(cx)$, and on replacing x by $I + \tilde{P}(cx)$ we get

$$(I-c)\big(I + \tilde{P}(cx)\big) = I - \tilde{Q}_0(cx). \tag{8.32}$$

Analogously, the equation $y - \tilde{Q}_0(yc) = I$ has a unique solution $y \in APW_{N\times N}$, and since $\tilde{Q}_0 = I - \tilde{P}$, it follows that

$$\big(I + \tilde{Q}_0(yc)\big)(I-c) = I - \tilde{P}(yc)$$

whence

$$\big(I - \tilde{P}(yc)\big)(I-c)^{-1} = I + \tilde{Q}_0(yc). \tag{8.33}$$

Multiplying (8.32) and (8.33) we obtain

$$\big(I - \tilde{P}(yc)\big)\big(I + \tilde{P}(cx)\big) = \big(I + \tilde{Q}_0(yc)\big)\big(I - \tilde{Q}_0(cx)\big). \tag{8.34}$$

As $\tilde{P}(yc), \tilde{P}(cx) \in APW^+_{N\times N}$ and $\tilde{Q}_0(yc), \tilde{Q}_0(cx) \in APW^-_{N\times N}$ and as $M(\tilde{Q}_0(yc)) = M(\tilde{Q}_0(cx)) = 0$, we see that (8.34) must be equal to the identity matrix. Taking the determinant of (8.34), we arrive at the conclusion that

$$\det\big(I + \tilde{P}(cx)\big) \in GAPW^+, \quad \det\big(I - \tilde{Q}_0(cx)\big) \in GAPW^-,$$

which implies that

$$I + \tilde{P}(cx) \in GAPW^+_{N\times N}, \quad I - \tilde{Q}_0(cx) \in GAPW^-_{N\times N}.$$

Thus, from (8.32) we get the right APW factorization

$$I - c = \Phi_-(I-c)\,\Phi_+(I-c)$$

with

$$\Phi_-(I-c) := I - \tilde{Q}_0(cx), \quad \Phi_+(I-c) := \big(I + \tilde{P}(cx)\big)^{-1}, \quad x = (I-L_c)^{-1}I.$$

Clearly, $\Phi_\pm$ are continuous. Finally, from (8.31) we infer that

$$x = I + \tilde{P}(cx) = (I - L_c)^{-1}I = \sum_{j=0}^{\infty} L_c^j I$$

Since

$$\mathbf{d}(I-c) = M\big(I - \tilde{Q}_0(cx)\big)\Big(M\big(I + \tilde{P}(cx)\big)\Big)^{-1} = \big(M(x)\big)^{-1},$$

we finally obtain (8.30). □

Theorem 8.24. *Let $a \in GAPW_{N\times N}$ have a right APW factorization with the right AP indices $\lambda_1, \ldots, \lambda_N$. Then the following are equivalent:*

(i) *there is an $\varepsilon > 0$ such that every right APW factorable $b \in GAPW_{N\times N}$ with $\|b-a\|_W < \varepsilon$ has the same right AP indices as a;*

(ii) *there is an $\varepsilon > 0$ such that every $b \in GAPW_{N\times N}$ with $\|b-a\|_W < \varepsilon$ has a right APW factorization with the same right AP indices as a;*

(iii) $\lambda_1 = \ldots = \lambda_N$.

Proof. The implication (ii) $\Rightarrow$ (i) is trivial.

Proceeding exactly as in the proof of Theorem 8.20, we see that if $(\mu_j) \prec (\lambda_j)$, then for every $\varepsilon > 0$ there is a $b \in GAPW_{N\times N}$ such that $\|b-a\|_W < \varepsilon$ and b has a right APW factorization with the right AP indices $\mu_1, \ldots, \mu_N$. This gives the implication (i) $\Rightarrow$ (iii).

To prove the implication (iii) $\Rightarrow$ (ii) , let $a = a_- d a_+$, $d = e_\lambda I$ be a right AP factorization of a. Suppose $b \in GAPW_{N\times N}$ and

$$\|d^{-1}a_-^{-1}\|_W \|b-a\|_W \|a_+^{-1}\|_W < 1. \tag{8.35}$$

We have

$$b = a + b - a = a_- d\big[I + d^{-1}a_-^{-1}(b-a)a_+^{-1}\big]a_+,$$

and from (8.35) and Proposition 8.23 we infer that there is a canonical right APW factorization

$$I + d^{-1}a_-^{-1}(b-a)a_+^{-1} = c_- c_+.$$

It follows that $b = a_- d c_- c_+ a_+ = (a_- c_-) d (c_+ a_+)$ is a right APW factorization of b with the AP indices $\lambda, \ldots, \lambda$. □

We conclude this section with two results that will be needed later. Let $\mathcal{G}_{N\times N}$ stand for the set of all matrix functions in $GAPW_{N\times N}$ which have a canonical right APW factorization.

Proposition 8.25. (a) *The set $\mathcal{G}_{N\times N}$ is an open subset of $GAPW_{N\times N}$ which contains $I - B_1$.*

(b) *The map $\mathbf{d} : \mathcal{G}_{N\times N} \to \mathbf{C}^{N\times N}$, $a \mapsto \mathbf{d}(a)$ is continuous.*

(c) *If $a = I - c \in I - B_1$ and the set $\Omega(a) \setminus \{0\}$ is linearly independent over $\mathbf{Z}_+$, then $\mathbf{d}(a) = M(a)$.*

Proof. (a) This is immediate from Proposition 8.23 and Theorem 8.24.

(b) Fix $a \in \mathcal{G}_{N\times N}$ and let $a = a_- a_+$ be a canonical right APW factorization. Let $b \in APW_{N\times N}$ satisfy the inequality

$$\|a_-^{-1}\|_W \|b-a\|_W \|a_+^{-1}\|_W < 1.$$

Then, as in the proof of Theorem 8.24,

$$b = a_-\left(I + a_-^{-1}(b-a)a_+^{-1}\right)a_+,$$

and from Proposition 8.23 we know that there is a canonical right APW factorization

$$I + a_-^{-1}(b-a)a_+^{-1} = c_-c_+$$

such that $c_\pm$ depend continuously on b. As

$$\mathbf{d}(b) = M(a_-)M(c_-)M(c_+)M(a_+),$$

it follows that $\mathbf{d}(b)$ is also a continuous function of b.

(c) By (8.30), $\mathbf{d}(a) = (\sum_{j=0}^{\infty} M(L_c^j I))^{-1}$. Suppose first that $M(c) = 0$, so that $M(a) = I$. Let Δ be the collection of all finite sums

$$\sum n_k\lambda_k \text{ where } n_k \in \mathbf{Z}_+, \quad \sum n_k > 0, \ \lambda_k \in \Omega(c).$$

Clearly, $\Omega(L_c I) = \Omega(\tilde{P}c) \subset \Delta$. If $\Omega(L_c^j I) \subset \Delta$, then

$$\Omega(L_c^{j+1} I) = \Omega\big(\tilde{P}(cL_c^j I)\big) \subset \Delta,$$

which shows that $\Omega(L_c^j I) \subset \Delta$ for all $j \geq 1$. Since $\Omega(c) = \Omega(a) \setminus \{0\}$ is linearly independent over $\mathbf{Z}_+$, the set Δ does not contain the origin. Hence, $M(L_c^j I) = 0$ for all $j \geq 1$ and therefore

$$\mathbf{d}(a) = \left(\sum_{j=0}^{\infty} M(L_c^j I)\right)^{-1} = (M(I))^{-1} = I = M(a).$$

Now assume $c_0 := M(c) \neq 0$. Put $a_0 := I - c_0$. Because $\|c_0\| \leq \|c\|_W < 1$, the matrix a_0 is invertible. Consider $b := a_0^{-1}a$. We have

$$b = a_0^{-1}\big(I - c_0 - (c - c_0)\big) = I - a_0^{-1}(c - c_0)$$

with $M(a_0^{-1}(c-c_0)) = 0$. As $\|c_0\| + \|c - c_0\|_W = \|c\|_W < 1$, we obtain that $\|c - c_0\|_W < 1 - \|c_0\|$, whence

$$\begin{aligned} \|a_0^{-1}(c-c_0)\|_W &\leq \|a_0^{-1}\| \, \|c - c_0\|_W = \|(I - c_0)^{-1}\| \, \|c - c_0\|_W \\ &< \left(\sum_{j=0}^{\infty} \|c_0\|^j\right)(1 - \|c_0\|) = 1. \end{aligned}$$

Thus, by what was proved in the preceding paragraph, $\mathbf{d}(b) = I$. Since $\mathbf{d}(a) = \mathbf{d}(a_0 b) = a_0\mathbf{d}(b)$, it follows that $\mathbf{d}(a) = a_0 = M(a)$. □

Proposition 8.26. *If $a \in GAP_{N\times N}$ has a right AP factorization, then all right AP indices belong to $[\inf \Omega(a), \sup \Omega(a)]$, the closed convex hull of the Bohr-Fourier spectrum $\Omega(a)$ of a.*

Proof. Let $a = a_- d a_+$ be a right AP factorization. Put $\omega := \inf \Omega(a)$. If $\omega = -\infty$, then clearly $\omega \le \lambda_j$ for all right AP indices λ_j. If $\omega > -\infty$, we can multiply the factorization $a = a_- d a_+$ by $e_{-\omega}$ to reduce the general case to the special case where $\omega = 0$. We have $aa_+^{-1} = a_- d$, the condition $\omega = 0$ implies that $\Omega(a) \subset [0, \infty)$, Corollary 7.6 gives $a \in AP^+_{N\times N}$, whence $aa_+^{-1} \in AP^+_{N\times N}$ and thus $a_- d \in AP^+_{N\times N}$. If some index λ_j is negative, then the jth column of $a_- d$ is in AP_N^- and has zero mean value. Consequently, the jth column of $aa_+^{-1} \in AP^+_{N\times N}$ also lies in AP_N^- and has mean value zero. Because $H_+^\infty \cap H_-^\infty = \mathbf{C}$, this is only possible if this column is identically zero, which contradicts the invertibility of aa_+^{-1}. Thus, $\lambda_j \ge 0 = \omega$ for all j. Analogously one can show that $\lambda_j \le \sup \Omega(a)$. □

Notes

The bulk of the results of this chapter was obtained in the eighties and announced in [128]. The detailed manuscript of that paper was deposited at VINITI as [130], and a shortened version of the latter manuscript was published later as [132]. In particular, Theorems 8.1, 8.14 and 8.21 were not included in [132]. Theorem 8.16 is borrowed from [147]; see also [76, Lemma 4.2]. Propositions 8.23, 8.25, and Theorem 8.24 are from [135], [212]. The material of Section 8.3 is of course well known and was mainly included for illustrative purposes. In particular, Theorem 8.11 follows from the general results on factorization in decomposing Banach algebras exhibited in [57], [92], [155].

The stability of the partial indices of Wiener-Hopf factorization was already studied by Gohberg and Krein [94]; they showed that the stable sets are just the sets in which the indices (which are integers) differ from each other by at most 1. The results of the last section concern stability properties of AP factorization under *uniformly* small perturbations. In fact, much stronger results can be proved. Namely, the canonical *APW* factorization, if it exists, is stable under simultaneous perturbations of the Bohr-Fourier coefficients *and* the Bohr-Fourier spectrum. To formulate the result, let us first fix some notation.

Let $a \in APW_{N\times N}$ be given in the form

$$a = \sum_{\lambda \in \Omega(a)} a_\lambda e_\lambda, \tag{8.36}$$

where $a_\lambda \in \mathbf{C}_{N\times N}$, $\|a\|_W = \sum \|a_\lambda\| < \infty$, and the set $\Omega(a)$ is contained in an Abelian group $\Sigma = \Sigma(a)$ generated by an at most countable set $\Gamma(a)$ of positive real numbers that are linearly independent over $\mathbf{Z}$. (We are greatful to Sergei Favorov for pointing out an inaccuracy we made with $\Omega(a)$ in [123] and in an earlier version of the manuscript of this book.)

We associate with each $\gamma \in \Gamma(a)$ a positive real number γ' and consider the Abelian group Σ' generated by $\Gamma'(a) = \{\gamma' : \gamma \in \Gamma(a)\}$. The map $\nu : \Gamma(a) \to \Gamma'(a)$, $\gamma \mapsto \gamma'$ can be extended to a homomorphism $\nu : \Sigma \to \Sigma'$ which, in its turn, induces a Banach algebra homomorphism

$$\psi_\nu : [APW_\Sigma]_{N\times N} \to [APW_{\Sigma'}]_{N\times N}, \quad \sum_{\lambda\in\Sigma} a_\lambda e_\lambda \mapsto \sum_{\lambda\in\Sigma'} a_\lambda e_{\nu(\lambda)},$$

where $[APW_\Sigma]_{N\times N} = \{G \in APW_{N\times N} : \Omega(G) \subset \Sigma\}$.

Theorem 8.27. *Let* $a \in [APW_\Sigma]_{N\times N}$,

$$a = \sum_{\lambda\in\Omega(a)} a_\lambda e_\lambda,$$

and let a *have a canonical APW factorization. Then for each* $\lambda \in \Omega(a)$ *and for each* $\gamma \in \Gamma(a)$, *there is a neighborhood* $U(a_\lambda)$ *of* a_λ *in* $\mathbf{C}_{N\times N}$ *and a neighborhood* U_γ *of* γ *in* $\mathbf{R}$ *such that, for every* $b_\lambda \in U(a_\lambda)$ *and for every* $\gamma' \in U_\gamma$, *the matrix function*

$$b = \sum_{\lambda\in\Omega(a)} b_\lambda e_{\nu(\lambda)}, \tag{8.37}$$

has a canonical APW factorization. □

In the case $b_\lambda = a_\lambda$, Theorem 8.27 was announced in [123]. It was published in the present form with full a proof in [23]; its particular case for *AP* polynomials is contained in [134]. Scrutinizing the proofs in [23], [134], it is possible to see in addition that, under the conditions of Theorem 8.27, the transformation $a \mapsto \mathbf{d}(a)$ is continuous. Moreover, if a is an *AP* polynomial and in (8.37) $b_\lambda = a_\lambda$, then $\mathbf{d}(a) = \mathbf{d}(b)$. Note that [23], [123] contain applications of this result to canonical *AP* factorization not included in the present book.

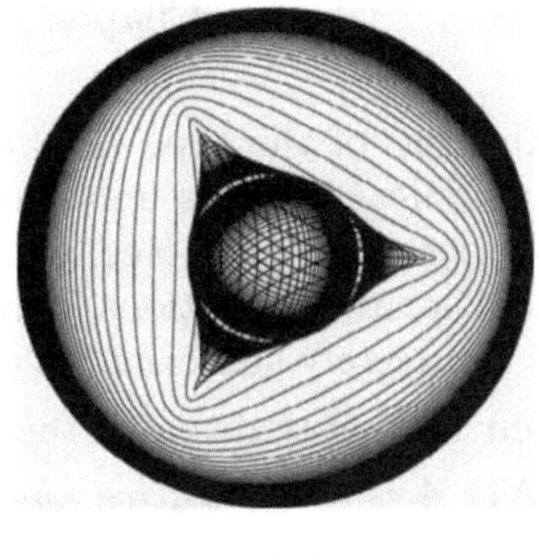

Chapter 9

Matrix Wiener-Hopf Operators with APW Symbols

With this chapter we reach the first big goal on our trek. If a matrix function $a \in APW_{N\times N}$ has a canonical right APW factorization $a = a_- a_+$, then $W(a)$ is obviously invertible on $L^2_N(\mathbf{R})$. We here prove that the converse is also true, that is, the symbol a necessarily has a canonical right APW factorization if $W(a)$ is invertible. The proof of this result requires some heavy tools. We first establish an isomorphism theorem for C^*-dynamical systems, which allows us to show that the invertibility of $W(a)$ on $L^2_N(\mathbf{R}_+)$ is equivalent to the invertibility of a certain operator on the vector-valued Besicovitch space B^2_N. After that we employ the Bochner-Phillips theorem on the inverse closedness of certain matrix algebras in order to prove that the invertibility of the latter operator implies the canonical right APW factorability of a.

9.1 An Isomorphism Theorem for C^*-Dynamical Systems

Let H be a Hilbert space, let A be a C^*-subalgebra of $\mathcal{B}(H)$ which contains the identity operator I, let G be a discrete group, and let

$$u : G \to \mathcal{B}(H),\ g \mapsto u_g$$

be a unitary representation of G (= a group homomorphism of G into the group of all unitary operators on H). We denote by $D := \operatorname{alg}(A, u_G)$ the smallest C^*-subalgebra of $\mathcal{B}(H)$ that contains $A \cup \{u_g : g \in G\}$.

Algebras of the type $D = \text{alg}(A, u_G)$ are called C^*-algebras associated with C^*-dynamical systems and are of great interest in connection with many problems of operator theory (see, e.g., [4], [5], [173]). They have been studied under several additional assumptions for a long time. We here consider a very special case, which is completely sufficient for our purposes.

Our first assumption is the following requirement:

(A1) *A is commutative and G is Abelian.*

Furthermore, we assume that

(A2) *if $a \in A$, then $u_g a u_g^* \in A$ for every $g \in G$.*

By (A2), the map $\alpha_g : a \mapsto u_g a u_g^*$ maps A onto itself. It is easily seen that α_g is in fact a C^*-algebra automorphism of A. Since A is a unital commutative C^*-algebra, we may identify A with $C(M)$, where M is the maximal ideal space (Theorem 1.7(a)). As every automorphism of $C(M)$ is induced by a homeomorphism of M, we can for each $g \in G$ find a homeomorphism $\beta_g : M \to M$ such that

$$(\alpha_g a)(m) = a\big(\beta_g(m)\big) \quad \text{for all } a \in C(M) \text{ and all } m \in M.$$

Our third assumption is that

(A3) *G acts topologically free on M,*

which, by definition, means that for every finite set $\{g_1, \dots, g_k\} \subset G$ and every open set $U \subset M$ there is a point $m_0 \in U$ such that the points $\beta_{g_1}(m_0), \dots, \beta_{g_k}(m_0)$ are pairwise distinct.

Let $\tilde{H}$ be a second Hilbert space, let $\tilde{A}$ be a C^*-subalgebra of $\mathcal{B}(\tilde{H})$ which contains the identity operator, and let $\tilde{u} : G \to \mathcal{B}(\tilde{H}),\ g \mapsto \tilde{u}_g$ be a unitary representation of the group G. Put $\tilde{D} := \text{alg}\,(\tilde{A}, \tilde{u}_G)$. We finally assume that

(A4) *there is a C^*-algebra isomorphism $\varphi : A \to \tilde{A}$ such that*

$$\varphi(u_g a u_g^*) = \tilde{u}_g \varphi(a) \tilde{u}_g^*$$

for all $g \in G$ and all $a \in A$.

Theorem 9.1. *Under the assumptions* (A1) – (A4) *there is a (unique) C^*-algebra isomorphism $\Phi : D \to \tilde{D}$ such that*

$$\Phi(u_g) = \tilde{u}_g \ \textit{for}\ g \in G \ \textit{and}\ \Phi|A = \varphi. \tag{9.1}$$

Proof. Let D^0 and $\tilde{D}^0$ be the set of all finite sums of the form

$$\sum_{g\in G} a_g u_g \ (a_g \in A) \quad \text{and} \quad \sum_{g\in G} \tilde{a}_g \tilde{u}_g \ (\tilde{a}_g \in \tilde{A}),$$

respectively. From (A2) we infer that D is the closure of D^0. Conditions (A2) and (A4) imply that $\tilde{\alpha}_g : \tilde{a} \mapsto \tilde{u}_g \tilde{a} \tilde{u}_g^*$ is a C^*-algebra automorphism of $\tilde{A}$, which, in turn, shows that $\tilde{D}^0$ is dense in $\tilde{D}$. Our aim is to show that every $b \in D^0$ can be uniquely written in the form $b = \sum a_g u_g$ ($a_g \in A$) and that the map

$$\Phi : D^0 \to \tilde{D}^0, \sum a_g u_g \mapsto \sum \varphi(a_g)\tilde{u}_g \tag{9.2}$$

extends to a C^*-algebra isomorphism of D onto $\tilde{D}$.

Let G_0 be a finite subset of G and suppose e, the identity of G, belongs to G_0. Let a_g ($g \in G_0$) be arbitrary elements of A and put $b = \sum_{g \in G_0} a_g u_g$. We claim that

$$\|b\| = \left\|\sum a_g u_g\right\| \geq \|a_e\|. \tag{9.3}$$

Pick $\varepsilon \in (0, \|a_e\|/2)$. There is an $m_0 \in M$ such that $\|a_e\| = |a_e(m_0)|$, and if V is a sufficiently small open neighborhood of m_0, then $|a_e(m)| \geq \|a_e\| - \varepsilon$ for all $m \in V$. Condition (A3) guarantees the existence of a point $m_1 \in V$ such that the points $\beta_g(m_1)\,(g \in G_0)$ are pairwise distinct. Hence, if $W \subset V$ is a sufficiently small open neighborhood of m_1, then $\beta_g(W) \cap \beta_h(W) = \emptyset$ whenever $g, h \in G_0$ and $g \neq h$. By Uryson's lemma, there is a $\psi \in A$ such that $0 \leq \psi \leq 1$ on M, $\psi | M \setminus W = 0$, and $\psi(m_1) = 1$. Consequently, $\|\psi a_e\| \geq |a_e(m_1)| \geq \|a_e\| - \varepsilon$.

Now recall that the elements of A are bounded operators on H. From what was proved in the preceding paragraph, we see that there is a $\xi_0 \in H$ such that

$$\|\xi_0\| = 1, \quad \|\psi a_e \xi_0\| \geq \|a_e\| - 2\varepsilon.$$

We have $(\alpha_g(\psi)\alpha_h(\psi))(m) = \psi(\beta_g(m))\psi(\beta_h(m))$. If this is nonzero, then $\beta_g(m) \in W = \beta_e(W)$ and $\beta_h(m) \in W = \beta_e(W)$, which, by the construction of W and the equality $\beta_h(m) = \beta_{hg^{-1}}(\beta_g(m))$, implies that $h = g$ whenever $g, h \in G_0$. Thus, $\alpha_g(\psi)\alpha_h(\psi) = 0$ whenever $g, h \in G_0$ and $g \neq h$. It results that in the sum $\sum a_g u_g \psi \xi_0$ the summands are pairwise orthogonal:

$$\begin{aligned}
&(a_g u_g \psi \xi_0, a_h u_h \psi \xi_0) = (\psi u_h^* a_h^* a_g u_g \psi \xi_0, \xi_0) \\
&= \big(u_h^*(u_h \psi u_h^*) a_h^* a_g (u_g \psi u_g^*) u_g \xi_0, \xi_0\big) \\
&= \big(u_h^*(u_h \psi u_h^*)(u_g \psi u_g^*) a_h^* a_g u_g \xi_0, \xi_0\big) \quad \text{(by (A1),(A2))} \\
&= \big(u_h^* \alpha_h(\psi)\alpha_g(\psi) a_h^* a_g u_g \xi_0, \xi_0\big) = 0.
\end{aligned}$$

Thus,

$$\begin{aligned}
\|b\psi\xi_0\|^2 &= \left\|\sum a_g u_g \psi \xi_0\right\|^2 = \sum \|a_g u_g \psi \xi_0\|^2 \\
&\geq \|a_e u_e \psi \xi_0\|^2 = \|a_e \psi \xi_0\|^2 \geq \big(\|a_e\| - 2\varepsilon\big)^2,
\end{aligned}$$

and as $\|\psi\xi_0\| \leq \|\psi\|\,\|\xi_0\| = 1$ and $\varepsilon \in (0, \|a_e\|/2)$ was arbitrary, we arrive at our claim (9.3).

From (9.3) we obtain that for every $h \in G_0$,

$$\|a_h\| \leq \left\|\sum a_g u_g u_h^*\right\| = \left\|\left(\sum a_g u_g\right) u_h^*\right\| = \left\|\sum a_g u_g\right\|.$$

Consequently, for each $h \in G$ the map

$$N_h : D^0 \to A, \sum a_g u_g \mapsto a_h$$

extends to a bounded linear operator of norm 1 of D onto A. Since, by (A4), for the algebra $\tilde{D} = \operatorname{alg}(\tilde{A}, \tilde{u}_G)$ the analogues of (A1) to (A3) are valid, we can repeat the above reasoning to show that for each $h \in G$ the map

$$\tilde{N}_h : \tilde{D}^0 \to \tilde{A}, \sum \tilde{a}_g \tilde{u}_g \mapsto \tilde{a}_h$$

extends to a bounded linear operator of norm 1 of $\tilde{D}$ onto $\tilde{A}$.

Now define Φ by (9.2). It is clear that Φ is linear and that Φ preserves adjoints. From the equalities

$$\begin{aligned}\Phi(a_g u_g a_h u_h) &= \Phi(a_g u_g a_h u_g^* u_{gh}) \\ &= \varphi(a_g u_g a_h u_g^*)\tilde{u}_{gh} \quad \text{(by (A2))} \\ &= \varphi(a_g)\tilde{u}_g \varphi(a_h)\tilde{u}_g^* \tilde{u}_g \tilde{u}_h \quad \text{(by (A4))} \\ &= \Phi(a_g u_g)\Phi(a_h u_h)\end{aligned}$$

we see that Φ is multiplicative. We now show that Φ is bounded. Fix an operator $b = \sum_{g \in F} a_g u_g$ in D^0. We have

$$\begin{aligned}\|\Phi(b)\|^{2n} &= \|\Phi(b)\Phi(b^*)\|^n \\ &= \big\|\big(\Phi(b)\Phi(b^*)\big)^n\big\| \quad \text{(because } \Phi(b)\Phi(b^*) \text{ is selfadjoint)} \\ &= \big\|\Phi\big((bb^*)^n\big)\big\|.\end{aligned} \tag{9.4}$$

Write $(bb^*)^n = \sum_{g \in F_n} c_g^{(n)} u_g$. Let $|F_n|$ be the number of elements in F_n. Then

$$\begin{aligned}\big\|\Phi\big((bb^*)^n\big)\big\| &\le \sum_{g \in F_n} \|c_g^{(n)}\| = \sum_{g \in F_n} \big\|N_g\big((bb^*)^n\big)\big\| \\ &\le \sum_{g \in F_n} \|(bb^*)^n\| = |F_n|\,\|(bb^*)^n\| \\ &= |F_n|\,\|bb^*\|^n \quad \text{(because } bb^* \text{ is selfadjoint)} \\ &= |F_n|\,\|b\|^{2n}.\end{aligned} \tag{9.5}$$

On the other hand, because $\|\tilde{N}_h\| = 1$, we get

$$\begin{aligned}\big\|\Phi\big((bb^*)^n\big)\big\| &\ge \max_{g \in F_n} \|\varphi(c_g^{(n)})\| \ge \max_{g \in F_n} \|c_g^{(n)}\| \\ &= \max_{g \in F_n} \big\|N_g\big((bb^*)^n\big)\big\| \ge |F_n|^{-1}\|(bb^*)^n\| = |F_n|^{-1}\|b\|^{2n}.\end{aligned} \tag{9.6}$$

Combining (9.4), (9.5), and (9.6) we obtain

$$|F_n|^{-1/n}\|b\|^2 \le \|\Phi(b)\|^2 \le |F_n|^{1/n}\|b\|^2 \quad \text{for all } n \in \mathbf{N}.$$

Suppose $|F| = k$. By conditions (A1) and (A2), we have $|F_n| \le \pi_k(n)$ where $\pi_k(n)$ is the number of all partitions

$$n = l_1 + \ldots + l_k \text{ with } l_1, \ldots, l_k \in \mathbf{Z}_+.$$

Clearly, $\pi_1(n) = 1$ and

$$\pi_k(n) = \pi_{k-1}(n) + \pi_{k-1}(n-1) + \ldots + \pi_{k-1}(0) < (n+1)\pi_{k-1}(n),$$

whence $\pi_k(n) < (n+1)^{k-1}$ and thus $|F_n|^{1/n} \to 1$ as $n \to \infty$. It follows that $\|\Phi(b)\| = \|b\|$ as desired.

The map Φ has a unique continuous extension from D^0 to all of D. We denote this extension also by Φ. By what was already proved, Φ is an isometric C^*-algebra homomorphism of D into $\tilde{D}$ satisfying (9.1). We are therefore left to prove that Φ is surjective. But since the image of a C^*-algebra homomorphism is always closed (Proposition 1.5(c)) and since $\tilde{D}^0 \subset \Phi(D^0)$ is dense in $\tilde{D}$, we see that Φ is surjective. □

9.2 Passage to the Besicovitch Space

Recall that P, Q are the projections given by (2.10) to (2.14).

Lemma 9.2. *Let $a \in L^\infty_{N\times N}(\mathbf{R}) \setminus \{0\}$. Then the operator $W(a)$ on $L^2_N(\mathbf{R}_+)$ is strongly Φ-equivalent to the operator $aP + Q$ on $L^2_N(\mathbf{R})$.*

Proof. Obviously, $W(a) \in \mathcal{B}(L^2_N(\mathbf{R}+))$ is strongly Φ-equivalent to the operator $\chi_+ W^0(a)\chi_+ + I - \chi_+$ on $L^2_N(\mathbf{R})$, and the latter operator is strongly Φ-equivalent to

$$\begin{aligned}&F(\chi_+ W^0(a)\chi_+ + I - \chi_+)F^{-1}\\ &= F\chi_+F^{-1}aF\chi_+F^{-1} + I - F\chi_+F^{-1} = PaP + Q.\end{aligned}$$

The assertion now follows from the identity

$$aP + Q = (PaP + Q)(I + QaP)$$

and the fact that $I + QaP$ has the inverse $I - QaP$. □

Let $\mathcal{D} := \text{alg}\,(P, AP_{N\times N})$ be the smallest closed C^*-subalgebra of $\mathcal{B}(L^2_N(\mathbf{R}))$ which contains P and the set $\{aI : a \in AP_{N\times N}\}$. Put

$$\chi_+ := \chi_{[0,\infty)}, \quad \chi_+^0 := \chi_{(0,\infty)}, \quad \chi_- := \chi_{(-\infty,0]}, \quad \chi_-^0 := \chi_{(-\infty,0)},$$

let $F_B : l^2_N(\mathbf{R}) \to B^2_N$ be the Bohr-Fourier transform, and define

$$\tilde{P} := F_B\chi_+F_B^{-1}, \quad \tilde{P}_0 := F_B\chi_+^0F_B^{-1}, \quad \tilde{Q} := F_B\chi_-F_B^{-1}, \quad \tilde{Q}_0 := F_B\chi_-^0F_B^{-1}.$$

These are four bounded projections on the $\mathbf{C}^N$-valued Besicovitch space B^2_N. If $a \in AP_{N\times N} = C_{N\times N}(\mathbf{R}_B)$, then the operator aI of multiplication by a is obviously bounded on $B^2_N = L^2_N(\mathbf{R}_B, d\mu)$. Let $\tilde{\mathcal{D}}$ (resp. $\tilde{\mathcal{D}}_0$) be the smallest C^*-subalgebra of $\mathcal{B}(B^2_N)$ which contains $\tilde{P}$ (resp. $\tilde{P}_0$) and the set $\{aI : a \in AP_{N\times N}\}$.

Theorem 9.3. *The C^*-algebras $\mathcal{D}, \tilde{\mathcal{D}}, \tilde{\mathcal{D}}_0$ are (isometrically) isomorphic. To be more precise, there are (unique) C^*-algebra isomorphisms $\Phi : \mathcal{D} \to \tilde{\mathcal{D}}$ and $\Phi_0 : \mathcal{D} \to \tilde{\mathcal{D}}_0$ such that*

$$\Phi(P) = \tilde{P}, \quad \Phi_0(P) = \tilde{P}_0, \quad \Phi(aI) = \Phi_0(aI) = aI \quad (a \in AP_{N\times N}).$$

Proof. We only prove the isomorphy of $\mathcal{D}$ and $\tilde{\mathcal{D}}$ (the isomorphy of $\mathcal{D}$ and $\tilde{\mathcal{D}}_0$ can be shown analogously). Furthermore, it suffices to consider the case $N = 1$.

Denote by $\mathcal{A}$ (resp. $\tilde{\mathcal{A}}$) the smallest C^*-subalgebra of $\mathcal{D}$ (resp. $\tilde{\mathcal{D}}$) which contains the set

$$\{I\} \cup \{e_\lambda P e_{-\lambda} I : \lambda \in \mathbf{R}\} \subset \mathcal{B}(L^2(\mathbf{R}))$$
$$\big(\text{resp. } \{I\} \cup \{e_\lambda \tilde{P} e_{-\lambda} I : \lambda \in \mathbf{R}\} \subset \mathcal{B}(B^2)\big).$$

Clearly, the maps

$$u : \mathbf{R} \to \mathcal{B}(L^2(\mathbf{R})), \quad \lambda \mapsto u_\lambda := e_\lambda I,$$
$$\tilde{u} : \mathbf{R} \to \mathcal{B}(B^2), \quad \lambda \mapsto \tilde{u}_\lambda := e_\lambda I$$

are unitary representations. It is easily seen that

$$\mathcal{D} = \operatorname{alg}(\mathcal{A}, u_{\mathbf{R}}), \quad \tilde{\mathcal{D}} = \operatorname{alg}(\tilde{\mathcal{A}}, \tilde{u}_{\mathbf{R}}).$$

To establish that $\mathcal{D}$ and $\tilde{\mathcal{D}}$ are isomorphic, we will employ Theorem 9.1.

The map $\Psi : \mathcal{A} \to \mathcal{B}(L^2(\mathbf{R}))$, $A \mapsto F^{-1}AF$ is an injective C^*-algebra homomorphism. Clearly, $\Psi(I) = I$, and a direct computation shows that for $\lambda \in \mathbf{R}$,

$$\Psi(e_\lambda P e_{-\lambda} I) = U_\lambda \chi_+ U_{-\lambda} = \chi_{[\lambda,\infty)} I,$$

where $(U_\mu f)(t) := f(t-\mu)$. The C^*-algebra generated by $\{\chi_{[\lambda,\infty)} : \lambda \in \mathbf{R}\}$ is PC (recall Section 1.5), and on identifying the multiplication operator $\chi_{[\lambda,\infty)} I$ with the function $\chi_{[\lambda,\infty)}$, we see that Ψ is a C^*-algebra isomorphism of $\mathcal{A}$ onto PC. This tells us in particular that (A1) is satisfied.

If $\lambda, \mu \in \mathbf{R}$, then $\alpha_\mu(I) = e_\mu I e_{-\mu} I = I$ and

$$\alpha_\mu(e_\lambda P e_{-\lambda} I) = e_{\lambda+\mu} P e_{-(\lambda+\mu)} I$$

which reveals that (A2) is valid. Moreover, since

$$\Psi(e_\lambda P e_{-\lambda} I) = \chi_{[\lambda,\infty)} I, \quad \Psi\big(\alpha_\mu(e_\lambda P e_{-\lambda} I)\big) = \chi_{[\lambda+\mu,\infty)},$$

we see that

$$\big(\Psi(\alpha_\mu A)\big)(t) = \big(\Psi(A)\big)(t-\mu), \quad t \in \mathbf{R},\ \mu \in \mathbf{R} \tag{9.7}$$

for all $A \in \mathcal{A}$. The maximal ideal space M of PC may be identified with $\dot{\mathbf{R}} \times \{0,1\}$ (with the topology induced by the exotic topology described after Theorem 2.21) in the following sense: for $a \in PC$ we have

$$\begin{aligned} a(t,0) &= a(t-0), \quad a(t,1) = a(t+0) \quad (t \in \mathbf{R}),\\ a(\infty,0) &= a(+\infty), \quad a(\infty,1) = a(-\infty). \end{aligned}$$

From (9.7) we infer that for $\mu \in \mathbf{R}$ the homeomorphism $\beta_\mu : M \to M$ acts by the rule

$$\begin{aligned} \beta(t,0) &= (t-\mu,0), \quad \beta(t,1) = (t-\mu,1) \quad (t \in \mathbf{R}),\\ \beta(\infty,0) &= (\infty,0), \quad \beta(\infty,1) = (\infty,1). \end{aligned}$$

Since every open set $U \subset M$ contains a set of the form $\gamma \times \{0,1\}$ where γ is an open subset of $\mathbf{R}$, it is obvious that $\mathbf{R}$ acts topologically free on M. Thus, (A3) is also true.

We now consider $\tilde{\mathcal{D}} = \operatorname{alg}(\tilde{\mathcal{A}}, \tilde{u}_{\mathbf{R}})$. In this case the map

$$\tilde{\Psi} : \tilde{\mathcal{A}} \to \mathcal{B}\big(l^2(\mathbf{R})\big),\ A \mapsto F_B^{-1} A F_B$$

is an injective C^*-algebra homomorphism. As above, one can straightforwardly verify that

$$\tilde{\Psi}(e_\lambda \tilde{P} e_{-\lambda} I) = U_\lambda \chi_+ U_\lambda = \chi_{[\lambda,\infty)} I,$$

and after identifying $\chi_{[\lambda,\infty)} I$ and $\chi_{[\lambda,\infty)}$, we arrive at the conclusion that $\tilde{\Psi}$ is a C^*-algebra isomorphism of $\tilde{\mathcal{D}}$ onto PC. Thus, $\varphi := \tilde{\Psi}^{-1}\Psi : \mathcal{A} \to \tilde{\mathcal{A}}$ is a C^*-algebra isomorphism. We have $\varphi(e_\lambda P e_{-\lambda} I) = e_\lambda \tilde{P} e_{-\lambda} I$ and therefore,

$$\varphi(e_\mu e_\lambda P e_{-\lambda} e_{-\mu} I) = e_\mu \varphi(e_\lambda P e_{-\lambda} I) e_{-\mu} I,$$

which shows that (A4) holds. The assertion is now an immediate consequence of Theorem 9.1. □

Corollary 9.4. *If $a \in AP_{N\times N}$, then*

$$\begin{aligned} W(a) \in G\mathcal{B}\big(L^2_N(\mathbf{R}_+)\big) &\iff aP + Q \in G\mathcal{B}\big(L^2_N(\mathbf{R})\big)\\ &\iff a\tilde{P} + \tilde{Q}_0 \in G\mathcal{B}(B^2_N) \iff a\tilde{P}_0 + \tilde{Q} \in G\mathcal{B}(B^2_N). \end{aligned}$$

Proof. The first equivalence follows from Lemma 9.2 and the other two equivalences result from Theorem 9.3 in conjunction with Proposition 1.4(a). □

9.3 Application of the Bochner-Phillips Theorem

Let G be a locally compact Abelian group, let G^* be the dual group, and let $G_B = (G_d^*)^*$ be the Bohr compactification of G. Further, suppose E is a unital Banach algebra. The uniform closure of the set of all E-valued trigonometric polynomials on G,

$$p : G \to E, \quad p(g) = \sum_j p_j \chi_j(g), \quad p_j \in E, \quad \chi_j \in G^*,$$

is denoted by $AP(G, E)$ (recall Section 7.1 for $E = \mathbf{C}$). One can show that $AP(G, E)$ may be identified with $C(G_B, E)$, i.e., with the continuous functions of G_B into E. Thus, $AP(G, E) = C(G_B, E)$. Let $APW(G, E)$ stand for the set of all functions $a \in AP(G, E)$ with absolutely convergent Fourier series: a is in $APW(G, E)$ if and only if

$$a(g) = \sum_j a_j \chi_j(g), \quad a_j \in E, \quad \chi_j \in G^*, \quad \|a\|_W := \sum_j \|a_j\|_E < \infty.$$

On identifying G^{**} and G via the equalities $g(\chi) = \chi(g)$ ($g \in G$, $\chi \in G^*$), we can write functions $a \in APW(G, E)$ and their extensions to G_B in the form

$$a(g) = \sum_j a_j \chi_j(g) = \sum_j a_j g(\chi_j) \quad (g \in G),$$

$$a(x) = \sum_j a_j \chi_j(x) = \sum_j a_j x(\chi_j) \quad (x \in G_B = (G_d^*)^*),$$

respectively. The set $APW(G, E)$ is frequently also denoted by $W(G_B, E)$. With pointwise operations and the norm $\|\cdot\|_W$, the set $APW(G, E) = W(G_B, E)$ is a Banach algebra.

Theorem 9.5 (Bochner and Phillips). *The algebra $W(G_B, E)$ is inverse closed in $C(G_B, E)$, that is, if $a \in W(G_B, E)$ is invertible in $C(G_B, E)$, then the inverse a^{-1} belongs to $W(G_B, E)$.*

We denote by $\mathcal{R}^W_{N\times N}$ the set of all operators $A \in \mathcal{B}(B^2_N)$ which can be represented in the form

$$A = \sum_{\lambda \in \mathbf{R}} \psi(a_\lambda) e_\lambda I, \quad a_\lambda \in l^\infty_{N\times N}(\mathbf{R}), \quad \sum_{\lambda \in \mathbf{R}} \|a_\lambda\|_\infty < \infty, \tag{9.8}$$

where $a_\lambda \neq 0$ for at most countably many $\lambda \in \mathbf{R}$ and $\psi(a) = F_B a F_B^{-1}$ is as in Section 7.4. Since

$$e_\lambda \psi(a) = \psi(U_\lambda a) e_\lambda I, \quad (U_\lambda a)(\mu) := a(\mu - \lambda),$$

it follows that $\mathcal{R}^W_{N\times N}$ is an algebra. With the norm

$$\|A\|_W := \sum_{\lambda\in\mathbf{R}} \|\psi(a_\lambda)\| = \sum_{\lambda\in\mathbf{R}} \|a_\lambda\|_\infty,$$

$\mathcal{R}^W_{N\times N}$ is a Banach algebra.

Theorem 9.6. *The algebra $\mathcal{R}^W_{N\times N}$ is inverse closed in $\mathcal{B}(B^2_N)$, i.e., if an operator in $\mathcal{R}^W_{N\times N}$ is invertible, then its inverse belongs to $\mathcal{R}^W_{N\times N}$.*

Proof. Let $A \in \mathcal{R}^W_{N\times N}$ be given by (9.8) and suppose $A \in G\mathcal{B}(B^2_N)$. Put $E = \mathcal{B}(B^2_N)$ and consider the function

$$a : \mathbf{R}_B \to E,\ a(\chi) = \sum_{\lambda\in\mathbf{R}} \chi(\lambda)\psi(a_\lambda)e_\lambda I.$$

Obviously, $a \in W(\mathbf{R}_B, E)$. Let $T_h := F_B e_h F_B^{-1}$ be as in Section 7.4. From (7.8) we see that

$$T_h e_\lambda T_h^{-1} = e^{i\lambda h} e_\lambda I. \tag{9.9}$$

As $T_h\psi(a_\lambda) = \psi(a_\lambda)T_h$ (Proposition 7.10), we therefore get for $h \in \mathbf{R}$,

$$\begin{aligned} a(e_h) &= \sum_{\lambda\in\mathbf{R}} e^{i\lambda h}\psi(a_\lambda)e_\lambda I = \sum_{\lambda\in\mathbf{R}} \psi(a_\lambda)T_h e_\lambda T_h^{-1} \\ &= T_h\Big(\sum_{\lambda\in\mathbf{R}} \psi(a_\lambda)e_\lambda I\Big)T_h^{-1} = T_h A T_h^{-1}. \end{aligned}$$

Since A is invertible, it follows that $a(e_h)$ is invertible and that $\|a^{-1}(e_h)\| = \|A^{-1}\|$ for all $h \in \mathbf{R}$. As $\mathbf{R}$ is dense in $\mathbf{R}_B$, we see that a is invertible in $C(\mathbf{R}_B, E)$. Consequently, by Theorem 9.5, a is invertible in $W(\mathbf{R}_B, E)$. Thus, we have

$$a^{-1}(\chi) = \sum_{\lambda\in\mathbf{R}} \chi(\lambda)C_\lambda, \quad C_\lambda \in E, \quad \sum_{\lambda\in\mathbf{R}} \|C_\lambda\| < \infty,$$

and $C_\lambda \neq 0$ for at most countably many $\lambda \in \mathbf{R}$. This implies that

$$A^{-1} = a^{-1}(e_0) = \sum_{\lambda\in\mathbf{R}} e_0(\lambda)C_\lambda = \sum_{\lambda\in\mathbf{R}} C_\lambda,$$

and we are left with showing that there are $d_\lambda \in l^\infty_{N\times N}(\mathbf{R})$ such that

$$C_\lambda = \psi(d_\lambda)e_\lambda I \ \text{ for all } \ \lambda \in \mathbf{R}. \tag{9.10}$$

We have

$$\begin{aligned}
T_h a(e_s) T_h^{-1} &= \sum_{\lambda \in \mathbf{R}} T_h e^{i\lambda s} \psi(a_\lambda) e_\lambda T_h^{-1} \\
&= \sum_{\lambda \in \mathbf{R}} e^{i\lambda s} \psi(a_\lambda) T_h e_\lambda T_h^{-1} \quad \text{(translation invariance)} \\
&= \sum_{\lambda \in \mathbf{R}} e^{i\lambda s} e^{i\lambda h} \psi(a_\lambda) e_\lambda I \quad \text{(by (9.9))} \\
&= a(e_{s+h})
\end{aligned}$$

and hence

$$T_h a^{-1}(e_s) T_h^{-1} = a^{-1}(e_{s+h}). \tag{9.11}$$

Letting $\nu = s$ and $\nu = s + h$ in $a^{-1}(e_\nu) = \sum_\lambda e^{i\lambda\nu} C_\lambda$, we obtain from (9.11) that

$$\sum_{\lambda \in \mathbf{R}} e^{i\lambda s} T_h C_\lambda T_h^{-1} = \sum_{\lambda \in \mathbf{R}} e^{i\lambda s} e^{i\lambda h} C_\lambda. \tag{9.12}$$

Considering (9.12) as a function of s and comparing the coefficients of $e^{i\lambda s}$ we get $T_h C_\lambda T_h^{-1} = e^{i\lambda h} C_\lambda$ and thus

$$T_h C_\lambda T_h^{-1} e_{-\lambda} I = e^{i\lambda h} C_\lambda e_{-\lambda} I.$$

As $T_h^{-1} e_{-\lambda} I = e^{i\lambda h} e_{-\lambda} T_h^{-1}$ by (9.9), it results that

$$T_h (C_\lambda e_{-\lambda} I) T_h^{-1} = C_\lambda e_{-\lambda} I.$$

This shows that $C_\lambda e_{-\lambda} I$ is translation invariant, and (9.10) therefore follows from Proposition 7.10. □

9.4 Invertibility of Operators with *APW* Symbols

We are now in a position to prove Theorem 5.16.

Lemma 9.7. *Let $a \in APW_{N \times N}$. Then $a\tilde{P} + \tilde{Q}_0$ is invertible on B_N^2 if and only if a has a canonical right APW factorization.*

Proof. Throughout this proof, we omit the tildes. Thus, $\tilde{P} =: P$, $\tilde{P}_0 =: P_0$, $\tilde{Q} =: Q$, $\tilde{Q}_0 =: Q_0$.

The "if" part of the lemma is trivial: if $a = a_- a_+$ is a canonical APW factorization, then $aP + Q_0 = a_-(a_+ P + a_-^{-1} Q_0)$, and $a_-^{-1} I$ is the inverse of $a_- I$, while a direct computation shows that $a_+^{-1} P + a_- Q_0$ is the inverse of $a_+ P + a_-^{-1} Q_0$.

To prove the "only if" portion, suppose $aP + Q_0$ is invertible. Let $I_{N\times N}$ stand for the $N \times N$ identity matrix. The invertibility of $aP + Q_0$ implies that there is an $f \in B^2_{N\times N}$ such that

$$(aP + Q_0)f = I_{N\times N}. \tag{9.13}$$

Letting $f_+ := Pf$ and $f_- := I_{N\times N} - Q_0 f$, we get

$$af_+ = f_-. \tag{9.14}$$

We will show that $a = f_- f_+^{-1}$ is the desired APW factorization.

We first prove that $a^T Q + P_0$ is also invertible. The identity

$$PaI + Q_0 = (I + PaQ_0)(aP + Q_0)(I - Q_0 aP)$$

and the invertibilty of the outer factors,

$$(I + PaQ_0)^{-1} = I - PaQ_0, \quad (I - Q_0 aP)^{-1} = I + Q_0 aP,$$

shows that $PaI + Q_0$ is invertible together with $aP + Q_0$. Therefore $a^*P + Q_0 = (PaI + Q_0)^*$ is also invertible. Let $C : \varphi \mapsto \overline{\varphi}$ be the operator of complex conjugation on B^2_N. Since

$$Ca^*C = a^T I, \quad CPC = Q, \quad CQ_0C = P_0,$$

it follows that $a^T Q + P_0 = C(a^*P + Q_0)C$ is invertible.

Let $g \in B^2_{N\times N}$ be the solution of the equation $(a^T Q + P_0)g = I_{N\times N}$ and put $g_-^T := Qg$, $g_+^T := I_{N\times N} - P_0 g$. Then $a^T g_-^T = g_+^T$ and hence

$$g_- a = g_+. \tag{9.15}$$

We claim that $aP + Q_0$ and $a^T Q + P_0$ belong to the algebra $\mathcal{R}^W_{N\times N}$. Because

$$aP + Q_0 = aP + I - P, \quad a^T Q + P_0 = a^T I - a^T P_0 + P_0,$$

it suffices to show that $e_\lambda I$, $e_\lambda P$, $e_\lambda P_0$ belong to $\mathcal{R}^W$ $(N = 1)$. As $e_\lambda I = \psi(1)e_\lambda I$, it is clear that $e_\lambda I \in \mathcal{R}^W$. The equalities

$$\begin{aligned} e_\lambda P &= e_\lambda P e_{-\lambda} e_\lambda I = \psi(\chi_{[\lambda,\infty)})e_\lambda I, \\ e_\lambda P_0 &= e_\lambda P_0 e_{-\lambda} e_\lambda I = \psi(\chi_{(\lambda,\infty)})e_\lambda I \end{aligned}$$

imply that $e_\lambda P, e_\lambda P_0 \in \mathcal{R}^W$. This proves out claim.

From Theorem 9.6 we deduce that $(aP + Q_0)^{-1} \in \mathcal{R}^W_{N\times N}$. Thus,

$$(aP + Q_0)^{-1} = \sum_\lambda \psi(d_\lambda)e_\lambda I, \quad d_\lambda \in l^\infty_{N\times N}(\mathbf{R}), \quad \sum_\lambda \|d_\lambda\|_\infty < \infty.$$

Equation (9.13) therefore gives

$$\begin{aligned} f &= \sum_\lambda \psi(d_\lambda) e_\lambda I_{N\times N} = \sum_\lambda F_B d_\lambda F_B^{-1} e_\lambda I_{N\times N} \\ &= \sum_\lambda F_B d_\lambda \chi_{\{\lambda\}} I_{N\times N} = \sum_\lambda F_B d_\lambda(\lambda) \chi_{\{\lambda\}} I_{N\times N} \\ &= \sum_\lambda d_\lambda(\lambda) e_\lambda I_{N\times N}, \end{aligned}$$

which shows that $f \in APW_{N\times N}$. Consequently,

$$f_+ \in APW^+_{N\times N}, \quad f_- \in APW^-_{N\times N}, \quad M(f_-) = I_{N\times N}.$$

It can be verified analogously that

$$g_+ \in APW^+_{N\times N}, \quad g_- \in APW^-_{N\times N}, \quad M(g_+) = I_{N\times N}.$$

Combining (9.14) and (9.15) we get

$$g_- f_- = g_-(af_+) = (g_- a) f_+ = g_+ f_+,$$

and since $g_\pm f_\pm \in APW^\pm_{N\times N}$, it results that

$$g_- f_- = g_+ f_+ = c \in \mathbf{C}^{N\times N}. \tag{9.16}$$

We prove that $\det c \neq 0$. So assume $\det c = 0$. Then $\det(g_- f_-) = 0$ on $\mathbf{R}$ and hence in $\mathbf{C}_-$. Because

$$\det f_-(x+iy) \to \det I_{N\times N} = 1 \ \text{ as } \ y \to -\infty,$$

we see that $\det g_-$ must vanish identically in a neighborhood of $-i\infty$, and the analyticity of $\det g_-$ implies that $\det g_- = 0$ on $\mathbf{R}$. From (9.15) we see that $\det g_+ = 0$ on $\mathbf{R}$, and hence $\det g_+$ must vanish identically in $\mathbf{C}_+$. However, this is impossible because

$$\det g_+(x+iy) \to \det I_{N\times N} = 1 \ \text{ as } \ y \to +\infty.$$

Thus, $\det c \neq 0$. From (9.16) we now obtain that $f_\pm$ belong to $GAPW^\pm_{N\times N}$, which shows that $a = f_- f_+^{-1}$ is a canonical right APW factorization. □

With the following result, we achieved the goal of this and the preceding four chapters.

Corollary 9.8. *If $a \in APW_{N\times N}$, then for a to have a canonical right APW factorization it is necessary and sufficient that $W(a)$ is invertible on $L^2_N(\mathbf{R}_+)$.*

Proof. Combine Corollary 9.4 and Lemma 9.7. □

9.5 *APW* Symbols with Good Numerical Range

In this and the following section we investigate *APW* factorization for two classes of matrix functions.

The *numerical range* (= Hausdorff set) of a matrix $A \in \mathbf{C}^{N\times N}$ is defined as

$$\mathcal{H}(A) := \{(A\eta, \eta) : \eta \in \mathbf{C}^N, \ \|\eta\| = 1\},$$

where $\|(\eta_1, \dots, \eta_N)\|^2 := |\eta_1|^2 + \dots + |\eta_n|^2 = 1$. It is well known that $\mathcal{H}(A)$ is a closed convex set which contains the spectrum (eigenvalues) of A. If A is normal, i.e., if $AA^* = A^*A$, then $\mathcal{H}(A)$ is the convex hull of the spectrum of A.

Let $a \in APW_{N\times N}$. Then $\mathcal{H}(a(x))$ is a well-defined subset of $\mathbf{C}$ for every $x \in \mathbf{R}$. We say that *the numerical range of a is bounded away from zero* if

$$\inf_{x\in\mathbf{R}} \operatorname{dist}\big(\mathcal{H}(a(x)), 0\big) > 0$$

or, equivalently, if there is an $\varepsilon > 0$ such that

$$\big|(a(x)\eta, \eta)\big| \geq \varepsilon\|\eta\|^2 \ \text{ for all } \ x \in \mathbf{R} \ \text{ and all } \ \eta \in \mathbf{C}^N. \tag{9.17}$$

In the scalar case ($N = 1$) we have $\mathcal{H}(a(x)) = a(x)$, and therefore the numerical range of a is bounded away from zero if and only if a is invertible in $L^\infty(\mathbf{R})$ and thus in APW.

Given $a \in APW_{N\times N}$ and $\eta \in \mathbf{C}^N \setminus \{0\}$, we denote by $(a\eta, \eta)$ the function $x \mapsto (a(x)\eta, \eta)$. Clearly, if the numerical range of a is bounded away from zero, then $(a\eta, \eta) \in GAPW$ for every $\eta \in \mathbf{C}^N \setminus \{0\}$ and hence the mean motion $\kappa((a\eta, \eta))$ is defined for every $\eta \in \mathbf{C}^N \setminus \{0\}$. The following theorem shows in particular that $\kappa((a\eta, \eta))$ is independent of $\eta \in \mathbf{C} \setminus \{0\}$.

Theorem 9.9 (Babadzhanyan and Rabinovich). *If $a \in GAPW_{N\times N}$ and the numerical range of a is bounded away from zero, then a has a right APW factorization and all right AP indices are equal to $\kappa((a\eta, \eta))$ where η is an arbitrary vector in $\mathbf{C}^N \setminus \{0\}$.*

Proof. Consider a as a continuous $\mathbf{C}^{N\times N}$-valued function on the Bohr compactification $\mathbf{R}_B$. As $\mathbf{R}$ is dense in $\mathbf{R}_B$, we deduce from (9.17) that

$$\big|(a(\chi)\eta, \eta)\big| \geq \varepsilon\|\eta\|^2 \ \text{ for all } \ \chi \in \mathbf{R}_B \ \text{ and all } \ \eta \in \mathbf{C}^N. \tag{9.18}$$

For $\chi \in \mathbf{R}_B$, denote by $S(\chi)$ the smallest sector with the vertex at the origin which contains the convex hull of $\{0\} \cup \mathcal{H}(a(\chi))$. Since $\mathcal{H}(a(\chi))$ is convex and $0 \notin \mathcal{H}(a(\chi))$, the set $S(\chi)$ is a closed sector of opening α less than π. Let $\beta(\chi)$ be the point of the bisector of $S(\chi)$ whose distance to the origin is 1. Because $\mathcal{H}(a(\chi))$ and thus also $S(\chi)$ depend continuously on $\chi \in \mathbf{R}_B$ in the Hausdorff metric (see, e.g., [104, Problem 175]), it follows that $\beta : \mathbf{R}_B \to \mathbf{T}$ is continuous. By the construction of β and by (9.18),

$$\operatorname{Re}\left(\overline{\beta(\chi)}a(\chi)\eta, \eta\right) \geq \varepsilon \left(\cos\frac{\alpha}{2}\right) \|\eta\|^2 \text{ for all } \chi \in \mathbf{R}_B \text{ and all } \eta \in \mathbf{C}^N,$$

which implies that

$$\operatorname{Re}\left(\overline{\beta(x)}a(x)\eta, \eta\right) \geq \varepsilon \left(\cos\frac{\alpha}{2}\right) \|\eta\|^2 \text{ for all } x \in \mathbf{R} \text{ and all } \eta \in \mathbf{C}^N.$$

The function $\overline{\beta} : \mathbf{R} \to \mathbf{T}$ belongs to $AP = C(\mathbf{R}_B)$, and hence we can approximate it by a function $\gamma \in APW$ such that

$$\operatorname{Re}\big(\gamma(x)a(x)\eta, \eta\big) \geq \frac{\varepsilon}{2} \left(\cos\frac{\alpha}{2}\right) \|\eta\|^2 \text{ for all } x \in \mathbf{R} \text{ and all } \eta \in \mathbf{C}^N.$$

Consequently, $\gamma a =: s$ is sectorial, and from Theorem 6.17 we obtain that $W(s)$ is invertible. Because $s = \gamma a \in APW_{N\times N}$, Corollary 9.8 implies that γa has a canonical right APW factorization $\gamma a = b_- b_+$. Theorems 2.25 and 8.11 give a right APW factorization $\gamma = \gamma_- e_{\kappa(\gamma)} \gamma_+$. In summary,

$$a = (\gamma_-^{-1} b_-)\, (e_{-\kappa(\gamma)} I)\, (\gamma_+^{-1} b_+)$$

is a right APW factorization of a. By the construction of β, there are continuous arguments such that

$$\left|\arg\big(a(x)\eta, \eta\big) + \arg\gamma(x)\right| < \pi/2, \quad \left|\arg\big(a(x)\eta, \eta\big) - \arg\beta(x)\right| < \pi/2$$

for every $\eta \in \mathbf{C}^N \setminus \{0\}$. Since

$$\arg\big(a(x)\eta, \eta\big) = \kappa\big((a\eta, \eta)\big)x + \varphi_\eta(x), \quad \arg\gamma(x) = \kappa(\gamma)x + \psi(x)$$

with real-valued $\varphi_\eta, \psi \in AP \subset L^\infty(\mathbf{R})$ by Theorem 2.25, we arrive at the conclusion that $\kappa((a\eta, \eta)) = -\kappa(\gamma)$. ☐

Corollary 9.10. *If $a \in GAPW_{N\times N}$ and the numerical range of a is bounded away from zero, then*

$$\begin{aligned}
\kappa\big((a\eta, \eta)\big) = 0 \quad &\Longrightarrow \quad W(a) \textit{ is invertible,}\\
\kappa\big((a\eta, \eta)\big) > 0 \quad &\Longrightarrow \quad W(a) \textit{ is properly } n\textit{-normal and left-invertible,}\\
\kappa\big((a\eta, \eta)\big) < 0 \quad &\Longrightarrow \quad W(a) \textit{ is properly } d\textit{-normal and right-invertible.}
\end{aligned}$$

Proof. Combine Theorems 2.28 and 9.9. ☐

9.6 Hermitian Matrix Symbols in *APW*

A matrix function $a \in L^\infty_{N\times N}(\mathbf{R})$ is *Hermitian* if $a(x) = a^*(x)$ for almost all $x \in \mathbf{R}$. Suppose a is a Hermitian matrix function in $G[BC(\mathbf{R})]_{N\times N}$. We denote by $\xi_1(x), \dots, \xi_N(x)$ the eigenvalues of $a(x)$. Since the eigenvalues are real numbers that depend continuously on x and

$$\det a(x) = \xi_1(x)\dots\xi_N(x) \neq 0 \quad \text{for all } x \in \mathbf{R},$$

it is clear that the *signature* of $a(x)$, i.e., the number of positive eigenvalues minus the number of negative eigenvalues, is independent of x. We therefore simply speak of the signature of a.

The purpose of this section is to prove that if a Hermitian matrix function $a \in GAPW_{N\times N}$ admits a right APW factorization, then it has a right $APW(u)$ factorization $a = b_+^* u b_+$ with $b_+ \in GAPW^+_{N\times N}$ and a special unitary factor u. This result has several consequences and will play a key role in the next section.

Let $a \in GAP_{N\times N}$ ($\subset G[BC(\mathbf{R})]_{N\times N}$) be Hermitian and let

$$a = a_- d a_+, \quad d = \operatorname{diag}(e_{\mu_j}), \quad \mu_1 \geq \dots \geq \mu_N$$

be a right AP factorization. Then

$$a = a^* = a_+^* d^{-1} a_-^*, \quad d^{-1} = \operatorname{diag}(e_{-\mu_j})$$

is also a right AP factorization. Theorem 8.2 therefore implies that μ_j equals $-\mu_{N-j+1}$ for every j. In other words (recall Definition 8.3),

$$\kappa(a) = (\lambda_m, \dots, \lambda_1, 0, \dots, 0, -\lambda_1, \dots, -\lambda_m)$$

where $\lambda_m \geq \dots \geq \lambda_1$ are the positive right AP indices of a.

Given square matrices $A_1, \dots, A_n$, we put

$$\operatorname{codiag}(A_1, \dots, A_n) := \begin{pmatrix} & & & A_n \\ & & \cdots & \\ & A_2 & & \\ A_1 & & & \end{pmatrix}.$$

Theorem 9.11. *Let $a \in GAPW_{N\times N}$ be Hermitian with the signature σ and suppose a admits a right APW factorization. Denote by $\lambda_m \geq \dots \geq \lambda_1 (> 0)$ the positive right AP indices of a. Then*

$$a = b_+^* d_0 b_+ \tag{9.19}$$

where $b_+ \in GAPW^+_{N\times N}$ and

$$d_0(x) = \operatorname{codiag}\left(e^{-i\lambda_m x}, \dots, e^{-i\lambda_1 x}, \ J, e^{i\lambda_1 x}, \dots, e^{i\lambda_m x}\right),$$
$$J := \operatorname{diag}(I_{l_+}, -I_{l_-}), \quad l_+ + l_- = N - 2m, \quad l_+ - l_- = \sigma,$$

and I_l stands for the $l \times l$ identity matrix.

Proof. Let $a = a_- d a_+$ with

$$d(x) = \operatorname{diag}\left(e^{-i\lambda_m x}, \ldots, e^{-i\lambda_1 x}, 1, \ldots, 1, e^{i\lambda_1 x}, \ldots, e^{i\lambda_m x}\right)$$

be a right *APW* factorization. Denote by $\nu_1 < \ldots < \nu_k$ the different positive right *AP* indices and let $l_1, \ldots, l_k$ be their multiplicities ($l_1 + \ldots + l_k = m$). Also put $\nu_{-s} := -\nu_s$ ($s = 1, \ldots, k$) and $\nu_0 := 0$. Since $a = a_- d a_+ = a_+^* d^{-1} a_-^*$, we get

$$\varphi d = d^{-1} \varphi^* \tag{9.20}$$

with $\varphi := (a_+^*)^{-1} a_- \in GAPW_{N\times N}^-$. Clearly,

$$a = a_+^* \varphi d a_+. \tag{9.21}$$

Suppose we had a matrix function $\psi \in GAPW_{N\times N}^-$ such that

$$\varphi d = \psi d_0 \psi^*. \tag{9.22}$$

Then we would obtain the theorem with $b_+ := \psi^* a_+$. Thus, we must prove that the equation (9.22) has a solution $\psi \in GAPW_{N\times N}^-$.

Write φ and ψ in block form:

$$\varphi = (\varphi_{rs})_{r,s=-k}^k, \quad \psi = (\psi_{rs})_{r,s=-k}^k$$

where φ_{rr} and ψ_{rr} are $l_{|r|} \times l_{|r|}$ matrix functions and the enumeration of the blocks is from left to right and from below to above. The equality (9.20) is equivalent to the equalities

$$\varphi_{rs}(x) = \varphi_{-s,-r}^*(x) e^{i(\nu_r - \nu_s)x}. \tag{9.23}$$

Hence $\varphi_{rs} = 0$ for $r > s$ and

$$\Omega(\varphi_{rs}) \subset [\nu_r - \nu_s, 0] \ \text{ for } \ r \le s. \tag{9.24}$$

In particular, the blocks φ_{rr} are constant. Since $\varphi = (a_+^*)^{-1} a_-$ and φ is block triangular, it follows that φ_{rr} is an invertible matrix for each r.

Equation (9.22) is equivalent to the system

$$\sum_{\alpha=-k}^{k} \psi_{r,-\alpha}(x) J_\alpha \psi_{-s,\alpha}^*(x) e^{i\nu_\alpha x} = \varphi_{rs}(x) e^{i\nu_s x} \tag{9.25}$$

where $J_0 := J$ and J_α is the $l_{|\alpha|} \times l_{|\alpha|}$ matrix $\operatorname{codiag}(1, \ldots, 1)$ for $\alpha \ne 0$. Since φd is Hermitian by (9.20) and since $\psi d_0 \psi^*$ is obviously Hermitian, it suffices to solve (9.25) for $r + s \ge 0$.

We look for a matrix function ψ such that $\psi_{\alpha\beta} = 0$ whenever $\alpha + \beta > 0$. This requirement guarantees that the summands on the left side of (9.25) may differ from zero only when $r \leq \alpha \leq s$. Accordingly, both sides of (9.25) are zero if $r > s$. Thus, we are left to solve (9.25) for $r \leq s$. We divide the equations (9.25) into groups, letting the jth group consist of the equations with $s = r + j$ ($j = 0, \ldots, 2k$). Since $r + s \geq 0$, it is sufficient to consider only the equations with $r = -[j/2], -[j/2]+1, \ldots, k-j$ in the jth group, where $[x]$ denotes the integer part of x.

The equations of the 0th group are

$$\psi_{r,-r} J_r \psi^*_{-r,r} = \varphi_{rr} \; (r = 1, \ldots, k), \tag{9.26}$$
$$\psi_{00} J \psi^*_{00} = \varphi_{00}. \tag{9.27}$$

It is obvious that the equations (9.26) have solutions $\psi_{r,-r}$ and $\psi_{-r,r}$ in $G\mathbf{C}^{l_{|r|} \times l_{|r|}}$ (e.g. $\psi_{-r,r} = \varphi^*_{rr}$, $\psi_{r,-r} = J_r$). From (9.21) we infer that a and φd have the same signature σ. On the other hand, (9.22) implies that φd and d_0 also have the same signature. Furthermore, the signature of d_0 coincides with the signature of J and, by (9.27), with the signature of φ_{00}. This implies that φ_{00} also has the signature σ and thus shows that (9.27) has a solution $\psi_{00} \in G\mathbf{C}^{l_0 \times l_0}$ where $l_0 = N - 2m$.

Put $\psi_{r,-\alpha} := 0$ for $\alpha \in \{|r|+1, \ldots, k\}$. Assume that we have solved the equations of the groups j with $j \leq j_0$ and that for the solutions $\psi_{\alpha,\beta}$ ($0 \geq \alpha + \beta \geq -j_0$) the inclusion

$$\Omega(\psi_{\alpha,\beta}) \subset [\nu_\alpha - \nu_{-\beta}, 0] \tag{9.28}$$

is satisfied. Every equation of the group $j_0 + 1$ can be written in the form

$$\begin{aligned} &\psi_{r,-r} J_r \psi^*_{-s,r}(x) e^{i\nu_r x} + \psi_{r,-s}(x) J_s \psi^*_{-s,s} e^{i\nu_s x} \\ &= \varphi_{rs}(x) e^{i\nu_s x} - \sum_{\alpha=r+1}^{s-1} \psi_{r,-\alpha}(x) J_\alpha \psi^*_{-s,\alpha}(x) e^{i\nu_\alpha x}. \end{aligned} \tag{9.29}$$

If $r + s \neq 0$, then $\psi_{r,-\alpha} := 0$ for $|r|+1 \leq \alpha \leq s$, and we determine $\psi_{-s,r}(x)$ as the solution of

$$\psi_{r,-r} J_r \psi^*_{-s,r}(x) = \varphi_{rs}(x) e^{i(\nu_s - \nu_r)x} - \sum_{\alpha=r+1}^{|r|} \psi_{r,-\alpha}(x) J_\alpha \psi^*_{-s,\alpha}(x) e^{i(\nu_\alpha - \nu_r)x}. \tag{9.30}$$

Here we take into account that $r \in \{-[j/2], -[j/2]+1, \ldots, k-j\}$. By (9.24) and (9.28), the Bohr-Fourier spectrum of the matrix function on the right of (9.30) is contained in $[0, \nu_s - \nu_r]$, which implies that

$$\Omega(\psi_{-s,r}) \subset [\nu_r - \nu_s, 0] = [\nu_{-s} - \nu_{-r}, 0].$$

If $r+s=0$, then $r<0$ (since $r<s$ for $j>0$), and (9.29) takes the form

$$\begin{aligned}
&2\mathrm{Re}\left(\psi_{r,-r}J_r\psi_{rr}^*(x)e^{i\nu_r x}\right)\\
&=\varphi_{r,-r}(x)e^{-i\nu_r x}-\sum_{\alpha=r+1}^{|r|-1}\psi_{r,-\alpha}(x)J_\alpha\psi_{r,\alpha}^*(x)e^{i\nu_\alpha x}\\
&=\varphi_{r,-r}(x)e^{-i\nu_r x}-\psi_{r0}(x)J\psi_{r0}^*(x)\\
&\quad-\sum_{\alpha=r+1}^{-1}\left(\psi_{r,-\alpha}(x)J_\alpha\psi_{r,\alpha}^*(x)e^{i\nu_\alpha x}+\psi_{r,\alpha}(x)J_\alpha\psi_{r,-\alpha}^*(x)e^{-i\nu_\alpha x}\right). \qquad (9.31)
\end{aligned}$$

From (9.23) we see that $\varphi_{r,-r}(x)e^{-i\nu_r x}$ is Hermitian. Hence, the right-hand side of (9.31) is also Hermitian. Since the Bohr-Fourier spectra of all terms on the right of (9.31) are contained in $[\nu_r,-\nu_r]=[-\nu_s,\nu_s]$ (recall (9.24) and (9.28)), the right-hand side of (9.31) can be represented in the form $2\mathrm{Re}\,\eta$ with $\eta\in APW_{N\times N}^+$ and $\Omega(\eta)\subset[0,\nu_s]$. Obviously,

$$\psi_{rr}(x)=e^{i\nu_r x}\eta^*(x)(\psi_{r,-r}^{-1})^*J_r$$

is a solution of (9.31), and

$$\Omega(\psi_{rr})\subset[\nu_r-\nu_s,\nu_r]\subset[\nu_r-\nu_{-r},0]$$

because $\nu_{-r}=-\nu_r=\nu_s>0$.

Thus, we have shown that there exists a matrix function ψ in $APW_{N\times N}^-$ such that (9.22) is satisfied. As

$$\det\psi(x)=\prod_{r=-k}^{k}\det\psi_{r,-r}$$

is a nonzero constant, it follows that $\psi\in GAPW_{N\times N}^-$. □

Corollary 9.12. *Let $a\in GAPW_{N\times N}$ be Hermitian and suppose a has a right AP factorization. Let $0<\lambda_1\le\ldots\le\lambda_m$ be the positive right AP indices. If the signature of a is 0, then N is even and there exists a $b_+\in GAPW_{N\times N}^+$ such that $a=b_+^*d_0b_+$ with*

$$\begin{aligned}
&d_0(x)=\mathrm{codiag}\,(e^{-i\lambda_m x},\ldots,e^{-i\lambda_1 x},J,e^{i\lambda_1 x},\ldots,e^{i\lambda_m x}),\\
&J=\mathrm{diag}\,(I_{N/2-m},-I_{N/2-m}).
\end{aligned}$$

Proof. Immediate from Theorem 9.11. □

Corollary 9.13. *Let $a\in GAPW_{N\times N}$ be Hermitian and suppose the signature of a is σ. Then $W(a)$ is invertible on $L_N^2(\mathbf{R}_+)$ if and only if there exists a matrix* *function* $b_+ \in GAPW^+_{N\times N}$ *such that*

$$a = b_+^* \begin{pmatrix} I_{l_+} & 0 \\ 0 & -I_{l_-} \end{pmatrix} b_+$$

where $l_+ + l_- = N$ *and* $l_+ - l_- = \sigma$.

Proof. Combine Corollary 9.8 and Theorem 9.11. □

A matrix function $a \in L^\infty_{N\times N}(\mathbf{R})$ is said to be *positive definite* if $a(x) = a^*(x)$ for almost all $x \in \mathbf{R}$ and if there exists an $\varepsilon > 0$ such that

$$\big(a(x)\eta, \eta\big) \geq \varepsilon\|\eta\|^2 \text{ for all } \eta \in \mathbf{C}^N \text{ and almost all } x \in \mathbf{R}.$$

Theorem 9.14. *For a matrix function* $a \in L^\infty_{N\times N}(\mathbf{R})$, *the following are equivalent:*

(i) *a is positive definite;*

(ii) $a = u_+^* u_+$ *with* $u_+ \in G[H^\infty_+]_{N\times N}$;

(iii) $a = v_+ v_+^*$ *with* $v_+ \in G[H^\infty_+]_{N\times N}$.

Proof. We first prove the implication (i)⇒(ii). Let a be positive definite. Then for $f \in L^2_N(\mathbf{R}_+)$,

$$\begin{aligned}\big(W(a)f, f\big) &= \big(\chi_+(af), f\big) = (af, \chi_+ f) = (af, f) \\ &= \int_{\mathbf{R}} \big(a(x)f(x), f(x)\big)\,dx \geq \varepsilon \int_{\mathbf{R}} \|f(x)\|^2\,dx = \varepsilon\|f\|^2,\end{aligned}$$

which shows that the self-adjoint operator $W(a)$ is invertible. By Theorem 6.5, a has a right WH factorization $a = a_- a_+$. Since $a = a^* = a_+^* a_-^*$ is also a right WH factorization, we deduce from Theorem 6.4 that $a_- = a_+^* h^{-1}$ and $a_+ = h a_-^*$ with some $h \in G\mathbf{C}^{N\times N}$. Thus, $a = a_- a_+ = a_+^* h^{-1} a_+$. The matrix h^{-1} is positive definite together with a. Therefore $h^{-1} = g^* g$ with $g \in G\mathbf{C}^{N\times N}$, and we get $a = u_+^* u_+$ with $u_+ = g a_+$. Consideration of the diagonal entries in the equalities $a = u_+^* u_+$ and $a^{-1} = u_+^{-1}(u_+^{-1})^*$ finally shows that u_+ and u_+^{-1} are bounded.

The implication (ii)⇒(i) is trivial, and the equivalence (i)⇔(iii) follows from the equivalence (i)⇔(ii) applied to a^{-1}. □

Obviously, a matrix function in $G[BC(\mathbf{R})]_{N\times N}$ is positive definite if and only if it is Hermitian with the signature N.

Corollary 9.15. *Let* $a \in GAPW_{N\times N}$ *be positive definite. Then*

(a) $W(a)$ *is invertible on* $L^2_N(\mathbf{R}_+)$;

(b) *there is a* $b_+ \in GAPW^+_{N\times N}$ *such that* $a = b_+^* b_+$.

Proof. Part (a) follows from Theorem 6.17, and part (b) results from part (a) along with Corollary 9.8 and Theorem 9.11. □

9.7 One-Sided Invertibility of Operators with *APW* Symbols

In this section, we study one-sided invertibility of Wiener-Hopf operators with symbols in $APW_{N\times N}$. By Theorem 5.1, we may restrict our attention to symbols in $GAPW_{N\times N}$. An invertibility criterion is given by Corollary 9.8. In the spirit of this chapter, Corollary 9.8 can be rephrased as follows.

Theorem 9.16. *If $a \in GAPW_{N\times N}$, then for $W(a)$ to be invertible on $L^2_N(\mathbf{R}_+)$ it is necessary and sufficient that a have a canonical right APW factorization.* □

By Proposition 1.4(b), a Hilbert space operator A is left-invertible (resp. right-invertible) if and only if A^*A (resp. AA^*) is invertible. If $A = W(a)$ is a Wiener-Hopf operator, then in general neither A^*A nor AA^* are Wiener-Hopf operators. The following simple observation reduces one-sided invertibility (resp. semi-Fredholmness) of $A = W(a)$ to invertibility (resp. Fredholmness) of a Wiener-Hopf operator $\hat{A} = W(\hat{a})$ with a double-size matrix function $\hat{a}$.

Lemma 9.17. *Let H be a Hilbert space and $A \in \mathcal{B}(H)$. The operator A is left-invertible (resp. n-normal) if and only if*

$$\hat{A} := \begin{pmatrix} I & A \\ A^* & 0 \end{pmatrix}$$

is invertible (resp. Fredholm) on $H_2 := H \oplus H$.

Proof. The identity

$$\begin{pmatrix} I & A_2 \\ A_1 & 0 \end{pmatrix} = \begin{pmatrix} I & 0 \\ A_1 & I \end{pmatrix} \begin{pmatrix} I & 0 \\ 0 & A_1A_2 \end{pmatrix} \begin{pmatrix} I & A_2 \\ 0 & -I \end{pmatrix}$$

and the invertibility of the outer factors,

$$\begin{pmatrix} I & 0 \\ A_1 & I \end{pmatrix}^{-1} = \begin{pmatrix} I & 0 \\ -A_1 & I \end{pmatrix}, \quad \begin{pmatrix} I & A_2 \\ 0 & -I \end{pmatrix}^{-1} = \begin{pmatrix} I & A_2 \\ 0 & -I \end{pmatrix},$$

show that $\begin{pmatrix} I & A_2 \\ A_1 & 0 \end{pmatrix}$ is invertible (resp. Fredholm) if and only if so is A_1A_2. The assertion now follows from Proposition 1.4(b). □

Theorem 9.18. *For $a \in GAPW_{N\times N}$ the following are equivalent:*

(i) *$W(a)$ is left-invertible;*

(ii) *there are $a_\pm \in GAP_{N\times N}$ such that*
$$a = a_-a_+, \quad a_-^{-1} \in AP^-_{N\times N}, \quad a_+ \in AP^+_{N\times N};$$

(iii) *there exist $a_\pm \in GAPW_{N\times N}$ such that*
$$a = a_-a_+, \quad a_-^{-1} \in APW^-_{N\times N}, \quad a_+ \in APW^+_{N\times N};$$

(iv) *a admits a right* $APW(u)$ *factorization with* $u \in APW^+_{N\times N}$, *i.e.,*

$$a = c_- u c_+, \quad c_\pm \in GAPW^\pm_{N\times N}, \quad u \in APW^+_{N\times N} \text{ is unitary;}$$

(v) *a has a right* $APW(u)$ *factorization as in* (iv) *with the additional property that*

$$u = \begin{pmatrix} v & 0 \\ 0 & I \end{pmatrix} \quad \text{and} \quad M(v) = 0.$$

Proof. The implications (v)⇒(iv)⇒(iii)⇒(ii)⇒(i) are trivial (note that we have $W(a)W(b_+) = W(ab_+)$ and $W(b_-)W(a) = W(b_-a)$ whenever $b_\pm \in AP^\pm_{N\times N}$).

(i)⇒(iii). Suppose $W(a)$ is left-invertible. Then, by Lemma 9.17, the operator $W(\hat{a})$, $\hat{a} := \begin{pmatrix} I & a \\ a^* & 0 \end{pmatrix}$, is invertible. From Theorem 9.16 we infer that $\hat{a}$ has a canonical right APW factorization. Since $\hat{a}$ is Hermitian and has the signature 0 due to the equality

$$\begin{pmatrix} I & a \\ a^* & 0 \end{pmatrix} = \begin{pmatrix} I & 0 \\ a^* & a^* \end{pmatrix}\begin{pmatrix} I & 0 \\ 0 & -I \end{pmatrix}\begin{pmatrix} I & a \\ 0 & a \end{pmatrix},$$

we deduce from Corollary 9.13 that

$$\hat{a} = b_+^* \begin{pmatrix} I & 0 \\ 0 & -I \end{pmatrix} b_+ \text{ with } b_+ \in GAPW^+_{2N\times 2N}.$$

In terms of the partition $b_+ = (b_{jk})_{j,k=1}^2$ into $N \times N$ blocks, this factorization is equivalent to the equalities

$$\begin{aligned} b_{11}^* b_{11} - b_{21}^* b_{21} &= I, \\ b_{11}^* b_{12} - b_{21}^* b_{22} &= a, \qquad (9.32) \\ b_{12}^* b_{12} - b_{22}^* b_{22} &= 0. \qquad (9.33) \end{aligned}$$

The polar decompositions $b_{12} = y_1(b_{12}^* b_{12})^{1/2}$, $b_{22} = y_2(b_{22}^* b_{22})^{1/2}$ (y_1, y_2 unitary) and (9.33) imply that

$$b_{22} = y b_{12} \text{ where } y \text{ is unitary-valued.} \qquad (9.34)$$

Hence

$$b_+ = \begin{pmatrix} 0 & I \\ b_{21} - y b_{11} & y \end{pmatrix}\begin{pmatrix} I & 0 \\ b_{11} & I \end{pmatrix}\begin{pmatrix} I & 0 \\ 0 & b_{12} \end{pmatrix}. \qquad (9.35)$$

Since $\det b_+$ is invertible in $GAPW$, we see from (9.35) that

$$b_{21} - y b_{11} \in GL^\infty_{N\times N}(\mathbf{R}), \quad b_{12} \in GL^\infty_{N\times N}(\mathbf{R}).$$

Since $b_{12} \in APW_{N\times N}$, it follows that actually $b_{12} \in GAPW_{N\times N}$. This together with (9.34) shows that $y \in APW_{N\times N}$. Consequently,

$$b_{22} \in APW^+_{N\times N} \cap GAPW_{N\times N}, \quad b_{21} - yb_{11} \in APW_{N\times N}. \tag{9.36}$$

Again by (9.35),

$$\begin{aligned}&\begin{pmatrix} -(b_{21}-yb_{11})^{-1}y & (b_{21}-yb_{11})^{-1} \\ I & 0 \end{pmatrix} \\ &= \begin{pmatrix} 0 & I \\ b_{21}-yb_{11} & y \end{pmatrix}^{-1} = \begin{pmatrix} I & 0 \\ b_{11} & I \end{pmatrix}\begin{pmatrix} I & 0 \\ 0 & b_{12} \end{pmatrix} b_+^{-1},\end{aligned}$$

which gives

$$(b_{21} - yb_{11})^{-1} \in APW^+_{N\times N}. \tag{9.37}$$

From (9.32) and (9.34) we finally get

$$a = b_{11}^* y^* b_{22} - b_{21}^* b_{22} = (-b_{21} + yb_{11})^* b_{22},$$

and letting $a_- := (-b_{21} + yb_{11})^*$, $a_+ := b_{22}$, we arrive at the desired factorization by virtue of (9.36) and (9.37).

(iii)⇒(iv). The matrix function $a_+^* a_+$ is positive definite. Corollary 9.15 therefore ensures a factorization

$$a_+^* a_+ = c_+^* c_+ \quad \text{with} \quad c_+ \in GAPW^+_{N\times N}. \tag{9.38}$$

From (9.38) we obtain that $u_1 := a_+ c_+^{-1}$ is unitary, and since a_+ and c_+^{-1} belong to $APW^+_{N\times N}$, it follows that $u_1 \in APW^+_{N\times N}$. Analogously,

$$a_- a_-^* = c_- c_-^* \quad \text{with} \quad c_- \in GAPW^-_{N\times N},$$

and $u_2 := a_-^{-1} c_-$ is a unitary-valued matrix function in $APW^-_{N\times N}$. In summary,

$$a = c_- c_-^{-1} a_- a_+ c_+^{-1} c_+ = c_- u c_+,$$

where $u = u_2^{-1} u_1 = u_2^* u_1 \in APW^+_{N\times N}$ is unitary-valued.

(iv)⇒(v). Let $a = c_- u c_+$ be a factorization as in (iv). Since $\|u(x)\| = 1$ for all $x \in \mathbf{R}$, we get

$$\|M(u)\| \le \lim_{T\to\infty} \frac{1}{2T} \int_{-T}^{T} \|u(x)\|\, dx = 1.$$

Hence, in the singular value decomposition $M(u) = W_1 D W_2$, where W_1, W_2 in $\mathbf{C}^{N\times N}$ are unitary and $D = \operatorname{diag}(s_1, \ldots, s_N)$, we have $s_j \in [0,1]$ for all j. By changing c_- to $c_- W_1$ and c_+ to $W_2 c_+$, we can (and will) assume that $M(u)$ itself equals D.

Suppose $0 \le s_1 \le \ldots \le s_m < 1 = s_{m+1} = \ldots = s_N$. Then

$$D = \begin{pmatrix} D_0 & 0 \\ 0 & I \end{pmatrix} \quad \text{where } D_0 := \operatorname{diag}(s_1, \ldots, s_m).$$

Let $u = (u_{jk})_{j,k=0}^{1}$ be the corresponding decomposition of u into blocks. Since $M(u_{11}) = I$, the Fourier series of u_{11} is of the form

$$u_{11}(x) = I + \sum_k f_k e^{i\lambda_k x}, \quad \lambda_k > 0.$$

Let

$$u_{10}(x) = \sum_l g_l e^{i\lambda_l x}, \quad \lambda_l \ge 0$$

be the Fourier series of u_{10}. The unitarity of u implies that $u_{11}u_{11}^* + u_{10}u_{10}^* = I$, whence

$$I = M(u_{11}u_{11}^*) + M(u_{10}u_{10}^*) = I + \sum_k f_k f_k^* + \sum_l g_l g_l^*.$$

Consequently, $f_k = 0$ and $g_l = 0$ for all k and l. It results that $u_{11}(x) = I$ and $u_{10}(x) = 0$ for all $x \in \mathbf{R}$. Analogously one can show that $u_{01}(x) = 0$ for all $x \in \mathbf{R}$. Thus,

$$u(x) = \begin{pmatrix} u_{00}(x) & 0 \\ 0 & I \end{pmatrix} \quad \text{with } M(u_{00}) = D_0.$$

Since $u_{00} - D_0 = (I - D_0 u_{00}^*)u_{00}$ and $\|D_0 u_{00}^*(x)\| \le \|D_0\| < 1$ for all $x \in \mathbf{R}$, it follows that $u_{00} - D_0$ is invertible and that

$$(u_{00} - D_0)^{-1} = \sum_{k=0}^{\infty} u_{00}^*(D_0 u_{00}^*)^k \in APW^-_{m\times m}.$$

Therefore $\varphi_- := I + D_0(u_{00} - D_0)^{-1}$ also belongs to $APW^-_{m\times m}$. As

$$\varphi_- = (u_{00} - D_0)(u_{00} - D_0)^{-1} + D_0(u_{00} - D_0)^{-1} = u_{00}(u_{00} - D_0)^{-1},$$

we see that φ_- is invertible and that

$$\varphi_-^{-1} = (u_{00} - D_0)u_{00}^* = I - D_0 u_{00}^* \in APW^-_{m\times m}.$$

Hence,

$$u_{00} = \varphi_-(u_{00} - D_0) \quad \text{with } \varphi_- \in GAPW^-_{m\times m}. \tag{9.39}$$

The Hermitian matrix function $h = (u_{00}-D_0)^*(u_{00}-D_0)$ is invertible in $APW_{m\times m}$ and therefore positive definite. Corollary 9.15 so implies that $h = \varphi_+^*\varphi_+$ with $\varphi_+ \in GAPW_{m\times m}^+$. Letting $v := (u_{00} - D_0)\varphi_+^{-1}$, we obtain

$$u_{00} - D_0 = v\varphi_+ \quad \text{with} \quad \varphi_+ \in GAPW_{m\times m}^+. \tag{9.40}$$

Clearly, $v \in APW_{m\times m}^+$. Because

$$v^*v = (\varphi_+^{-1})^*(u_{00} - D_0)^*(u_{00} - D_0)\varphi_+^{-1} = (\varphi_+^*)^{-1}h\varphi_+^{-1} = I,$$

the matrix function v is unitary valued. We have

$$M(v) = M(u_{00} - D_0)M(\varphi_+^{-1}) = \big(M(u_{00}) - D_0\big)M(\varphi_+^{-1}) = 0.$$

From (9.39) and (9.40) we obtain that

$$a = c_-\begin{pmatrix} u_{00} & 0 \\ 0 & I \end{pmatrix}c_+ = c_-\begin{pmatrix} \varphi_- & 0 \\ 0 & I \end{pmatrix}\begin{pmatrix} v & 0 \\ 0 & I \end{pmatrix}\begin{pmatrix} \varphi_+ & 0 \\ 0 & I \end{pmatrix}c_+$$

with

$$c_-\begin{pmatrix} \varphi_- & 0 \\ 0 & I \end{pmatrix} \in GAPW_{N\times N}^-, \quad \begin{pmatrix} \varphi_+ & 0 \\ 0 & I \end{pmatrix}c_+ \in GAPW_{N\times N}^+,$$

which proves the implication. □

Theorem 9.19. *For $a \in GAPW_{N\times N}$ the following are equivalent:*

(i) *$W(a)$ is right-invertible;*

(ii) *there are $a_\pm \in GAP_{N\times N}$ such that*
$$a = a_-a_+, \quad a_- \in AP_{N\times N}^-, \quad a_+^{-1} \in AP_{N\times N}^+;$$

(iii) *there exist $a_\pm \in GAPW_{N\times N}$ such that*
$$a = a_-a_+, \quad a_- \in APW_{N\times N}^-, \quad a_+^{-1} \in APW_{N\times N}^+;$$

(iv) *a admits a right $APW(u)$ factorization with $u \in APW_{N\times N}^-$, i.e.,*
$$a = c_-uc_+, \quad c_\pm \in GAPW_{N\times N}^\pm, \quad u \in APW_{N\times N}^- \text{ is unitary};$$

(v) *a has a right $APW(u)$ factorization as in* (iv) *with the additional property that*
$$u = \begin{pmatrix} v & 0 \\ 0 & I \end{pmatrix} \quad \text{and} \quad M(v) = 0.$$

Proof. This follows from Theorem 9.18 by passage to adjoints. □

Notes

In Section 9.1, we follow the approach of Antonevich; see [4, Chapter 2] and references therein. More general isomorphism theorems for non-commutative amenable groups can be found in [5], [119], [121]. The equivalence of invertibility for operators on Lebesgue and Besicovitch spaces (Section 9.2) was established in [122], [121]; a complete proof is published here for the first time. Results of this type, though for different kind of operators, had previously been known from [4], [201]. The original source of Theorem 9.5 is Bochner and Phillips' paper [27]. A full proof of that theorem can also be found in [149, Section 1.7.10].

Theorem 9.6 was first published in [11], but similar results in different settings appeared earlier in [8], [139], [149, Section 5.2.5], [199]. The exposition in Section 9.4 follows [120], [121], [122]. A full proof of Lemma 9.7 is also in [133]. The construction of the factorization there is similar to the scheme used for the classical factorization in Lebesgue spaces; see, for example, Theorem 3.4 in [155]. An alternative approach to the main result of this Section (Corollary 9.8), based on the norm formulas for Hankel operators on Besikovitch spaces, was announced in [126]; we refer to [121] and to Chapter 20 for the full account.

The results of Section 9.5 were obtained by Babadzhanyan and Rabinovich [7], [8]. Theorem 9.11 and its Corollaries 9.12, 9.13 were proved in [212]; see [57, Chapter 5] and [155, Chapter 7] for the corresponding results on the L^2 factorization of Hermitian L^∞ matrices. Section 9.7 is based on [136].

In [185], [215], the results of Sections 9.5 and 9.6 are applied to completion problems for positive and contractive APW functions. A generalization of the material of this chapter to the case of several variables is in [186], and further applications to completion problems for (matrix) functions of several variables, periodic or almost periodic, are in [9], [10], [187]. The K-theory associated with almost periodic Toeplitz operators is discussed in [61], [116], and [233]. Finally, some applications of AP factorization and related properties of Wiener-Hopf operators in mathematical physics and control theory can be found in [3], [51], [52].

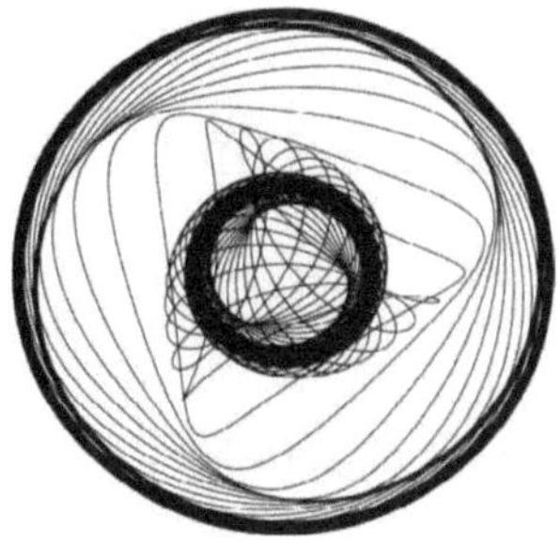

Chapter 10

Matrix Wiener-Hopf Operators with SAPW Symbols

In this chapter we prove a Fredholm criterion and an index formula for Wiener-Hopf operators with matrix symbols in $SAPW$, where $SAPW$ denotes the algebra of all semi-almost periodic functions whose almost periodic representatives a_l and a_r belong to APW. By using an argument familiar from the theory of so-called limit operators, we first show that $W(a_l)$ and $W(a_r)$ are invertible if $W(a)$ is Fredholm. Then we construct a factorization $a = f_- b f_+$ with the outer factors in $C + H^\infty_\mp$ and the middle factor in PC. In this way we can employ what we know about Wiener-Hopf operators with matrix-valued symbols in $C + H^\infty_\mp$ and PC in order to dispose of the Fredholm theory of operators with matrix symbols in $SAPW$.

As a by-product, we will see in Section 10.2 that if $a \in AP_{N\times N}$, then the operator $W(a)$ is Fredholm if and only if it is invertible (notice that $a_l = a_r = a$ in this case). This result in conjunction with the invertibility criterion of Section 9.4 settles the Fredholm theory of matrix Wiener-Hopf operators with APW symbols.

10.1 Invertibility of the Almost Periodic Representatives

Theorem 1.21 and its proof extend to the matrix case. Thus, given $u \in C(\overline{\mathbf{R}})$ such that $u(-\infty) = 0$ and $u(+\infty) = 1$, we can represent every matrix function $a \in SAP_{N\times N}$ in the form

$$a = (1-u)a_l + ua_r + a_0$$

where $a_0 \in [C_0(\dot{\mathbf{R}})]_{N\times N}$ and $a_l, a_r \in AP_{N\times N}$ are uniquely determined. The maps

$$SAP_{N\times N} \to AP_{N\times N},\ a \mapsto a_l, \tag{10.1}$$

$$SAP_{N\times N} \to AP_{N\times N},\ a \mapsto a_r, \tag{10.2}$$

are C^*-algebra homomorphisms. We call a_l and a_r the *almost periodic representatives* of a at $-\infty$ and $+\infty$, respectively.

Let $\mathcal{A}$ and $\mathcal{F}$ stand for the smallest closed subalgebras of $\mathcal{B}(L^2_N(\mathbf{R}))$ that contain the operators $\chi_\pm I$ and the sets $\{W^0(a) : a \in AP_{N\times N}\}$ and $\{W^0(a) : a \in SAP_{N\times N}\}$, respectively:

$$\mathcal{A} := \operatorname{alg}\big(\chi_\pm I, W^0(AP_{N\times N})\big), \quad \mathcal{F} := \operatorname{alg}\big(\chi_\pm I, W^0(SAP_{N\times N})\big).$$

Since $I \in W^0(AP)$, we could in these definitions replace $\chi_\pm I$ by only $\chi_+ I$. The purpose of this section is to show that the homomorphisms (10.1), (10.2) extend in a natural way to C^*-algebra homomorphisms of $\mathcal{F}$ onto $\mathcal{A}$. This will have some remarkable consequences.

It is well known that $\operatorname{alg}(\chi_\pm I, W^0(C(\overline{\mathbf{R}})))$ contains all compact operators. Since $C(\overline{\mathbf{R}}) \subset SAP$, it follows that $\mathcal{F}$ also contains all compact operators. Recall that $e_h(t) := e^{ith}$ $(t \in \mathbf{R})$.

Lemma 10.1. *If $K \in \mathcal{K}(L^2_N(\mathbf{R}))$ and $h_n \to +\infty$, then $e_{h_n} K e_{-h_n} I$ converges strongly to zero.*

Proof. It suffices to show the assertion for $N = 1$. If $f, g \in L^2(\mathbf{R})$, then

$$(e_{-h_n} f, g) = \int_{\mathbf{R}} e^{-ith_n} f(t)\overline{g(t)}\, dt = (f\overline{g})\hat{}\,(-h_n),$$

and as $f\overline{g} \in L^1(\mathbf{R})$ and therefore $(f\overline{g})\hat{} \in C_0(\dot{\mathbf{R}})$, it follows that $(f\overline{g})\hat{}\,(-h_n) \to 0$. Hence, $e_{-h_n} I$ converges weakly to zero. It results that $K e_{-h_n} I$ and also $e_{h_n} K e_{-h_n} I$ converges strongly to zero. □

For $a \in L^\infty_{N\times N}(\mathbf{R})$ and $h \in \mathbf{R}$, we define $a_h \in L^\infty_{N\times N}(\mathbf{R})$ by $a_h(x) := a(x+h)$, $x \in \mathbf{R}$. The simultaneous use of a_h and e_h should not cause confusion: we will never denote a matrix function in $L^\infty_{N\times N}$ by e, that is, e_h is always the function $e_h(t) = e^{iht}$, e_{-h} is always the function $e_{-h}(t) = e^{-iht}$, etc. A direct computation gives

$$e_h W^0(a) e_{-h} I = W^0(a_h). \tag{10.3}$$

Lemma 10.2. *If $a_1, \dots, a_M \in AP^0_{N\times N}$ is a finite collection of almost periodic matrix polynomials, then there exists a sequence $\{h_n\} \subset \mathbf{R}$ such that $h_n \to +\infty$ and $\|(a_m)_{h_n} - a_m\|_\infty \to 0$ for each $m \in \{1, \dots, M\}$ as $n \to \infty$.*

Proof. Let $a_m(x) = \sum_j a_j^{(m)} e^{i\lambda_j^{(m)} x}$. By Lemma 1.14, there are numbers $\beta_1, \dots, \beta_s$ which are linearly independent over $\mathbf{Z}$ and integers $k_1^{(j,m)}, \dots, k_s^{(j,m)}$ such that

$$\lambda_j^{(m)} = k_1^{(j,m)} \beta_1 + \ldots + k_s^{(j,m)} \beta_s \text{ for all } j, m.$$

We have

$$\begin{aligned} \|(a_m)_h - a_m\|_\infty &\le \sum_j \|a_j^{(m)}\| \, |e^{i\lambda_j^{(m)} h} - 1| \\ &= \sum_j \|a_j^{(m)}\| \, |(e^{i\beta_1 h})^{k_1^{(j,m)}} \ldots (e^{i\beta_s h})^{k_s^{(j,m)}} - 1|, \end{aligned} \tag{10.4}$$

and due to Corollary 1.13, we can find $h_n \to +\infty$ such that

$$(e^{i\beta_1 h_n}, \dots, e^{i\beta_s h_n}) \to (1, \dots, 1) \text{ as } n \to \infty,$$

which implies that for $h = h_n$ the right-hand side of (10.4) becomes as small as desired. □

Theorem 10.3. *The maps*

$$\begin{aligned} \mu_r &: \chi_\pm I \mapsto \chi_\pm I, \; W^0(a) \mapsto W^0(a_r) \quad (a \in SAP_{N\times N}), \\ \mu_l &: \chi_\pm I \mapsto \chi_\pm I, \; W^0(a) \mapsto W^0(a_l) \quad (a \in SAP_{N\times N}) \end{aligned}$$

extend to C^*-algebra homomorphisms of $\mathcal{F}$ onto $\mathcal{A}$ such that

$$\|\mu_r(A)\| \le \|A\|_{\text{ess}}, \quad \|\mu_l(A)\| \le \|A\|_{\text{ess}}$$

for all $A \in \mathcal{F}$.

Proof. It is sufficient to prove the theorem for $N = 1$. A dense subset of $\mathcal{F}$ is constituted by the operators $A = \sum_j \prod_k A_{jk}$ where the sum and the (ordered) products are finite and each A_{jk} is either of the form $\chi_\pm I$ or of the form

$$W^0(a_{jk}) = W^0\big((1-u)a_{jk}^l + u a_{jk}^r + a_{jk}^0\big) \tag{10.5}$$

with $a_{jk}^0 \in C_0(\dot{\mathbf{R}})$ and almost periodic polynomials a_{jk}^l, a_{jk}^r. For such operators we define

$$\mu_r(A) := \sum_j \prod_k \mu_r(A_{jk})$$

where $\mu_r(\chi_\pm I) := \chi_\pm I$ and μ_r takes the value $W^0(a_{jk}^r)$ at the operator (10.5). The assertion will follow once we have shown that

$$\|\mu_r(A)\| \le \|A + K\| \text{ for every } K \in \mathcal{K}(L^2(\mathbf{R})). \tag{10.6}$$

By Lemma 10.2, we can find a sequence $h_n \to +\infty$ such that

$$\|(a^r_{jk})_{h_n} - a^r_{jk}\|_\infty \to 0 \text{ for all } j, k. \tag{10.7}$$

We claim that for each $K \in \mathcal{K}(L^2(\mathbf{R}))$ the strong limit

$$\text{s-lim}_{n\to\infty}\, e_{h_n}(A+K)e_{-h_n}I$$

exists and equals $\mu_r(A)$. Clearly, this gives (10.6). By virtue of Lemma 10.1, we are left to show that

$$\text{s-lim}_{n\to\infty}\, e_{h_n}Ae_{-h_n}I = \mu_r(A). \tag{10.8}$$

Obviously,

$$\text{s-lim}_{n\to\infty}\, e_{h_n}Ae_{-h_n}I = \text{s-lim}_{n\to\infty} \sum_j \prod_k e_{h_n}A_{jk}e_{-h_n}I.$$

If $A_{jk} = \chi_\pm I$, then

$$e_{h_n}A_{jk}e_{-h_n}I = e_{h_n}\chi_\pm e_{-h_n}I = \chi_\pm I = A_{jk} = \mu_r(A_{jk}).$$

So let A_{jk} be the operator (10.5). For brevity, we omit the subscripts j, k. In view of (10.3), we have to show that

$$\|W^0(a_{h_n})f - W^0(a_r)f\|_2 \to 0 \text{ as } n \to \infty \tag{10.9}$$

for all f in a dense subset of $L^2(\mathbf{R})$. If $f \in L^2(\mathbf{R})$, then $Ff \in L^2(\mathbf{R})$ can be approximated by functions $g_n \in L^2(\mathbf{R})$ with compact support as closely as desired. Clearly, $\|f - F^{-1}g_n\|_2 \to 0$. Thus, the set of all $f \in L^2(\mathbf{R})$ for which Ff has compact support is dense in $L^2(\mathbf{R})$, and it remains to verify (10.9) if $Ff \in L^2(\mathbf{R})$ has finite support. For such f we have

$$\begin{aligned}
&\|W^0(a_{h_n})f - W^0(a_r)f\|_2 \\
&\le \|F^{-1}\|\, \big\|\big((1-u)a_l + ua_r + a_0\big)_{h_n}Ff - a_rFf\big\|_2 \\
&\le \|F^{-1}\| \sup_{x\in\operatorname{supp} Ff} \big|(1-u(x+h_n))a_l(x+h_n)\big|\, \|Ff\|_2 \\
&\quad + \|F^{-1}\| \sup_{x\in\operatorname{supp} Ff} \big|u(x+h_n)a_r(x+h_n) - a_r(x)\big|\, \|Ff\|_2 \\
&\quad + \|F^{-1}\| \sup_{x\in\operatorname{supp} Ff} |a_0(x+h_n)|\, \|Ff\|_2.
\end{aligned} \tag{10.10}$$

It is clear that the first and third terms on the right of (10.10) go to zero as $h_n \to +\infty$. Further,

$$\begin{aligned}
&\sup_{x\in\operatorname{supp} Ff} \big|u(x+h_n)a_r(x+h_n) - a_r(x)\big| \\
&\le \sup_{x\in\operatorname{supp} Ff} \big|(u(x+h_n)-1)a_r(x+h_n)\big| + \sup_{x\in\operatorname{supp} Ff} \big|a_r(x+h_n) - a_r(x)\big|,
\end{aligned}$$

the first term on the right goes obviously to zero, and the second term approaches zero due to (10.7). This gives (10.9) and completes the proof for μ_r. The proof is analogous for μ_l. □

Corollary 10.4. *If $A \in \mathcal{F}$ is Fredholm (n-normal, resp. d-normal), then $\mu_r(A)$ and $\mu_l(A)$ are invertible (left-invertible, resp. right-invertible).*

Proof. Let $\mathcal{K} := \mathcal{K}(L^2_N(\mathbf{R}))$. If $A \in \mathcal{F}$ is Fredholm, then $A + \mathcal{K}$ is invertible in the Calkin algebra $\mathcal{B}/\mathcal{K}$ and hence in the C^*-subalgebra $\mathcal{F}/\mathcal{K}$ of $\mathcal{B}/\mathcal{K}$. Consequently, there is a $B \in \mathcal{F}$ such that $BA = I + K_1$ and $AB = I + K_2$ with $K_1, K_2 \in \mathcal{K}$. From Theorem 10.3 we deduce that

$$\mu_r(B)\mu_r(A) = I = \mu_r(A)\mu_r(B), \quad \mu_l(B)\mu_l(A) = I = \mu_l(A)\mu_l(B),$$

which shows that $\mu_r(A)$ and $\mu_l(A)$ are invertible. If $A \in \mathcal{F}$ is n-normal, then A^*A is Fredholm by (2.1), hence

$$\mu_r(A^*A) = \big(\mu_r(A)\big)^*\mu_r(A), \quad \mu_l(A^*A) = \big(\mu_l(A)\big)^*\mu_l(A)$$

are invertible, which implies that $\mu_r(A)$ and $\mu_l(A)$ are left-invertible. Analogously one can treat the case of d-normality. □

Corollary 10.5. *Let $a \in SAP_{N\times N}$. If $W(a)$ is Fredholm (n-normal, resp. d-normal), then $W(a_l)$ and $W(a_r)$ are invertible (left-invertible, resp. right-invertible).*

Proof. This follows from Corollary 10.4 and the fact that $W(b)$ is strongly Φ-equivalent to the operator $\chi_- I + W^0(b)\chi_+ I$ (recall the proof of Lemma 9.2). □

10.2 Fredholmness of Operators with *APW* Symbols

We are now in a position to complete the results of Section 9.4 and to prove Theorem 5.15.

Corollary 10.6. *If $a \in AP_{N\times N}$, then*

$$\begin{aligned} W(a) &\text{ is Fredholm} &\iff W(a) &\text{ is invertible},\\ W(a) &\text{ is } n\text{-normal} &\iff W(a) &\text{ is left-invertible},\\ W(a) &\text{ is } d\text{-normal} &\iff W(a) &\text{ is right-invertible}.\end{aligned}$$

Proof. Since $a_l = a_r = a$ whenever $a \in AP_{N\times N}$, this is immediate from Corollary 10.5. □

In particular, for $a \in APW_{N\times N}$ we now obtain the following.

Corollary 10.7. *For $a \in GAPW_{N\times N}$ the following are equivalent:*

(i) $W(a)$ *is Fredholm;*

(ii) $W(a)$ *is invertible;*

(iii) a *has a canonical right AP factorization;*

(iv) a *has a canonical right APW factorization.*

Proof. The implications (iv)⇒(iii)⇒(ii)⇒(i) are trivial, the implication (i)⇒(ii) results from Corollary 10.6, and the implication (ii)⇒(iv) can be deduced from Corollary 9.8. □

Here is a rather curious consequence of Corollary 10.6 and Theorems 9.18, 9.19.

Corollary 10.8. *Let $a \in AP_{N\times N}$ and suppose $W(a)$ is semi-Fredholm. Then*

$$W(a) \text{ is invertible} \Longleftrightarrow W(\det a) \text{ is invertible.}$$

Proof. The semi-Fredholmness of $W(a)$ implies that $a \in GAP_{N\times N}$ (Theorem 5.1). For the sake of definiteness, suppose $W(a)$ is n-normal.

Let us first assume that $a \in GAPW_{N\times N}$. The implication "⇒" is immediate from Theorem 9.16. To prove the implication "⇐", let $W(\det a)$ be invertible. Then $\det a = \varphi_-\varphi_+$ with $\varphi_\pm \in GAPW^\pm$ due to Theorems 2.28 and 8.11. On the other hand, Corollary 10.6 and Theorem 9.18 show that $a = a_-a_+$ with

$$a_\pm \in GAPW_{N\times N}, \quad a_-^{-1} \in APW_{N\times N}^-, \quad a_+ \in APW_{N\times N}^+,$$

whence $\det a = (\det a_-)(\det a_+)$ with

$$\det a_\pm \in GAPW, \quad (\det a_-)^{-1} \in APW^-, \quad \det a_+ \in APW^+. \tag{10.11}$$

It follows that $(\det a_-)^{-1}\varphi_- = \varphi_+^{-1}(\det a_+)$, and since the left-hand side is in APW^- and the right-hand side belongs to APW^+, both of them must be a nonzero constant c:

$$(\det a_-)^{-1}\varphi_- = \varphi_+^{-1}(\det a_+) = c.$$

Hence,

$$\det a_- = c^{-1}\varphi_- \in APW^-, \quad (\det a_+)^{-1} = c^{-1}\varphi_+^{-1} \in APW^+. \tag{10.12}$$

From (10.11) and (10.12) we see that $a_\pm \in GAPW_{N\times N}^\pm$, which gives the invertibility of $W(a)$.

Now assume $a \in GAP_{N\times N}$. If $b \in GAPW_{N\times N}$ is sufficiently close to a, then $W(b)$ is n-normal,

$$\operatorname{Ind} W(a) = \operatorname{Ind} W(b), \tag{10.13}$$

and, by Proposition 8.18, $\kappa(\det a) = \kappa(\det b)$. If $W(a)$ is invertible, then so is $W(b)$, and from what was already proved, we know that $W(\det b)$ is invertible. Hence $\kappa(\det a) = \kappa(\det b) = 0$, which, by Theorem 2.28, gives the invertibility of $W(\det a)$. Conversely, the invertibility of $W(\det a)$ implies the invertibility of $W(\det b)$. Again by what was already proved, $W(b)$ is invertible. From (10.13) we therefore obtain that $W(a)$ is Fredholm, and Corollary 10.6 finally shows that $W(a)$ is invertible. □

10.3 Reduction of SAP Symbols to PC Symbols

In this section we establish a matrix analogue of Lemma 3.11. This will allow us to reduce questions about $SAP_{N\times N}$ symbols to problems on symbols in $PC_{N\times N}$.

Theorem 10.9. *Let $a, b \in GSAP_{N\times N}$ and suppose the almost periodic representatives a_l, b_l, a_r, b_r satisfy*

$$a_l = \varphi_l^- b_l \varphi_l^+ \quad \textit{with} \quad \varphi_l^\pm \in GAP^\pm_{N\times N},$$
$$a_r = \varphi_r^- b_r \varphi_r^+ \quad \textit{with} \quad \varphi_r^\pm \in GAP^\pm_{N\times N},$$
$$M(\varphi_l^-) = M(\varphi_r^-), \quad M(\varphi_l^+) = M(\varphi_r^+).$$

Then $a = f_- b f_+$ with certain matrix functions

$$f_\pm \in GSAP_{N\times N} \cap G\big[C(\dot{\mathbf{R}}) + H^\infty_\pm\big]_{N\times N}.$$

Proof. Put $c_0 := M(\varphi_l^+) = M(\varphi_r^+)$ and choose almost periodic matrix polynomials

$$\psi_l^+(x) = c_0 + \sum r_j e^{i\lambda_j x}, \quad \psi_r^+(x) = c_0 + \sum s_j e^{i\lambda_j x}$$

such that $\lambda_j > 0$ for all j and

$$\|\psi_l^+ - \varphi_l^+\|_\infty < q/\|(\varphi_l^+)^{-1}\|_\infty, \quad \|\psi_r^+ - \varphi_r^+\|_\infty < q/\|(\varphi_r^+)^{-1}\|_\infty \tag{10.14}$$

with some $q \in (0,1)$. Put

$$v := I + \big(\psi_r^+(\varphi_r^+)^{-1} - I\big)u + \big(\psi_l^+(\varphi_l^+)^{-1} - I\big)(1-u),$$

where $u \in C(\overline{\mathbf{R}})$ is any fixed function such that $0 \le u \le 1$ and $u(-\infty) = 0$, $u(+\infty) = 1$. Since $M(\psi_r^+(\varphi_r^+)^{-1} - I) = M(\psi_r^+)(M(\varphi_r^+))^{-1} - I = 0$, we deduce from Lemma 3.5 that

$$\big(\psi_r^+(\varphi_r^+)^{-1} - I\big)u \in \big[C(\dot{\mathbf{R}}) + H^\infty_+\big]_{N\times N}.$$

Analogously,

$$\big(\psi_l^+(\varphi_l^+)^{-1} - I\big)(1-u) \in \big[C(\dot{\mathbf{R}}) + H^\infty_+\big]_{N\times N},$$

and thus, $v \in SAP_{N\times N} \cap [C(\dot{\mathbf{R}}) + H^\infty_+]_{N\times N}$. By (10.14),

$$\|\psi_r^+(\varphi_r^+)^{-1} - I\|_\infty \le \|\psi_r^+ - \varphi_r^+\|_\infty \|(\varphi_r^+)^{-1}\|_\infty \le q < 1,$$
$$\|\psi_l^+(\varphi_l^+)^{-1} - I\|_\infty \le \|\psi_l^+ - \varphi_l^+\|_\infty \|(\varphi_l^+)^{-1}\|_\infty \le q < 1,$$

whence $\|v(x) - I\|_\infty \le qu(x) + q(1-u(x)) = q < 1$ for all $x \in \mathbf{R}$. This shows that

$$v \in GSAP_{N\times N} \cap G\big[C(\dot{\mathbf{R}}) + H^\infty_+\big]_{N\times N}. \tag{10.15}$$

In what follows, we let $e^{-\infty} := 0$. Put

$$w(x) := c_0 + \sum p_j(x) e^{i\lambda_j x}$$

where

$$p_j(x) := \begin{cases} s_j e^{-\lambda_j/x} & \text{for} \quad x \ge 0, \\ r_j e^{\lambda_j/x} & \text{for} \quad x < 0. \end{cases}$$

Notice that $p_j(0-0) = p_j(0+0) = p_j(0) = 0$ and $p_j(-\infty) = r_j$, $p_j(+\infty) = s_j$. Thus, $p_j \in C_{N\times N}(\overline{\mathbf{R}})$. Again by Lemma 3.5,

$$w \in SAP_{N\times N} \cap \big[C(\dot{\mathbf{R}}) + H^\infty_+\big]_{N\times N}. \tag{10.16}$$

From (10.14) we infer that ψ_l^+ and ψ_r^+ are invertible in $AP^+_{N\times N}$. As ψ_l^+ and ψ_r^+ are almost periodic matrix polynomials, they are even invertible in $APW^+_{N\times N}$. Thus,

$$\big(\psi_l^+(x)\big)^{-1} = \tilde{c}_0 + \sum \tilde{r}_j e^{i\tilde{\lambda}_j x}, \quad \tilde{\lambda}_j > 0, \quad \sum \|\tilde{r}_j\|_\infty < \infty,$$
$$\big(\psi_r^+(x)\big)^{-1} = \tilde{c}_0 + \sum \tilde{s}_j e^{i\tilde{\lambda}_j x}, \quad \tilde{\lambda}_j > 0, \quad \sum \|\tilde{s}_j\|_\infty < \infty.$$

Put

$$\tilde{w}(x) := \tilde{c}_0 + \sum \tilde{p}_j(x) e^{i\tilde{\lambda}_j x}$$

where

$$\tilde{p}_j(x) := \begin{cases} \tilde{s}_j e^{-\tilde{\lambda}_j/x} & \text{for} \quad x \ge 0, \\ \tilde{r}_j e^{\tilde{\lambda}_j/x} & \text{for} \quad x < 0. \end{cases}$$

As above, we see that

$$\tilde{w} \in SAP_{N\times N} \cap \big[C(\dot{\mathbf{R}}) + H^\infty_+\big]_{N\times N}. \tag{10.17}$$

If $x > 0$, then

$$w(x) = c_0 + \sum s_j e^{-\lambda_j/x} e^{i\lambda_j x} = c_0 + \sum s_j e^{i\lambda_j(x+i/x)} = \psi_r^+\Big(x + \frac{i}{x}\Big),$$
$$\tilde{w}(x) = \tilde{c}_0 + \sum \tilde{s}_j e^{-\tilde{\lambda}_j/x} e^{i\tilde{\lambda}_j x} = \tilde{c}_0 + \sum \tilde{s}_j e^{i\tilde{\lambda}_j(x+i/x)} = \Big[\psi_r^+\Big(x + \frac{i}{x}\Big)\Big]^{-1}.$$

Analogously, for $x < 0$,

$$w(x) = \psi_l^+\Big(x - \frac{i}{x}\Big), \quad \tilde{w}(x) = \Big[\psi_l^+\Big(x - \frac{i}{x}\Big)\Big]^{-1}.$$

Consequently, $w\tilde{w} = \tilde{w}w = I$ on $\mathbf{R} \setminus \{0\}$. From (10.16) and (10.17) we therefore get

$$w \in GSAP_{N\times N} \cap G\big[C(\dot{\mathbf{R}}) + H_+^\infty\big]_{N\times N}. \tag{10.18}$$

Combining (10.15) and (10.18) we obtain that

$$g^+ := v^{-1}w \in GSAP_{N\times N} \cap G\big[C(\dot{\mathbf{R}}) + H_+^\infty\big]_{N\times N}. \tag{10.19}$$

Clearly,

$$(g^+)_l = v_l^{-1}w_l = \Big(I + \big(\psi_l^+(\varphi_l^+)^{-1} - I\big)\Big)^{-1}\psi_l^+ = \varphi_l^+, \tag{10.20}$$

$$(g^+)_r = v_r^{-1}w_r = \Big(I + \big(\psi_r^+(\varphi_r^+)^{-1} - I\big)\Big)^{-1}\psi_r^+ = \varphi_r^+. \tag{10.21}$$

Analogously one can construct

$$g^- \in GSAP_{N\times N} \cap G\big[C(\dot{\mathbf{R}}) + H_-^\infty\big]_{N\times N} \tag{10.22}$$

such that

$$(g^-)_l = \varphi_l^-, \quad (g^-)_r = \varphi_r^-. \tag{10.23}$$

Now put $h := g^-bg^+$. We claim that the theorem is true for $f_- := g^-$ and $f_+ := g^+h^{-1}a$. First of all, we have

$$f_-bf_+ = g^-bg^+(g^-bg^+)^{-1}a = a. \tag{10.24}$$

By (10.22),

$$f_- \in GSAP_{N\times N} \cap G\big[C(\dot{\mathbf{R}}) + H_-^\infty\big]_{N\times N}. \tag{10.25}$$

Since $g^-, b, g^+ \in GSAP_{N\times N}$, it follows that $h \in GSAP_{N\times N}$ and thus

$$f_+ \in GSAP_{N\times N}. \tag{10.26}$$

From (10.20), (10.21), (10.23) we get

$$h_l = (g^-)_l b_l (g^+)_l = \varphi_l^- b_l \varphi_l^+ = a_l,$$
$$h_r = (g^-)_r b_r (g^+)_r = \varphi_r^- b_r \varphi_r^+ = a_r,$$

and hence $a - h \in [C_0(\dot{\mathbf{R}})]_{N\times N}$. This implies that

$$h^{-1}a = I + h^{-1}(a-h) \in C_{N\times N}(\dot{\mathbf{R}}), \quad a^{-1}h = I - a^{-1}(a-h) \in C_{N\times N}(\dot{\mathbf{R}})$$

and shows that

$$f_+ = g^+ h^{-1} a \in G\big[C(\dot{\mathbf{R}}) + H_+^\infty\big]_{N\times N}. \tag{10.27}$$

With (10.24), (10.25), (10.26), (10.27) the proof is complete. □

Corollary 10.10. *If $a, b \in GSAP_{N\times N}$ are subject to the hypotheses of Theorem 10.9, then $W(a)$ and $W(b)$ are weakly Φ-equivalent. Moreover, if $a = f_- b f_+$ is the factorization ensured by Theorem 10.9 and one of the operators $W(a)$ and $W(b)$ is semi-Fredholm, then*

$$\operatorname{Ind} W(a) = \operatorname{Ind} W(f_-) + \operatorname{Ind} W(b) + \operatorname{Ind} W(f_+). \tag{10.28}$$

Proof. Using (5.1) to (5.4) and Theorem 5.7(c), one can proceed as in the proof of Lemma 3.10, but with one difference: one now cannot have recourse to Corollary 2.6(a). Instead of this, one can use the fact that if A is normally solvable and B_1, B_2 are Fredholm, then $B_1 A B_2$ is also normally solvable (see, e.g., [99, Vol. 1, Corollaries 2.2 and 2.3 of Chapter 4]). □

10.4 Fredholmness and Index of Operators with $SAPW$ Symbols

Recall that if $c \in GAP_{N\times N}$ has a right AP factorization with the right AP indices $\lambda_1 \ge \ldots \ge \lambda_N$, then $\kappa(c)$ is defined as the vector $(\lambda_1, \ldots, \lambda_N)$. In case $\kappa(c) = (0, \ldots, 0)$, i.e., if c has a canonical right AP factorization $c = c_- c_+$, we put $\mathbf{d}(c) := M(c_-)M(c_+)$ (Proposition 8.4 and Definition 8.5).

Theorem 10.11. *Let $a \in SAP_{N\times N}$ and assume the almost periodic representatives a_l, a_r admit a right AP factorization. Then $W(a)$ is Fredholm if and only if*

$$a \in GSAP_{N\times N}, \quad \kappa(a_l) = \kappa(a_r) = (0, \ldots, 0), \tag{10.29}$$

$$\operatorname{sp}\left(\mathbf{d}^{-1}(a_r)\mathbf{d}(a_l)\right) \cap (-\infty, 0] = \emptyset, \tag{10.30}$$

where $\operatorname{sp}(\mathbf{d}^{-1}(a_r)\mathbf{d}(a_l))$ stands for the set of the eigenvalues of the matrix

$$\mathbf{d}^{-1}(a_r)\mathbf{d}(a_l) := (\mathbf{d}(a_r))^{-1}\mathbf{d}(a_l).$$

Proof. If $W(a)$ is Fredholm, then $a \in GSAP_{N\times N}$ by Theorem 5.1 and $W(a_l)$, $W(a_r)$ are invertible by Corollary 10.5. As already observed after Definition 6.8, this implies that $\kappa(a_l)$ and $\kappa(a_r)$ are the zero vectors.

Now suppose (10.29) holds. Let $a_l = a_l^- a_l^+$ and $a_r = a_r^- a_r^+$ be canonical right AP factorizations. We can write

$$a_l = a_l^- M^{-1}(a_l^-)\mathbf{d}(a_l)M^{-1}(a_l^+)a_l^+,$$

$$a_r = a_r^- M^{-1}(a_r^-)\mathbf{d}(a_r)M^{-1}(a_r^+)a_r^+.$$

Define $b \in C_{N\times N}(\overline{\mathbf{R}})$ by $b = (1-u)\mathbf{d}(a_l) + u\mathbf{d}(a_r)$. Then $b_l = \mathbf{d}(a_l)$, $b_r = \mathbf{d}(a_r)$, and hence

$$a_l = \varphi_l^- b_l \varphi_l^+, \quad a_r = \varphi_r^- b_r \varphi_r^+$$

with

$$\begin{aligned}\varphi_l^- &= a_l^- M^{-1}(a_l^-), \quad \varphi_l^+ = M^{-1}(a_l^+)a_l^+,\\ \varphi_r^- &= a_r^- M^{-1}(a_r^-), \quad \varphi_r^+ = M^{-1}(a_r^+)a_r^+.\end{aligned}$$

Clearly, $\varphi_l^-, \varphi_r^- \in GAP_{N\times N}^-$, $\varphi_l^+, \varphi_r^+ \in GAP_{N\times N}^+$, and

$$M(\varphi_l^-) = M(\varphi_r^-) = M(\varphi_l^+) = M(\varphi_r^+) = I.$$

Thus, by Corollary 10.10, $W(a)$ and $W(b)$ are weakly Φ-equivalent. This and Theorem 5.10 show that (10.30) holds if $W(a)$ is Fredholm and also show that $W(a)$ must be Fredholm whenever (10.29) and (10.30) are satisfied. □

If (10.29) is valid, then $\det a \in GSAP_{0,0}$ (recall Section 3.4) and hence the Cauchy index $\operatorname{ind}(\det a)$ is well-defined (Definition 3.13).

Theorem 10.12. *If $a \in SAP_{N\times N}$, if the almost periodic representatives a_l, a_r admit right AP factorizations, and if $W(a)$ is Fredholm, then*

$$\operatorname{Ind} W(a) = -\operatorname{ind}(\det a) - \sum_{k=1}^{N}\left(\frac{1}{2} - \left\{\frac{1}{2} - \frac{1}{2\pi}\arg\xi_k\right\}\right) \tag{10.31}$$

where $\xi_1, \ldots, \xi_N \in \mathbf{C}\setminus(-\infty, 0]$ are the eigenvalues of the matrix $\mathbf{d}^{-1}(a_r)\mathbf{d}(a_l)$ and $\{\cdot\}$ stands for the fractional part. When choosing $\arg\xi_k$ in $(-\pi,\pi)$, we have

$$\operatorname{Ind} W(a) = -\operatorname{ind}(\det a) - \frac{1}{2\pi}\sum_{k=1}^{N}\arg\xi_k. \tag{10.32}$$

Proof. Let $a_l^\pm, a_r^\pm, b, \varphi_l^\pm, \varphi_r^\pm$ be as in the proof of Theorem 10.11 and write $a = f_- b f_+$ as in Theorem 10.9. Then (10.28) holds. Theorems 5.7 and 2.19 give

$$\operatorname{Ind} W(f_\pm) = \operatorname{Ind} W(\det f_\pm). \tag{10.33}$$

From Theorem 5.10 we infer that

$$\operatorname{Ind} W(b) = -\operatorname{ind}(\det b) - \frac{1}{2\pi}\sum_{k=1}^{N}\arg\xi_k \tag{10.34}$$

if only the arguments $\arg\xi_k$ are taken in $(-\pi,\pi)$. Condition (10.30) also guarantees that $\det(\mathbf{d}^{-1}(a_r)\mathbf{d}(a_l)) = \xi_1 \ldots \xi_N \neq 0$.

We know from Example 5.11 that the Fredholmness of $W(b)$ does not necessarily imply the Fredholmness of $W(\det b)$. Thus, let us for a moment assume that

$$\det\left(\mathbf{d}^{-1}(a_r)\mathbf{d}(a_l)\right) = \xi_1 \dots \xi_N \notin (-\infty, 0). \tag{10.35}$$

Since $\det b = \det f_- \det a \det f_+$ is nonzero on $\mathbf{R}$ and

$$\det b(-\infty) = \det \mathbf{d}(a_l), \quad \det b(+\infty) = \det \mathbf{d}(a_r),$$

condition (10.35) and Theorem 5.10 imply that $W(\det b)$ is Fredholm and

$$\operatorname{Ind} W(\det b) = -\operatorname{ind}(\det b) - \frac{1}{2\pi} \arg \frac{\det \mathbf{d}(a_l)}{\det \mathbf{d}(a_r)}, \tag{10.36}$$

provided $\arg(\det \mathbf{d}(a_l)/\det \mathbf{d}(a_r))$ is taken in $(-\pi, \pi)$. Inserting (10.33), (10.34), (10.36) in (10.28) we get

$$\begin{aligned} \operatorname{Ind} W(a) &= \operatorname{Ind} W(\det f_-) + \operatorname{Ind} W(\det b) + \operatorname{Ind} W(\det f_+) \\ &\quad + \frac{1}{2\pi} \arg \frac{\det \mathbf{d}(a_l)}{\det \mathbf{d}(a_r)} - \frac{1}{2\pi} \sum_{k=1}^{N} \arg \xi_k. \end{aligned} \tag{10.37}$$

By Theorem 3.14 and (3.25),

$$\operatorname{Ind} W(\det a) = -\operatorname{ind}(\det a) - \frac{1}{2\pi} \arg \frac{\det \mathbf{d}(a_l)}{\det \mathbf{d}(a_r)}. \tag{10.38}$$

It is clear that

$$\operatorname{Ind} W(\det f_-) + \operatorname{Ind} W(\det b) + \operatorname{Ind} W(\det f_+) = \operatorname{Ind} W(\det a). \tag{10.39}$$

Substituting (10.38) and (10.39) into (10.37), we arrive at (10.32).

It remains to remove the temporary assumption (10.35). So suppose

$$\det\left(\mathbf{d}^{-1}(a_r)\mathbf{d}(a_l)\right) = \xi_1 \dots \xi_N \in (-\infty, 0).$$

We have

$$a = (1-u)a_l^- a_l^+ + u a_r^- a_r^+ + a_0$$

and we consider the perturbation

$$a^\varepsilon := (1-u)a_l^-(a_l^+ + \varepsilon I) + u a_r^- a_r^+ + a_0 \tag{10.40}$$

where $\varepsilon \in \mathbf{C}$. Obviously, $\mathbf{d}((a^\varepsilon)_r) = \mathbf{d}(a_r)$ and

$$\mathbf{d}((a^\varepsilon)_l) = M(a_l^-)M(a_l^+ + \varepsilon I) = \mathbf{d}(a_l) + \varepsilon M(a_l^-).$$

The function

$$\begin{aligned} f(\varepsilon) &:= \det\left[\mathbf{d}^{-1}((a^{\varepsilon})_r)\mathbf{d}((a^{\varepsilon})_l)\right] \\ &= \det\left[\mathbf{d}^{-1}(a_r)\mathbf{d}(a_l)+\varepsilon\mathbf{d}^{-1}(a_r)M(a_l^{-})\right] \end{aligned}$$

is a polynomial of degree N with the leading coefficient

$$\det\left[\mathbf{d}^{-1}(a_r)M(a_l)\right]=\det\mathbf{d}^{-1}(a_r)\det M(a_l^{-})\neq 0.$$

Hence, there exists a sequence $\varepsilon_n\to 0$ such that $f(\varepsilon_n)\notin(-\infty,0)$ for all n. Changing a to a^{ε_n} does not violate (10.29) and (10.30) if only $|\varepsilon_n|$ is sufficiently small (note that the almost periodic representatives of a^{ε} have right AP factorizations by (10.40)). Since a^{ε_n} satisfies (10.35), we get

$$\operatorname{Ind}W(a^{\varepsilon_n})=-\operatorname{ind}(\det a^{\varepsilon_n})-\frac{1}{2\pi}\sum_{k=1}^{N}\arg\xi_k(\varepsilon_n),$$

and the limit $\varepsilon_n\to 0$ gives formula (10.32). □

For symbols $a\in SAPW_{N\times N}$ the requirement that the almost periodic representatives a_l, a_r have a right AP factorization is redundant in the hypothesis of Theorems 10.11 and 10.12. In this case the theorems can be formulated as follows.

Theorem 10.13. *Let $a\in SAPW_{N\times N}$. Then $W(a)$ is Fredholm if and only if $a\in GSAPW_{N\times N}$, the matrix functions a_l, a_r have canonical right APW factorizations, and* (10.30) *holds. In the case of Fredholmness, the index of $W(a)$ is given by* (10.31) *and* (10.32).

Proof. If $W(a)$ is Fredholm, we deduce from Corollaries 10.5 and 10.7 that a_l and a_r have canonical right APW factorizations. The assertion now follows from Theorems 10.11 and 10.12. □

10.5 Two Results from Linear Algebra

In this section we prove two lemmas that will be needed in the following section.

Lemma 10.14. *Let $C=(c_{ij})_{i,j=1}^{2}$ be a block matrix with blocks $c_{ij}\in\mathbf{C}^{m_i\times m_j}$. If $\det C\neq 0$, then the set*

$$\left\{x\in\mathbf{C}^{m_1\times m_2}:\det(c_{22}+c_{21}x)\neq 0\right\}$$

is dense in $\mathbf{C}^{m_1\times m_2}$.

Proof. Fix an arbitrary $x_0\in\mathbf{C}^{m_1\times m_2}$ and put

$$D:=C\begin{pmatrix} I & x_0\\ 0 & I\end{pmatrix}=\begin{pmatrix} c_{11} & c_{11}x_0+c_{12}\\ c_{21} & c_{21}x_0+c_{22}\end{pmatrix}.$$

Denote by $\xi_1, \dots, \xi_{m_1}$ the columns of c_{21} and by $\eta_1, \dots, \eta_{m_2}$ the columns of $d_{22} := c_{21}x_0 + c_{22}$. Since $\det D \neq 0$, the matrix $(c_{21}d_{22})$ has the rank m_2. Let r be the rank of d_{22} and put $k = m_2 - r$. We can find r linearly independent columns $\eta_{l_1}, \dots, \eta_{l_r}$ and k linearly independent columns $\xi_{i_1}, \dots, \xi_{i_k}$ such that the m_2 columns $\xi_{i_1}, \dots, \xi_{i_k}, \eta_{l_1}, \dots, \eta_{l_r}$ are linearly independent. Let

$$\{\eta_{j_1}, \dots, \eta_{j_k}\} := \{\eta_1, \dots, \eta_{m_2}\} \setminus \{\eta_{l_1}, \dots, \eta_{l_r}\}.$$

Denote by $w \in \mathbf{C}^{m_1 \times m_2}$ the matrix whose (i_s, j_s) entry $(s = 1, \dots, k)$ is 1 and the remaining entries of which are zero. Finally, consider $x := x_0 + \varepsilon w$. We have $y := c_{22} + c_{21}x = d_{22} + \varepsilon c_{21} w$. A direct computation shows that the jth column of y is η_j for $j \notin \{j_1, \dots, j_k\}$ and $\eta_{j_s} + \varepsilon \xi_{i_s}$ for $j = j_s$. Since each of the columns $\eta_1, \dots, \eta_{m_2}$ is a linear combination of the columns $\eta_{l_1}, \dots, \eta_{l_r}$, the linear hull of the columns of y coincides with the linear hull of $\{\eta_{l_1}, \dots, \eta_{l_r}\} \cup \{\varepsilon\xi_{i_1}, \dots, \varepsilon\xi_{i_k}\}$. Thus, for $\varepsilon \neq 0$, the linear hull of the columns of y has the dimension m_2. This shows that $\det y = \det(c_{22} + c_{21}x) \neq 0$ and completes the proof. □

Lemma 10.15. *Suppose E is a proper subset of $\mathbf{C}$, $\mathcal{L}$ is a finite-dimensional linear space, $\mathcal{L}_1$ is a subspace of $\mathcal{L}$, and $T_1 : \mathcal{L}_1 \to \mathcal{L}$ is a linear operator. Then T_1 can be extended to a linear operator $T : \mathcal{L} \to \mathcal{L}$ such that* $\operatorname{sp} T \cap E = \emptyset$ *if and only if $\mathcal{L}_1 \cap \operatorname{Ker}(T_1 - \lambda I) = \{0\}$ for all $\lambda \in E$.*

Proof. The "only if" part is trivial. To prove the "if" portion, we can without loss of generality assume that T_1 is injective, since otherwise we can replace T_1 by $T_1 - \xi I$ and E by $E - \xi$ with an appropriate $\xi \in \mathbf{C}$.

We first show that the space $\mathcal{L}_1 + T_1\mathcal{L}_1$ can be decomposed into a direct sum

$$\mathcal{L}_1 + T_1\mathcal{L}_1 = \mathcal{M}_0 \dot{+} \mathcal{M}_1 \dot{+} \dots \dot{+} \mathcal{M}_l$$

such that $T_1\mathcal{M}_0 = \mathcal{M}_0 \subset \mathcal{L}_1$ and, for $j = 1, \dots, l$, there is a basis $f_1^{(j)}, \dots, f_{m_j}^{(j)}$ in $\mathcal{M}_j$ with the following property: $f_i^{(j)} \in \mathcal{L}_1$ and $T_1 f_i^{(j)} = f_{i+1}^{(j)}$ for $i < m_j$, $f_{m_j}^{(j)} \notin \mathcal{L}_1$.

Put $\mathcal{L}_{j+1} := T_1\mathcal{L}_j \cap \mathcal{L}_1$ $(j = 1, 2, \dots)$. It is easily seen that $\mathcal{L}_j \supset \mathcal{L}_{j+1}$ and that the equality $\mathcal{L}_j = \mathcal{L}_{j+1}$ implies that $\mathcal{L}_j = \mathcal{L}_s$ for all $s > j$. As $\mathcal{L}_1$ is finite-dimensional, there exists a j such that $\mathcal{L}_j = \mathcal{L}_{j+1}$. Let $k+1$ be the smallest j with this property and put $\mathcal{L}_{k+1} =: \mathcal{M}_0$. Thus, we have

$$\mathcal{L}_1 \supset \mathcal{L}_2 \supset \dots \supset \mathcal{L}_k \supset \mathcal{L}_{k+1} = \mathcal{L}_{k+2} = \dots = \mathcal{M}_0,$$

the inclusions "$\supset$" being proper. Since $T_1\mathcal{M}_0 \cap \mathcal{L}_1 = \mathcal{M}_0$, we see that $T_1\mathcal{M}_0 \supset \mathcal{M}_0$. As $\dim T_1\mathcal{M}_0 \leq \dim \mathcal{M}_0$, it follows that actually $T_1\mathcal{M}_0 = \mathcal{M}_0$.

If $k = 0$, then $\mathcal{M}_0 = \mathcal{L}_1$ and the remaining $\mathcal{M}_j$ do not emerge. So assume $k \geq 1$. Choose a maximal set $T_1^{k-1}f_{11}, \dots, T_1^{k-1}f_{n_1,1}$ of vectors in $\mathcal{L}_k$ which are linearly independent over $\mathcal{M}_0$. Then (provided $k \geq 2$) extend $T_1^{k-2}f_{11}, \dots, T_1^{k-2}f_{n_1,1}$ by vectors $T_1^{k-2}f_{12}, \dots, T_1^{k-2}f_{n_2,2}$ in $\mathcal{L}_{k-1}$ to a maximal set of vectors which are linearly independent over $\mathcal{L}_k$. Continuing in this way, we obtain a set $\mathcal{Z} = \{z_j\}_{j=1}^N$ which equals the set $\tilde{\mathcal{Z}}$ of all the vectors in Table $(*)$ minus the vectors in brackets.

$$
\begin{array}{cccccc}
f_{11} & T_1 f_{11} & \ldots & T_1^{k-2} f_{11} & T_1^{k-1} f_{11} & \left[\begin{array}{c} T_1^k f_{11} \\ \vdots \\ T_1^k f_{n_1,1} \end{array}\right] \\
\vdots & \vdots & & \vdots & \vdots & \\
f_{n_1,1} & T_1 f_{n_1,1} & \ldots & T_1^{k-2} f_{n_1,1} & T_1^{k-1} f_{n_1,1} & \\
 & & & & & \\
f_{12} & T_1 f_{12} & \ldots & T_1^{k-2} f_{12} & \left[\begin{array}{c} T_1^{k-1} f_{12} \\ \vdots \\ T_1^{k-1} f_{n_2,2} \end{array}\right] & \\
\vdots & \vdots & & \vdots & & \\
f_{n_2,2} & T_1 f_{n_2,2} & \ldots & T_1^{k-2} f_{n_2,2} & & \\
\vdots & \vdots & & & & \\
f_{1k} & \left[\begin{array}{c} T_1 f_{1k} \\ \vdots \\ T_1 f_{n_k,k} \end{array}\right] & & & & \\
\vdots & & & & & \\
f_{n_k,k} & & & & & \text{Table } (*)
\end{array}
$$

We denote the vectors in the first column of Table $(*)$ by ξ_i $(i = 1, \ldots, n_1 + \ldots + n_k)$. The remaining vectors of the table can be written in the form $T_1 z_j$ $(z_j \in \mathcal{Z})$. We claim that the elements of $\tilde{\mathcal{Z}}$ are linearly independent over $\mathcal{M}_0$. Indeed, if

$$\sum \alpha_i \xi_i + T_1 \Big(\sum \beta_i z_j \Big) \in \mathcal{M}_0,$$

then

$$\sum \alpha_i \xi_i \in (T_1 \mathcal{L}_1 + \mathcal{M}_0) \cap \mathcal{L}_1 = T_1 \mathcal{L}_1 \cap \mathcal{L}_1 = \mathcal{L}_2.$$

Since the vectors $\xi_i \in \mathcal{L}_1$ are linearly independent over $\mathcal{L}_2$, it follows that $\alpha_i = 0$ for all i. Consequently, $T_1 \left(\sum \beta_j z_j\right) \in \mathcal{M}_0$. As $\mathcal{M}_0 = T_1 \mathcal{M}_0$, this implies that $T_1 \left(\sum \beta_j z_j\right) = T_1 g$ with some $g \in \mathcal{M}_0$, and because T_1 is injective, we arrive at the conclusion that $\sum \beta_j z_j \in \mathcal{M}_0$. Taking into account that, by construction, the vectors in $\mathcal{Z}$ are linearly independent over $\mathcal{M}_0$, we obtain that $\beta_j = 0$ for all j. This proves our claim.

Obviously, $\tilde{\mathcal{Z}} = \mathcal{Z} \cup T_1 \mathcal{Z}$. Since the linear hull of $\mathcal{Z}$ is a direct complement of $\mathcal{M}_0$ in $\mathcal{L}_1$, the linear hull of $\tilde{\mathcal{Z}}$ must be a direct complement of $\mathcal{M}_0$ in $\mathcal{L}_1 + T_1 \mathcal{L}_1$. Moreover, since the vectors in $\mathcal{Z}$ and $\tilde{\mathcal{Z}}$ are linearly independent over $\mathcal{M}_0$, we have $(\tilde{\mathcal{Z}} \setminus \mathcal{Z}) \cap \mathcal{L}_1 = \emptyset$. Thus, denoting by $\mathcal{M}_j$ the linear hull of the elements in the jth row of Table $(*)$ $(j = 1, \ldots, n_1 + \ldots + n_k)$, we get the desired decomposition.

We now extend $T_1 : \mathcal{L}_1 \to \mathcal{L}$ to a linear operator $T : \mathcal{L} \to \mathcal{L}$. We already know that $T f_i^{(j)} = f_{i+1}^{(j)}$ for $i < m_j$. Put

$$T f_{m_j}^{(j)} := -\alpha_{m_j}^{(j)} f_1^{(j)} - \ldots - \alpha_1^{(j)} f_{m_j}^{(j)};$$

the α's will be specified in the following. Clearly, $\mathcal{M}_j$ is an invariant subspace of T, and the matrix representation of T in the basis $f_1^{(j)}, \dots, f_m^{(j)}$ is the companion matrix

$$A_j := \begin{pmatrix} 0 & \cdots & 0 & 0 & -\alpha_{m_j}^{(j)} \\ 1 & \cdots & 0 & 0 & -\alpha_{m_j-1}^{(j)} \\ \vdots & \ddots & \vdots & \vdots & \vdots \\ 0 & \cdots & 1 & 0 & -\alpha_2^{(j)} \\ 0 & \cdots & 0 & 1 & -\alpha_1^{(j)} \end{pmatrix}.$$

We have

$$\det(A_j - \lambda I) = (-1)^{m_j} \left(\lambda^{m_j} + \alpha_1^{(j)} \lambda^{m_j - 1} + \ldots + \alpha_{m_j}^{(j)}\right),$$

and hence we can choose $\alpha_1^{(j)}, \dots, \alpha_{m_j}^{(j)}$ so that sp $(T|\mathcal{M}_j) \cap E = \emptyset$ for $j = 1, \dots, l$. Because

$$\mathcal{L}_1 \cap \Big(\cup_{\lambda \in E} \operatorname{Ker}(T_1 - \lambda I) \Big) = \{0\},$$

by assumption, we get sp $(T|\mathcal{M}_0) \cap E = \emptyset$. It therefore results that

$$\operatorname{sp}(T|(\mathcal{L}_1 + T_1 \mathcal{L}_1)) \cap E = \emptyset.$$

If $\mathcal{L}_1 + T_1\mathcal{L}_1 \neq \mathcal{L}$, we choose any direct complement $\mathcal{M}_{l+1}$ of $\mathcal{L}_1 + T_1\mathcal{L}_1$ in $\mathcal{L}$ and extend T from its invariant subspace $\mathcal{L}_1 + T_1\mathcal{L}_1$ to all of $\mathcal{L}$ by any operator $\mathcal{M}_{l+1} \to \mathcal{M}_{l+1}$ whose spectrum does not contain points of E. □

10.6 Semi-Fredholm Theory

Let $a \in SAP_{N\times N}$. We say that a has a *pair of pre-factorizations* if the almost periodic representatives a_l, a_r can be represented in the form

$$a_l = a_l^- \begin{pmatrix} v_l & 0 \\ 0 & I \end{pmatrix} a_l^+, \quad a_r = a_r^- \begin{pmatrix} v_r & 0 \\ 0 & I \end{pmatrix} a_r^+ \tag{10.41}$$

where $a_l^\pm, a_r^\pm \in GAP_{N\times N}^\pm$ and v_l, v_r are unitary-valued functions such that

$$v_l \in APW_{m_l\times m_l}^- \cup APW_{m_l\times m_l}^+, \quad M(v_l) = 0, \tag{10.42}$$

$$v_r \in APW_{m_r\times m_r}^- \cup APW_{m_r\times m_r}^+, \quad M(v_r) = 0, \tag{10.43}$$

and $m_l, m_r \in \{0, 1, \dots, N\}$. Let, for example, a_l and a_r have right AP factorizations:

$$a_l = a_l^- d_l a_l^+, \quad d_l(x) = \operatorname{diag}(e^{i\lambda_j^{(l)} x}), \tag{10.44}$$

$$a_r = a_r^- d_r a_r^+, \quad d_r(x) = \operatorname{diag}(e^{i\lambda_j^{(r)} x}). \tag{10.45}$$

The factorizations (10.44), (10.45) are a pair of pre-factorizations for a if and only if all nonzero $\lambda_j^{(l)}$ have the same sign and all nonzero $\lambda_j^{(r)}$ also have equal signs. The following proposition shows that the existence of a pair of pre-factorizations is ensured in many interesting cases.

Proposition 10.16. (a) *Let $a \in SAP_{N\times N}$. If a_l and a_r have right $AP(u_l)$ and $AP(u_r)$ factorizations whose unitary middle factors u_l and u_r belong to $APW_{N\times N}$ and if $W(a)$ is semi-Fredholm, then a has a pair of pre-factorizations.*

(b) *If $a \in SAPW_{N\times N}$ and $W(a)$ is semi-Fredholm, then a possesses a pair of pre-factorizations.*

Proof. By Corollary 10.5, the semi-Fredholmness of $W(a)$ implies the one-sided invertibility of $W(a_l)$ and $W(a_r)$. If

$$a = c_l^- u_l c_l^+, \quad a_r = c_r^- u_r c_r^+$$

where $c_l^\pm, c_r^\pm \in GAP_{N\times N}^\pm$ and $u_l, u_r \in APW_{N\times N}$ are unitary-valued, then $W(u_l)$ and $W(u_r)$ are also one-sided invertible. Applying Theorems 9.18 and 9.19 to $W(u_l), W(u_r)$ we obtain factorizations

$$u_l = \varphi_l^- \begin{pmatrix} v_l & 0 \\ 0 & I \end{pmatrix} \varphi_l^+, \quad u_r = \varphi_r^- \begin{pmatrix} v_r & 0 \\ 0 & I \end{pmatrix} \varphi_r^+$$

with $\varphi_l^\pm, \varphi_r^\pm \in GAPW_{N\times N}^\pm$ and unitary-valued v_l, v_r satisfying (10.42), (10.43). Clearly,

$$a_l = c_l^- \varphi_l^- \begin{pmatrix} v_l & 0 \\ 0 & I \end{pmatrix} \varphi_l^+ c_l^+, \quad a_r = c_r^- \varphi_r^- \begin{pmatrix} v_r & 0 \\ 0 & I \end{pmatrix} \varphi_r^+ c_r^+$$

is a pair of pre-factorizations for a. In case a belongs even to $SAPW_{N\times N}$, we can immediately use Theorems 9.18 and 9.19 to deduce that a has a pair of pre-factorizations. □

Now let $a \in SAP_{N\times N}$ and suppose (10.41), (10.42), (10.43) is any pair of pre-factorizations. Consider the constant $N \times N$ matrices Φ^-, Φ^+ defined by

$$\Phi^- := M^{-1}(a_r^-)M(a_l^-), \quad \Phi^+ := M(a_l^+)M^{-1}(a_r^+), \tag{10.46}$$

where $M^{-1}(\varphi) := (M(\varphi))^{-1}$, and partition them into blocks as follows:

$$\Phi^- = \begin{pmatrix} \Phi_{11}^- & \Phi_{10}^- \\ \Phi_{01}^- & \Phi_{00}^- \end{pmatrix} \text{ with } \Phi_{11}^- \in \mathbf{C}^{m_r \times m_l}, \tag{10.47}$$

$$\Phi^+ = \begin{pmatrix} \Phi_{11}^+ & \Phi_{10}^+ \\ \Phi_{01}^+ & \Phi_{00}^+ \end{pmatrix} \text{ with } \Phi_{11}^+ \in \mathbf{C}^{m_l \times m_r}. \tag{10.48}$$

Theorem 10.17. *Let $a \in SAP_{N\times N}$ have the pair of pre-factorizations* (10.41). *Then $W(a)$ is n-normal if and only if*

(a) $a \in GSAP_{N\times N}$,

(b) *the entries of* v_l, v_r *are in* AP^+,

(c) *the columns of the matrix*

$$\begin{pmatrix} \Phi_{00}^-\Phi_{00}^+ - \mu I \\ \Phi_{10}^+ \\ \Phi_{10}^-\Phi_{00}^+ \end{pmatrix} \in \mathbf{C}^{(N+m_l)\times(N-m_r)}$$

are linearly independent for all $\mu \in (-\infty, 0] \cap \operatorname{sp}(\Phi_{00}^-\Phi_{00}^+)$.

Theorem 10.18. *Let* $a \in SAP_{N\times N}$ *and suppose* (10.41) *is a pair of pre-factorizations of* a. *Then* $W(a)$ *is d-normal if and only if*

(a) $a \in GSAP_{N\times N}$,

(b) *the entries of* v_l, v_r *lie in* AP^-,

(c) *the rows of the matrix*

$$(\Phi_{00}^-\Phi_{00}^+ - \mu I,\ \Phi_{01}^-,\ \Phi_{00}^-\Phi_{01}^+) \in \mathbf{C}^{(N-m_r)\times(N+m_l)} \tag{10.49}$$

are linearly independent for every $\mu \in (-\infty, 0] \cap \operatorname{sp}(\Phi_{00}^-\Phi_{00}^+)$.

Note that we consider condition (c) of Theorems 10.17 and 10.18 as satisfied if $m_r = N$.

Since each of the above two theorems follows from the other one by passage to adjoints, it suffices to prove one of them. We prove Theorem 10.18. Our proof is rather long and is divided into five parts: a lemma, a prologue, the proof of the sufficiency portion, the proof of the necessity, and an epilogue.

Lemma 10.19. *Let* $a \in GSAP_{N\times N}$ *have the pair of pre-factorizations*

$$a_l = a_l^- \begin{pmatrix} v_l & 0 \\ 0 & I \end{pmatrix} a_l^+, \quad a_r = \begin{pmatrix} v_r & 0 \\ 0 & I \end{pmatrix}$$

(*i.e.*, $a_r^- = a_r^+ = I$). *Let further* X_1, X_2, X_3, X_4 *be any constant* $N \times N$ *matrices of the form*

$$X_j = \begin{pmatrix} I & x_j \\ 0 & I \end{pmatrix}, \quad x_j \in \mathbf{C}^{m\times(N-m)}. \tag{10.50}$$

Then there is a matrix function $b \in GSAP_{N\times N}$ *such that* $W(b)$ *is weakly* Φ-*equivalent to* $W(a)$ *and*

$$b_l = X_1 a_l^- X_2 \begin{pmatrix} v_l & 0 \\ 0 & I \end{pmatrix} X_3 a_l^+ X_4, \quad b_r = \begin{pmatrix} v_r & 0 \\ 0 & I \end{pmatrix}$$

Proof. We have

$$\begin{aligned} X_1^{-1} a_r X_4^{-1} &= \begin{pmatrix} I & -x_1 \\ 0 & I \end{pmatrix} \begin{pmatrix} v_r & 0 \\ 0 & I \end{pmatrix} \begin{pmatrix} I & -x_4 \\ 0 & I \end{pmatrix} \\ &= \begin{pmatrix} I & -v_r x_4 \\ 0 & I \end{pmatrix} \begin{pmatrix} v_r & 0 \\ 0 & I \end{pmatrix} \begin{pmatrix} I & -v_r^{-1} x_1 \\ 0 & I \end{pmatrix}, \end{aligned}$$

and because $M(v_r) = 0$, it follows that

$$M \begin{pmatrix} I & -v_r x_4 \\ 0 & I \end{pmatrix} = M \begin{pmatrix} I & -v_r^{-1} x_1 \\ 0 & I \end{pmatrix} = \begin{pmatrix} I & 0 \\ 0 & I \end{pmatrix}.$$

Let $b^{(1)} \in GSAP_{N\times N}$ be any matrix function such that

$$b_l^{(1)} = a_l^- \begin{pmatrix} v_l & 0 \\ 0 & I \end{pmatrix} a_l^+, \quad b_r^{(1)} = X_1^{-1} \begin{pmatrix} v_r & 0 \\ 0 & I \end{pmatrix} X_4^{-1}.$$

From Theorem 10.9 (also recall Lemma 3.10 and Corollary 10.10) we deduce that $W(a)$ and $W(b^{(1)})$ are weakly Φ-equivalent. Clearly, $W(b^{(1)})$ is weakly Φ-equivalent to $W(b^{(2)})$ where

$$b^{(2)} := (a_l^-)^{-1} b^{(1)} (a_l^+)^{-1}.$$

We have

$$b_l^{(2)} = \begin{pmatrix} v_l & 0 \\ 0 & I \end{pmatrix}, \quad b_r^{(2)} = (a_l^-)^{-1} X_1^{-1} \begin{pmatrix} v_r & 0 \\ 0 & I \end{pmatrix} X_4^{-1} (a_l^+)^{-1},$$

and an argument analogous to the one employed above shows that there is a $b^{(3)} \in GSAP_{N\times N}$ such that $W(b^{(2)})$ is weakly Φ-equivalent to $W(b^{(3)})$ and

$$b_l^{(3)} = X_2 \begin{pmatrix} v_l & 0 \\ 0 & I \end{pmatrix} X_3, \quad b_r^{(3)} = (a_l^-)^{-1} X_1^{-1} \begin{pmatrix} v_r & 0 \\ 0 & I \end{pmatrix} X_4^{-1} (a_l^+)^{-1}.$$

Letting $b := X_1 a_l^- b_l^{(3)} a_l^+ X_4$ we get the assertion of the lemma. □

Prologue. By virtue of Theorem 5.1, we can a-priori assume that condition (a) is satisfied, i.e., that $a \in GSAP_{N\times N}$. Let (10.41) be a pair of pre-factorizations. Then

$$\begin{aligned} a_l &= a_l^- \begin{pmatrix} v_l & 0 \\ 0 & I \end{pmatrix} a_l^+ = \tilde{a}_l^- M(a_l^-) \begin{pmatrix} v_l & 0 \\ 0 & I \end{pmatrix} M(a_l^+) \tilde{a}_l^+, \\ a_r &= a_r^- \begin{pmatrix} v_r & 0 \\ 0 & I \end{pmatrix} a_r^+ = \tilde{a}_r^- M(a_r^-) \begin{pmatrix} v_r & 0 \\ 0 & I \end{pmatrix} M(a_r^+) \tilde{a}_r^+ \end{aligned}$$

where $M(\tilde{a}_l^\pm) = M(\tilde{a}_r^\pm) = \begin{pmatrix} I & 0 \\ 0 & I \end{pmatrix}$. By Theorem 10.9, there is a matrix function $a^{(1)} \in GSAP_{N\times N}$ such that $W(a)$ and $W(a^{(1)})$ are weakly Φ-equivalent and

$$a_l^{(1)} = M(a_l^-) \begin{pmatrix} v_l & 0 \\ 0 & I \end{pmatrix} M(a_l^+), \quad a_r^{(1)} = M(a_r^-) \begin{pmatrix} v_r & 0 \\ 0 & I \end{pmatrix} M(a_r^+).$$

Obviously, $W(a^{(1)})$ is (strongly) Φ-equivalent to $W(a^{(2)})$ where

$$a^{(2)} = M^{-1}(a_r^-)a^{(1)}M^{-1}(a_r^+).$$

With $\Phi^\pm$ defined by (10.46), it follows that

$$a_l^{(2)} = \Phi^- \begin{pmatrix} v_l & 0 \\ 0 & I \end{pmatrix} \Phi^+, \quad a_r^{(2)} = \begin{pmatrix} v_r & 0 \\ 0 & I \end{pmatrix}.$$

We now assume that $m_l = m_r =: m$. In the epilogue, we will show that the general case can be reduced to this case. Let X_1, X_2, X_3, X_4 be any matrices of the form (10.50). From Lemma 10.19 we infer that there is a $b \in GSAP_{N\times N}$ such that $W(b)$ is weakly Φ-equivalent to $W(a^{(2)})$ and

$$b_l = X_1\Phi^- X_2 \begin{pmatrix} v_l & 0 \\ 0 & I \end{pmatrix} X_3\Phi^+ X_4, \quad b_r = \begin{pmatrix} v_r & 0 \\ 0 & I \end{pmatrix}.$$

We have

$$X_1\Phi^- X_2 = \begin{pmatrix} \Phi_{11}^- + x_1\Phi_{01}^- & \Phi_{10}^- + \Phi_{11}^- x_2 + x_1(\Phi_{00}^- + \Phi_{01}^- x_2) \\ \Phi_{01}^- & \Phi_{00}^- + \Phi_{01}^- x_2 \end{pmatrix}.$$

Applying Lemma 10.14 to $C = \Phi^-$, we see that there is an x_2 such that $\Phi_{00}^- + \Phi_{01}^- x_2$ is invertible. Taking such an x_2, we can then find an x_1 such that

$$\Phi_{10}^- + \Phi_{11}^- x_2 + x_1(\Phi_{00}^- + \Phi_{01}^- x_2) = 0.$$

Thus,

$$X_1\Phi^- X_2 = \begin{pmatrix} \Phi_{11}^- + x_1\Phi_{01}^- & 0 \\ \Phi_{01}^- & \Phi_{00}^- + \Phi_{01}^- x_2 \end{pmatrix}.$$

Analogously, there are x_3 and x_4 such that $\Phi_{00}^+ + \Phi_{01}^+ x_4$ is invertible and

$$X_3\Phi^+ X_4 = \begin{pmatrix} \Phi_{11}^+ + x_3\Phi_{01}^+ & 0 \\ \Phi_{01}^+ & \Phi_{00}^+ + \Phi_{01}^+ x_4 \end{pmatrix}.$$

In summary, $W(a)$ is weakly Φ-equivalent to $W(b)$ where $b \in GSAP_{N\times N}$ and

$$b_l = \begin{pmatrix} (b_{11})_l & 0 \\ (b_{01})_l & (b_{00})_l \end{pmatrix}, \quad b_r = \begin{pmatrix} (b_{11})_r & 0 \\ (b_{01})_r & (b_{00})_r \end{pmatrix},$$

$$\begin{aligned}
(b_{11})_l &= (\Phi_{11}^- + x_1\Phi_{01}^-)v_l(\Phi_{11}^+ + x_3\Phi_{01}^+), \\
(b_{01})_l &= \Phi_{01}^- v_l(\Phi_{11}^+ + x_3\Phi_{01}^+) + (\Phi_{00}^- + \Phi_{01}^- x_2)\Phi_{01}^+, \\
(b_{00})_l &= (\Phi_{00}^- + \Phi_{01}^- x_2)(\Phi_{00}^+ + \Phi_{01}^+ x_4), \\
(b_{11})_r &= v_r, \quad (b_{01})_r = 0, \quad (b_{00})_r = I.
\end{aligned}$$

By virtue of Theorem 10.9, we can without loss of generality assume that b in $GSAP_{N\times N}$ itself is of the form

$$b = \begin{pmatrix} b_{11} & 0 \\ b_{01} & b_{00} \end{pmatrix}$$

and that

$$\begin{aligned} b_{01} &= (b_{01})_l(1-u) + (b_{01})_r u \\ &= (\Phi_{01}^- v_l(\Phi_{11}^+ + x_3\Phi_{01}^+) + (\Phi_{00}^- + \Phi_{01}^- x_2)\Phi_{01}^+)(1-u). \end{aligned} \tag{10.51}$$

Since $(b_{00})_l$ is invertible by construction, Theorem 10.9 allows us also to assume that

$$\begin{aligned} b_{00} &= (b_{00})_l(1-u) + (b_{00})_r u \\ &= (\Phi_{00}^- + \Phi_{01}^- x_2)(\Phi_{00}^+ + \Phi_{01}^+ x_4)(1-u) + uI. \end{aligned} \tag{10.52}$$

□

Sufficiency. Suppose conditions (a),(b),(c) of Theorem 10.18 are satisfied. Put $k = N - m$ and $\mathcal{L} = \mathbf{C}^k$. We think of the elements of $\mathcal{L}$ as columns. Let $T_1 : \mathcal{L} \to \mathcal{L}$ be given by the matrix $(\Phi_{00}^- \Phi_{00}^+)^*$ and let

$$\mathcal{L}_1 := (\operatorname{Im} \Phi_{01}^- + \operatorname{Im} \Phi_{00}^- \Phi_{01}^+)^\perp,$$

where $^\perp$ stands for the orthogonal complement in $\mathcal{L}$. Condition (c) implies that if $\mu \in (-\infty, 0]$, then

$$\operatorname{Im}(T_1^* - \mu I) + \mathcal{L}_1^\perp = \mathcal{L}$$

whence $\operatorname{Ker}(T_1 - \mu I) \cap \mathcal{L}_1 = \{0\}$ (note that this is trivial for points μ in the set $(-\infty, 0] \setminus \operatorname{sp}(\Phi_{00}^- \Phi_{00}^+)$). By Lemma 10.15, there is a linear operator $T : \mathcal{L} \to \mathcal{L}$ such that

$$T|\mathcal{L}_1 = T_1|\mathcal{L}_1 \ \text{ and } \ \operatorname{sp} T \cap (-\infty, 0] = \emptyset.$$

Obviously, $\operatorname{Ker}(T - T_1) \supset \mathcal{L}_1$. This shows that

$$\operatorname{Im}(T - T_1)^* \subset \mathcal{L}_1^\perp = \operatorname{Im} \Phi_{01}^- + \operatorname{Im} \Phi_{00}^- \Phi_{01}^+,$$

and hence the columns of $(T - T_1)^*$ are linear combinations of the columns of $(\Phi_{01}^-, \Phi_{00}^- \Phi_{01}^+)$:

$$(T - T_1)^* = \Phi_{01}^- z_1 + \Phi_{00}^- \Phi_{01}^+ z_2, \quad z_1, z_2 \in \mathbf{C}^{m\times k}.$$

If $\varepsilon > 0$ is small enough, then $\operatorname{sp} T_0 \cap (-\infty, 0] = \emptyset$ for all $T_0 : \mathcal{L} \to \mathcal{L}$ such that $\|T_0 - T\| < \varepsilon$. By Lemma 10.14, the x_4 chosen in the prologue can be picked so that

$$\|\Phi_{00}^- \Phi_{01}^+ (x_4 - z_2)\| < \varepsilon/2.$$

Again by Lemma 10.14, we can find an x_2 as in the prologue so that

$$\left\|\Phi_{01}^-\left(x_2 - z_1(\Phi_{00}^+ + \Phi_{01}^+ x_4)^{-1}\right)\right\| < \|\Phi_{00}^+ + \Phi_{01}^+ x_4\|^{-1}\frac{\varepsilon}{2}.$$

Put

$$T_0 := T_1 + \left(\Phi_{01}^- x_2(\Phi_{00}^+ + \Phi_{01}^+ x_4) + \Phi_{00}^- \Phi_{01}^+ x_4\right)^*.$$

Then

$$\begin{aligned}
&\|(T_0 - T)^*\| = \|(T_0 - T_1)^* + (T_1 - T)^*\| \\
&= \|\Phi_{01}^- x_2(\Phi_{00}^+ + \Phi_{01}^+ x_4) + \Phi_{00}^- \Phi_{01}^+ x_4 - \Phi_{01}^- z_1 - \Phi_{00}^- \Phi_{01}^+ z_2\| \\
&\leq \|\Phi_{00}^- \Phi_{01}^+ (x_4 - z_2)\| + \|\Phi_{01}^- x_2(\Phi_{00}^+ + \Phi_{01}^+ x_4) - \Phi_{01}^- z_1\| \\
&< \varepsilon/2 + \|\Phi_{01}^- x_2(\Phi_{00}^+ + \Phi_{01}^+ x_4) - \Phi_{01}^- z_1\|
\end{aligned}$$

and the second term on the right is

$$\begin{aligned}
&\|\Phi_{01}^- x_2(\Phi_{00}^+ + \Phi_{01}^+ x_4) - \Phi_{01}^- z_1(\Phi_{00}^+ + \Phi_{01}^+ x_4)^{-1}(\Phi_{00}^+ + \Phi_{01}^+ x_4)\| \\
&\leq \|\Phi_{01}^- x_2 - \Phi_{01}^- z_1(\Phi_{00}^+ + \Phi_{01}^+ x_4)^{-1}\| \, \|\Phi_{00}^+ + \Phi_{01}^+ x_4\| < \varepsilon/2.
\end{aligned}$$

This proves that sp $T_0 \cap (-\infty, 0] = \emptyset$.

By (10.52), the matrix function b_{00} belongs to $[C(\overline{\mathbf{R}})]_{k\times k}$. We have

$$\begin{aligned}
b_{00}(+\infty) &= (b_{00})_r = I, \\
b_{00}(-\infty) &= (b_{00})_l = (\Phi_{00}^- + \Phi_{01}^- x_2)(\Phi_{00}^+ + \Phi_{01}^+ x_4) = T_0^*,
\end{aligned}$$

and hence $b_{00}^{-1}(+\infty)b_{00}(-\infty) = T_0^*$. Therefore

$$\text{sp } \left(b_{00}^{-1}(+\infty)b_{00}(-\infty)\right) \cap (-\infty, 0] = \emptyset,$$

and since b_{00} is invertible due to the invertibility of b, we obtain from Theorem 5.10 that $W(b_{00})$ is Fredholm.

Since $M(v_l) = M(v_r) = 0$, Lemma 3.5 and condition (b) imply that the entries of b_{11} are in $C(\dot{\mathbf{R}}) + H_-^\infty$. The invertibility of b yields the invertibility of b_{11}. Consequently, $W(b_{11})$ is d-normal due to Theorem 5.7.

In summary, we have proved that the diagonal entries G and H of

$$W(b) = \begin{pmatrix} G & 0 \\ U & H \end{pmatrix}$$

have right regularizers $G^{(-1)}$ and $H^{(-1)}$. From the identity

$$\begin{pmatrix} G & 0 \\ U & H \end{pmatrix}\begin{pmatrix} G^{(-1)} & 0 \\ -H^{(-1)}UG^{(-1)} & H^{(-1)} \end{pmatrix} = \begin{pmatrix} GG^{(-1)} & 0 \\ UG^{(-1)} - HH^{(-1)}UG^{(-1)} & HH^{(-1)} \end{pmatrix}$$

we finally obtain that $W(b)$ also has a right regularizer. Thus, the operator $W(b)$ is d-normal. □

Necessity. Now suppose $W(a)$ is d-normal. Then $a \in GSAP_{N\times N}$ by Theorem 5.1. Proceeding as in the proof of Proposition 10.16, we see that condition (b) is also satisfied. Thus, we are left to prove (c).

Put $A = \Phi_{00}^{-}\Phi_{00}^{+}$ and $B = (\Phi_{01}^{-}, \Phi_{00}^{-}\Phi_{01}^{+})$. Let $A \in \mathbf{C}^{k\times k}$ have r Jordan cells. Then there are

$$\mu_j \in \operatorname{sp} A, \quad g_{j1}, \dots, g_{jk_j} \in \mathbf{C}^k \quad (j = 1, \dots, r)$$

such that all the g_{js} form a basis in $\mathbf{C}^k$ and

$$(A - \mu_j I)g_{js} = g_{j,s-1} \quad (s = 2, \dots, k_j), \qquad (A - \mu_j I)g_{j1} = 0.$$

Note that $k_1 + \ldots + k_r = k$. We denote by $\operatorname{Col} M$ the column space of a matrix M, that is, the linear span of its columns.

Fix $\mu \in \operatorname{sp} A$. Let $\mathcal{N}_\mu$ be the linear span of the columns g_{jk_j} for which $\mu_j = \mu$. Further, put $\mathcal{M}_\mu = \operatorname{Col}(A - \mu I, B)$. Obviously, $\mathcal{M}_\mu$ is the linear span of $\operatorname{Col} B$ and the columns

$$\{(A - \mu I)g_{js} : j = 1, \dots, r;\ s \le k_j\}.$$

Since

$$\begin{aligned}(A - \mu I)g_{js} &= (A - \mu_j I)g_{js} + (\mu_j - \mu)g_{js} \\ &= \begin{cases} g_{j,s-1} + (\mu_j - \mu)g_{js} & \text{for} \quad s \ge 1, \\ 0 + (\mu_j - \mu)g_{js} & \text{for} \quad s = 1, \end{cases}\end{aligned}$$

we see that, for each fixed j,

$$\operatorname{span}\{(A - \mu I)g_{js} : s \le k_j\} = \begin{cases} \operatorname{span}\{g_{js} : s \le k_j\} & \text{if} \quad \mu \ne \mu_j, \\ \operatorname{span}\{g_{js} : s < k_j\} & \text{if} \quad \mu = \mu_j. \end{cases}$$

Because $\mathcal{N}_\mu$ is the linear span of all those g_{js} for which $s = k_j$ and $\mu_j = \mu$, it follows that $\mathcal{M}_\mu + \mathcal{N}_\mu$ contains all the columns g_{js}. Thus,

$$\mathcal{M}_\mu + \mathcal{N}_\mu = \mathbf{C}^k. \tag{10.53}$$

Contrary to what we want, let us assume that there is a point $\mu \in \operatorname{sp} A \cap (-\infty, 0]$ such that the rows of $(A - \mu I, B)$ are linearly dependent. Then $\mathcal{M}_\mu \ne \mathbf{C}^k$, and from (10.53) we obtain the existence of a nonzero subspace $\mathcal{R}_\mu \subset \mathcal{N}_\mu$ such that

$$\mathcal{R}_\mu \dot{+} \mathcal{M}_\mu = \mathbf{C}^k, \tag{10.54}$$

where $\dot{+}$ stands for the direct sum.

Given a subset Z of $\mathbf{C}^N$, we denote by $L^2[Z]$ the space of all functions in $L^2_N(\mathbf{R}_+)$ whose values lie in the linear span of Z. In particular,

$$L^2_N(\mathbf{R}_+) = L^2[e_1, \dots, e_N] \quad (:= L^2[\{e_1, \dots, e_N\}])$$

where $\{e_1, \ldots, e_N\}$ is the standard basis of $\mathbf{C}^N$. Recall that $m + k = N$ and put

$$X := L^2[e_1, \ldots, e_m], \quad Y := L^2[e_{m+1}, \ldots, e_N].$$

Define $b \in GSAP_{N\times N}$ as in the prologue. Then $W(b)$ is d-normal, and hence $\operatorname{Im} W(b)$ has finite codimension. Because

$$\operatorname{Im} W(b) = \begin{pmatrix} W(b_{11}) & 0 \\ W(b_{01}) & W(b_{00}) \end{pmatrix} \begin{pmatrix} X \\ Y \end{pmatrix} = \begin{pmatrix} W(b_{11})X \\ W(b_{01})X + W(b_{00})Y \end{pmatrix},$$

the space $W(b_{01})X + W(b_{00})Y$ has finite codimension in Y.

The columns of the product AB are linear combinations of the columns of A. Consequently, the columns of

$$\Phi_{01}^- W\big(v_l(\Phi_{11}^+ + x_3\Phi_{01}^+)(1-u)\big)X \text{ and } \Phi_{01}^- W\big(x_2\Phi_{01}^+(1-u)\big)X$$

all lie in $L^2[\operatorname{Col}\Phi_{01}^-]$. Analogously, the columns of

$$\Phi_{00}^-\Phi_{01}^+ W(1-u)X$$

all belong to $L^2[\operatorname{Col}\Phi_{00}^-\Phi_{01}^+]$. From (10.51) we therefore obtain that

$$W(b_{01})X \subset L^2[\operatorname{Col} B].$$

This implies that $L^2[\operatorname{Col} B] + W(b_{00})Y$ has finite codimension in Y.

By (10.52), we have $b_{00} = h_1 + h_2$ where

$$h_1 := \Phi_{00}^-\Phi_{00}^+(1-u) + uI,$$
$$h_2 := \Phi_{00}^-\Phi_{01}^+x_4 + \Phi_{01}^-x_2(\Phi_{00}^+ + \Phi_{01}^+x_4)(1-u).$$

From (10.54) we infer that $Y = L^2[\mathcal{R}_\mu]\dot{+}L^2[\mathcal{M}_\mu]$. As the columns of $W(h_2)Y$ lie in

$$L^2[\operatorname{Col}\Phi_{00}^-\Phi_{01}^+] \cup L^2[\operatorname{Col}\Phi_{01}^-] \subset L^2[\operatorname{Col} B],$$

it follows that

$$W(b_{00})Y \subset W(h_1)Y + W(h_2)Y \subset W(h_1)Y + L^2[\operatorname{Col} B].$$

Hence, $L^2[\operatorname{Col}B] + W(h_1)Y$ has finite codimension in Y. Letting $\varphi := \mu(1-u)+u$, we get

$$W(h_1) = AW(1-u) + IW(u) = (A - \mu I)W(1-u) + IW(\varphi).$$

Because $(A-\mu I)W(1-u)Y \subset L^2[\mathcal{M}_\mu]$, it results that

$$W(h_1)Y \subset L^2[\mathcal{M}_\mu] + W(\varphi)Y,$$

and as $L^2[\mathrm{Col}\, B] \subset L^2[\mathcal{M}_\mu]$, we obtain that

$$L^2[\mathrm{Col}\, B] + W(h_1)Y \subset L^2[\mathcal{M}_\mu] + W(\varphi)Y.$$

Thus, $L^2[\mathcal{M}_\mu] + W(\varphi)Y$ has finite codimension in Y. Since $W(\varphi)$ leaves both $L^2[\mathcal{M}_\mu]$ and $L^2[\mathcal{R}_\mu]$ invariant, it results that

$$L^2[\mathcal{M}_\mu]\dot{+}W(\varphi)L^2[\mathcal{R}_\mu]$$

has finite codimension in Y. This shows that $W(\varphi)L^2[\mathcal{R}_\mu]$ has finite codimension in $L^2[\mathcal{R}_\mu]$. Consequently, $W(\varphi)$ is d-normal.

The function φ belongs to $C(\overline{\mathbf{R}})$. By Theorem 2.20, the d-normality of $W(\varphi)$ implies that $0 \notin [\varphi(+\infty), \varphi(-\infty)] = [1, \mu]$, which is impossible for $\mu \in (-\infty, 0]$. This contradiction proves that condition (c) must be satisfied. □

Epilogue. It remains to show that the general case can be reduced to the case where $m_l = m_r = m$. So assume $m_l < m_r$ and put $d = m_r - m_l$. Let

$$g(x) := e^{i\alpha x}(1 - u(x)) + u(x) \text{ where } \alpha < 0$$

and set

$$c := \begin{pmatrix} a & 0 \\ 0 & gI_d \end{pmatrix}.$$

Since $W(gI_d)$ is d-normal by virtue of Theorem 3.9(d), we see that $W(a)$ is d-normal if and only if $W(c)$ is d-normal. Let I_n denote the $n \times n$ identity matrix. By what was said in the prologue, we can assume that

$$\begin{aligned} c_l &= \begin{pmatrix} \Phi^- \begin{pmatrix} v_l & 0 \\ 0 & I_{N-m_l} \end{pmatrix} \Phi^+ & 0 \\ 0 & e_\alpha I_d \end{pmatrix} \\ &= \begin{pmatrix} \Phi_{11}^- & 0 & \Phi_{10}^- \\ \Phi_{01}^- & 0 & \Phi_{00}^- \\ 0 & I_d & 0 \end{pmatrix} \begin{pmatrix} v_l & 0 & 0 \\ 0 & e_\alpha I_d & 0 \\ 0 & 0 & I_{N-m_l} \end{pmatrix} \begin{pmatrix} \Phi_{11}^+ & \Phi_{10}^+ & 0 \\ 0 & 0 & I_d \\ \Phi_{01}^+ & \Phi_{00}^+ & 0 \end{pmatrix} \end{aligned}$$

and

$$c_r = \begin{pmatrix} v_r & 0 & 0 \\ 0 & I_{N-m_r} & 0 \\ 0 & 0 & I_d \end{pmatrix}.$$

Since $m_l + d = m_r$, the criterion for d-normality we have already proved can be applied to $W(c)$. It results that $W(c)$ is d-normal if and only if (a) and (b) are satisfied and if the rows of the matrix

$$\begin{aligned} &\left(\begin{pmatrix} \Phi_{00}^- \\ 0 \end{pmatrix} (\Phi_{00}^+ \; 0) - \mu \begin{pmatrix} I & 0 \\ 0 & I \end{pmatrix}, \begin{pmatrix} \Phi_{01}^- & 0 \\ 0 & I \end{pmatrix}, \begin{pmatrix} \Phi_{00}^- \\ 0 \end{pmatrix} \Phi_{01}^+ \right) \\ &= \begin{pmatrix} \Phi_{00}^-\Phi_{00}^+ - \mu I & 0 & \Phi_{01}^- & 0 & \Phi_{00}^-\Phi_{01}^+ \\ 0 & -\mu I & 0 & I & 0 \end{pmatrix} \end{aligned}$$

or, equivalently, of the matrix $(\Phi_{00}^{-}\Phi_{00}^{+}-\mu I, \Phi_{01}^{-}, \Phi_{00}^{-}\Phi_{01}^{+})$ are linearly independent for every $\mu \in \mathrm{sp}\,(\Phi_{00}^{-}\Phi_{00}^{+}) \cap (-\infty, 0]$. □

At this point the proof of Theorems 10.17 and 10.18 is complete.

Combining Theorems 10.17 and 10.18, we get a slight generalization of the Fredholm criterion contained in Theorem 10.11.

Corollary 10.20. *Let $a \in SAP_{N\times N}$ and suppose a_l, a_r have a pair of pre-factorizations* (10.41). *Then $W(a)$ is Fredholm if and only if*

(a) $a \in GSAP_{N\times N}$;

(b) $m_l = m_r = 0$;

(c) $\mathrm{sp}\,\left(\mathbf{d}^{-1}(a_r)\mathbf{d}(a_l)\right) \cap (-\infty, 0] = \emptyset$.

Proof. Conditions (b) of Theorems 10.17 and 10.18 are equivalent to the requirement that v_l and v_r are constant matrices, and since $M(v_l) = M(v_r) = 0$, this means that $v_l = v_r = 0$. In that case $m_l = m_r = 0$, and hence $\Phi^{\pm}$ reduce to the blocks $\Phi_{00}^{\pm}$. Thus, conditions (c) of Theorems 10.17 and 10.18 are satisfied if and only if $\Phi_{00}^{-}\Phi_{00}^{+} - \mu I$ is invertible for all $\mu \in \mathrm{sp}\,(\Phi_{00}^{-}\Phi_{00}^{+}) \cap (-\infty, 0]$, which happens if and only if $\mathrm{sp}\,(\Phi_{00}^{-}\Phi_{00}^{+}) \cap (-\infty, 0] = \emptyset$. Since

$$\begin{aligned}\mathrm{sp}\,(\Phi_{00}^{-}\Phi_{00}^{+}) &= \mathrm{sp}\,\left(M^{-1}(a_r^{-})M(a_l^{-})M(a_l^{+})M^{-1}(a_r^{+})\right)\\ &= \mathrm{sp}\,\left(M^{-1}(a_r^{+})M^{-1}(a_r^{-})M(a_l^{-})M(a_l^{+})\right) = \mathrm{sp}\,\left(\mathbf{d}^{-1}(a_r)\mathbf{d}(a_l)\right),\end{aligned}$$

we arrive at condition (c) of the corollary. □

Notes

The results of Sections 10.1 and 10.2 were announced in [122], and a detailed exposition (for the more general case of Wiener-Hopf operators on L^p, $1 < p < \infty$) was given in [133]. Observe that the proof of Theorem 10.3 is based on the so called limit operators technique. This technique was originated by Favard [75] and then used extensively in [150], [151], [159], [182], [202]. The equivalence of (iii) and (iv) in Corollary 10.7 was earlier established in [212]. The rest of the chapter follows [132]. However, we remark that when considering semi-Fredholmness in [132], we required that the middle factor of the pre-factorizations (10.41) of the almost periodic representatives were diagonal.

Chapter 11

Left Versus Right Wiener-Hopf Factorization

In this chapter we show that if κ_l and κ_r are the left and right total indices of some matrix function in $L^\infty_{N\times N}$ with at most a finite number d of discontinuities on $\mathbf{R}\cup\{\infty\}$, then $|\kappa_l+\kappa_r|\le d(N-1)$. Conversely, given any integers λ and ϱ satisfying $|\lambda+\varrho|\le d(N-1)$, there exist matrix functions in $L^\infty_{N\times N}$ with at most d discontinuities whose left and right total indices are just λ and ϱ. We remark that such matrix functions cannot be found in $[C+H^\infty_\pm]_{N\times N}$, $PC_{N\times N}$ or $AP_{N\times N}$, but that one can find them in $SAPW_{N\times N}$.

11.1 The Associated Operator

Given a measurable subset Ω of the real line $\mathbf{R}$, we denote by $L^p_{N\times N}(\Omega)$ and $L^p_N(\Omega)$ the $N\times N$ matrix functions with entries in $L^p(\Omega)$ and the column vectors of height N with components in $L^p(\Omega)$, respectively. For $a\in L^\infty_{N\times N}(\mathbf{R})$, the Wiener-Hopf operator $W(a)$ is the compression of the convolution operator $W^0(a)$ to $L^2_N(\mathbf{R}_+)$. Given $a\in L^\infty_{N\times N}(\mathbf{R})$, we define $\widetilde{a}\in L^\infty_{N\times N}(\mathbf{R})$ by

$$\widetilde{a}(x):=a(-x)\quad (x\in\mathbf{R}).$$

The Wiener-Hopf operator $W(\widetilde{a})$ is called the *associated operator* of $W(a)$. A moment's thought reveals that $W(\widetilde{a})$ is unitarily equivalent to the compression of $W^0(a)$ to $L^2_N(\mathbf{R}_-)$, where $\mathbf{R}_-=(-\infty,0)$.

If a is the Fourier transform of an L^1 matrix function, $a=Fk$ with k in $L^1_{N\times N}(\mathbf{R})$, then $W(a)$ and $W(\widetilde{a})$ can be written in the form

$$(W(a)f)(t) = \int_0^\infty k(t-s)f(s)ds \quad (t>0),$$

$$(W(\widetilde{a})f)(t) = \int_0^\infty k(s-t)f(s)ds \quad (t>0).$$

We know that the symbol $a(x) = -\operatorname{sgn} x$ $(N=1)$ induces the Cauchy singular integral operator on $\mathbf{R}_+$,

$$(W(a)f)(t) = \frac{1}{\pi i}\int_0^\infty \frac{f(s)}{s-t}ds \quad (t>0),$$

and in the corresponding formula for $W(\widetilde{a})$ we have to replace $s-t$ by $t-s$. Finally, if $a(x) = e^{ix}$ $(N=1)$, then

$$(W(a)f)(t) = \begin{cases} f(t-1) & \text{for } t>1, \\ 0 & \text{for } 0<t<1, \end{cases} \tag{11.1}$$

$$(W(\widetilde{a})f)(t) = f(t+1) \text{ for } t>0. \tag{11.2}$$

It is well known that the answers to many questions on the operator $W(a)$ depend not only on the properties of $W(a)$ itself but also on the properties of the associated operator $W(\widetilde{a})$. This is, for instance, the case when studying the finite section method for $W(a)$ (see [46], [92]) or the Fredholm determinants of the truncations of $W(a)$ (see [228]). Actually, questions on the relation between certain properties of $W(a)$ and $W(\widetilde{a})$ are always questions on the connection between left and right Wiener-Hopf factorizations of the matrix function a (recall Section 6.2).

11.2 The Index of the Associated Operator

In this section we tell a nice story. We cite several results without proofs - these will be given in the remaining sections of this chapter.

In the scalar case $(N=1)$, $W(\widetilde{a})$ is the transpose operator of $W(a)$, that is, $W(\widetilde{a}) = (W(a))^T := CW^*(a)C$, where $C\varphi = \overline{\varphi}$ for $\varphi \in L^2(\mathbf{R}_+)$. Therefore $W(a)$ is Fredholm of index κ if and only if $W(\widetilde{a})$ is Fredholm of index $-\kappa$. This chapter concerns the connection between the Fredholm indices of $W(a)$ and $W(\widetilde{a})$, or equivalently, the connection between the right and left total indices of a in the matrix case $(N \geq 2)$. To be more precise, we consider the following questions:

(a) Does the Fredholmness of $W(a)$ imply that $W(\widetilde{a})$ is also Fredholm?

(b) If $W(a)$ and $W(\widetilde{a})$ are Fredholm, does it follow that $\operatorname{Ind} W(a) = -\operatorname{Ind} W(\widetilde{a})$?

The answers to both questions are known to be in the affirmative in case a belongs to certain classes of symbols, for example, if

$$a \in [C+H^\infty_+]_{N\times N} \cup [C+H^\infty_-]_{N\times N} \cup PC_{N\times N}$$

(see Sections 5.2 and 5.3).

We will show that the two questions have nevertheless negative answers for general $a \in L^\infty_{N\times N}(\mathbf{R})$ $(N \geq 2)$.

Proposition 11.1. *There exist $a \in AP_{2\times 2}$ such that $W(a)$ is invertible but $W(\widetilde{a})$ is not Fredholm.*

Proof. Example 6.7. □

Thus, the answer to question (a) is negative within $APW_{N\times N}$ $(N \geq 2)$. If f is in $AP_{N\times N}$ and $W(f)$ is Fredholm, then $W(f)$ is automatically invertible (Corollary 10.6). Hence, for $a \in AP_{N\times N}$ the answer to question (b) remains positive.

We know from Chapter 4 that much evil with Wiener-Hopf operators begins with semi-almost periodic symbols. This experience is supported by the following result, which shows that the answer to question (b) is already negative within $SAPW_{N\times N}$ $(N \geq 2)$.

Proposition 11.2. *There are $a \in SAPW_{2\times 2}$ such that $W(a)$ is Fredholm of index 0 and $W(\widetilde{a})$ is Fredholm of index 1.*

This proposition suggests to sharpen question (b) as follows:

(c) Given two integers κ and ν, is there a matrix function a such that $W(a)$ is Fredholm of index κ and $W(\widetilde{a})$ is Fredholm of index ν?

An answer is given by the following theorem. Notice that $a \in L^\infty_{N\times N}(\mathbf{R})$ is said to have at most k discontinuities on $\mathbf{R}\cup\{\infty\}$ if there are $x_1, \ldots, x_k \in \mathbf{R}\cup\{\infty\}$ such that a is continuous on $(\mathbf{R}\cup\{\infty\}) \setminus \{x_1, \ldots, x_k\}$. We call a point $x_0 \in \mathbf{R}\cup\{\infty\}$ a *harmless discontinuity* of a if the matrix function a is discontinuous at x_0 and if there exist an open neighborhood U of x_0 and a matrix function

$$b \in [C+H^\infty_+]_{N\times N} \cup [C+H^\infty_-]_{N\times N} \cup PC_{N\times N} \cup AP_{N\times N}$$

such that b is continuous on $U \setminus \{x_0\}$ and $a|U = b|U$. A discontinuity which is not harmless will be called a *severe discontinuity.*

Theorem 11.3. *If $a \in L^\infty_{N\times N}(\mathbf{R})$ has at most a finite number m of harmless discontinuities and at most a finite number d of severe discontinuities on $\mathbf{R}\cup\{\infty\}$ and both $W(a)$ and $W(\widetilde{a})$ are Fredholm, then*

$$|\operatorname{Ind} W(a) + \operatorname{Ind} W(\widetilde{a})| \leq d(N-1).$$

Conversely, given integers κ and ν satisfying $|\kappa + \nu| \le d(N-1)$, there exist matrix functions $a \in L^\infty_{N\times N}(\mathbf{R})$ with at most d severe discontinuities and no other discontinuities on $\mathbf{R} \cup \{\infty\}$ such that $W(a)$ is Fredholm of index κ and $W(\widetilde{a})$ is Fredholm of index ν.

Here is an obvious consequence of Theorem 11.3.

Corollary 11.4. *Given any natural number $N \ge 2$ and any two integers κ and ν, there exist $a \in L^\infty_{N\times N}(\mathbf{R})$ with at most finitely many discontinuities such that $W(a)$ is Fredholm of index κ and $W(\widetilde{a})$ is Fredholm of index ν.* □

As functions in SAP have at most a discontinuity at ∞, we also arrive at the following straightforward consequence of Theorem 11.3, which shows that Proposition 11.2 cannot be "improved".

Corollary 11.5. *There is no $a \in SAP_{2\times 2}$ such that $W(a)$ is Fredholm of index 0 and $W(\widetilde{a})$ is Fredholm of index 2.* □

Of course, Theorem 11.3 even implies that if $a \in SAP_{N\times N}$, $W(a)$ is Fredholm of index κ and $W(\widetilde{a})$ is Fredholm of index ν, then necessarily $|\kappa+\nu| \le N-1$. This leads to the following question:

(d) Given integers κ and ν satisfying $|\kappa+\nu| \le N-1$, is there a matrix function $a \in SAP_{N\times N}$ such that $W(a)$ is Fredholm of index κ and $W(\widetilde{a})$ is Fredholm of index ν?

The answer is yes.

Theorem 11.6. *If κ and ν are any integers subject to the condition $|\kappa+\nu| \le N-1$, then there exist $a \in SAPW_{N\times N}$ such that $W(a)$ is Fredholm of index κ and $W(\widetilde{a})$ is Fredholm of index ν.*

The rest of this chapter is devoted to the proofs of the results quoted above. To prove Theorem 11.3, we will have recourse to Wiener-Hopf factorization, and in this context passage from the real line to the unit circle makes things more comfortable. In Section 11.4 we therefore translate Theorem 11.3 into the language of Toeplitz operators, and then we give a proof of the Toeplitz analogue of the first assertion of Theorem 11.3.

We give two proofs of the second part of Theorem 11.3. The first of these proofs is in Section 11.5. The approach of Section 11.5 is based on the Fredholm criterion for Wiener-Hopf operators with piecewise continuous symbols. We there show even more: we prove that we can find symbols a which are analytic in the plane minus a finite number of points such that $W(a)$ and $W(\widetilde{a})$ have prescribed indices. In other words, the symbols we are looking for are in fact not as exotic as one might expect at the first glance.

In Section 11.6 we use the Fredholm criterion and the index formula for Wiener-Hopf operators with $SAPW$ symbols established in Chapter 10 to prove Theorem 11.6. Clearly, Theorem 11.6 and standard localization techniques yield a second proof of the second assertion of Theorem 11.3. Notice, however, that the machinery employed in Section 11.6 is much heavier than the tools we are using in Section 11.5. Finally, Proposition 11.2 is an obvious consequence of Theorem 11.6.

11.3 Introduction to Toeplitz Operators

It will be more convenient to pass from the real line $\mathbf{R}$ to the complex unit circle $\mathbf{T}$ or, equivalently, from Wiener-Hopf integral operators to Toeplitz operators.

Let g be a matrix function in $L^\infty_{N\times N}(\mathbf{T})$. Denote by $g_k \in \mathbf{C}^{N\times N}$ $(k \in \mathbf{Z})$ the Fourier coefficients of g,

$$g_k = \frac{1}{2\pi}\int_0^{2\pi} g(e^{i\theta})e^{-ik\theta}\,d\theta.$$

Further, let $l^2(\mathbf{Z}_+, \mathbf{C}^N)$ stand for the $\mathbf{C}^N$-valued l^2 space over $\mathbf{Z}_+ := \{0, 1, 2, \dots\}$. The operator induced on $l^2(\mathbf{Z}_+, \mathbf{C}^N)$ by the matrix

$$\begin{pmatrix} g_0 & g_{-1} & g_{-2} & \dots \\ g_1 & g_0 & g_{-1} & \dots \\ g_2 & g_1 & g_0 & \dots \\ \dots & \dots & \dots & \dots \end{pmatrix} \tag{11.3}$$

is called the *Toeplitz operator with the symbol* g and is denoted by $T(g)$.

Given $a \in L^\infty_{n\times n}(\mathbf{R})$, define $B_0 a \in L^\infty_{N\times N}(\mathbf{T})$ by (6.5):

$$(B_0 a)(t) := a\Big(i\frac{1+t}{1-t}\Big), \quad t \in \mathbf{T} \setminus \{1\}.$$

It is easily verified that $W(a)$ on $L^2_N(\mathbf{R}_+)$ and $T(B_0 a)$ on $l^2(\mathbf{Z}_+, \mathbf{C}^N)$ are unitarily equivalent, see, e.g., [46, p. 400, 9.5(e)]. This implies the following.

Lemma 11.7. *The Wiener-Hopf integral operator $W(a)$ is Fredholm of index κ if and only if the Toeplitz operator $T(B_0 a)$ is Fredholm of index κ.* □

If $a \in L^\infty_{N\times N}(\mathbf{R})$, we have $(B_0\widetilde{a})(t) = (B_0 a)(1/t)$ for $t \in \mathbf{T}$. Thus, the associated operator of the Toeplitz operator $T(g)$ is defined as $T(\widetilde{g})$ with $\widetilde{g}(t) := g(1/t)$ $(t \in \mathbf{T})$. Notice that if $T(g)$ is given by the matrix $(g_{j-k})_{j,k=0}^\infty$, then $T(\widetilde{g})$ has the matrix $(g_{k-j})_{j,k=0}^\infty$.

The following two theorems summarize a few basic and well known properties of Toeplitz operators, which can be proved as their analogues for Wiener-Hopf operators.

Theorem 11.8. (a) *If* $a \in [C(\mathbf{T})]_{N\times N}$, *then* $T(g)$ *is Fredholm if and only if* $\det g$ *has no zeros on* $\mathbf{T}$. *In that case,* $\operatorname{Ind} T(g)$ *is minus the winding number of* $\det g$ *about the origin.*
(b) *If* $g, h \in L^\infty_{N\times N}(\mathbf{T})$ *and if at each point of* $\mathbf{T}$ *at least one of the matrix functions* g *and* h *is continuous, then* $T(gh) - T(g)T(h)$ *is a compact operator.*
(c) *Let* $g \in L^\infty_{N\times N}(\mathbf{T})$ *and suppose for each* $\tau \in \mathbf{T}$ *there exist an open subarc* $\Gamma_\tau \subset \mathbf{T}$ *containing* τ *and a matrix function* $g_\tau \in L^\infty_{N\times N}(\mathbf{T})$ *such that* $g = g_\tau$ *on* Γ_τ *and* $T(g_\tau)$ *is Fredholm. Then* $T(g)$ *is also Fredholm.*

A matrix function $g \in L^\infty_{N\times N}(\mathbf{T})$ is said to have m harmless and d severe discontinuities on $\mathbf{T}$ if $g = B_0 a$ where a has m harmless and d severe discontinuities on $\mathbf{R} \cup \{\infty\}$.

Theorem 11.9. *If* $g \in L^\infty_{N\times N}(\mathbf{T})$ *has at most finitely many harmless discontinuities and both* $T(g)$ *and* $T(\widetilde{g})$ *are Fredholm, then* $\operatorname{Ind} T(g) = -\operatorname{Ind} T(\widetilde{g})$.

In accordance with Remark 6.2, a right Wiener-Hopf (WH) factorization of a matrix function $g \in GL^\infty_{N\times N}(\mathbf{T})$ is a representation

$$g(t) = g_-(t)m(t)g_+(t), \quad m(t) = \operatorname{diag}(t^{\varrho_1}, \dots, t^{\varrho_N})$$

for almost all $t \in \mathbf{T}$ such that $\varrho_1, \dots, \varrho_N$ are integers, $g_\pm \in G[H^2_\pm(\mathbf{T})]_{N\times N}$, and the operator $g_+^{-1} S_0 g_+ I$ is bounded on $L^2_N(\mathbf{T})$. The sum $\varrho_1 + \dots + \varrho_N$ is referred to as the total index of the given right WH factorization.

The following theorem is the Toeplitz analogue of Theorem 6.5.

Theorem 11.10 (Simonenko). *Let* $g \in L^\infty_{N\times N}(\mathbf{T})$. *The operator* $T(g)$ *is Fredholm of index* κ *if and only if* g *is invertible in* $L^\infty_{N\times N}(\mathbf{T})$ *and* g *has a right* WH *factorization with the total index* $-\kappa$.

11.4 The Inequality for the Total Indices

We begin with two refinements of Theorem 11.10.

Proposition 11.11. *Let* g *be a matrix function in* $GL^\infty_{N\times N}(\mathbf{T})$ *and suppose* g *extends to an analytic matrix function in some open set* $U \subset \mathbf{C} \cup \{\infty\}$ *such that* $U \cap \mathbf{T} \neq \emptyset$. *If* $g = g_- m g_+$ *is a right* WH *factorization, then* g_+ *is analytic in* $U \setminus \{\infty\}$ *and* g_- *is analytic in* $U \setminus \{0\}$. *Moreover, if* m *is the identity matrix, then* $g_\pm$ *are analytic in all of* U.

Proof. We have $g_+ = m^{-1} g_-^{-1} g$ almost everywhere on $\mathbf{T}$. Define

$$\psi(z) := \begin{cases} g_+(z) & \text{for } z \in (\mathbf{D}_+ \cup \mathbf{T}) \cap U, \\ m^{-1}(z) g_-^{-1}(z) g(z) & \text{for } z \in (\mathbf{D}_- \cup \mathbf{T}) \cap U. \end{cases}$$

The assertion for g_+ will follow once we have shown that ψ is analytic in $U \setminus \{\infty\}$; the analyticity of g_- can be proved analogously.

To show that ψ is analytic in $U \setminus \{\infty\}$, we employ Morera's theorem. Thus let δ be the positively oriented circumference of a triangle $\Delta \subset U \setminus \{\infty\}$. If $\delta \subset \mathbf{D}_+$ or $\delta \subset \mathbf{D}_-$, then $\int_\delta \psi(z)dz = 0$ because ψ is analytic in $\mathbf{D}_+ \cap U$ and $\mathbf{D}_- \cap U$. If δ intersects $\mathbf{T}$, we can write $\int_\delta \psi(z)\,dz = \int_{\delta_+} \psi(z)\,dz + \int_{\delta_-} \psi(z)\,dz$, where

$$\delta_+ = (\delta \cap \mathbf{D}_+) \cup \gamma, \quad \delta_- = (\delta \cap \mathbf{D}_-) \cup (-\gamma),$$

γ is the positively oriented arc $\mathbf{T} \cap \overline{\Delta}$ of $\mathbf{T}$, and $-\gamma$ is γ with the opposite orientation. Clearly,

$$\int_{\delta_+} \psi(z)dz = \int_{\delta_+} g_+(z)dz. \tag{11.4}$$

The matrix function $g_+ \in [H_+^2(\mathbf{T})]_{N\times N} \subset [H_+^1(\mathbf{T})]_{N\times N}$ can be approximated by analytic polynomials P_k^+ in the $H_+^1(\mathbf{T})$ norm as closely as desired, and because

$$\int_{\delta_+} P_k^+(z)dz = 0,$$

it follows that (11.4) also equals zero. Analogously,

$$\int_{\delta_-} \psi(z)dz = \int_{\delta_-} m^{-1}(z)g_-^{-1}(z)g(z)dz, \tag{11.5}$$

and $g_-^{-1} \in [H_-^2(\mathbf{T})]_{N\times N} \subset [H_-^1(\mathbf{T})]_{N\times N}$ can be approximated in $[H_-^1(\mathbf{T})]_{N\times N}$ by matrix functions of the form

$$P_k^-(z) = q_0 + q_1 z^{-1} + \ldots + q_k z^{-k}$$

as closely as wanted. As $m^{-1}P_k^- g$ is analytic in $U \setminus \{0, \infty\}$ and $\delta_- \subset U \setminus \{0, \infty\}$, we get

$$\int_{\delta_-} m^{-1}(z)P_k^-(z)g(z)dz = 0,$$

which implies that (11.5) is also zero.

In summary, $\int_\delta \psi(z)dz = 0$ for every triangle curve $\delta = \partial\Delta$ with $\Delta \subset U \setminus \{\infty\}$. Hence, by Morera's theorem, ψ is analytic in $U \setminus \{\infty\}$. Clearly, if $m = I_N$, then g_+ is analytic in all of U, and the same is true for g_-. □

Every real number $\alpha > 0$ can be uniquely written in the form $\alpha = [\alpha] + \{\alpha\}$ where $[\alpha]$ is an integer and $\{\alpha\} \in [0, 1)$. We denote by C^α the $[\alpha]$ times continuously differentiable functions on $\mathbf{T}$ whose $[\alpha]$-th derivative satisfies a Hölder condition with the exponent $\{\alpha\}$.

Proposition 11.12. *Let $g \in L^\infty_{N\times N}(\mathbf{T})$, let $\Gamma \subset \mathbf{T}$ be an open arc, and suppose g is C^α $(\alpha > 0)$ on Γ. If $T(g)$ is Fredholm, then g possesses a right WH factorization $g = g_- m g_+$ such that $g_\pm$ and $g_\pm^{-1}$ are also C^α on Γ.*

Proof. The result is well known in the case where $\Gamma = \mathbf{T}$. So assume that Γ is a proper subarc of $\mathbf{T}$, that is, $\Gamma = \{e^{i\theta} : \eta < \theta < \beta\}$ with $\beta - \eta < 2\pi$. Let $\gamma \subset \Gamma$ be any arc of the form $\gamma = \{e^{i\theta} : \eta_0 < \theta < \beta_0\}$ with $\eta < \eta_0 < \beta_0 < \beta$. We must show that g has a right WH factorization $g = g_- m g_+$ such that $g_\pm$, $g_\pm^{-1}$ are C^α on γ.
Since $T(g)$ is Fredholm, the matrix function g is invertible in $L^\infty_{N\times N}(\mathbf{T})$. Hence, we can find an $N \times N$ matrix function h whose entries are C^α on all of $\mathbf{T}$ such that $h|\gamma = g|\gamma$ and $\det h$ has no zeros on $\mathbf{T}$. As $T(h)$ is Fredholm and as the proposition is true for $\Gamma = \mathbf{T}$, there is a right WH factorization $h = h_- n h_+$ with $h_\pm$, $h_\pm^{-1}$ in C^α on $\mathbf{T}$. Theorem 11.10 gives us a right WH factorization $g = g_- m g_+$ with $g_\pm$, $g_\pm^{-1}$ in $H^2_\pm(\mathbf{T})$. Because $h = g$ on γ, it follows that

$$m g_+ h_+^{-1} = g_-^{-1} h_- n \quad \text{on} \quad \gamma. \tag{11.6}$$

Put

$$\psi(z) := \begin{cases} m(z) g_+(z) h_+^{-1}(z) & \text{for } z \in \mathbf{D}_+ \setminus \{0\}, \\ g_-^{-1}(z) h_-(z) n(z) & \text{for } z \in \mathbf{D}_- \setminus \{\infty\}, \end{cases}$$

and let ψ be given by (11.6) on γ.
The matrix functions $g_+, g_-^{-1}, h_+^{-1}, h_-$ can be approximated in L^2 by matrix functions of the form

$$P_k^+(z) = p_0^+ + p_1^+ z + \ldots + p_k^+ z^k, \quad P_k^-(z) = p_0^- + p_1^- z^{-1} + \ldots + p_k^- z^{-k},$$

$$Q_k^+(z) = q_0^+ + q_1^+ z + \ldots + q_k^+ z^k, \quad Q_k^-(z) = q_0^- + q_1^- z^{-1} + \ldots + q_k^- z^{-k},$$

respectively, as closely as desired. Since $m P_k^+ Q_k^+$ and $P_k^- Q_k^- n$ are analytic in $\mathbf{D}_+ \setminus \{0\}$ and $\mathbf{D}_- \setminus \{\infty\}$ and converge in L^1 to $m g_+ h_+^{-1}$ and $g_-^{-1} h_- n$, respectively, we can employ Morera's theorem as in the proof of Proposition 11.11 to conclude that ψ is analytic in $\mathbf{C} \setminus \{0\}$.
Because $g_+ = m^{-1} \psi h_+$ on γ, we see that g_+ is C^α on γ. Since $\det g_+$ cannot have zeros on γ (Theorem 11.8(a)), it follows that g_+^{-1} is also C^α on γ.
Analogously, using the representation

$$h_-^{-1} g_- m = n h_+ g_+^{-1} \text{ on } \gamma$$

in place of (11.6), one can prove that g_- is C^α on γ. Invoking Theorem 11.8(a) again, we conclude that $\det g_-$ does not have zeros on γ, so that g_-^{-1} is C^α on γ as well. □

The following proposition is the Toeplitz analogue of the first part of Theorem 11.3. Recall that $\widetilde{g}(t) = g(1/t)$ for $t \in \mathbf{T}$.

Proposition 11.13. *If $g \in L^\infty_{N\times N}(\mathbf{T})$ has at most a finite number m of harmless discontinuities and at most a finite number d of severe discontinuities on $\mathbf{T}$ and if $T(g)$ and $T(\widetilde{g})$ are Fredholm, then*

$$|\operatorname{Ind} T(g) + \operatorname{Ind} T(\widetilde{g})| \le d(N-1).$$

Proof. By Theorem 11.9, there is nothing to prove for $d = 0$. Let us now assume that g has exactly one discontinuity on $\mathbf{T}$, namely a severe discontinuity at $t_0 \in \mathbf{T}$. We can approximate g in $L^\infty_{N\times N}(\mathbf{T})$ as closely as desired by matrix functions which are C^1 on $\mathbf{T} \setminus \{t_0\}$. Since both the Fredholmness and the index are stable under uniform approximation, we may a priori assume that g is C^1 on $\mathbf{T}\setminus\{t_0\}$. Applying Theorem 11.10 and Proposition 11.12 to g and to $\widetilde{g}$, we get factorizations

$$g = g_- u g_+ = h_+ v h_- \tag{11.7}$$

where $g_+^{\pm 1}, h_+^{\pm 1} \in [H^2_+(\mathbf{T})]_{N\times N}$, $g_-^{\pm 1}, h_-^{\pm 1} \in [H^2_-(\mathbf{T})]_{N\times N}$,

$$\begin{aligned} u(t) &= \operatorname{diag}(t^{\varrho_1}, \ldots, t^{\varrho_N}), \qquad & \varrho &:= \varrho_1 + \ldots + \varrho_N = -\operatorname{Ind} T(g),\\ v(t) &= \operatorname{diag}(t^{\lambda_1}, \ldots, t^{\lambda_N}), \qquad & \lambda &:= \lambda_1 + \ldots + \lambda_N = \operatorname{Ind} T(\widetilde{g}), \end{aligned}$$

and $g_+^{\pm 1}, h_+^{\pm 1}, g_-^{\pm 1}, h_-^{\pm 1}$ are C^1 on $\mathbf{T} \setminus \{t_0\}$. Taking the determinant of (11.7) we obtain

$$\det g_-(t)\, t^{\varrho} \det g_+(t) = \det h_+(t)\, t^{\lambda} \det h_-(t),$$

whence

$$t^{\varrho-\lambda}\varphi_+(t) = \varphi_-(t), \quad \varphi_+ := \frac{\det g_+}{\det h_+}, \quad \varphi_- := \frac{\det h_-}{\det g_-}. \tag{11.8}$$

Obviously, $\det g_+^{\pm 1} \in H^{2/N}_+(\mathbf{T})$ and $\det h_-^{\pm 1} \in H^{2/N}_-(\mathbf{T})$. This shows that $\varphi_\pm$ is in $H^{1/N}_\pm(\mathbf{T})$. Since $\varphi_\pm(z) \neq 0$ for $z \in \mathbf{D}_\pm$ and $\varphi_\pm$ are nonzero and continuous on $\mathbf{T} \setminus \{t_0\}$, there are functions $\psi_\pm$ such that

$$(\psi_\pm(z))^N = \varphi_\pm(z) \quad \text{for} \quad z \in \mathbf{D}_\pm \cup (\mathbf{T} \setminus \{t_0\}) \tag{11.9}$$

and $\psi_\pm$ are analytic in $\mathbf{D}_\pm$ and continuous on $\mathbf{T} \setminus \{t_0\}$. Because $\varphi_\pm \in H^{1/N}_\pm(\mathbf{T})$ and hence

$$\sup_{0<r^{\pm1}<1} \int_0^{2\pi} |\psi_\pm(re^{i\theta})| d\theta = \sup_{0<r^{\pm1}<1} \int_0^{2\pi} |\varphi_\pm(re^{i\theta})|^{1/N} d\theta < \infty,$$

it results that $\psi_\pm \in H^1_\pm(\mathbf{T})$. Analogously one can show that $\psi_\pm^{-1} \in H^1_\pm(\mathbf{T})$.

Our aim is to prove that $|\varrho - \lambda| \le N - 1$. Contrary to what we want, let us assume that $|\varrho - \lambda| \ge N$. For the sake of definiteness, suppose $\varrho - \lambda = N + \alpha$ with an integer $\alpha \ge 0$ (the case $\varrho - \lambda \le -N$ can be reduced to the case considered here by passage to adjoint operators). Let $t^{\alpha/N}$ be any branch of the function which is continuous on $\mathbf{T} \setminus \{t_0\}$. From (11.8) and (11.9) we obtain

$$(t\, t^{\alpha/N} \psi_+(t))^N = (\psi_-(t))^N \quad \text{for} \quad t \in \mathbf{T} \setminus \{t_0\}.$$

Taking into account that $t\,t^{\alpha/N}\psi_+(t)$ and $\psi_-(t)$ are continuous on $\mathbf{T}\setminus\{t_0\}$, we therefore get

$$t\,t^{\alpha/N}\psi_+(t) = c\psi_-(t) \quad \text{for} \quad t\in\mathbf{T}\setminus\{t_0\},$$

where $c\in\mathbf{C}$ satisfies $c^N=1$. We now employ an old trick, namely, we write

$$t^{\alpha/N} = (t-t_0)^{\alpha/N}(1-t_0/t)^{-\alpha/N},$$

which is valid with appropriately chosen branches of $(t-t_0)^{\alpha/N}$ and $(1-t_0/t)^{-\alpha/N}$. It follows that

$$t(t-t_0)^{\alpha/N}\psi_+(t) = c(1-t_0/t)^{\alpha/N}\psi_-(t) \quad \text{for} \quad t\in\mathbf{T}\setminus\{t_0\}. \tag{11.10}$$

Since $\psi_\pm\in H^1_\pm(\mathbf{T})$ and

$$(t-t_0)^{\alpha/N}\in H^\infty_+(\mathbf{T}), \quad (t-t_0/t)^{\alpha/N}\in H^\infty_-(\mathbf{T}),$$

we see that the left-hand side of (11.10) belongs to $H^1_+(\mathbf{T})$, while the right-hand side of (11.10) is a function in $H^1_-(\mathbf{T})$. As $H^1_+(\mathbf{T})\cap H^1_-(\mathbf{T})$ is the set of all constant functions, the function

$$t(t-t_0)^{\alpha/N}\psi_+(t)\in H^1_+(\mathbf{T})$$

must be constant. This function vanishes at the origin, whence

$$t(t-t_0)^{\alpha/N}\psi_+(t) = 0 \quad \text{for} \quad t\in\mathbf{T}\setminus\{t_0\},$$

which is impossible because $\psi_+^{-1}\in H^1_+(\mathbf{T})$. This contradiction completes the proof in the case where g has exactly 1 severe discontinuity and no other discontinuities.

If g has exactly d severe discontinuities and exactly m harmless discontinuities on $\mathbf{T}$, we can write

$$g = g_1\cdots g_d g_{d+1}\cdots g_{d+m}$$

where each g_j has exactly 1 discontinuity on $\mathbf{T}$, which is a severe discontinuity for $1\le j\le d$ and a harmless discontinuity for $d+1\le j\le d+m$. A simple application of Theorems 11.8 and 11.9 shows that all $T(g_j)$ and $T(\widetilde{g}_j)$ are Fredholm together with $T(g)$ and $T(\widetilde{g})$. Also by Theorem 11.8, the operators

$$T(g)-\prod_{j=1}^{d+m}T(g_j) \quad \text{and} \quad T(\widetilde{g})-\prod_{j=1}^{d+m}T(\widetilde{g}_j)$$

are compact. Consequently, from what was already proved we get

$$
\begin{aligned}
&|\operatorname{Ind} T(g) + \operatorname{Ind} T(\widetilde{g})| \\
&\quad = \left| \sum_{j=1}^{d} \operatorname{Ind} T(g_j) + \sum_{j=1}^{d} \operatorname{Ind} T(\widetilde{g}_j) + \sum_{j=d+1}^{d+m} \operatorname{Ind} T(g_j) + \sum_{j=d+1}^{d+m} \operatorname{Ind} T(\widetilde{g}_j) \right| \\
&\quad = \left| \sum_{j=1}^{d} \operatorname{Ind} T(g_j) + \sum_{j=1}^{d} \operatorname{Ind} T(\widetilde{g}_j) \right| \\
&\quad \le \sum_{j=1}^{d} |\operatorname{Ind} T(g_j) + \operatorname{Ind} T(\widetilde{g}_j)| \le d(N-1). \qquad \square
\end{aligned}
$$

Combining Lemma 11.7 and Proposition 11.13 we obtain the first assertion of Theorem 11.3.

11.5 Bounded Symbols with Prescribed Total Indices

We now proceed to the proof of the second part of Theorem 11.3.

We know from Lemma 11.7 and Theorem 11.9 that if $a \in [C(\overline{\mathbf{R}})]_{N\times N}$ and $W(a)$ as well as $W(\widetilde{a})$ are Fredholm, then

$$\operatorname{Ind} W(a) + \operatorname{Ind} W(\widetilde{a}) = 0. \tag{11.11}$$

It should be mentioned that this can also be easily deduced from Theorem 5.10. Indeed, let η_j be the eigenvalues of $a^{-1}(-\infty)a(+\infty)$. Since

$$\widetilde{a}^{-1}(-\infty)\widetilde{a}(+\infty) = a^{-1}(+\infty)a(-\infty),$$

the eigenvalues $\widetilde{\eta}_j$ of $\widetilde{a}^{-1}(-\infty)\widetilde{a}(+\infty)$ are $\widetilde{\eta}_j = 1/\eta_j$, whence $\arg \eta_j = -\arg \widetilde{\eta}_j$. From Theorem 5.10 we therefore obtain

$$
\begin{aligned}
\operatorname{Ind} W(a) &= -\frac{1}{2\pi}\{\arg \det a\}_{-\infty}^{\infty} + \frac{1}{2\pi}\sum_{j=1}^{N} \arg \eta_j, \\
\operatorname{Ind} W(\widetilde{a}) &= \frac{1}{2\pi}\{\arg \det a\}_{-\infty}^{\infty} - \frac{1}{2\pi}\sum_{j=1}^{N} \arg \eta_j,
\end{aligned}
$$

and addition of these two equalities gives (11.11).

One of the main ingredients of our proof is the following beautiful result by Ballantine [12], [13] (also see [112, Problem 10 on p. 295]).

Theorem 11.14 (Ballantine). *A matrix $M \in \mathbf{C}^{N\times N}$ is the product of four positive definite matrices in $\mathbf{C}^{N\times N}$ if and only if*

$$\det M > 0 \quad \textit{and} \quad M \notin \bigcup_{\lambda<0}\{\lambda I_N\},$$

where I_N is the $N\times N$ identity matrix.

The next result is the key to the proof of the second assertion of Theorem 11.3.

Proposition 11.15. *Let $N \ge 2$ be a natural number and let ν be a number in $\{-(N-1),\dots,N-1\}$. Then there exists an $a \in L^\infty_{N\times N}(\mathbf{R})$ which is continuous on $\mathbf{R}$ and extends to an analytic function in $\mathbf{C}\setminus\Sigma$, $\Sigma = i(-\infty,-1]\cup i[1,\infty)$, such that $W(a)$ is invertible and $W(\widetilde{a})$ is Fredholm of index ν.*

Proof. Without loss of generality assume $\nu \in \{1,\dots,N-1\}$ (the case $\nu = 0$ is trivial and the case $\nu \in \{-1,\dots,-(N-1)\}$ can be disposed of by passage to adjoint operators). Let $k := N-\nu$ and let M be the $N\times N$ diagonal matrix

$$M := e^{-2\pi ik/N}\operatorname{diag}(r_1,\dots,r_N) \text{ with positive } r_1,\dots,r_N.$$

To guarantee the applicability of Theorem 11.14, we impose the condition that not all $r_1,\dots,r_N$ are the same (which of course is relevant only for even N and $\nu = N/2$). Then there exist four positive definite matrices $b(\pm\infty), c(\pm\infty)$ such that

$$M = c^{-1}(+\infty)b^{-1}(+\infty)b(-\infty)c(-\infty).$$

Define $b, c \in [C(\overline{\mathbf{R}})]_{N\times N}$ by

$$b(x) = (1-u(x))b(-\infty) + u(x)b(+\infty),$$

$$c(x) = (1-u(x))c(-\infty) + u(x)c(+\infty),$$

where $u = (1+\Phi)/2$ and Φ is the function defined by (3.9). Clearly, b and c extend to analytic functions in $\mathbf{C}$. Since $0 \le u \le 1$ on $\mathbf{R}$, we also see that there is an $\varepsilon > 0$ such that

$$(b(x)\zeta,\zeta) \ge \varepsilon\|\zeta\|^2 \quad \text{and} \quad (c(x)\zeta,\zeta) \ge \varepsilon\|\zeta\|^2$$

for all $x \in \mathbf{R}$ and all $\zeta \in \mathbf{C}^n$. From Theorem 9.14 we therefore deduce the existence of $b_+, c_+ \in G[H^\infty_+]_{N\times N}$ such that $b = b_+b_+^*$ and $c = c_+c_+^*$. Taking into account Lemma 11.7 and Proposition 11.11, it is easily seen that b_+, c_+ extend to analytic functions in $\mathbf{C}$.

For a real number $\delta \in (0,1/2)$, define $\Lambda \in [C(\overline{\mathbf{R}})]_{N\times N}$ by

$$\Lambda(x) := \operatorname{diag}\left(\left(\frac{x-i}{x+i}\right)^\delta,\dots,\left(\frac{x-i}{x+i}\right)^\delta\right) \quad (x \in \mathbf{R}).$$

The branch of $((x-i)/(x+i))^\delta$ is chosen so that

$$\Lambda(-\infty) = I_N \quad \text{and} \quad \Lambda(+\infty) = e^{2\pi i\delta} I_N.$$

The operator $W(((x-i)/(x+i))^\delta)$ is known to be invertible, and hence $W(\Lambda)$ is also invertible.

Put $a := b_+^* \Lambda c_+$. Clearly, $W(a) = W(b_+^*)W(\Lambda)W(c_+)$ is invertible, a is continuous on $\mathbf{R}$, and a extends to an analytic function in $\mathbf{C} \setminus \Sigma$. Since

$$\begin{aligned} W(\widetilde{b}_+)W(\widetilde{a})W(\widetilde{c}_+^*) &= W(\widetilde{b}_+)W(\widetilde{b}_+^* \widetilde{\Lambda}\widetilde{c}_+)W(\widetilde{c}_+^*) \\ &= W(\widetilde{b}_+\widetilde{b}_+^* \widetilde{\Lambda}\widetilde{c}_+\widetilde{c}_+^*) = W(\widetilde{b\Lambda c}), \end{aligned}$$

$W(\widetilde{a})$ is Fredholm if and only if $W(\widetilde{b\Lambda c})$ is Fredholm, in which case both operators have the same index.

As $\widetilde{b\Lambda c} \in [C(\overline{\mathbf{R}})]_{N\times N}$, the Fredholm properties of $W(\widetilde{b\Lambda c})$ can be studied with the help of Theorem 5.10. It is clear that $\det(\widetilde{b\Lambda c})$ does not vanish on $\mathbf{R} \cup \{\pm\infty\}$. Moreover,

$$\begin{aligned} -\frac{1}{2\pi}\{\arg\det(\widetilde{b\Lambda c})\}_{-\infty}^{\infty} &= \frac{1}{2\pi}\{\arg\det(b\Lambda c)\}_{-\infty}^{\infty} \\ &= \frac{1}{2\pi}\{\arg\det b\}_{-\infty}^{\infty} + \frac{1}{2\pi}\{\arg\det \Lambda\}_{-\infty}^{\infty} + \frac{1}{2\pi}\{\arg\det c\}_{-\infty}^{\infty} \\ &= 0 + \frac{1}{2\pi}N \cdot 2\pi\delta + 0 = N\delta. \end{aligned} \tag{11.12}$$

We have

$$\begin{aligned} (\widetilde{b\Lambda c})^{-1}(-\infty)(\widetilde{b\Lambda c})(+\infty) &= (b\Lambda c)^{-1}(+\infty)(b\Lambda c)(-\infty) \\ &= c^{-1}(+\infty)\Lambda^{-1}(+\infty)b^{-1}(+\infty)b(-\infty)\Lambda(-\infty)c(-\infty) \\ &= c^{-1}(+\infty)e^{-2\pi i\delta}I_N b^{-1}(+\infty)b(-\infty)I_N c(-\infty) \\ &= e^{-2\pi i\delta}M = \operatorname{diag}\left(r_1 e^{2\pi i(-k/N-\delta)}, \dots, r_N e^{2\pi i(-k/N-\delta)}\right). \end{aligned}$$

The eigenvalues of this matrix are

$$\xi_j = r_j e^{2\pi i(-k/N-\delta)}, \quad j = 1, \dots, N.$$

Now choose $\delta \in (0, 1/2)$ so that $-1 < -k/N - \delta < -1/2$ (recall that k is in $\{1, \dots, N-1\}$). Then $0 < 2\pi(-k/N-\delta) + 2\pi < \pi$, and hence the argument $\arg \xi_j$ lying in $(-\pi, \pi)$ is $2\pi(-k/N-\delta) + 2\pi$. This together with (11.12) and Theorem 5.10 yields that $W(\widetilde{b\Lambda c})$ is Fredholm and that

$$\operatorname{Ind} W(\widetilde{b\Lambda c}) = N\delta + N(-k/N-\delta) + N = N - k = \nu. \qquad \square$$

We are now in a position to complete the proof of Theorem 11.3. Combining Proposition 11.15 and Lemma 11.7, we get matrix functions $g \in L^\infty_{N\times N}(\mathbf{T})$ with at most a discontinuity at $t_0 = 1$ such that $T(g)$ is invertible and thus Fredholm of index zero, whereas $T(\widetilde{g})$ is Fredholm with any prescribed index in the set $\{-(N-1), \dots, N-1\}$. On replacing $g(t)$ by $g(t/t_0)$ with $t_0 \in \mathbf{T}$, which amounts to replacing the matrix (11.3) by

$$\begin{pmatrix} g_0 & t_0 g_{-1} & t_0^2 g_{-2} & \dots \\ t_0^{-1} g_1 & g_0 & t_0 g_{-1} & \dots \\ t_0^{-2} g_2 & t_0^{-1} g_1 & g_0 & \dots \\ \dots & \dots & \dots & \dots \end{pmatrix} = D_{t_0}^{-1} \begin{pmatrix} g_0 & g_{-1} & g_{-2} & \dots \\ g_1 & g_0 & g_{-1} & \dots \\ g_2 & g_1 & g_0 & \dots \\ \dots & \dots & \dots & \dots \end{pmatrix} D_{t_0}$$

with $D_{t_0} := \operatorname{diag}(1, t_0, t_0^2, \dots)$, we can achieve that g has its discontinuity at an arbitrarily given point $t_0 \in \mathbf{T}$. Using Lemma 11.7 to go back to the real line, we obtain $a \in L^\infty_{N\times N}(\mathbf{R})$ with at most a discontinuity at a given point $x_0 \in \mathbf{R} \cup \{\infty\}$ such that $W(a)$ is Fredholm of index zero and $W(\widetilde{a})$ is Fredholm of prescribed index $\nu \in \{-(N-1), \dots, N-1\}$. Considering products $a := a_1 \dots a_d$ in which each a_j has at most one discontinuity and a_j, a_k have no common discontinuity for $j \neq k$ and taking into account Theorem 11.8, we arrive at an $a \in L^\infty_{N\times N}(\mathbf{R})$ with at most d discontinuities on $\mathbf{R} \cup \{\infty\}$ for which $W(a)$ is Fredholm of index zero and $W(\widetilde{a})$ is Fredholm of prescribed index $\nu \in \{-d(N-1), \dots, d(N-1)\}$. Replacing $a(x)$ by

$$a(x) \operatorname{diag}\left(\left(\frac{x+i}{x-i}\right)^\kappa, 1, \dots, 1\right),$$

we get the desired a's with

$$\operatorname{Ind} W(a) = \kappa \quad \text{and} \quad \operatorname{Ind} W(\widetilde{a}) \in \{-\kappa - d(N-1), \dots, -\kappa + d(N-1)\}.$$

11.6 *SAP* Symbols with Prescribed Total Indices

In this section we prove Theorem 11.6.

We remark that Theorem 10.13 immediately gives the *SAP* version of the first assertion of Theorem 11.3, that is, the inequality

$$|\operatorname{Ind} W(a) + \operatorname{Ind} W(\widetilde{a})| \leq n - 1.$$

Indeed, since $SAPW_{N\times N}$ is dense in $SAP_{N\times N}$, it suffices to prove the inequality for a in $SAPW_{N\times N}$. In that case we obtain from Theorem 10.13 (see also Theorem 5.10) that

$$\begin{aligned} \operatorname{Ind} W(a) &= -\frac{1}{2\pi}\{\arg\det a\}_{-\infty}^{\infty} + \frac{1}{2\pi}\sum_{j=1}^N \arg\eta_j, \\ \operatorname{Ind} W(\widetilde{a}) &= \frac{1}{2\pi}\{\arg\det a\}_{-\infty}^{\infty} + \frac{1}{2\pi}\sum_{j=1}^N \arg\widetilde{\eta}_j, \end{aligned}$$

where η_i (resp. $\widetilde{\eta}_i$) are the eigenvalues of the matrix $\mathbf{d}(a_l)^{-1}\mathbf{d}(a_r)$ (resp. of the matrix $\mathbf{d}((\widetilde{a})_l)^{-1}\mathbf{d}((\widetilde{a})_r)$) and $\arg\eta_j$, $\arg\widetilde{\eta}_j$ are chosen in $(-\pi,\pi)$. Adding these two equalities we get

$$|\operatorname{Ind} W(a) + \operatorname{Ind} W(\widetilde{a})| < \frac{1}{2\pi}(N\pi + N\pi) = N,$$

which completes the proof.

If $f \in APW_{N\times N}$ is positive definite, then f has both a left and a right canonical APW factorization (apply Corollary 9.15 to $a = f$ and $a = f^T$), the geometric means $\mathbf{d}_l(f)$ and $\mathbf{d}_r(f)$ (see Definition 8.5) are positive definite matrices, and, obviously,

$$\det \mathbf{d}_l(f) = \det \mathbf{d}_r(f) = \exp\left(M(\log\det f)\right).$$

The following result shows that the equality of the determinants is the only general relation between the left and right geometric means of positive definite matrix functions in $APW_{N\times N}$.

Lemma 11.16. *Given any two positive definite matrices D_l and D_r in $\mathbf{C}^{N\times N}$ for which* $\det D_l = \det D_r$, *there exists a 2π-periodic positive definite matrix function f in $APW_{N\times N}$ such that*

$$\mathbf{d}_l(f) = D_l \quad \textit{and} \quad \mathbf{d}_r(f) = D_r. \tag{11.13}$$

Proof. It is again more convenient to pass from the real line to the unit circle $\mathbf{T}$. Let W denote the Wiener algebra of all functions g on $\mathbf{T}$ with absolutely convergent Fourier series and put $W^\pm := W \cap H^\infty_\pm(\mathbf{T})$. A right (resp. left) canonical W factorization of $g \in W_{N\times N}$ is a representation $g = g_- g_+$ (resp. $g = k_+ k_-$) with $g_\pm, g_\pm^{-1}$ in $W^\pm_{N\times N}$ (resp. with $k_\pm, k_\pm^{-1}$ in $W^\mp_{N\times N}$). Given a right canonical W factorization of $g = g_- g_+$, we define the right geometric mean $\mathbf{d}_r(g)$ as $g_-(\infty)g_+(0)$. Notice that functions in $W^\pm$ extend to analytic functions in $\mathbf{D}_\pm$. We also remark that the definition of $\mathbf{d}_r(g)$ is independent of the particular W factorization $g = g_- g_+$. Analogously, if $g = k_+ k_-$ is a left canonical W factorization, the left geometric mean is defined by $\mathbf{d}_l(g) = k_+(0)k_-(\infty)$.

We will construct a positive definite matrix function $g \in W_{N\times N}$ that has left and right canonical W factorizations $g = k_+ k_- = g_- g_+$ such that

$$\mathbf{d}_l(g) = k_+(0)k_-(\infty) = D_l \quad \text{and} \quad \mathbf{d}_r(g) = g_-(\infty)g_+(0) = D_r. \tag{11.14}$$

Letting $f(x) := g(e^{ix})$, we get a 2π-periodic and positive definite matrix function $f \in APW_{N\times N}$. Clearly,

$$f(x) = k_+(e^{ix})k_-(e^{ix}) = g_-(e^{ix})g_+(e^{ix})$$

are canonical APW factorizations of f, and since

$$\begin{aligned}\mathbf{d}_l(f) &= M(k_+(e^{ix}))M(k_-(e^{ix})) = k_+(0)k_-(\infty),\\ \mathbf{d}_r(f) &= M(g_-(e^{ix}))M(g_+(e^{ix})) = g_-(\infty)g_+(0),\end{aligned}$$

we arrive at (11.13).

To construct g, let U be a unitary matrix diagonalizing $D_l^{-1/2}D_rD_l^{-1/2}$, that is,

$$D_l^{-1/2}D_rD_l^{-1/2} = U^*\Lambda U.$$

Suppose we have a positive definite matrix function ϕ on $\mathbf{T}$ for which

$$\mathbf{d}_l(\phi) = I, \quad \mathbf{d}_r(\phi) = \Lambda. \tag{11.15}$$

Then $g := D_l^{1/2}U^*\phi U D_l^{1/2}$ is a positive definite matrix function satisfying (11.14). Indeed,

$$\begin{aligned}\mathbf{d}_l(g) &= D_l^{1/2}U^*\mathbf{d}_l(\phi)UD_l^{1/2} = D_l^{1/2}U^*UD_l^{1/2} = D_l,\\ \mathbf{d}_r(g) &= D_l^{1/2}U^*\mathbf{d}_r(\phi)UD_l^{1/2} = D_l^{1/2}U^*\Lambda UD_l^{1/2}\\ &= D_l^{1/2}D_l^{-1/2}D_rD_l^{-1/2}D_l^{1/2} = D_r.\end{aligned}$$

In other words, it suffices to construct a positive definite matrix function ϕ having property (11.15).

Observe that $\Lambda = \operatorname{diag}(\lambda_1, \dots, \lambda_N)$ with $\lambda_j > 0$ and

$$\lambda_1 \dots \lambda_N = \det\Lambda = \det(D_l^{-1/2}D_rD_l^{-1/2}) = 1.$$

We will be looking for ϕ in the form

$$\phi(z) = (I + zA)(I + zA)^* \quad (|z| = 1), \tag{11.16}$$

where A is a matrix all eigenvalues of which belong to $\mathbf{D}_+$. Then, of course, (11.16) is a left canonical W factorization of ϕ and $\mathbf{d}_l(\phi) = I$. Suppose that at the same time

$$\phi(z) = (\Lambda^{1/2} + zB)^*(\Lambda^{1/2} + zB) \quad (|z| = 1), \tag{11.17}$$

and $\Lambda^{-1/2}B$ has all its eigenvalues in $\mathbf{D}_+$. Then (11.17) delivers a right canonical W factorization of ϕ with $\mathbf{d}_r(\phi) = \Lambda^{1/2}\Lambda^{1/2} = \Lambda$.

For (11.16) and (11.17) to hold simultaneously it is necessary and sufficient that

$$I + AA^* = \Lambda + B^*B \quad \text{and} \quad A = \Lambda^{1/2}B,$$

which, in turn, leads to the equation

$$I + \Lambda^{1/2}BB^*\Lambda^{1/2} = \Lambda + B^*B. \tag{11.18}$$

Let B be the matrix of a weighted shift transformation,

$$B = \begin{pmatrix} 0 & b_1 & 0 & \dots & 0 \\ 0 & 0 & b_2 & \dots & 0 \\ & & & \ddots & \\ 0 & 0 & 0 & \dots & b_{N-1} \\ b_N & 0 & 0 & \dots & 0 \end{pmatrix},$$

with positive weights b_j to be specified later. Then

$$BB^* = \operatorname{diag}(b_1^2, \dots, b_N^2), \quad B^*B = \operatorname{diag}(b_N^2, b_1^2, \dots, b_{N-1}^2),$$

so that (11.18) takes the form

$$1 + \lambda_{j+1} b_{j+1}^2 = \lambda_{j+1} + b_j^2 \ (j = 1, \dots, N-1), \quad 1 + \lambda_1 b_1^2 = \lambda_1 + b_N^2. \tag{11.19}$$

Choose a number $\varepsilon \in (0,1)$ smaller than all the products $\prod_{j=2}^k \lambda_j$ $(k = 2, \dots, N)$ and define b_j by the recursive relation

$$b_1 = (1-\varepsilon)^{1/2}, \quad b_{j+1} = \left(1 - \lambda_{j+1}^{-1}(1 - b_j^2)\right)^{1/2}, \ j = 1, \dots, N. \tag{11.20}$$

An easy induction argument shows that, with our choice of ε, we have $1 - b_j^2 < \lambda_{j+1}$ for $j = 1, \dots, N-1$, and $1 - b_N^2 = \lambda_N^{-1} \dots \lambda_2^{-1} \varepsilon < 1$. Therefore, formula (11.20) indeed defines a sequence of positive weights b_j. Also, (11.20) implies the first $N-1$ equalities in (11.19); the last equality in (11.19) holds because $1 - b_N^2 = (\lambda_2 \cdots \lambda_N)^{-1}(1 - b_1^2)$ and $\lambda_1 \cdots \lambda_N = 1$.

Thus, we have found a solution B of (11.18). Letting $A = \Lambda^{1/2} B$, we get the two factorizations (11.16) and (11.17).

It remains to show that the eigenvalues of $\Lambda^{\pm 1/2} B$ are all located in $\mathbf{D}_+$. To this end, notice that $\Lambda^{\pm 1/2} B$ are weighted shifts simultaneously with B and that

$$\det(\Lambda^{\pm 1/2} B) = \det B = (-1)^{N-1} b_1 \cdots b_N.$$

Therefore, the eigenvalues of $\Lambda^{\pm 1/2} B$ lie on the circle centered at the origin and having radius $r = (b_1 \cdots b_N)^{1/N}$. But, according to (11.20), all b_j are smaller than 1. Hence, $r < 1$ as well. □

Proof of Theorem 11.6. Pick any $k \in \{-(N-1), \dots, N-1\}$. Let J be the $N \times N$ Jordan block with the eigenvalue $e^{i\pi k/N}$. By Ballantine's Theorem 11.14, there are four positive definite matrices H_1, H_2, H_3, H_4 such that

$$J^2 = H_1^{-1} H_2 H_3^{-1} H_4.$$

Clearly,

$$1 = \det J^2 = \det H_1^{-1} \det H_2 \det H_3^{-1} \det H_4.$$

This implies that we can find $\alpha > 0$ and $\beta > 0$ such that

$$\det(\alpha H_1) = \det(\beta H_4), \quad \det(\alpha H_2) = \det(\beta H_3).$$

Put $D_1 = \alpha H_1, D_2 = \alpha H_2, D_3 = \beta H_3, D_4 = \beta H_4$. We then have

$$J^2 = D_1^{-1} D_2 D_3^{-1} D_4, \quad \det D_1 = \det D_4, \quad \det D_2 = \det D_3, \tag{11.21}$$

and letting $X := D_3^{-1} D_4 J^{-1}$, we therefore get

$$J = D_1^{-1} D_2 X, \quad J = X^{-1} D_3^{-1} D_4. \tag{11.22}$$

As the matrices D_1, D_2, D_3, D_4 are positive definite, we deduce from the last two equalities of (11.21) and from Lemma 11.16 that there are 2π-periodic matrix functions $f_l, f_r \in GAPW_{N\times N}$ such that

$$\mathbf{d}_r(f_l) = D_1, \quad \mathbf{d}_l(f_l) = D_4, \quad \mathbf{d}_l(f_r) = D_3, \quad \mathbf{d}_r(f_r) = D_2.$$

Put $a_l = f_l$, $a_r = f_r X$, choose any $a_0 \in [C(\dot{\mathbf{R}})]_{N\times N}$ such that $a_0(\infty)$ is the zero matrix, and consider

$$a(x) = (1 - u(x)) a_l(x) + u(x) a_r(x) + a_0(x).$$

It is easily seen that a_0 can be chosen so that $\det a(x)$ is nonzero for all $x \in \mathbf{R}$. By construction and by (11.22),

$$(\mathbf{d}_r(a_l))^{-1} \mathbf{d}_r(a_r) = D_1^{-1} D_2 X = J \tag{11.23}$$

$$(\mathbf{d}_r((\widetilde{a})_l))^{-1} \mathbf{d}_r((\widetilde{a})_r) = (\mathbf{d}_l(a_r))^{-1} \mathbf{d}_l(a_l) = X^{-1} D_3^{-1} D_4 = J. \tag{11.24}$$

The eigenvalues of J are all equal to $e^{i\pi k/N}$. Thus, all eigenvalues η_j and $\widetilde{\eta}_j$ of the matrices on the left-hand sides of (11.23) and (11.24), respectively, are also equal to $e^{i\pi k/N}$. Now Theorem 10.13 implies that $W(a)$ and $W(\widetilde{a})$ are Fredholm and that

$$\begin{aligned} &\operatorname{Ind} W(a) + \operatorname{Ind} W(\widetilde{a}) \\ &= -\frac{1}{2\pi}\{\arg\det a\}_{-\infty}^{\infty} + \frac{1}{2\pi}\sum_{j=1}^{N} \arg \eta_j \\ &\quad + \frac{1}{2\pi}\{\arg\det a\}_{-\infty}^{\infty} + \frac{1}{2\pi}\sum_{j=1}^{N} \arg \widetilde{\eta}_j \\ &= \frac{1}{2\pi} N \cdot \pi \frac{k}{N} + \frac{1}{2\pi} N \cdot \pi \frac{k}{N} = k. \end{aligned}$$

Consequently, $\operatorname{Ind} W(a) + \operatorname{Ind} W(\widetilde{a})$ may assume any prescribed value k between $-(N-1)$ and $N-1$. On replacing $a(x)$ by

$$a(x) \operatorname{diag}\left(\left(\frac{x-i}{x+i}\right)^{\mu}, 1, \ldots, 1\right)$$

with appropriate integers μ we can finally obtain a in $SAP_{N\times N}$ (and even in $SAPW_{N\times N}$) such that $\operatorname{Ind} W(a)$ and $\operatorname{Ind} W(\widetilde{a})$ are any integers κ and ν subject to the constraint $|\kappa+\nu| \le N-1$. □

Notes

The material of this chapter is taken from [29]. Proposition 11.1 was proved earlier in [210] by means of an example different from Example 6.7.

Question (b) of Section 11.2 can be sharpened as follows.

(e) Given two vectors $\varkappa, \nu \in \mathbf{Z}^N$, is there a matrix function $a \in GL^{\infty}_{N\times N}(\mathbf{R})$ which possesses a right WH factorization with the right partial indices $\varkappa$ and a left WH factorization with the left partial indices ν ?

For matrix functions

$$a \in [C+H^{\infty}_{+}]_{N\times N} \cup [C+H^{\infty}_{-}]_{N\times N} \cup PC_{N\times N}$$

the right and left total indices $\sum \varkappa_j$ and $\sum \nu_j$ always coincide (Theorems 6.5, 5.7, 5.9). The following result shows that even for very nice matrix functions the coincidence of the right and left total indices is all we can state without further information.

Theorem 11.17 (Feldman and Markus). *Let $N \ge 2$. Given any two vectors $\varkappa, \nu \in \mathbf{Z}^N$ such that $\sum \varkappa_j = \sum \nu_j$, there exist $N\times N$ matrix functions with rational entries whose vectors of right and left partial indices are $\varkappa$ and ν, respectively.*

For a proof see [78].

The following theorem was established in [30]. It shows that the answer to question (e) is yes.

Theorem 11.18. *Let $N \ge 2$ and let $\varkappa$ and ν be any two vectors in $\mathbf{Z}^N$. Then there exist $a \in GL^{\infty}_{N\times N}(\mathbf{R})$ which admit a right WH factorization with the right partial indices $\varkappa$ and a left WH factorization with the left partial indices ν.*

Paper [30] contains a full proof of this theorem.

Chapter 12

Corona Theorems

The results of Chapters 9 and 10 leave us with the question of deciding whether a given almost periodic matrix function has a canonical AP factorization and with computing the geometric mean if the answer is yes. In the following chapters we will study this set of problems for several special classes of matrix functions. The unifying idea behind these investigations is the Portuguese transformation, which will be introduced in the next chapter. The Portuguese transformation requires the solution of a corona problem whose data are almost periodic functions. Therefore the present chapter is devoted to the general theory of the corona problem for almost periodic functions.

Carleson's corona theorem says that the algebras $H^\infty_\pm$ have no corona. We here consider algebras of analytic almost periodic functions whose Bohr-Fourier spectrum is supported in some additive semigroup $\Sigma \subset [0,\infty)$. It will turn out that the corona is empty for some Σ's and nonempty for others.

12.1 The Corona Problem for AP and APW

Let $\Sigma \subset [0,\infty)$ be an additive semigroup (i.e., $\lambda, \mu \in \Sigma$ implies that $\lambda + \mu \in \Sigma$) and suppose $0 \in \Sigma$. Recall that AP^0 stands for the set of all almost periodic polynomials and put

$$AP^0_\Sigma := \{a \in AP^0 : \Omega(a) \subset \Sigma\}.$$

Let AP_Σ and APW_Σ denote the closures of AP^0_Σ in AP and APW, respectively. Corollary 7.6 shows that

$$AP_\Sigma = \{a \in AP : \Omega(a) \subset \Sigma\}, \quad APW_\Sigma = \{a \in APW : \Omega(a) \subset \Sigma\}.$$

Since Σ is a semigroup, the sets AP_Σ and APW_Σ are closed subalgebras of AP^+ and APW^+, respectively. As $0 \in \Sigma$, the algebras AP^+ and APW^+ have a unit element.

In what follows we let $AP(W)$ stand for one of the algebras AP and APW. Thus, an implication of the type $f \in AP(W) \Rightarrow g \in AP(W)$ means that if $f \in AP$ then $g \in AP$ and if $f \in APW$ then $g \in APW$.

The *corona problem* for $AP(W)_\Sigma$ consists in the following: given $f_1, \ldots, f_m$ in $AP(W)_\Sigma$, find $g_1, \ldots, g_m$ in $AP(W)_\Sigma$ such that

$$f_1 g_1 + \cdots + f_m g_m = 1. \tag{12.1}$$

The functions $f_1, \ldots, f_m$ are referred to as the *corona data* and any vector $(g_1, \ldots, g_m)$ satisfying (12.1) is called a *corona solution.*

Because $\Sigma \subset [0, \infty)$, every function $a \in AP_\Sigma$ can be extended to an analytic function $\hat{a}$ in the upper half-plane $\mathbf{C}_+$ by the Poisson integral (2.17). Obviously,

$$\sup_{z \in \mathbf{C}_+} |\hat{a}(z)| \le \|a\|_\infty \quad \bigl(:= \sup_{t \in \mathbf{R}} |a(t)| \bigr). \tag{12.2}$$

If $\lambda, \mu \ge 0$ and $x + iy \in \mathbf{C}_+$, then

$$\begin{aligned}(e_\lambda e_\mu)\hat{}\,(x+iy) &= \hat{e}_{\lambda+\mu}(x+iy) = e^{i(\lambda+\mu)x} e^{-(\lambda+\mu)y} \\ &= e^{i\lambda x} e^{-\lambda y} e^{i\mu x} e^{-\mu y} = \hat{e}_\lambda(x+iy)\hat{e}_\mu(x+iy).\end{aligned}$$

Hence, the map $a \mapsto \hat{a}$ is linear and multiplicative on AP_Σ^0. Because of (12.2), the map $a \mapsto \hat{a}$ is linear and multiplicative on all AP_Σ. In particular, if (12.1) holds then

$$\hat{f}_1(z)\hat{g}_1(z) + \cdots + \hat{f}_m(z)\hat{g}_m(z) = 1 \ \text{ for all } \ z \in \mathbf{C}_+,$$

which, by virtue of (12.2), implies that necessarily

$$\inf_{z \in \mathbf{C}_+} \sum_{j=1}^m |\hat{f}_j(z)| > 0. \tag{12.3}$$

We now restate the corona problem in *Banach algebraic language.* Let $M(AP_\Sigma)$ and $M(APW_\Sigma)$ be the maximal ideal space of AP_Σ and APW_Σ, respectively. In what follows, we think of the elements of $M(AP(W)_\Sigma)$ as nonzero multiplicative linear functionals on $AP(W)_\Sigma$.

The map $\tau : \mathbf{R} \times (0, \infty) \to \mathbf{C}_+$, $(x, y) \mapsto x + iy$ identifies $\mathbf{R} \times (0, \infty)$ and $\mathbf{C}_+$. For $(x, y) \in \mathbf{R} \times (0, \infty)$, define

$$\varphi_{x,y} : AP_\Sigma^0 \to \mathbf{C}, \ \sum a_\lambda e_\lambda \mapsto \sum a_\lambda e^{i\lambda x} e^{-\lambda y}. \tag{12.4}$$

Since

$$\varphi_{x,y}\left(\sum a_\lambda e_\lambda\right) = \sum a_\lambda e^{i\lambda(x+iy)} = \left(\sum a_\lambda e_\lambda\right)\hat{}\,(x+iy), \tag{12.5}$$

we see that $\varphi_{x,y}(a)$ is the value of the analytic extension of a at $x+iy \in \mathbf{C}_+$. Thus, by (12.2), $\varphi_{x,y}$ extends to a nonzero multiplicative linear functional on AP_Σ. From (12.4) we also get

$$\left|\varphi_{x,y}\left(\sum a_\lambda e_\lambda\right)\right| \le \sum |a_\lambda| = \left\|\sum a_\lambda e_\lambda\right\|_W,$$

which shows that $\varphi_{x,y}$ extends to a nonzero multiplicative linear functional on APW_Σ, too. In summary, we have a map

$$\sigma : \mathbf{R} \times (0,\infty) \to M\big(AP(W)_\Sigma\big), \ (x,y) \mapsto \varphi_{x,y}.$$

If $\Sigma = [0,\infty)$, then σ is injective: $\varphi_{x_1,y_1}(e_\lambda) = \varphi_{x_2,y_2}(e_\lambda)$ for all $\lambda \in [0,\infty)$ implies that $x_1 = x_2$ and $y_1 = y_2$. However, in general σ need not be injective. For example, if $\Sigma = \mathbf{Z}_+ := \{0,1,2,\dots\}$, then $\varphi_{x_1,y_1} = \varphi_{x_2,y_2}$ if and only if

$$e^{i\lambda x_1}e^{-\lambda y_1} = e^{i\lambda x_2}e^{-\lambda y_2} \text{ for all } \lambda \in \mathbf{Z}_+,$$

which happens exactly if $y_1 = y_2$ and $x_1 - x_2 \in 2\pi\mathbf{Z}$.
The map $\sigma\tau^{-1} : \mathbf{C}_+ \to M(AP(W)_\Sigma)$ sends each point of $\mathbf{C}_+$ to a maximal ideal of $AP(W)_\Sigma$. If σ is injective, this is an embedding of $\mathbf{C}_+$ into $M(AP(W)_\Sigma)$. In case σ is not injective, we think of $\mathbf{C}_+$ as being contained in $M(AP(W)_\Sigma)$ in "rolled up" form. In what follows, when saying that $\mathbf{C}_+$ is dense in $M(AP(W)_\Sigma)$ we always mean that $\sigma\tau^{-1}(\mathbf{C}_+)$ is dense in $M(AP(W)_\Sigma)$ (in the Gelfand topology).

Theorem 12.1. *Let Σ be an additive semi-subgroup of $[0,\infty)$ and suppose $0 \in \Sigma$. Then the following are equivalent:*

(i) *for every $f_1,\dots,f_m \in AP(W)_\Sigma$ satisfying* (12.3) *there exist $g_1,\dots,g_m$ in $AP(W)_\Sigma$ such that* (12.1) *holds;*

(ii) *$\mathbf{C}_+$ is dense in $M(AP(W)_\Sigma)$.*

Proof. (i)$\Rightarrow$(ii). Let $\varphi \in M(AP(W)_\Sigma)$ and suppose φ is not in the closure of $\sigma\tau^{-1}(\mathbf{C}_+)$. Then, by the definition of the Gelfand topology (= weak $*$-topology), there exist $h_1,\dots,h_m \in AP^0_\Sigma$ and $\varepsilon > 0$ such that for each $(x,y) \in \mathbf{R}\times(0,\infty)$ at least one of the inequalities $|\varphi_{x,y}(h_j) - \varphi(h_j)| \ge \varepsilon$ is satisfied. Put $f_k := h_k - \varphi(h_k)$. Then $f_k \in AP^0_\Sigma$ and for each $(x,y) \in \mathbf{R}\times(0,\infty)$ there is a j such that

$$|\hat{f}_j(x+iy)| = |\varphi_{x,y}(h_j) - \varphi(h_j)| \ge \varepsilon.$$

Hence, by (i), we can find $g_1,\dots,g_m \in AP(W)_\Sigma$ such that $\Sigma f_k g_k = 1$. It follows that $\sum \varphi(f_k)\varphi(g_k) = 1$, which is impossible because $\varphi(f_k) = 0$ for all k.

(ii)$\Rightarrow$(i). Assume there are $f_1, \dots, f_m \in AP(W)_\Sigma$ subject to (12.3) such that (12.1) is not satisfied by any $g_1, \dots, g_m \in AP(W)_\Sigma$. Then the set $\{\sum f_j g_j : g_j \in AP(W)_\Sigma\}$ is a proper ideal of $AP(W)_\Sigma$ and is therefore contained in some maximal ideal of $AP(W)_\Sigma$. Consequently, there exists a nonzero multiplicative linear functional φ on $AP(W)_\Sigma$ such that $\varphi(f_j) = 0$ for all $j = 1, \dots, m$. Since $\sigma\tau^{-1}(\mathbf{C}_+)$ is dense in $M(AP(W)_\Sigma)$, we can find $(x_n, y_n) \in \mathbf{R} \times (0, \infty)$ such that $|\varphi(f_j) - \varphi_{x_n,y_n}(f_j)| < 1/n$ for all j. It results that

$$|\hat{f}_j(x_n + iy_n)| = |\varphi_{x_n,y_n}(f_j)| < 1/n$$

for all j, which contradicts (12.3). □

The set $M(AP(W)_\Sigma) \setminus \overline{\mathbf{C}}_+$ is called the corona of $\mathbf{C}_+$. Thus, Theorem 12.1 states that (i) holds if and only if the corona of $\mathbf{C}_+$ is empty. This accounts for the name "corona problem". Figures 12.1 and 12.2 show real coronas.

12.2 The Arens-Singer Theorem

Our next objective is to describe $M(AP(W)_\Sigma)$, after which we will turn to the question whether $\mathbf{C}_+$ is dense in $M(AP(W)_\Sigma)$. Let $\Gamma \subset \mathbf{R}$ be the smallest additive group that contains the semigroup Σ,

$$\Gamma := \Sigma - \Sigma := \{\lambda - \mu : \lambda, \mu \in \Sigma\}.$$

Figure 12.1: The total solar eclipse of 11 August 1999. The two pictures were taken by one of the authors (A. B.) with a usual autofocus camera in Ensheim (Saar, Germany). It's no professional quality, but the corona is seen.

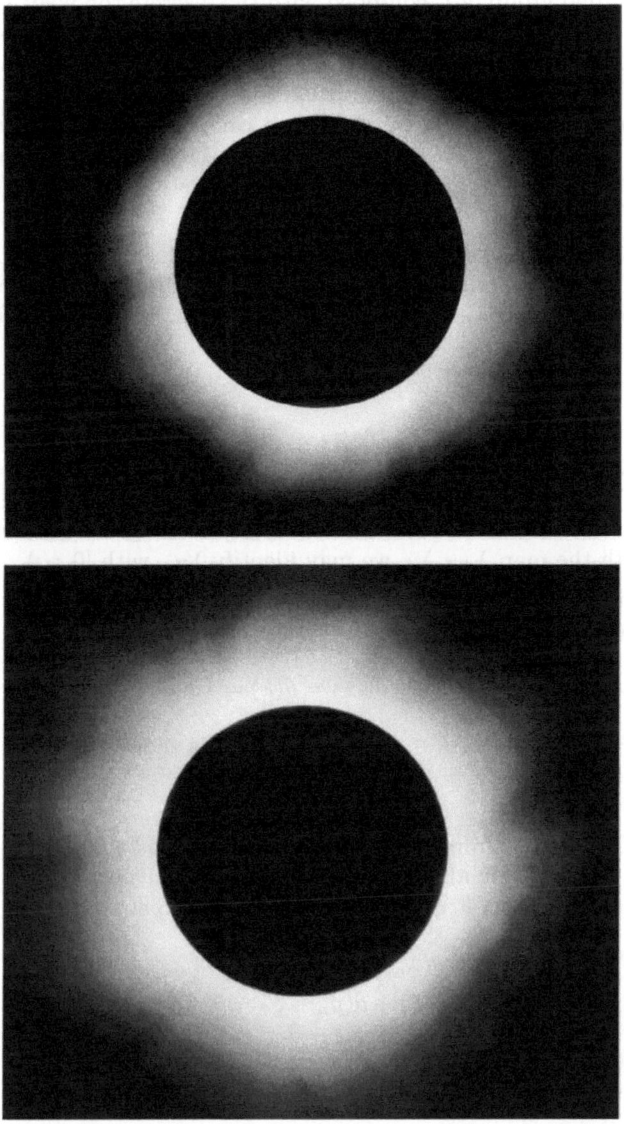

Figure 12.2: And here is professional quality. These two pictures of the total solar eclipse of 11 August 1999 were taken by Thomas Löger und Michael Mentschl in Rechnitz (Burgenland, Austria). The photos are reproduced here by courtesy of the photographers.

Let Γ_d^* be the dual group of Γ equipped with the discrete topology. Clearly, $\mathbf{R}_d^* \subset \Gamma_d^*$. In accordance with Section 7.1, the Bohr compactifications $\mathbf{R}_B$ and Γ_B are defined by $\mathbf{R}_B = (\mathbf{R}_d^*)^*$ and $\Gamma_B = (\Gamma_d^*)^*$. Now put $X_\Gamma := \{e_\lambda : \lambda \in \Gamma\}$. Then $X_\Gamma \subset \mathbf{R}_d^* \subset \Gamma_d^*$. Of course, we think of X_Γ as a group with the discrete topology. The dual group X_Γ^* is denoted by $\mathbf{R}_{B,\Gamma}$ and called the *Bohr compactification of* $\mathbf{R}$ *relative to* Γ (see, e.g., [171, p. 21] or [107, 26.11–26.12]). Thus, an element of $\mathbf{R}_{B,\Gamma}$ is a map $\chi : X_\Gamma \to \mathbf{T}$ such that $\chi(e_{\lambda+\mu}) = \chi(e_\lambda)\chi(e_\mu)$ for all $\lambda, \mu \in \Gamma$. If $\Gamma = \mathbf{R}$, then $\mathbf{R}_{B,\Gamma} = \mathbf{R}_B$.

One can show (see, for example, [189, Theorem 2.1.4]) that every $\chi \in \mathbf{R}_{B,\Gamma}$ can be extended (not necessarily in a unique way) to a character $\tilde{\chi}$ on all of $\mathbf{R}_d^*$ and even on all of Γ_d^*. Hence, if $\chi \in \mathbf{R}_{B,\Gamma}$ then there is a $\tilde{\chi} \in \Gamma_B \subset \mathbf{R}_B$ such that $\tilde{\chi}|X_\Gamma = \chi$.

We denote by Y_Σ the set of all maps $\theta : \Sigma \to [0,\infty)$ satisfying $\theta(\lambda+\mu) = \theta(\lambda)+\theta(\mu)$ for all $\lambda, \mu \in \Sigma$.

Example 12.2. **(a)** Let $\mathbf{Z}_+ := \{0,1,2,\ldots\}$. If $\theta \in Y_{\mathbf{Z}_+}$, then $\theta(\lambda) = \lambda\theta(1)$, and denoting $\theta(1) \in [0,\infty)$ by y, we get $\theta(\lambda) = \lambda y$ for all $\lambda \in \mathbf{Z}_+$. Conversely, if $\theta(\lambda) = \lambda y$ for some $y \in [0,\infty)$, then obviously $\theta \in Y_{\mathbf{Z}_+}$. Thus, on identifying $y \in [0,\infty)$ with the map $\lambda \mapsto \lambda y$, we may identify $Y_{\mathbf{Z}_+}$ with $[0,\infty)$.

(b) Let $\mathbf{Q}_+$ be the set of the nonnegative rational numbers. If $\lambda = k/l \in \mathbf{Q}_+$ and $\theta \in Y_{\mathbf{Q}_+}$, then

$$l\theta(\lambda) = \theta(l\lambda) = \theta(k) = k\theta(1),$$

whence $\theta(\lambda) = \lambda\theta(1)$ for all $\lambda \in \mathbf{Q}_+$. As in (a), this shows that $Y_{\mathbf{Q}_+}$ consists just of the maps $\lambda \mapsto \lambda y$ with $y \in [0,\infty)$ and that therefore $Y_{\mathbf{Q}_+}$ may be identified with $[0,\infty)$.

(c) Suppose $\Sigma = [0,\infty)$. From part (b) we deduce that if $\theta \in Y_{[0,\infty)}$, then $\theta(y) = \lambda y$ with some $y \in [0,\infty)$ for all $\lambda \in \mathbf{Q}_+$. Let $\xi \in (0,\infty)$ be irrational and choose $\lambda_n, \mu_n \in \mathbf{Q}_+$ such that $\lambda_n < \xi < \mu_n$, $\lambda_n \to \xi$, $\mu \to \xi$. Since

$$\begin{aligned} \lambda_n y &= \theta(\lambda_n) \le \theta(\lambda_n) + \theta(\xi - \lambda_n) = \theta(\xi) \\ &\le \theta(\xi) + \theta(\mu_n - \xi) = \theta(\mu_n) = \mu_n y, \end{aligned}$$

it follows that $\theta(\xi) = \xi y$. Conversely, for every $y \in [0,\infty)$ the map $\lambda \mapsto \lambda y$ belongs to $Y_{[0,\infty)}$. Thus, again we can identify $Y_{[0,\infty)}$ with $[0,\infty)$.

(d) Let $\Sigma := \{k + l\sqrt{2} : k \in \mathbf{Z}_+,\ l \in \mathbf{Z}_+\}$. If $\theta \in Y_\Sigma$, then

$$\theta(k + l\sqrt{2}) = k\theta(1) + l\theta(\sqrt{2}) =: k\eta_1 + l\eta_2 \tag{12.6}$$

with $(\eta_1, \eta_2) \in [0,\infty)^2$. Conversely, for every pair $(\eta_1,\eta_2) \in [0,\infty)^2$ the map θ given by (12.6) is in Y_Σ. Hence, we can identify the sets Y_Σ and $[0,\infty)^2$ in a natural way. □

We define $\overline{Y}_\Sigma$ as the collection of all maps $\theta : \Sigma \to [0,\infty]$ such that

$$\theta(0) = 0 \text{ and } \theta(\lambda+\mu) = \theta(\lambda) + \theta(\mu) \text{ for all } \lambda, \mu \in \Sigma.$$

Here, of course, $\nu + \infty = \infty + \nu = \infty$ for $\nu \in [0,\infty]$. We also make the convention that $\nu \cdot \infty = \infty \cdot \nu = \infty$ for $\nu \in (0,\infty]$ and $0 \cdot \infty = \infty \cdot 0 = 0$. The map $\theta \in \overline{Y}_\Sigma$ given by $\theta(0) := 0$ and $\theta(\lambda) := \infty$ for $\lambda \in \Sigma \setminus \{0\}$ will simply be denoted by ∞.

Example 12.3. **(a)** If Δ is an additive subgroup of $\mathbf{R}$ and $\Sigma := \Delta \cap [0,\infty)$, then $\overline{Y}_\Sigma = Y_\Sigma \cup \{\infty\}$. Indeed, let $\theta \in \overline{Y}_\Sigma$ and suppose $\theta(\lambda_0) < \infty$ for some $\lambda_0 \in \Sigma \setminus \{0\}$. Given any $\lambda \in \Sigma$, there is a natural number k such that $k\lambda_0 > \lambda$. As $k\lambda_0 - \lambda \in \Delta$ and $k\lambda_0 - \lambda > 0$, it results that $k\lambda_0 - \lambda \in \Sigma$. Therefore $\theta(\lambda) + \theta(k\lambda_0 - \lambda) = k\theta(\lambda_0)$, and since $\theta(\lambda_0) < \infty$, we see that $\theta(\lambda) < \infty$. Thus, either $\theta(\lambda) = \infty$ for all $\lambda \in \Sigma \setminus \{0\}$ or $\theta(\lambda) < \infty$ for all $\lambda \in \Sigma \setminus \{0\}$.

(b) From part (a) and Example 12.2(a),(b),(c) we deduce that if Σ is $[0,\infty), \mathbf{Z}_+$, or $\mathbf{Q}_+$, then $\overline{Y}_\Sigma = [0,\infty) \cup \{\infty\} = [0,\infty]$.

(c) Let $\Sigma := \{k + l\sqrt{2} : k \in \mathbf{Z}_+, l \in \mathbf{Z}_+\}$. For every $\theta \in \overline{Y}_\Sigma$, the equalities (12.6) hold with $(\eta_1,\eta_2) \in [0,\infty]^2$. Conversely, given $(\eta_1,\eta_2) \in [0,\infty]^2$, the map θ defined by (12.6) belongs to $\overline{Y}_\Sigma$. Notice that if $\eta_1, \eta_2 \in (0,\infty)$, then

$$\begin{aligned} k\infty + l\eta_2 &:= \begin{cases} l\eta_2 & \text{for } k = 0 \\ \infty & \text{for } k \ge 1, \end{cases} \\ k\eta_1 + l\infty &:= \begin{cases} k\eta_1 & \text{for } l = 0 \\ \infty & \text{for } l \ge 1, \end{cases} \\ k\infty + l\infty &:= \infty. \end{aligned}$$

In summary, we may identify $\overline{Y}_\Sigma$ with $[0,\infty]^2$. □

For $\theta \in \overline{Y}_\Sigma$, we put

$$\Sigma_\theta := \{\lambda \in \Sigma : \theta(\lambda) < \infty\}, \tag{12.7}$$

we denote by Γ_θ the smallest additive subgroup of $\mathbf{R}$ which contains Σ_θ, and we let Γ_B^θ stand for the Bohr compactification of $\mathbf{R}$ relative to Γ_θ:

$$\Gamma_\theta := \Sigma_\theta - \Sigma_\theta, \quad \Gamma_B^\theta := \mathbf{R}_{B,\Gamma_\theta}. \tag{12.8}$$

Theorem 12.4 (Arens and Singer). *Let $\Sigma \subset [0,\infty)$ be an additive semigroup and suppose $0 \in \Sigma$. Then the maximal ideal space of $AP(W)_\Sigma$ can be identified with*

$$M_\Sigma := \bigcup_{\theta \in \overline{Y}_\Sigma} (\Gamma_B^\theta \times \{\theta\})$$

in the following sense: φ is a nonzero multiplicative linear functional on $AP(W)_\Sigma$ if and only if there are $\theta \in \overline{Y}_\Sigma$ and $\chi \in \Gamma_B^\theta$ such that $\varphi = \varphi_{\chi,\theta}$, where

$$\varphi_{\chi,\theta}\Big(\sum_{\lambda\in\Sigma} a_\lambda e_\lambda\Big) := \sum_{\lambda\in\Sigma_\theta} a_\lambda \chi(e_\lambda) e^{-\theta(\lambda)} \tag{12.9}$$

for $\sum_{\lambda\in\Sigma} a_\lambda e_\lambda \in AP_\Sigma^0$.

Proof. It is obvious that $\varphi_{\chi,\theta}$ is a nonzero multiplicative linear functional on APW_Σ. To prove that $\varphi_{\chi,\theta}$ is a nonzero multiplicative linear functional on AP_Σ, it suffices to show that $\varphi_{\chi,\theta}$ is bounded. This in turn will follow once we have shown that $|\varphi_{\chi,\theta}(a)| \le 2\|a\|_\infty$ for all $a \in AP_\Sigma^0$. So let $a = \sum_{j=1}^m a_{\lambda_j} e_{\lambda_j} \in AP_\Sigma^0$. Put

$$a^N = \Big(\sum_{j=1}^m a_{\lambda_j} e_{\lambda_j}\Big)^N =: \sum_{\lambda\in F_N} a_\lambda^{(N)} e_\lambda$$

The argument employed in the proof of Theorem 9.1 (after (9.5)) shows that the number $|F_N|$ of elements in F_N is at most $(N+1)^{m-1}$. Hence $|F_N|^{1/N} \to 1$ as $N \to \infty$. Since $\varphi_{\chi,\theta}$ has norm 1 on APW_Σ, we get

$$\begin{aligned} |\varphi_{\chi,\theta}(a)| &= |\varphi_{\chi,\theta}(a^N)|^{1/N} \le \|a^N\|_W^{1/N} = \Big(\sum_{\lambda\in F_N} |a_\lambda^{(N)}|\Big)^{1/N} \\ &\le |F_N|^{1/(2N)} \Big(\sum_{\lambda\in F_N} |a_\lambda^{(N)}|^2\Big)^{1/(2N)} \le 2\Big(\sum_{\lambda\in F_N} |a_\lambda^{(N)}|^2\Big)^{1/(2N)} \end{aligned} \tag{12.10}$$

if only N is large enough. Parseval's equality says that

$$\sum_{\lambda\in F_N} |a_\lambda^{(N)}|^2 = \int_{\mathbf{R}_B} |a^N(\xi)|^2 \, d\mu(\xi) \tag{12.11}$$

where μ is the normalized Haar measure on $\mathbf{R}_B$ (Section 7.1). Since $|a(\xi)| \le \|a\|_\infty$ for all $\xi \in \mathbf{R}_B$, we obtain from (12.10) and (12.11) that

$$|\varphi_{\chi,\theta}(a)| \le 2\Big(\int_{\mathbf{R}_B} |a(\xi)|^{2N} \, d\mu(\xi)\Big)^{1/(2N)} \le 2\|a\|_\infty,$$

as desired.

Conversely, let φ be a nonzero multiplicative linear functional on $AP(W)_\Sigma$. Since $|\varphi(e_\lambda)| \le \|e_\lambda\| = 1$, it follows that $|\varphi(e_\lambda)| = e^{-\theta(\lambda)}$ with $\theta \in \overline{Y}_\Sigma$. Define Σ_θ and Γ_θ by (12.7) and (12.8). The map $\theta : \Sigma_\theta \to [0,\infty)$ admits a unique extension $\tilde{\theta} : \Gamma_\theta \to [0,\infty)$ such that $\tilde{\theta}(\lambda+\mu) = \tilde{\theta}(\lambda) + \tilde{\theta}(\mu)$ for all $\lambda, \mu \in \Gamma_\theta$:

$$\tilde{\theta}(\lambda - \mu) := \theta(\lambda) - \theta(\mu) \ \text{ for } \ \lambda, \mu \in \Sigma_\theta.$$

There is also a unique extension of the map

$$\{e_\lambda\}_{\lambda\in\Sigma_\theta} \to \mathbf{C}\setminus\{0\},\ e_\lambda \mapsto \varphi(e_\lambda)$$

to a map $\tilde{\varphi} : \{e_\lambda\}_{\lambda\in\Gamma_\theta} \to \mathbf{C}\setminus\{0\}$ such that $\tilde{\varphi}(e_{\lambda+\mu}) = \tilde{\varphi}(e_\lambda)\tilde{\varphi}(e_\mu)$ for all $\lambda, \mu \in \Gamma_\theta$, namely

$$\tilde{\varphi}(e_{\lambda-\mu}) := \varphi(e_\lambda)\big(\varphi(e_\mu)\big)^{-1} \text{ for } \lambda, \mu \in \Sigma_\theta.$$

For $\lambda \in \Gamma_\theta$, put $\chi(e_\lambda) := e^{\tilde{\theta}(\lambda)}\tilde{\varphi}(e_\lambda)$. Then $|\chi(e_\lambda)| = 1$ and $\chi(e_{\lambda+\mu}) = \chi(e_\lambda)\chi(e_\mu)$ for all $\lambda, \mu \in \Gamma_\theta$, whence $\chi \in \Gamma_B^\theta$. Since

$$\varphi(e_\lambda) = \begin{cases} \chi(e_\lambda)e^{-\theta(\lambda)} & \text{for } \lambda \in \Sigma_\theta, \\ e^{-\infty} = 0 & \text{for } \lambda \in \Sigma\setminus\Sigma_\theta, \end{cases}$$

and since φ is a multiplicative linear functional, it results that φ is of the form (12.9). □

As already said, every $\chi \in \Gamma_B^\theta$ can be extended (not necessarily in a unique way) to a $\tilde{\chi} \in \Gamma_B$. Theorem 12.4 therefore implies that $M_\Sigma \subset \Gamma_B \times \overline{Y}_\Sigma$ in the following sense: every nonzero multiplicative linear functional φ is of the form $\varphi = \varphi_{\chi,\theta}$ where $\tilde{\chi} \in \Gamma_B$, $\theta \in \overline{Y}_\Sigma$, and

$$\varphi_{\chi,\theta}\Big(\sum_{\lambda\in\Sigma} a_\lambda e_\lambda\Big) := \sum_{\lambda\in\Sigma} a_\lambda \tilde{\chi}(e_\lambda)e^{-\theta(\lambda)}$$

for $\sum_{\lambda\in\Sigma} a_\lambda e_\lambda \in AP_\Sigma^0$. Of course, the values of $\tilde{\chi}$ on $\Sigma\setminus\Sigma_\theta$ are irrelevant because $e^{-\theta(\lambda)} = e^{-\infty} = 0$ for $\lambda \in \Sigma\setminus\Sigma_\theta$.

If $\theta \in Y_\Sigma$, then $\Sigma_\theta = \Sigma$ and $\Gamma_\theta = \Gamma$. Thus, provided $\overline{Y}_\Sigma = Y_\Sigma \cup \{\infty\}$ (Example 12.3(a),(b)), we have

$$M_\Sigma = (\mathbf{R}_{B,\Gamma} \times Y_\Sigma) \cup \{\infty\}, \tag{12.12}$$

where the ∞ on the right of (12.12) stands for the $\varphi \in M_\Sigma$ defined by $\theta = \infty$, i.e., for the $\varphi \in M_\Sigma$ given by $\varphi\left(\sum a_\lambda e_\lambda\right) := a_0$.

Example 12.5. Example 12.3(a) shows that (12.12) always holds if $\Sigma = \Delta \cap [0, \infty)$ with some additive group $\Delta \subset \mathbf{R}$, and from Example 12.3(b) we infer that

$$\begin{aligned} M_{[0,\infty)} &= \big(\mathbf{R}_B \times [0,\infty)\big) \cup \{\infty\}, \\ M_{\mathbf{Z}_+} &= \big(\mathbf{R}_{B,\mathbf{Z}} \times [0,\infty)\big) \cup \{\infty\} \cong \big(\mathbf{T} \times [0,\infty)\big) \cup \{\infty\}, \\ M_{\mathbf{Q}_+} &= \big(\mathbf{R}_{B,\mathbf{Q}} \times [0,\infty)\big) \cup \{\infty\}. \end{aligned}$$

□

Example 12.6. Let $\Sigma = \{k + l\sqrt{2} : k \in \mathbf{Z}_+, l \in \mathbf{Z}_+\}$. For $t_1, t_2 \in \mathbf{T}$ and $\eta_1, \eta_2 \in [0, \infty]$, let $\varphi_{t_1,t_2,\eta_1,\eta_2} : AP(W)_\Sigma \to \mathbf{C}$ denote the functional given by

$$\varphi_{t_1,t_2,\eta_1,\eta_2}\Big(\sum a_{k+l\sqrt{2}}e_{k+l\sqrt{2}}\Big) = \sum a_{k+l\sqrt{2}} t_1^k t_2^l e^{-k\eta_1} e^{-l\eta_2}.$$

Combining Example 12.3(c) and Theorem 12.4 and taking into account that $\mathbf{R}_{B,\mathbf{Z}}$ may be identified with $\mathbf{T}$, we arrive at the conclusion that M_Σ can be decomposed in the form

$$M_\Sigma = M_{*,*} \cup M_{*,\infty} \cup M_{\infty,*} \cup M_{\infty,\infty}, \tag{12.13}$$

where

$$\begin{aligned} M_{*,*} &= \{\varphi_{t_1,t_2,\eta_1,\eta_2} : t_1 \in \mathbf{T}, t_2 \in \mathbf{T}, \eta_1 \in [0,\infty), \eta_2 \in [0,\infty)\}, \\ M_{*,\infty} &= \{\varphi_{t,1,\eta,\infty} : t \in \mathbf{T}, \eta \in [0,\infty)\}, \\ M_{\infty,*} &= \{\varphi_{1,t,\infty,\eta} : t \in \mathbf{T}, \eta \in [0,\infty)\}, \\ M_{\infty,\infty} &= \{\varphi_{1,1,\infty,\infty}\} \end{aligned}$$

and, of course,

$$\begin{aligned} \varphi_{t,1,\eta,\infty}\left(\sum a_{k+l\sqrt{2}}e_{k+l\sqrt{2}}\right) &= \sum a_k t^k e^{-k\eta}, \\ \varphi_{1,t,\infty,\eta}\left(\sum a_{k+l\sqrt{2}}e_{k+l\sqrt{2}}\right) &= \sum a_{l\sqrt{2}} t^l e^{-l\eta}, \\ \varphi_{1,1,\infty,\infty}\left(\sum a_{k+l\sqrt{2}}e_{k+l\sqrt{2}}\right) &= a_0. \end{aligned}$$

Let $\overline{\mathbf{D}} := \{z \in \mathbf{C} : |z| \le 1\}$ and $\overline{\mathbf{D}}_* = \overline{\mathbf{D}} \setminus \{0\}$. On making the identifications

$$\begin{aligned} M_{*,*} &\cong \overline{\mathbf{D}}_* \times \overline{\mathbf{D}}_*, \quad & M_{*,\infty} &\cong \overline{\mathbf{D}}_* \times \{0\}, \\ M_{\infty,*} &\cong \{0\} \times \overline{\mathbf{D}}_*, \quad & M_{\infty,\infty} &\cong \{0\} \times \{0\}, \end{aligned}$$

we can rewrite (12.13) in the form

$$M_\Sigma \cong \overline{\mathbf{D}} \times \overline{\mathbf{D}}. \qquad \square$$

12.3 Empty and Nonempty Coronas

We write $\overline{Y}_\Sigma \subset [0,\infty]$ if every $\theta \in \overline{Y}_\Sigma$ is of the form $\theta(\lambda) = \lambda y$ for some $y \in [0,\infty]$. Analogously, we write $Y_\Sigma \subset [0,\infty)$ in case every $\theta \in Y_\Sigma$ has the form $\theta(\lambda) = \lambda y$ with some $y \in [0,\infty)$. Clearly, if $y \in [0,\infty)$ then the map $\theta : \lambda \mapsto \lambda y$ belongs to Y_Σ, while if $y = \infty$ then the map $\theta : \lambda \mapsto \lambda y$ is the map we denote by ∞. Thus, $\overline{Y}_\Sigma \subset [0,\infty]$ if and only if $Y_\Sigma \subset [0,\infty)$ and $\overline{Y}_\Sigma = Y_\Sigma \cup \{\infty\}$.

Theorem 12.7. *Let $\Sigma \subset [0,\infty)$ be an additive semigroup and suppose $0 \in \Sigma$. Then $\mathbf{C}_+$ is dense in M_Σ if and only if $\overline{Y}_\Sigma \subset [0,\infty]$.*

Proof. Suppose $\overline{Y}_\Sigma \subset [0,\infty]$ and pick $(\chi,\theta) \in \Gamma_B^\theta \times \overline{Y}_\Sigma$ (Theorem 12.4). We must show that if we are given $f_1,\dots,f_m \in AP_\Sigma^0$ and $\varepsilon > 0$, then there is a point $(x,y) \in \mathbf{R} \times (0,\infty)$ such that

$$|\varphi_{\chi,\theta}(f_j) - \varphi_{x,y}(f_j)| < \varepsilon \text{ for } j = 1,\dots,m, \tag{12.14}$$

where $\varphi_{x,y}$ is defined by (12.4). Write $f_j = \sum_\lambda f_\lambda^{(j)} e_\lambda$.

If $\theta = \infty$, then $\varphi_{\chi,\theta}(f_j) = f_0^{(j)}$ and it is clear that (12.14) holds whenever $x = 0$ and $y \in (0, \infty)$ is sufficiently large. Thus, assume $\theta \neq \infty$. Then, by what was said above, $\theta(\lambda) = \lambda y_0$ with some $y_0 \in [0, \infty)$. For $\lambda \in \Gamma := \Sigma - \Sigma$ and $x \in \mathbf{R}$, define $\chi_x(e_\lambda) := e^{i\lambda x}$.

Since $\mathbf{R}$ is dense in $\mathbf{R}_{B,\Gamma}$ (see [107, Theorem 26.15]), there is an $x \in \mathbf{R}$ such that

$$|\varphi_{\chi,\theta}(f_j) - \varphi_{\chi_x,\theta}(f_j)| \leq \sum_\lambda |f_\lambda^{(j)}|\, |\chi(e_\lambda) - \chi_x(e_\lambda)| < \frac{\varepsilon}{2}. \tag{12.15}$$

Put $y := y_0$ if $y_0 \in (0, \infty)$ and let $y > 0$ be sufficiently small if $y_0 = 0$. Then

$$|\varphi_{\chi_x,\theta}(f_j) - \varphi_{x,y}(f_j)| \leq \sum_\lambda |f_\lambda^{(j)}|\, |e^{-\lambda y_0} - e^{-\lambda y}| < \frac{\varepsilon}{2}. \tag{12.16}$$

Adding (12.15) and (12.16), we get (12.14).

Conversely, suppose now that $\sigma\tau^{-1}(\mathbf{C}_+)$ is dense in M_Σ. Pick any $\theta \in \overline{Y}_\Sigma$. By Theorem 12.4, $\varphi_{\chi_0,\theta} \in M_\Sigma$ ($\chi_0(e_\lambda) = 1$ for all $\lambda \in \mathbf{R}$). The density of $\sigma\tau^{-1}(\mathbf{C}_+)$ in M_Σ implies that if we are given any $\lambda_1, \lambda_2 \in \Sigma$ and any $\varepsilon > 0$, we can find $(x, y) \in \mathbf{R} \times (0, \infty)$ such that

$$|\varphi_{\chi_0,\theta}(e_{\lambda_j}) - \varphi_{x,y}(e_{\lambda_j})| < \varepsilon \text{ for } j = 1, 2.$$

Equivalently,

$$|e^{-\theta(\lambda_j)} - e^{i\lambda_j x} e^{-\lambda_j y}| < \varepsilon \text{ for } j = 1, 2.$$

Passage to real and imaginary parts gives

$$|e^{-\theta(\lambda_j)} - \cos(\lambda_j x) e^{-\lambda_j y}| < \varepsilon, \ |\sin(\lambda_j x) e^{-\lambda_j y}| < \varepsilon,$$

and adding the squares of these inequalities we obtain

$$e^{-2\theta(\lambda_j)} - 2\cos(\lambda_j x) e^{-\theta(\lambda_j)} e^{-\lambda_j y} + e^{-2\lambda_j y} < 2\varepsilon^2.$$

Since $-\cos(\lambda_j x) \geq -1$, it results that

$$|e^{-\theta(\lambda_j)} - e^{-\lambda_j y}| < \sqrt{2}\,\varepsilon \text{ for } j = 1, 2. \tag{12.17}$$

If $\Sigma = \{0\}$, then $\overline{Y}_\Sigma = \{0\}$ is obviously contained in $[0, \infty]$. So assume $\Sigma \neq \{0\}$. Fix a point $\lambda_1 \in \Sigma \setminus \{0\}$ and let $\lambda_2 \in \Sigma$ be an arbitrary point. From (12.17) we infer that for every natural number N there exists a $y_N = y_N(\lambda_1, \lambda_2) \in (0, \infty)$ such that

$$|e^{-\theta(\lambda_j)} - e^{-\lambda_j y_N}| < 1/N \text{ for } j = 1, 2.$$

Hence $\lambda_j y_N \to \theta(\lambda_j)$ as $N \to \infty$. This implies that

$$y_N \to \frac{\theta(\lambda_1)}{\lambda_1} =: y \in [0, \infty] \text{ as } N \to \infty,$$

and therefore $\theta(\lambda_2) = \lambda_2 y$. As $\lambda_2 \in \Sigma$ was arbitrary, we arrive at the conclusion that $\theta(\lambda) = \lambda y$ for some $y \in [0, \infty]$. Thus, $\overline{Y}_\Sigma \subset [0, \infty]$. □

Example 12.8. If $\Sigma = [0, \infty)$, then $\overline{Y}_\Sigma = [0, \infty]$ by Example 12.3(b) and hence Theorem 12.7 shows that $\mathbf{C}_+$ is dense in M_Σ. We know from Example 12.5 that $M_\Sigma = (\mathbf{R}_B \times [0, \infty)) \cup \{\infty\}$. Since $\mathbf{R}$ is dense in $\mathbf{R}_B$, it is no surprise that $\mathbf{C}_+ \cong \mathbf{R} \times [0, \infty)$ is dense in M_Σ. Figure 12.3 illustrates things. □

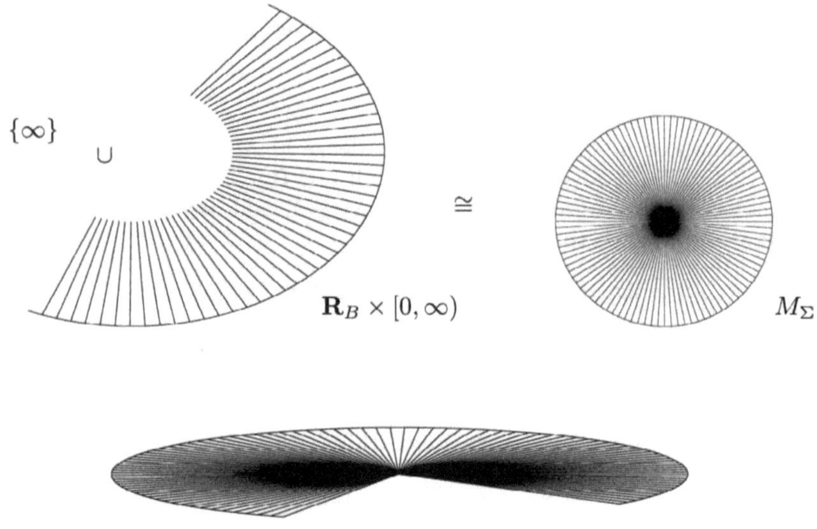

Figure 12.3: The "half-plane" $\mathbf{R}_B \times [0, \infty)$ can transformed into a "disk". Part of this "disk" is represented from another perspective in the bottom.

Example 12.9. Let $\Sigma = \{k + l\sqrt{2} : k \in \mathbf{Z}_+, l \in \mathbf{Z}_+\}$. Since $\overline{Y}_\Sigma = [0, \infty]^2$ by Example 12.3(c), Theorem 12.7 tells us that $\mathbf{C}_+$ is not dense in M_Σ. Here is another way to understand this phenomenon. The maximal ideal space M_Σ was completely identified in Example 12.6. A big part (even nearly all) of M_Σ is formed by $M_{*,*}$. Clearly, we can write

$$\mathbf{C}_+ = \mathbf{R} \times [0, \infty) \cong \mathbf{R} \times [0, 1), \tag{12.18}$$

$$M_{*,*} = \mathbf{T}^2 \times [0, \infty)^2 \cong \mathbf{T}^2 \times [0, 1)^2. \tag{12.19}$$

Why is (12.18) not dense in (12.19)? A functional $\varphi_{x+iy} \in \mathbf{C}_+$ acts by the rule

$$\varphi_{x+iy}\left(\sum a_{k+l\sqrt{2}}\, e_{k+l\sqrt{2}}\right) = \sum a_{k+l\sqrt{2}}\, e^{ixk} e^{ix\sqrt{2}\,l} e^{-yk} e^{-y\sqrt{2}\,l},$$

while the functionals in $M_{*,*}$ are given by

$$\varphi_{e^{ix_1},e^{ix_2},y_1,y_2}\left(\sum a_{k+l\sqrt{2}}e_{k+l\sqrt{2}}\right)=\sum a_{k+l\sqrt{2}}e^{ix_1k}e^{ix_2l}e^{-y_1k}e^{-y_2l}$$

with $x, x_1, x_2 \in \mathbf{R}$ and $y, y_1, y_2 \in [0,\infty)$. By Corollary 1.13, the image of the map

$$\mathbf{R}\to\mathbf{T}^2,\quad x\mapsto\left(e^{ix},e^{ix\sqrt{2}}\right)$$

is dense in $\mathbf{T}^2$. This easily implies that the $\mathbf{R}$ in (12.18) is a dense subset of the $\mathbf{T}^2$ of (12.19). Consequently, that (12.18) is not dense in (12.19) is due to the circumstance that the image of the map $[0,\infty)\to[0,\infty)^2$, $y\mapsto(y,y\sqrt{2})$ is not dense in $[0,\infty)$ or, equivalently, that $[0,1)$, the diagonal of the square $[0,1)^2$, is not dense in $[0,1)^2$. On representing the $\mathbf{T}^2$ in (12.19) by a circle (recall that $\mathbf{R}$ is dense in $\mathbf{T}^2$) and $[0,1)^2$ by a square, we may think of (12.19) as the body resulting from appropriate revolution of the square. Part of the body of revolution is shown in Figure 12.4. As $\mathbf{C}_+$ is contained in the equatorial disk of this body, we see that $M_{*,*}$ and thus also M_Σ consists almost entirely of corona. □

We conclude with a general result that describes a case in which the corona is empty.

Theorem 12.10. *If $\Delta\subset\mathbf{R}$ is an additive group and $\Sigma=\Delta\cap[0,\infty)$, then $\mathbf{C}_+$ is dense in the maximal ideal space of $AP(W)_\Sigma$.*

Proof. We know from Example 12.3(a) that $\overline{Y}_\Sigma=Y_\Sigma\cup\{\infty\}$. Thus, by Theorem 12.7, it suffices to show that $Y_\Sigma\subset[0,\infty)$. Obviously, $\Gamma:=\Sigma-\Sigma\subset\Delta$. Since

$$\Sigma\subset\Gamma\cap[0,\infty)\subset\Delta\cap[0,\infty)=\Sigma,$$

it follows that $\Sigma=\Gamma\cap[0,\infty)$. Let $\theta\in Y_\Sigma$. We extend θ from Σ to all of Γ by defining $\theta(\lambda-\mu):=\theta(\lambda)-\theta(\mu)$.

Fix $\lambda_0\in\Sigma\setminus\{0\}$. If $\lambda\in\Sigma$ and $\lambda\le\lambda_0$, then $\lambda_0-\lambda\in\Gamma$ and $\lambda_0-\lambda\in[0,\infty)$, whence $\lambda_0-\lambda\in\Sigma$ and thus

$$0\le\theta(\lambda)\le\theta(\lambda)+\theta(\lambda_0-\lambda)=\theta(\lambda_0).$$

Consequently, θ is bounded on $\{\lambda\in\Sigma:\lambda\le\lambda_0\}$. This implies that θ is also bounded on $\{\lambda\in\Sigma:|\lambda|\le\lambda_0\}$, which in turn shows that the (additive) function $\theta:\Gamma\to\mathbf{R}$ is continuous at the origin:

$$\theta(\lambda_N)\to0\ \text{ as }\ \lambda_N\in\Gamma\ \text{ and }\ \lambda_N\to0. \tag{12.20}$$

Every $\lambda\in\Sigma$ can be written as $\lambda=c\lambda_0$ with $c\in[0,\infty)$. By a theorem of Dirichlet (see, e.g., [109, p. 11]), for every natural number N there are a natural number $k_N\le N$ and a natural number l_N such that

$$\left|c-\frac{l_N}{k_N}\right|<\frac{1}{Nk_N}.$$

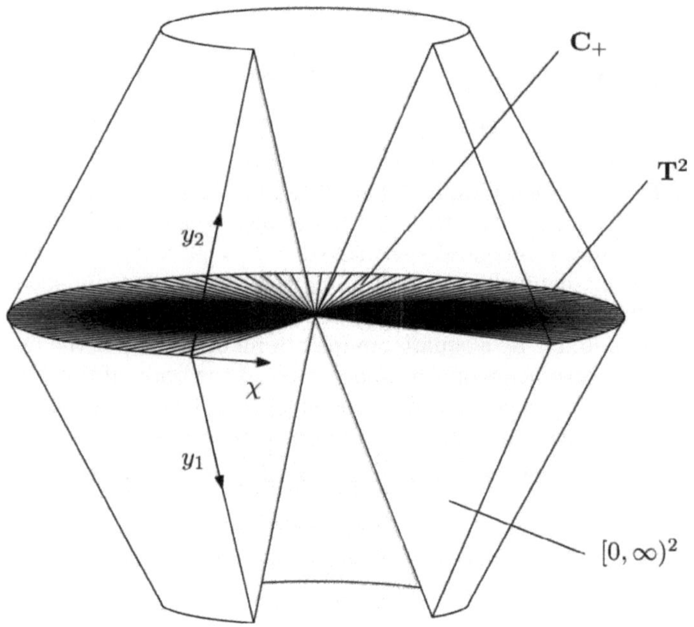

Figure 12.4: The set $M_{*,*}$ for $\Sigma = \{k + l\sqrt{2} : k \in \mathbf{Z}_+, l \in \mathbf{Z}_+\}$. As shown in Example 12.9, we here have a situation where the maximal ideal space consists almost entirely of corona.

Thus, $|k_N\lambda - l_N\lambda_0| = |(k_N c - l_N)\lambda_0| < \lambda_0/N$, and this approaches zero as N goes to infinity. From (12.20) we therefore deduce that $k_N\theta(\lambda) - l_N\theta(\lambda_0) \to 0$ as $N \to \infty$, whence

$$\theta(\lambda) = c\theta(\lambda_0) = \lambda\frac{\theta(\lambda_0)}{\lambda_0} =: \lambda y$$

with $y \in [0,\infty)$. This shows that $Y_\Sigma \subset [0,\infty)$. □

We remark that the converse of Theorem 12.10 is not true: if

$$\Sigma := \mathbf{Z}_+ \setminus \{1\} = \{0,2,3,4,\ldots\},$$

then $\overline{Y}_\Sigma \subset [0,\infty]$, hence $\mathbf{C}_+$ is dense in M_Σ, but there is no subgroup Δ of $\mathbf{R}$ such that $\Sigma = \Delta \cap [0,\infty)$.

Notes

The famous corona theorem for H^∞ was established by Carleson in [53], and it reads as follows.

Theorem 12.11 (Carleson). *The upper half-plane* $\mathbf{C}_+$ *is dense in the maximal ideal space of* $H^\infty(\mathbf{C}_+)$*. Equivalently, for every* $f_1, \dots, f_n \in H^\infty(\mathbf{C}_+)$ *subject to the condition*

$$\inf_{z \in \mathbf{C}_+} (|f_1(z) + \ldots + |f_n(z)|) > 0,$$

there exist $g_1, \dots, g_n \in H^\infty(\mathbf{C}_+)$ *such that*

$$f_1 g_1 + \cdots + f_n g_n = 1.$$

One can strengthen the theorem in the following way.
For every $n \geq 2$ *and every* $\varepsilon > 0$ *there exists a constant* $C(n, \varepsilon) < \infty$ *such that if* $f_1, \dots, f_n \in H^\infty(\mathbf{C}_+)$ *and*

$$\sup_{z \in \mathbf{C}_+} |f_j(z)| \leq 1, \quad \inf_{z \in \mathbf{C}_+} (|f_1(z)| + \ldots + |f_n(z)|) \geq \varepsilon,$$

then there exist $g_1, \dots, g_n \in H^\infty(\mathbf{C}_+)$ *such that*

$$\sup_{z \in \mathbf{C}_+} |g_j(z)| \leq C(n, \varepsilon), \quad f_1 g_1 + \cdots + f_n g_n = 1.$$

Nice presentations of the topic are in [84] and [142].

The almost periodic version of that theorem is much simpler and is, in principle, already in Arens and Singer's paper [6]. It was explicitly stated by Xia [232]. The latter paper and [219] also contain discussions of the corona problem in the abstract setting of analytic functions induced by flows. Generalizations of Carleson's theorem to so-called hyper-analytic functions with spectra in certain semigroups were established by Tonev [221].

The exposition of this chapter is a one-dimensional adaptation of [28]. Example 12.3 (a) was considered earlier in [184], where an AP version of the corona theorem was derived from Carleson's result.

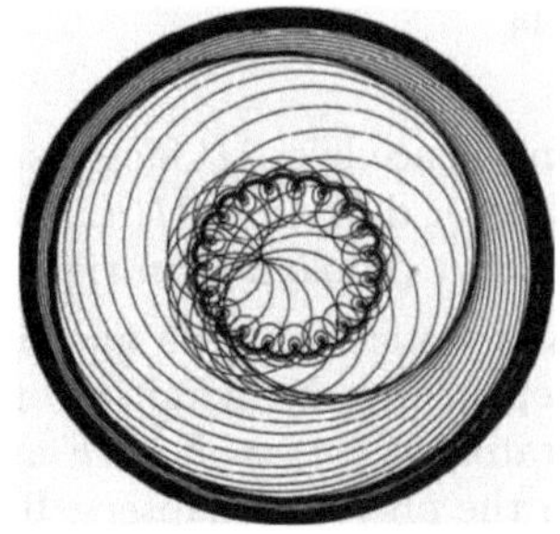

Chapter 13

The Portuguese Transformation

We now begin with studying AP factorization for special classes of almost periodic matrix functions. The Portuguese transformation is a procedure of reducing the factorization problem for one matrix function to another one, which is often simpler than the original matrix function. Repeated application of the Portuguese transformation may therefore lead to a matrix function which is easily factorable. The Portuguese transformation is based on the solution of a special corona theorem. This corona problem will be studied in detail in this chapter.

Unfortunately, we now switch from right factorizations to left factorizations. The first section tells the reader why this will be done.

13.1 Our Tower of Babel

As shown in the preceding chapters, right AP factorization, the right AP indices, and the value of $\mathbf{d}(a)$ $(= \mathbf{d}_r(a)) = M(a_-)M(a_+)$ are just the tools we need to tackle Wiener-Hopf operators $W(a)$. We also know (Lemma 9.2) that the theory of the operators $W(a)$ on $L^2_N(\mathbf{R}_+)$ is essentially equivalent to the theory of the operators $aP + Q$ on $L^2_N(\mathbf{R})$.

The *Riemann-Hilbert boundary value problem* consists in finding a pair (f_+, f_-) of vector functions $f_\pm \in [H^2_\pm(\mathbf{R})]_N$ such that

$$f_+(x) + a(x)f_-(x) = g(x) \text{ for almost all } x \in \mathbf{R},$$

where $a \in L^\infty_{N\times N}(\mathbf{R})$ and $g \in L^2_N(\mathbf{R})$ are given. The operator associated with this problem is the operator $P + aQ$ on $L^2_N(\mathbf{R})$. If $a \in AP_{N\times N}$ and $a = a_+a_-$ is a canonical left AP factorization of a, then

$$P + aQ = (P + a_+Q)(P + a_-Q),$$

and hence left AP factorization, the left AP indices, and the matrices defined by $\tilde{\mathbf{d}}(a)(=\mathbf{d}_l(a)) := M(a_+)M(a_-)$ are more natural in the study of the Riemann-Hilbert problem than right AP factorization.

Also notice that the question on which of the two factorizations is more natural depends on the definition of the Fourier transform. Definition (1.3) leads to the operators $W^0(a) := F^{-1}aF$ and their compressions $W(a), W_\lambda(a)$ we have considered in the previous chapters. If we would define the Fourier transform by

$$(\tilde{F}f)(x) := \int_{\mathbf{R}} f(t)e^{-ixt}\,dt, \quad x \in \mathbf{R},$$

we would arrive at the operators $\tilde{W}^0(a) := \tilde{F}^{-1}a\tilde{F}$ and their compressions $\tilde{W}(a)$, $\tilde{W}_\lambda(a)$. Since $\tilde{W}^0(a) = W^0(\tilde{a})$ where $\tilde{a}(x) := a(-x)$ and since $a = a_- d a_+$ is a right AP (resp. WH) factorization of a if and only if $\tilde{a} = \tilde{a}_- \tilde{d} \tilde{a}_+$ is a left AP (resp. WH) factorization of $\tilde{a}$, left factorizations do exactly the same for $\tilde{W}(a)$ and $\tilde{W}_\lambda(a)$ as right factorizations do for $W(a)$ and $W_\lambda(a)$.

Consequently, part of the literature expresses things in terms of right factorizations, whereas the other part uses the language of left factorizations. In Chapters 2 to 10, we preferred right factorizations because these are dominating in the Wiener-Hopf literature. However, almost all papers devoted to AP factorization of special matrix functions originated from the investigation of the Riemann-Hilbert problem and therefore deal with left AP factorization. We follow this tradition and concentrate upon left AP factorization throughout this and the next two chapters.

If one wants a right AP factorization of a given matrix function a, one can do this by simply applying the results of this chapter to $\tilde{a}$. For example, in Section 1.7 we encountered the matrix function

$$a(x) = \begin{pmatrix} e^{-i\lambda x}I & 0 \\ f(x) & e^{i\lambda x}I \end{pmatrix}, \tag{13.1}$$

and constructing a right AP factorization of this matrix function is equivalent to finding a left AP factorization of

$$\tilde{a}(x) = \begin{pmatrix} e^{i\lambda x}I & 0 \\ f(-x) & e^{-i\lambda x}I \end{pmatrix}. \tag{13.2}$$

Sometimes, another replacement of a is more convenient. Given a matrix function $a = (a_{jk}) \in L^\infty_{m\times m}(\mathbf{R})$, we denote by a^T and a^* the transpose and (Hermitian) adjoint of a:

$$a^T(x) := \big(a_{kj}(x)\big), \quad a^*(x) := \big(\overline{a_{kj}(x)}\big).$$

Also let

$$J := \begin{pmatrix} 0 & I \\ I & 0 \end{pmatrix}, \quad I = I_N.$$

Notice that if $a \in AP_{2N\times 2N}$ is written in the form

$$a = \begin{pmatrix} b & h \\ f & c \end{pmatrix} \quad \text{with } b, h, f, c \in AP_{N\times N}, \tag{13.3}$$

then

$$Ja^T J = \begin{pmatrix} c^T & h^T \\ f^T & b^T \end{pmatrix}, \quad Ja^* J = \begin{pmatrix} c^* & h^* \\ f^* & b^* \end{pmatrix}. \tag{13.4}$$

In particular, if a is given by (13.1), then

$$(Ja^T J)(x) = \begin{pmatrix} e^{i\lambda x} I & 0 \\ f^T(x) & e^{-i\lambda x} I \end{pmatrix}. \tag{13.5}$$

Note that in the $N = 1$ case, (13.1) and (13.5) have the same non-diagonal entry, while these entries are in general different in (13.1) and (13.2). Moreover, we have

$$\Omega(f^T) = \Omega(f) \ \text{ and } \ \Omega(\tilde{f}) = -\Omega(f). \tag{13.6}$$

Proposition 13.1. *Let $a \in AP_{2N\times 2N}$. Then*

$$Ja^T J = G_+ D G_- \tag{13.7}$$

is a left AP factorization of $Ja^T J$ if and only if

$$a = (JG_-^T J)(JDJ)(JG_+^T J) \tag{13.8}$$

is a right AP factorization of a. If (13.7) and (13.8) are AP factorizations, then the right AP indices of a coincide with the left AP indices of $Ja^T J$, and if all right AP indices of a (equivalently, all left AP indices of $Ja^T J$) are equal to each other, then

$$\mathbf{d}(a) = J\big(\tilde{\mathbf{d}}(Ja^T J)\big)^T J.$$

All assertions remain true with AP replaced by APW.

Proof. Let (13.7) be a left AP factorization. Then

$$\begin{aligned} a &= (JG_+DG_-J)^T = (JG_+JJDJJG_-J)^T \\ &= (JG_-^T J)(JDJ)(JG_+^T J), \end{aligned}$$

and since $G_\pm \in GAP^\pm_{N\times N}$ if and only if $JG_\pm^T J \in GAP^\pm_{N\times N}$, it is clear that (13.8) is a right AP factorization. Analogously one sees that (13.7) is a left AP factorization whenever (13.8) is a right AP factorization. The diagonal matrices D and JDJ differ only in the arrangement of the diagonal entries, and hence the left AP indices of $Ja^T J$ equal the right AP indices of a. Finally, if $\kappa(a) = \tilde{\kappa}(Ja^T J)$ is of the form $(\lambda, \dots, \lambda)$, then, by definition,

$$\begin{aligned} \mathbf{d}(a) &= M(JG_-^T J)M(JG_+^T J) = J\big(M(G_-)\big)^T \big(M(G_+)\big)^T J \\ &= J\big(M(G_+)M(G_-)\big)^T J = J\big(\tilde{\mathbf{d}}(Ja^T J)\big)^T J. \end{aligned}$$

□

13.2 Preliminary Observations

Throughout the rest of this chapter and of the next two chapters we omit the attribute "left". Thus, instead of left AP factorization and left AP indices, we simply speak of AP factorization and AP indices.

The following counterpart of Proposition 13.1 is often very useful.

Proposition 13.2. *Let $a \in AP_{2N\times 2N}$. Then*

$$Ja^*J = G_+DG_- \tag{13.9}$$

is an AP factorization of Ja^*J if and only if

$$a = (JG_-^*J)(JD^*J)(JG_+^*J) \tag{13.10}$$

is an AP factorization of a. If (13.9) *and* (13.10) *are AP factorizations, then the AP indices of Ja^*J are the negatives of the AP indices of a, and in case all AP indices of Ja^*J (equivalently, all AP indices of a) coincide, then*

$$\tilde{\mathbf{d}}(Ja^*J) = J\big(\tilde{\mathbf{d}}(a)\big)^*J.$$

All assertions remain valid with AP replaced by APW.

Proof. The proof is analogous to the proof of Proposition 13.1, the only difference being that $G_\pm \in GAP^\pm_{N\times N}$ if and only if $JG_\pm^*J \in GAP^\mp_{N\times N}$. □

In what follows, we restrict ourselves to matrix functions of the form

$$G_f(x) := G_f^{(\lambda)}(x) := \begin{pmatrix} e^{i\lambda x}I & 0 \\ f(x) & e^{-i\lambda x}I \end{pmatrix}, \quad x \in \mathbf{R}, \tag{13.11}$$

where $\lambda \in \mathbf{R}$ and $f \in AP_{N\times N}$. Recall that I here stands for the $N \times N$ identity matrix.

If a is the matrix function (13.11), then

$$(Ja^*J)(x) = \begin{pmatrix} e^{i\lambda x}I & 0 \\ f^*(x) & e^{-i\lambda x}I \end{pmatrix}. \tag{13.12}$$

The analogue of (13.6) is

$$\Omega(f^*) = -\Omega(f). \tag{13.13}$$

Theorem 1.25 indicates that the case $\lambda \leq 0$ is of no interest in connection with the operators $W_\lambda(a)$. The next result tells us that this case is also not interesting from the viewpoint of AP factorization.

Proposition 13.3. *If $\lambda \leq 0$ and $f \in APW_{N\times N}$, then the matrix function G_f has an APW factorization, the AP indices are $\lambda, -\lambda$ (N pairs), and if $\lambda = 0$ then*

$$\tilde{\mathbf{d}}(G_f) = \begin{pmatrix} I & 0 \\ M(f) & I \end{pmatrix}.$$

Proof. Write $f = f_+ + f_-$ with $f_\pm \in APW^\pm_{N\times N}$ and put $\varphi_+ := e_{-\lambda} f_+$, $\varphi_- := e_\lambda f_-$. Then $\varphi_\pm \in APW^\pm_{N\times N}$,

$$f = e_\lambda \varphi_+ + e_{-\lambda}\varphi_-$$

and it can be straightforwardly verified that

$$G_f = \begin{pmatrix} I & 0 \\ \varphi_+ & I \end{pmatrix} \begin{pmatrix} e_\lambda I & 0 \\ 0 & e_{-\lambda} I \end{pmatrix} \begin{pmatrix} I & 0 \\ \varphi_- & I \end{pmatrix}$$

is an APW factorization. If $\lambda = 0$, we have

$$\tilde{\mathbf{d}}(G_f) = \begin{pmatrix} I & 0 \\ M(\varphi_+) & I \end{pmatrix} \begin{pmatrix} I & 0 \\ M(\varphi_-) & I \end{pmatrix} = \begin{pmatrix} I & 0 \\ M(f) & I \end{pmatrix}.$$ □

Thus, we henceforth will always tacitly suppose that $\lambda > 0$.

Given a subset E of $\mathbf{R}$, we define the projection P_E on $APW_{N\times N}$ by

$$P_E : \sum a_\lambda e_\lambda \mapsto \sum_{\lambda \in E} a_\lambda e_\lambda.$$

If $\lambda > 0$, then every $f \in APW_{N\times N}$ decomposes into the sum

$$f = P_{(-\lambda,\lambda)} f + P_{(-\infty,-\lambda]} f + P_{[\lambda,\infty)} f, \tag{13.14}$$

and we can write

$$P_{(-\infty,-\lambda]} f = e_{-\lambda}\varphi_-, \quad P_{[\lambda,\infty)} f = e_\lambda \varphi_+ \tag{13.15}$$

with $\varphi_\pm \in APW^\pm_{N\times N}$. The following result shows that the problem of factorizing $G_f^{(\lambda)}$ can be reduced to the task of finding a factorization of $G^{(\lambda)}_{P_{(-\lambda,\lambda)}f}$.

Proposition 13.4. *Let $\lambda > 0$ and write $f \in APW_{N\times N}$ in the form* (13.14), (13.15). *Put $g := P_{(-\lambda,\lambda)}f$. Then*

$$G_f = \begin{pmatrix} I & 0 \\ \varphi_+ & I \end{pmatrix} G_g \begin{pmatrix} I & 0 \\ \varphi_- & I \end{pmatrix}, \tag{13.16}$$

and hence G_f has an APW factorization if and only if G_g possesses an APW factorization. In this case the AP indices of G_f and G_g coincide, and if all these indices are equal to each other then

$$\tilde{\mathbf{d}}(G_f) = \begin{pmatrix} I & 0 \\ M_\lambda(f) & I \end{pmatrix} \tilde{\mathbf{d}}(G_g) \begin{pmatrix} I & 0 \\ M_{-\lambda}(f) & I \end{pmatrix} \tag{13.17}$$

where $M_\mu(f) := M(fe_{-\mu})$.

Proof. The identity (13.16) can be verified directly, and (13.17) easily results from the equalities

$$M\begin{pmatrix} I & 0 \\ \varphi_\pm & I \end{pmatrix} = \begin{pmatrix} I & 0 \\ M(\varphi_\pm) & I \end{pmatrix} = \begin{pmatrix} I & 0 \\ M_{\pm\lambda}(f) & I \end{pmatrix}. \qquad \square$$

Consequently, in what follows we can limit ourselves to the case where $\lambda > 0$ and $\Omega(f) \subset (-\lambda, \lambda)$.

Proposition 13.5. *Let $\lambda > 0$ and $f \in APW_{N\times N}$. If G_f has an APW factorization, then all AP indices are contained in the segment $[-\lambda, \lambda]$ and their sum is zero. If, in particular, $N = 1$ and G_f is APW factorable, then the two AP indices are $\mu, -\mu$ with $\mu \in [0, \lambda]$. Moreover, all the $2N$ AP indices are equal to each other if and only if they are all zero.*

Proof. By Proposition 13.4, we may assume that $\Omega(f) \subset (-\lambda, \lambda)$. Suppose G_f has an *APW* factorization $G_f = G_+ D G_-$. From Proposition 8.26 we deduce that all *AP* indices belong to $[-\lambda, \lambda]$. Because $\det G_f = \det G_+ \det D \det G_-$ is an *APW*-factorization of $\det G_f = 1$, we infer from the scalar case of Theorem 8.2 that $\det D = 1$, which shows that the sum of the *AP* indices is zero. For $N = 1$, this is only possible if the *AP* indices are μ and $-\mu$. In case all *AP* indices coincide, their sum is zero if and only if each of them is zero. $\square$

13.3 The Idea of the Portuguese Transformation

As in Chapter 12, we let $AP(W)$ stand for one of the algebras *AP* and *APW*.

We are interested in finding or proving the non-existence of an $AP(W)$ factorization of the matrix function

$$G_f := G_f^{(\lambda)} := \begin{pmatrix} e_\lambda I & 0 \\ f & e_{-\lambda} I \end{pmatrix}, \tag{13.18}$$

where $f \in AP(W)_{N\times N}$. The idea of the Portuguese transformation is to construct functions u, v, g_1, g_2 in $AP(W)^+_{N\times N}$ such that

$$\begin{pmatrix} u & v \\ g_1 & g_2 \end{pmatrix} \begin{pmatrix} e_\lambda I & 0 \\ f & e_{-\lambda} I \end{pmatrix} \begin{pmatrix} 0 & I \\ I & 0 \end{pmatrix} = \begin{pmatrix} e_\nu I & 0 \\ f_1 & e_{-\nu} I \end{pmatrix} \tag{13.19}$$

where

$$\nu \in (-\infty, \lambda) \tag{13.20}$$

and

$$\begin{pmatrix} u & v \\ g_1 & g_2 \end{pmatrix} \in G\big[AP(W)^+_{2N\times 2N}\big]. \tag{13.21}$$

Condition (13.21) guarantees that $G_f^{(\lambda)}$ (the matrix function in the middle on the left of (13.19)) and $G_{f_1}^{(\nu)}$ (the matrix on the right of (13.19)) are $AP(W)$ factorable only simultaneously, i.e., $G_f^{(\lambda)}$ has an $AP(W)$ factorization if and only if $G_{f_1}^{(\nu)}$ has an $AP(W)$ factorization. As for condition (13.20), we remark that if $\nu \in (-\infty, 0]$ and $f_1 \in APW_{N\times N}$, then $G_{f_1}^{(\nu)}$ is APW factorable due to Proposition 13.3, while if $\nu \in (0, \lambda)$, Proposition 13.4 nourishes the hope that the problem of finding an $AP(W)$ factorization for $G_{f_1}^{(\nu)}$ is simpler than the same problem for $G_f^{(\lambda)}$. Of course, one also expects that repeated application of the Portuguese transformation makes ν so small that eventually an $AP(W)$ factorable matrix function $G_{f_n}^{(\nu_n)}$ emerges. Equation (13.19) is equivalent to the four equations

$$ve_{-\lambda} I = e_\nu I, \tag{13.22}$$
$$g_2 e_{-\lambda} I = f_1, \tag{13.23}$$
$$u e_\lambda I + v f = 0, \tag{13.24}$$
$$g_1 e_\lambda I + g_2 f = e_{-\nu} I. \tag{13.25}$$

Equations (13.22) and (13.24) are satisfied if and only if $v = e_{\lambda+\nu} I$ and $u = -e_{-\lambda} v f = -e_\nu f$. Equation (13.25) can be rewritten in the form

$$g_1 e_{\lambda+\nu} I + g_2 e_\nu f = I. \tag{13.26}$$

Thus, consider the matrix function

$$\begin{pmatrix} u & v \\ g_1 & g_2 \end{pmatrix} = \begin{pmatrix} -e_\nu f & e_{\lambda+\nu} I \\ g_1 & g_2 \end{pmatrix}. \tag{13.27}$$

We have to satisfy (13.21). Clearly, (13.27) belongs to $AP(W)^+_{2N\times 2N}$ if and only if

$$\Omega(e_\nu f) \subset [0, \infty), \tag{13.28}$$
$$\nu \in [-\lambda, \infty), \tag{13.29}$$
$$g_1, g_2 \in AP(W)^+_{N\times N} \tag{13.30}$$

(also recall Corollary 7.6). We simply *require* that $\Omega(e_\nu f) \subset [0,\infty)$ for some ν in $(-\lambda,\lambda)$ – otherwise the Portuguese transformation is not applicable. Then, of course, (13.20), (13.28), (13.29) are fulfilled. If, in addition, (13.30) is valid, then passage to the determinant in (13.19) gives

$$\det\begin{pmatrix} u & v \\ g_1 & g_2 \end{pmatrix} \cdot 1 \cdot (-1)^N = 1,$$

which shows that (13.21) is automatically invertible in $AP(W)^+_{2N\times 2N}$.
In summary, we arrive at the following. Given $f \in AP(W)_{N\times N}$ such that

$$\Omega(e_\nu f) \subset [0,\infty) \text{ for some } \nu \in (-\lambda,\lambda), \tag{13.31}$$

find

$$g_1, g_2 \in AP(W)^+_{N\times N} \text{ so that } g_1 e_{\lambda+\nu} I + g_2 e_\nu f = I \tag{13.32}$$

and replace $G_f^{(\lambda)}$ by $G_{f_1}^{(\nu)}$ where, by (13.23),

$$f_1 := g_2 e_{-\lambda}.$$

Then $G_f^{(\lambda)}$ is $AP(W)$ factorable if and only if $G_{f_1}^{(\nu)}$ is $AP(W)$ factorable. In the case of $AP(W)$ factorability, the AP indices of $G_f^{(\lambda)}$ and $G_{f_1}^{(\nu)}$ coincide, and if, in addition, all AP indices are zero, we have

$$\tilde{\mathbf{d}}(G_f^{(\lambda)}) = \begin{pmatrix} -M(e_\nu f) & 0 \\ M(g_1) & M(g_2) \end{pmatrix}^{-1} \tilde{\mathbf{d}}(G_{f_1}^{(\nu)}) \begin{pmatrix} 0 & I \\ I & 0 \end{pmatrix} \tag{13.33}$$

by virtue of (13.19) and (13.27). Notice that, by (13.23) and Proposition 13.4, we can further replace $f_1 = g_2 e_{-\lambda}$ with

$$f_1 := P_{(-\nu,\nu)}(g_2 e_{-\lambda}) = e_{-\lambda} P_{(\lambda-\nu,\lambda+\nu)} g_2.$$

The passage from $G_f^{(\lambda)}$ to $G_{f_1}^{(\nu)}$ will be called the *Portuguese transformation.*
The crucial step in the Portuguese transformation is the solution of problem (13.32) under the assumption (13.31). This problem is a special corona problem.

13.4 A Special Corona Problem: Scalar Case

We now turn to the corona problem which arises in connection with the Portuguese transformation. We first consider the scalar case, that is, $N = 1$.
Let $\Sigma \subset [0,\infty)$ be an additive semigroup and suppose $0 \in \Sigma$. Assume we are given

$$a \in APW_\Sigma, \quad \Omega(a) \subset [0,\mu), \quad a_0 := M(a) \neq 0. \tag{13.34}$$

We look for solutions $g_1, g_2 \in APW^+$ of the equation

$$g_1 e_\mu + g_2 a = 1. \tag{13.35}$$

In the context of Section 13.3, μ is $\lambda + \nu$ and a is $e_\nu f$.

Since $|e_\mu(x+iy)| = e^{-\mu y}$ and $|a(x+iy)| \to |a_0| \neq 0$ as $y \to \infty$, we see that

$$\inf_{z \in \mathbf{C}_+} \big(|e_\mu(z)| + |a(z)|\big) > 0.$$

As $\overline{Y}_{[0,\infty)} = [0, \infty]$ (Example 12.3(b)), we deduce from Theorems 12.1 and 12.7 that the problem (13.34), (13.35) always has solutions $g_1, g_2 \in APW^+$. Since in general $\overline{Y}_\Sigma$ is not contained in $[0, \infty]$, the results of Chapter 12 do not guarantee that the problem (13.34), (13.35) is solvable with $g_1, g_2 \in APW_\Sigma$. However, the problem (13.34), (13.35) is a very special corona problem, and we will show that it always possesses a solution $g_1, g_2 \in APW^+$ such that

$$P_{[0,\mu)} g_2 \in APW_\Sigma. \tag{13.36}$$

Note that (13.36) is just what we need for the Portuguese transformation.

Lemma 13.6. *The general solution $\tilde{g}_1, \tilde{g}_2 \in APW^+$ of problem* (13.34), (13.35) *is*

$$\tilde{g}_1 = g_1 + \varphi a, \quad \tilde{g}_2 = g_2 - \varphi e_\mu \tag{13.37}$$

where $g_1, g_2 \in APW^+$ is any particular solution and φ is an arbitrary function in APW^+.

Proof. If $g_1, g_2 \in APW^+$ satisfy (13.35) and $\varphi \in APW^+$, then

$$(g_1 + \varphi a)e_\mu + (g_2 - \varphi e_\mu)a = g_1 e_\mu + g_2 a = 1,$$

and hence (13.37) also gives a solution.

Conversely, assume $g_1, g_2, \tilde{g}_1, \tilde{g}_2 \in APW^+$ and

$$g_1 e_\mu + g_2 a = 1, \quad \tilde{g}_1 e_\mu + \tilde{g}_2 a = 1.$$

Then

$$(g_2 - \tilde{g}_2)a = e_\mu(\tilde{g}_1 - g_1). \tag{13.38}$$

Write

$$g_2 - \tilde{g}_2 =: \Sigma c_\lambda e_\lambda, \quad a = a_0 + \Sigma a_\lambda e_\lambda, \quad e_\mu(\tilde{g}_1 - g_1) =: \Sigma b_\lambda e_\lambda.$$

Equality (13.38) at $z = x + iy \in \mathbf{C}_+$ yields

$$(\Sigma c_\lambda e^{-\lambda y} e_\lambda)(a_0 + \Sigma a_\lambda e^{-\lambda y} e_\lambda) = \Sigma b_\lambda e^{-\lambda y} e_\lambda,$$

and since $\|\Sigma a_\lambda e^{-\lambda y} e_\lambda\|_W < |a_0|$ for sufficiently large y, we obtain

$$\Sigma c_\lambda e^{-\lambda y} e_\lambda = \Big(\sum_{\lambda \geq \mu} b_\lambda e^{-\lambda y} e_\lambda\Big)\Big(a_0^{-1} + \sum_{\lambda > 0} d_\lambda(y) e_\lambda\Big)$$

for all y large enough. This shows that $c_\lambda = 0$ for $\lambda < \mu$, and hence

$$\varphi := e_{-\mu}(g_2 - \tilde{g}_2) \in APW^+. \tag{13.39}$$

From (13.39) we get $\tilde{g}_2 = g_2 - \varphi e_\mu$, and (13.38) then implies that $\tilde{g}_1 = g_1 + \varphi a$. □

From the preceding lemma we see in particular that $P_{[0,\mu)} g_2$ is independent of the special choice of g_2.

Now suppose that, in addition, a is an almost periodic polynomial, $a \in AP_\Sigma^0$. By virtue of (13.34), we can write a in the form

$$a = a_0 + \sum_{k=1}^{m} a_k e_{\gamma_k}, \quad a_0 \neq 0, \quad 0 < \gamma_1 < \gamma_2 < \ldots < \gamma_m < \mu \tag{13.40}$$

with $\gamma_k \in \Sigma$ for all k. Put $b_k := -a_0^{-1} a_k$ $(k = 1, \ldots, m)$, and for $n = (n_1, \ldots, n_m)$ in $\mathbf{Z}_+^m$, set

$$c_n := \frac{(n_1 + \ldots + n_m)!}{n_1! \ldots n_m!} b_1^{n_1} \ldots b_m^{n_m}. \tag{13.41}$$

Further, let $\gamma := (\gamma_1, \ldots, \gamma_m)$ and define

$$g_2 := \sum_{(n,\gamma) < \mu} a_0^{-1} c_n e_{(n,\gamma)}, \tag{13.42}$$

$$g_1 := \sum_{k=1}^{m} b_k \sum_{\mu - \gamma_k \leq (n,\gamma) < \mu} c_n e_{(n,\gamma) + \gamma_k - \mu}, \tag{13.43}$$

where $(n, \gamma) := n_1\gamma_1 + \ldots + n_m\gamma_m$ and the n in (13.42) and (13.43) is taken from $\mathbf{Z}_+^m$. Obviously, $\Omega(g_2) \subset \Sigma \cap [0, \mu)$. However, $\Omega(g_1)$ need not be contained in Σ.

Theorem 13.7. *Let $a \in AP_\Sigma^0$ be given by (13.40) and define $g_2, g_1 \in AP_{[0,\infty)}^0$ by (13.42), (13.43). Then $g_1 e_\mu + g_2 a = 1$.*

Proof. We have

$$
\begin{aligned}
&g_1 e_\mu + g_2 a \\
&= \sum_{k=1}^{m} b_k \sum_{\mu-\gamma_k \le (n,\gamma) < \mu} c_n e_{(n,\gamma)+\gamma_k} \\
&\quad + \sum_{(n,\gamma)<\mu} c_n e_{(n,\gamma)} + \sum_{k=1}^{m} a_k \sum_{(n,\gamma)<\mu} a_0^{-1} c_n e_{(n,\gamma)+\gamma_k} \\
&= \sum_{(n,\gamma)<\mu} c_n e_{(n,\gamma)} + \sum_{k=1}^{m} b_k \Big[\sum_{\mu-\gamma_k \le (n,\gamma)<\mu} c_n e_{(n,\gamma)+\gamma_k} - \sum_{(n,\gamma)<\mu} c_n e_{(n,\gamma)+\gamma_k} \Big] \\
&= \sum_{(n,\gamma)<\mu} c_n e_{(n,\gamma)} - \sum_{k=1}^{m} b_k \sum_{(n,\gamma)<\mu-\gamma_k} c_n e_{(n,\gamma)+\gamma_k} \\
&= \sum_{(n,\gamma)<\mu} c_n e_{(n,\gamma)} - \sum_{k=1}^{m} b_k \sum_{(l+\delta_k,\gamma)<\mu} c_l e_{(l+\delta_k,\gamma)} \qquad (13.44)
\end{aligned}
$$

where $\delta_k \in \mathbf{Z}_+^m$ is the vector whose kth component is 1 and the remaining components of which are zero. For fixed $\alpha \in \{1, \dots, m\}$, the map $l \mapsto l + \delta_\alpha =: n$ is a bijection of the set of all $l \in \mathbf{Z}_+^m$ satisfying $(l+\delta_\alpha, \gamma) < \mu$ onto the set of all $n \in \mathbf{Z}_+^m$ for which $n_\alpha > 0$ and $(n, \gamma) < \mu$. Fix $n \in \mathbf{Z}_+^m \setminus \{0\}$ and denote by $n_{\alpha_1}, \dots, n_{\alpha_q}$ the positive components of n. Let $l_j = n - \delta_{\alpha_j}$ $(j = 1, \dots, q)$. Also put

$$|n| := n_1 + \ldots + n_m = n_{\alpha_1} + \ldots + n_{\alpha_q}.$$

Then

$$
\begin{aligned}
&b_{\alpha_1} c_{l_1} e_{(l_1+\delta_{\alpha_1},\gamma)} + \ldots + b_{\alpha_q} c_{l_q} e_{(l_q+\delta_{\alpha_q},\gamma)} \\
&= e_{(n,\gamma)} (b_{\alpha_1} c_{l_1} + \ldots + b_{\alpha_q} c_{l_q}) \\
&= e_{(n,\gamma)} \Big[b_{\alpha_1} \frac{(|n|-1)!}{(n_{\alpha_1}-1)! n_{\alpha_2}! \dots n_{\alpha_q}!} b_{\alpha_1}^{n_{\alpha_1}-1} b_{\alpha_2}^{n_{\alpha_2}} \dots b_{\alpha_q}^{n_{\alpha_q}} \\
&\qquad + \ldots + b_{\alpha_q} \frac{(|n|-1)!}{n_{\alpha_1}! \dots n_{\alpha_{q-1}}! (n_{\alpha_q}-1)!} b_{\alpha_1}^{n_{\alpha_1}} \dots b_{\alpha_{q-1}}^{n_{\alpha_{q-1}}} b_{\alpha_q}^{n_{\alpha_q}-1} \Big] \\
&= e_{(n,\gamma)} b_{\alpha_1}^{n_1} \dots b_{\alpha_q}^{n_q} \Big[\frac{(|n|-1)! n_{\alpha_1} + \ldots + (|n|-1)! n_{\alpha_q}}{n_{\alpha_1}! \dots n_{\alpha_q}!} \Big] \\
&= e_{(n,\gamma)} b_{\alpha_1}^{n_1} \dots b_{\alpha_q}^{n_q} \frac{(n_{\alpha_1} + \ldots + n_{\alpha_q})!}{n_{\alpha_1}! \dots n_{\alpha_q}!} = c_n e_{(n,\gamma)}. \qquad (13.45)
\end{aligned}
$$

Thus, except for the $n = (0, \dots, 0)$ term, every term in the first sum of (13.44) is compensated by a sum of corresponding terms in the second sum of (13.44). What results is that (13.44) equals $c_0 e_{(0,\gamma)} = 1$. ☐

Theorem 13.8. *The problem* (13.34), (13.35) *always has a solutions* g_2, g_1 *in* APW^+ *such that* $P_{[0,\mu)}g_2 \in APW_\Sigma$.

Proof. Theorem 13.7 shows that this is true if $a \in AP_\Sigma^0$ (and even shows that in this case we have a solution for which $g_2 = P_{[0,\mu)}g_2$ and $g_2 \in AP_\Sigma^0$).
Now let a be an arbitrary function subject to (13.34). Determine $g_1, g_2 \in APW^+$ so that $g_1 e_\mu + g_2 a = 1$; this is possible due to Theorems 12.1 and 12.7 with $\Sigma = [0, \infty)$. Choose $a_m \in AP_\Sigma^0$ so that $\|a - a_m\|_W \to 0$ as $m \to \infty$. We have

$$g_1 e_\mu + g_2 a_m = g_1 e_\mu + g_2 a + g_2(a_m - a) = 1 + g_2(a_m - a) =: \psi_m.$$

Since $\|\psi_m - 1\|_W \to 0$ as $m \to \infty$, it follows that ψ_m is invertible in APW^+ for all sufficiently large m and that $\|\psi_m^{-1} - 1\|_W \to 0$ as $m \to \infty$. Clearly,

$$g_1 \psi_m^{-1} e_\mu + g_2 \psi_m^{-1} a_m = 1. \tag{13.46}$$

By Theorem 13.7, there are $h_1^{(m)}, h_2^{(m)} \in APW^+$ such that

$$h_1^{(m)} e_\mu + h_2^{(m)} a_m = 1, \quad h_2^{(m)} \in AP_\Sigma^0. \tag{13.47}$$

Combining (13.46) and (13.47) with Lemma 13.6, we obtain the existence of functions $\varphi_m \in APW^+$ such that $g_2 \psi_m^{-1} = h_2^{(m)} - \varphi_m e_\mu$. Consequently,

$$P_{[0,\mu)}(g_2 \psi_m^{-1}) = P_{[0,\mu)} h_2^{(m)} \in AP_\Sigma^0,$$

and since $\|g_2 \psi_m^{-1} - g_2\|_W \to 0$ as $m \to \infty$, it results that $P_{[0,\mu)}g_2 \in APW_\Sigma$. □

13.5 A Special Corona Problem: Matrix Case

Our next concern is the matrix analogue of problem (13.34), (13.35). We begin with a simple general result.

Proposition 13.9. *Let* $\Sigma \subset [0, \infty)$ *be an additive semigroup, suppose* $0 \in \Sigma$, *and* $\mathbf{C}_+$ *is dense in* M_Σ. *If* $f_1, \ldots, f_m \in [AP(W)_\Sigma]_{N \times N}$ *and*

$$\inf_{z \in \mathbf{C}_+} \sum_{j=1}^{m} |\det f_j(z)| > 0, \tag{13.48}$$

then there exist $g_1, \ldots, g_m \in [AP(W)_\Sigma]_{N \times N}$ *such that*

$$g_1 f_1 + \ldots + g_m f_m = I \tag{13.49}$$

where I *is the* $N \times N$ *identity matrix.*

Proof. By the implication (ii)$\Rightarrow$(i) of Theorem 12.1, there are functions $h_1, \dots, h_m$ in $AP(W)_\Sigma$ such that

$$\sum_{j=1}^{m} h_j(\det f_j) = 1.$$

We have $(\det f_j)I = (\operatorname{adj} f_j) f_j$ where $\operatorname{adj} f_j$ is the adjugate matrix of f_j. Hence, letting $g_j := h_j(\operatorname{adj}\ f_j)$, we obtain

$$\sum_{j=1}^{m} g_j f_j = \sum_{j=1}^{m} h_j(\operatorname{adj} f_j) f_j = \sum_{j=1}^{m} h_j(\det f_j) I = I. \qquad \square$$

Note that (13.48) is not necessary for (13.49) to hold if $N > 1$: we have, for example,

$$g_1 f_1 + g_2 f_2 := \begin{pmatrix} 1 & 0 \\ 0 & 1 \end{pmatrix}\begin{pmatrix} 1 & 0 \\ 0 & 0 \end{pmatrix} + \begin{pmatrix} 1 & 0 \\ 0 & 1 \end{pmatrix}\begin{pmatrix} 0 & 0 \\ 0 & 1 \end{pmatrix} = I,$$

but $\det f_1$ and $\det f_2$ vanish identically.

Here is a matrix analogue of Theorem 13.8.

Theorem 13.10. *Let $\Sigma \subset [0, \infty)$ be an additive semigroup which contains the origin and suppose*

$$a \in [APW_\Sigma]_{N \times N}, \quad \Omega(a) \subset [0, \mu), \quad \det M(a) \neq 0.$$

Then there are $g_1, g_2 \in APW^+_{N\times N}$ such that

$$g_1 e_\mu I + g_2 a = I \tag{13.50}$$

and $P_{[0,\mu)} g_2 \in [APW_\Sigma]_{N\times N}$.

Proof. Since $|(\det(e_\mu I))(x + iy)| = e^{-N\mu y}$ and

$$|(\det a)(x + iy)| \to |\det M(a)| > 0 \ \text{ as } \ y \to \infty,$$

we deduce from Theorem 13.8 that there are $h_1, h_2 \in APW^+$ such that

$$h_1 \det(e_\mu I) + h_2 \det a = 1, \quad P_{[0,\mu)} h_2 \in APW_\Sigma.$$

The argument employed in the proof of Proposition 13.9 gives (13.50) with

$$g_1 = h_1 \operatorname{adj}(e_\mu I), \ g_2 = h_2 \operatorname{adj} a.$$

Writing $h_2 = P_{[0,\mu)} h_2 + e_\mu \varphi$ with $\varphi \in APW^+$, we get

$$\begin{aligned} P_{[0,\mu)} g_2 &= P_{[0,\mu)}\big((P_{[0,\mu)} h_2) \operatorname{adj} a + e_\mu \varphi \operatorname{adj} a\big) \\ &= P_{[0,\mu)}\big((P_{[0,\mu)} h_2) \operatorname{adj} a\big) \end{aligned} \tag{13.51}$$

and since $P_{[0,\mu)}h_2 \in APW_\Sigma$ and $\operatorname{adj} a \in [APW_\Sigma]_{N\times N}$, it is clear that (13.51) also belongs to $[APW_\Sigma]_{N\times N}$. □

Theorem 13.7 provides us with an explicit solution of problem (13.34), (13.35) in case a is a scalar almost periodic polynomial. If a is a matrix-valued almost periodic polynomial, then $\det a$ is a scalar almost periodic polynomial, and the construction of the proof of Theorem 13.10 again yields an explicit solution of problem (13.34), (13.35). However, one can also proceed in the following way.

Suppose $a \in [AP^0_\Sigma]_{N\times N}$ is of the form

$$a = a_0 + \sum_{k=1}^{m} a_k e_{\gamma_k}, \quad 0 < \gamma_1 < \gamma_2 < \ldots < \gamma_m < \mu \tag{13.52}$$

and

$$a_0 \text{ is invertible.} \tag{13.53}$$

Let $b_k := -a_0^{-1}a_k$ $(k = 1, \ldots, m)$, and for $n = (n_1, \ldots, n_m) \in \mathbf{Z}^m_+$ put

$$c_n := \Sigma b_{j_1} b_{j_2} \ldots b_{j_{|n|}}, \tag{13.54}$$

where $|n| := n_1 + \ldots + n_m$ and the sum is over all $(j_1, \ldots, j_{|n|}) \in \{1, 2, \ldots, m\}^{|n|}$ for which exactly n_k components are equal to $k(k = 1, \ldots, m)$. Clearly, if the matrices $b_1, \ldots, b_m$ commute pairwise, then (13.54) coincides with (13.41). Finally, define g_2 and g_1 by (13.42) and (13.43).

Theorem 13.11. *Let $a \in [AP^0_\Sigma]_{N\times N}$ be given by* (13.52), (13.53). *Then the functions $g_2, g_1 \in [AP^0_{[0,\infty)}]_{N\times N}$ defined by* (13.54), (13.42), (13.43) *satisfy the equation $g_1 e_\mu I + g_2 a = I$.*

Proof. This follows from calculations similar to those of the proof of Theorem 13.7. □

13.6 One More Approach

We here present another approach to Theorems 13.8 and 13.10. For $\varphi > 0$, consider the bounded linear operator

$$V_\varphi : [APW_\Sigma]_{N\times N} \to [APW_\Sigma]_{N\times N}, \ (V_\varphi g)(x) := g(x - i\log\varphi).$$

Lemma 13.12. *Let $\Sigma \subset [0,\infty)$ be an additive semigroup which contains the origin. Also suppose that $\Sigma \neq \{0\}$. If $\eta \in [APW_\Sigma]_{N\times N}$ and $M(\eta) = 0$, then the series $\sum_{k=0}^{\infty} P_{[0,\mu)}(\eta^k)$ converges in $[APW_\Sigma]_{N\times N}$ for every $\mu \in (0,\infty)$.*

Proof. Since $M(\eta)=0$, we can write the matrix function η in the form

$$\eta = \sum_{\lambda\in\Sigma\setminus\{0\}} c_\lambda e_\lambda$$

where $c_\lambda \in \mathbf{C}_{N\times N}$ and $\|\eta\|_W := \Sigma\|c_\lambda\| < \infty$.
Fix $\varepsilon > 0$. There exist $\lambda_\varepsilon \in \Sigma\setminus\{0\}$ and $\varphi \in (0,1)$ such that

$$\sum_{\lambda\in\Sigma,0<\lambda<\lambda_\varepsilon} \|c_\lambda\| < \varepsilon/2, \quad \varphi^{\lambda_\varepsilon}\|\eta\|_W < \varepsilon/2. \tag{13.55}$$

Put

$$\eta_\varphi := V_\varphi \eta = \sum_{\lambda\in\Sigma\setminus\{0\}} c_\lambda \varphi^\lambda e_\lambda.$$

By (13.55),

$$\begin{aligned}\|\eta_\varphi\|_W &\le \sum_{0<\lambda<\lambda_\varepsilon} \|c_\lambda\|\varphi^\lambda + \sum_{\lambda\ge\lambda_\varepsilon}\|c_\lambda\|\varphi^\lambda \\ &\le \sum_{0<\lambda<\lambda_\varepsilon}\|c_\lambda\| + \varphi^{\lambda_\varepsilon}\|\eta\|_W < \varepsilon < 1,\end{aligned}$$

which implies that $\sum_{k=0}^\infty \eta_\varphi^k$ converges in $[APW_\Sigma]_{N\times N}$. Consequently, the series

$$\sum_{k=0}^\infty V_{\varphi^{-1}} P_{[0,\mu)}\eta_\varphi^k = \sum_{k=0}^\infty V_{\varphi^{-1}}P_{[0,\mu)}V_\varphi(\eta^k)$$

is also convergent in $[APW_\Sigma]_{N\times N}$, and as $V_{\varphi^{-1}}P_{[0,\mu)}V_\varphi = P_{[0,\mu)}$, we arrive at the assertion. □

Theorem 13.13. *Let Σ, μ, and a be as in Theorem* 13.10. *Define $\eta \in [APW_\Sigma]_{N\times N}$ as $\eta = I - (M(a))^{-1}a$ and $g_1, g_2 \in APW^+_{N\times N}$ by*

$$g_2 = \Big(\sum_{k=0}^\infty P_{[0,\mu)}(\eta^k)\Big)\big(M(a)\big)^{-1}, \quad g_1 = (I-g_2 a)e_{-\mu}.$$

Then g_1, g_2 satisfy (13.50) *and $P_{[0,\mu)}g_2 \in [APW_\Sigma]_{N\times N}$.*

Proof. If $\Sigma = \{0\}$, then $a = M(a)$, $\eta = 0$, $g_2 = (M(a))^{-1}$, $g_1 = 0$, and the assertion is therefore trivial. So let $\Sigma \ne \{0\}$. Since $M(\eta) = 0$, the matrix function $g_2 = P_{[0,\mu)}g_2$ belongs to $[APW_\Sigma]_{N\times N}$ by virtue of Lemma 13.12. On the other hand, for every $k \in \{0,1,2,\dots\}$ there exists an $h_k \in APW^+_{N\times N}$ such that

$$\eta^k = P_{[0,\mu)}(\eta^k) + h_k e_\mu,$$

whence, for $k \in \{1,2,\dots\}$, $(P_{[0,\mu)}(\eta^{k-1}))\eta - P_{[0,\mu)}(\eta^k) = \psi_k e_\mu$ with $\psi_k = h_k - h_{k-1}\eta \in APW^+_{N\times N}$. It follows that

$$\begin{aligned} & I + \Big(\sum_{k=0}^{\infty} P_{[0,\mu)}(\eta^k)\Big)(\eta - I) \\ &= \Big(\sum_{k=1}^{\infty} P_{[0,\mu)}(\eta^{k-1})\Big)\eta - \sum_{k=1}^{\infty} P_{[0,\mu)}(\eta^k) = \Big(\sum_{k=1}^{\infty} \psi_k\Big)e_\mu. \end{aligned}$$

This shows that the series $\sum_{k=0}^{\infty} \psi_k e_\mu$ converges in $APW^+_{N\times N}$ and that

$$\begin{aligned} g_1 &= (I - g_2 a)e_{-\mu} = \Big(I + \Big(\sum_{k=0}^{\infty} P_{[0,\mu)}(\eta^k)\Big)(\eta - I)\Big)e_{-\mu} \\ &= \sum_{k=1}^{\infty} \psi_k \in APW^+_{N\times N}. \end{aligned}$$

Since $g_1 = (I - g_2 a)e_{-\mu}$, it is clear that (13.50) is satisfied. □

Obviously, Theorem 13.13 implies Theorems 13.8 and 13.10.

Notes

The elementary properties of AP factorization gathered in Sections 13.1 and 13.2 are taken from [130],[132]. For some particular cases, transformation (13.19) was already employed in [214] and [220], while in the explicit form it is used here it appeared in [24] for the first time. This paper received an essential impetus from joint work of its first three authors in Lisbon, by virtue of which we decided to attach the attribute Portuguese to the transformation.

To be exact, we remark that [24] is based on the additional assumption that the Fourier coefficients of the off diagonal block f of the matrix commute pairwise. Of course, this condition is automatically satisfied when f is a scalar-valued function. In the case of non-commuting coefficients, the corresponding formulas were given only in [177]. The full understanding of the relationship between the Portuguese transformation and the corona problem was achieved in [184] and [28].

For some f's, one step of the Portuguese transformation is all we need to construct an AP factorization. In these cases, the solvability of the corona problem is equivalent to explicit AP factorability. Up to this equivalence, Theorem 13.7 was in fact established in [24]; see the proof of Theorem 3.1 there. Its block matrix analogue (Theorem 13.11) follows from [177, Theorem 2.1]. The case of APW functions with infinite Fourier spectra (Theorems 13.8, 13.10) was first considered in [184]. The alternative approach of Section 13.6 is a variation of the method proposed in [20, Theorem 5.7]. The φ-transform used there was first introduced in [22].

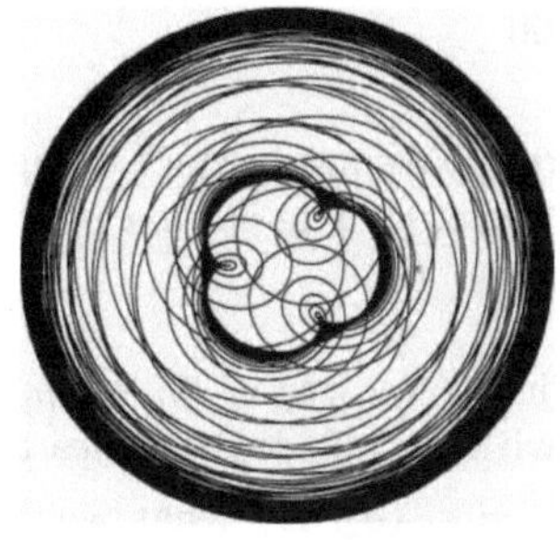

Chapter 14

Some Concrete Factorizations

Convolution operators on intervals lead to the problem of factorizing matrix functions of the form

$$\begin{pmatrix} e^{i\lambda x} & 0 \\ f(x) & e^{-i\lambda x} \end{pmatrix}.$$

In this chapter we compute the AP indices and the geometric mean of such matrix functions under certain additional assumptions. In the one-sided case it is assumed that $\Omega(f)$ is contained in the positive or negative half-lines, the big gap case concerns the situation where one point of $\Omega(f)$ is separated from the rest of $\Omega(f)$ by a big gap, and we will also be able to treat the case where the points of $\Omega(f)$ are commensurable. A fairly complete picture will be given if f is a scalar almost periodic binomial. All results are obtained by employing the Portuguese transformation.

14.1 The One-Sided Case

This case concerns the $2N \times 2N$ matrix functions

$$G_f := \begin{pmatrix} e_\lambda I & 0 \\ f & e_{-\lambda} I \end{pmatrix}, \quad f \in APW_{N\times N}$$

for which $\Omega(f) \subset [0,\infty)$ or $\Omega(f) \subset (-\infty, 0]$. By Propositions 13.3 and 13.4, it suffices to consider the situation in which $\lambda > 0$ and $\Omega(f) \subset [0,\lambda)$ or $\Omega(f) \subset (-\lambda, 0]$.

Theorem 14.1. *If $\Omega(f) \subset [\mu, \lambda)$ where $0 \le \mu < \lambda$ and $M_\mu(f) := M(fe_{-\mu})$ is invertible, then the matrix function G_f admits an APW factorization and the AP indices are $\pm\mu$ (N pairs).*

Proof. We apply the Portuguese transformation with $\nu := -\mu$ (note that $\Omega(e_{-\mu}f) \subset [0,\infty)$). The corona problem (13.32) now reads

$$g_1 e_{\lambda-\mu} I + g_2 e_{-\mu} f = I. \tag{14.1}$$

Since $\Omega(e_{-\mu}f) \subset [0, \lambda - \mu)$ and $M(e_{-\mu}f) = M_\mu(f)$ is invertible, Theorem 13.10 (with $\Sigma = [0,\infty)$) implies that there are $g_1, g_2 \in APW^+_{N\times N}$ satisfying (14.1). Put $f_1 = e_{-\lambda} g_2$. By what was said in Section 13.3, $G_f = G_f^{(\lambda)}$ is APW factorable if and only if so is $G_{f_1}^{(\nu)} = G_{f_1}^{(-\mu)}$. Since $\nu = -\mu \le 0$, we deduce from Proposition 13.3 that $G_f^{(\nu)}$ has an APW factorization and that the AP indices are $\nu = -\mu$, $-\nu = \mu$ (N pairs). □

Note that in the case $N = 1$ the requirement that $M_\mu(f)$ be invertible is equivalent to demanding that $M_\mu(f) \ne 0$. Thus, in the scalar case we have

$$\begin{aligned} &\Omega(f) \subset [\mu, \lambda) \text{ for some } 0 \le \mu < \lambda \text{ and } M_\mu(f) \ne 0 \\ \Longleftrightarrow\ &\Omega(f) \subset [0,\lambda) \text{ and } \Omega(f) \text{ has a minimal element.} \end{aligned}$$

Theorem 14.2. *Let $\Omega(f) \subset [0,\lambda)$. The $2N \times 2N$ matrix function G_f admits a canonical APW factorization if and only if $M(f)$ is invertible. In that case $\tilde{\mathbf{d}}(G_f)$ has the form*

$$\tilde{\mathbf{d}}(G_f) = \begin{pmatrix} 0 & -M^{-1}(f) \\ M(f) & * \end{pmatrix}, \tag{14.2}$$

where $M^{-1}(f) := (M(f))^{-1}$.

Proof. If $M(f)$ is invertible, then Theorem 14.1 implies that G_f has a canonical APW factorization. To compute $\tilde{\mathbf{d}}(G_f)$, we proceed as in Section 13.3. Let g_1, g_2 in $APW^+_{N\times N}$ be any solutions of the corona problem

$$g_1 e_\lambda I + g_2 f = I \tag{14.3}$$

(which is (13.32) with $\nu = 0$ or, equivalently, (14.1) with $\mu = 0$). From (13.33) we get

$$\tilde{\mathbf{d}}(G_f) = \begin{pmatrix} -M(f) & 0 \\ M(g_1) & M(g_2) \end{pmatrix}^{-1} \tilde{\mathbf{d}}(G_{f_1}^{(0)}) \begin{pmatrix} 0 & I \\ I & 0 \end{pmatrix}$$

where $f_1 := e_{-\lambda} g_2$. By Proposition 13.3,

$$\tilde{\mathbf{d}}(G_{f_1}^{(0)}) = \begin{pmatrix} I & 0 \\ M(f_1) & I \end{pmatrix}.$$

Taking the mean value of (14.3) we obtain $M(g_2)M(f) = I$, that is, $M(g_2) = M^{-1}(f)$. It can be easily verified that

$$\begin{pmatrix} -M(f) & 0 \\ M(g_1) & M^{-1}(f) \end{pmatrix}^{-1} = \begin{pmatrix} -M^{-1}(f) & 0 \\ M(f)M(g_1)M^{-1}(f) & M(f) \end{pmatrix}.$$

Putting things together we arrive at (14.2), the asterisk being

$$M(f)M(g_1)M^{-1}(f) + M(f)M(e_{-\lambda}g_2). \tag{14.4}$$

Now suppose G_f has a canonical APW factorization $G_f = G_+G_-$. Choose $f_m \in APW^+_{N\times N}$ such that $\|f - f_m\|_W \to 0$ and $M(f_m)$ is invertible. From what was already proved we know that G_{f_m} has a canonical APW factorization $G_{f_m} = G_m^+G_m^-$ and that

$$\tilde{\mathbf{d}}(G_{f_m}) = \begin{pmatrix} 0 & -M^{-1}(f_m) \\ M(f_m) & * \end{pmatrix} \tag{14.5}$$

Proposition 8.25(b) implies that $\tilde{\mathbf{d}}(G_{f_m}) \to \tilde{\mathbf{d}}(G_f)$ as $m \to \infty$, and therefore $\{M^{-1}(f_m)\}$ must be a convergent sequence. This shows that the matrix $M(f)$ is invertible. □

In the case where f is an almost periodic matrix polynomial, a simple formula for the asterisk in (14.2) is available.

Theorem 14.3. *Let $f \in AP^0_{N\times N}$ be of the form*

$$f = a_0 + \sum_{k=1}^{m} a_k e_{\gamma_k}, \quad 0 < \gamma_1 < \gamma_2 < \ldots < \gamma_m < \lambda.$$

Then G_f admits a canonical APW factorization if and only if a_0 is invertible. Under this condition,

$$\tilde{\mathbf{d}}(G_f) = \begin{pmatrix} 0 & -a_0^{-1} \\ a_0 & a_0\left(\sum_{(n,\gamma)=\lambda} c_n\right)a_0^{-1} \end{pmatrix}$$

where the n in the sum ranges over $\mathbf{Z}_+^m$, $\gamma := (\gamma_1, \ldots, \gamma_m)$, and c_n is given by (13.54) (or by (13.41) in the case $N = 1$).

Proof. Theorem 14.2 shows that G_f has an APW factorization if and only if a_0 is invertible and that

$$\tilde{\mathbf{d}}(G_f) = \begin{pmatrix} 0 & -a_0^{-1} \\ a_0 & * \end{pmatrix},$$

where $*$ is the matrix (14.4). By Theorem 13.11, we can assume that g_2 and g_1 are given by (13.42) and (13.43) with μ replaced by λ. Obviously, $M(e_{-\lambda}g_2) = 0$ and

$$M(g_1) = \sum_{k=1}^{m} b_k \sum_{(l,\gamma)+\gamma_k=\lambda} c_l. \tag{14.6}$$

Equality (13.45) shows that the right-hand side of (14.6) equals $\sum_{(n,\gamma)=\lambda} c_n$. □

Combining Proposition 13.2 with the above three theorems, we arrive at the following result.

Theorem 14.4. (a) *If $\Omega(f) \subset (-\lambda, -\mu]$ where $0 \le \mu < \lambda$ and $M_{-\mu}(f)$ is invertible, then G_f has an APW factorization with the AP indices $\pm\mu$ (N pairs).*
(b) *If $\Omega \subset (-\lambda, 0]$, then for G_f to have a canonical APW factorization it is necessary and sufficient that $M_0(f)$ is invertible. In that case*

$$\tilde{\mathbf{d}}(G_f) = \begin{pmatrix} * & -M_0(f) \\ M_0(f) & 0 \end{pmatrix}.$$

(c) *Let $f \in AP^0_{N\times N}$ be of the form*

$$f = a_0 + \sum_{k=1}^{m} a_k e_{-\gamma_k}, \quad 0 < \gamma_1 < \gamma_2 < \ldots < \gamma_m < \lambda$$

and assume a_0 is invertible. Then G_f admits a canonical APW factorization and

$$\tilde{\mathbf{d}}(G_f) = \begin{pmatrix} a_0^{-1}\left(\sum_{(n,\gamma)=\lambda} c_n\right) a_0 & -a_0^{-1} \\ a_0 & 0 \end{pmatrix},$$

where the n in the sum ranges over $\mathbf{Z}^m_+$, $\gamma := (\gamma_1, \ldots, \gamma_m)$, and c_n is given by (13.54) (or by (13.41) in the case $N = 1$) with $b_k := -a_k a_0^{-1}$ for $k = 1, \ldots, m$. □

14.2 The Big Gap Case

We now consider

$$G_f = \begin{pmatrix} e_\lambda I & 0 \\ f & e_{-\lambda} I \end{pmatrix}, \quad f \in APW_{N\times N}$$

under the assumption that $\Omega(f) \subset (-\lambda, \lambda)$ and one point of $\Omega(f)$ is separated from the rest of $\Omega(f)$ by a big gap, i.e., at least by λ. This type of matrix functions can be reduced to the one-sided case by a single Portuguese transformation.

Theorem 14.5. *Let $f \in APW_{N\times N}$ and suppose*

$$\Omega(f) \subset \{-\nu\} \cup [\mu, \lambda) \;\; \textit{where} \;\; -\nu < 0 < \lambda - \nu \le \mu < \lambda. \tag{14.7}$$

If $a_0 := M_{-\nu}(f)$ and $a_1 := M_\mu(f)$ are invertible, then $G_f^{(\lambda)}$ has an APW factorization and the AP indices are $\pm(\mu + \nu - \lambda)$ (N pairs).

Proof. The Portuguese transformation leads to the corona problem

$$g_1 e_{\lambda+\nu} I + g_2 e_\nu f = I. \tag{14.8}$$

Since $\Omega(e_\nu f) \subset \{0\} \cup [\mu+\nu, \lambda+\nu)$ and $M(e_\nu f) = a_0$ is invertible, Theorem 13.10 with $\Sigma := \{0\} \cup [\mu+\nu, \infty)$ gives that (14.8) has a solution $g_1, g_2 \in APW^+_{N\times N}$ such that

$$\Omega(P_{[0,\lambda+\nu)}g_2) \subset \{0\} \cup [\mu+\nu, \lambda+\nu). \tag{14.9}$$

It follows that $\Omega(P_{(-\nu,\nu)}(g_2 e_{-\lambda})) \subset [\mu+\nu-\lambda, \nu) \subset [0,\nu)$ and we are therefore in the one-sided case.

Put $f_1 := P_{(-\nu,\nu)}(g_2 e_{-\lambda})$. In order to make use of Theorem 14.1, we prove that

$$M_{\mu+\nu-\lambda}(f_1) = M_{\mu+\nu}(g_2) =: q$$

is invertible. From (14.8) we see that

$$M(g_2) = M^{-1}(e_\nu f) = a_0^{-1}. \tag{14.10}$$

This and (14.8) show that

$$g_2 = a_0^{-1} + q e_{\mu+\nu} + \sum_{\eta>\mu+\nu} q_\eta e_\eta .$$

From (14.8) we now obtain

$$g_1 e_{\lambda+\nu} I + a_0^{-1} e_\nu f + q e_{\mu+\nu} e_\nu f + \sum_{\eta>\mu+\nu} q_\eta e_\eta e_\nu f = I,$$

multiplying this by $e_{-\mu-\nu}$ gives

$$g_1 e_{\lambda-\mu} I + a_0^{-1} e_{-\mu} f + q e_\nu f + \sum_{\eta>\mu+\nu} q_\eta e_{\eta-\mu-\nu} e_\nu f = e_{-\mu-\nu} I,$$

and taking the mean value of the last equality we get

$$0 + a_0^{-1} a_1 + q a_0 + 0 = 0,$$

whence

$$q = -a_0^{-1} a_1 a_0^{-1}. \tag{14.11}$$

This shows that q is invertible.

Theorem 14.1 now implies that $G^{(\nu)}_{f_1}$ is APW factorable with the AP indices $\pm(\mu+\nu-\lambda)$ (N pairs), and from Section 13.3 we know that the same is true for $G^{(\lambda)}_f$. □

In the scalar case ($N = 1$), the hypothesis of Theorem 14.5 is equivalent to requiring that

$$-\nu \in \Omega(f) \subset \{-\nu\} \cup [\lambda-\nu, \lambda)$$

and that $\Omega(f)$ has a leftmost point μ in $[\lambda-\nu, \lambda)$.

Our next objective is to prove that, under the condition (14.7), the invertibility of a_0 and a_1 is necessary for $G_f^{(\lambda)}$ to have a canonical *APW* factorization.

Lemma 14.6. *Let $f \in APW_{N\times N}$ and suppose $\Omega(f)\cap(-\lambda,0]$ consists of at most one point. If $G_f^{(\lambda)}$ admits a canonical APW factorization, then $\Omega(f)\cap(-\lambda,0]$ consists of exactly one point, say $-\nu$, and $M_{-\nu}(f)$ is invertible.*

Proof. If $g \in AP^0_{N\times N}$, $P_{(-\lambda,0]}g = P_{(-\lambda,0]}f$, and $\|g-f\|_W$ is sufficiently small, then $G_g^{(\lambda)}$ has a canonical *APW* factorization together with $G_f^{(\lambda)}$ (Theorem 8.24 or Corollary 9.8). Thus, we may without loss of generality assume that f itself is an almost periodic polynomial.

If $\Omega(f)\cap(-\lambda,0]=\emptyset$, then $G_f^{(\lambda)}$ cannot have a canonical *APW* factorization by virtue of Theorem 14.2. So suppose $\Omega(f)\cap(-\lambda,0]=\{-\nu\}$, that is

$$f = M_{-\nu}(f)e_{-\nu}+h, \quad h\in AP^0_{N\times N}, \quad \Omega(h)\subset(0,\infty).$$

Contrary to what we want, let us assume that rank $M_{-\nu}(f)=m<N$. Then there are invertible constant $N\times N$ matrices A, B such that

$$AM_{-\nu}(f)B=\begin{pmatrix} 0 & 0 \\ 0 & I_m \end{pmatrix}.$$

Put

$$f_0 := \begin{pmatrix} 0 & 0 \\ 0 & I_m \end{pmatrix} e_{-\nu} + AhB.$$

Clearly, $AhB\in AP^0_{N\times N}$ and $\Omega(AhB)\subset(0,\infty)$. Because

$$\begin{pmatrix} B^{-1} & 0 \\ 0 & A \end{pmatrix} G_f^{(\lambda)} \begin{pmatrix} B & 0 \\ 0 & A^{-1} \end{pmatrix} = G_{f_0}^{(\lambda)},$$

the matrix function $G_{f_0}^{(\lambda)}$ has a canonical *APW* factorization together with $G_f^{(\lambda)}$. Let $\varepsilon>0$ be the leftmost point of $\Omega(AhB)$ and choose any scalar-valued function $g_-\in AP\backslash\{0\}$ such that $\Omega(g_-)\subset(-\varepsilon,0)\cap(-\lambda,0)$. Denote the column $(g_-\,0\ldots0)^T$ by φ_-. A straightforward computation gives

$$\varphi_+ := G_{f_0}^{(\lambda)}\varphi_- = (e_\lambda g_-\,0\ldots0\,h_{11}g_-\ldots h_{N1}g_-)^T,$$

where $(h_{11}\ldots h_{N1})^T$ is the first column of AhB.

Now let $G_{f_0}^{(\lambda)}=G_+G_-$ be a canonical *APW* factorization of $G_{f_0}^{(\lambda)}$. We then have

$$G_+^{-1}\varphi_+ = G_-\varphi_-,$$

and since $\Omega(G_+^{-1}\varphi_+)\subset(0,\infty)$, $\Omega(G_-\varphi_-)\subset(-\infty,0)$, it results that $G_+^{-1}\varphi_+ = G_-\varphi_- = 0$. In particular, $\varphi_-=0$, which is impossible because g_- was chosen from $AP\setminus\{0\}$. Consequently, rank $M_{-\nu}(f)=N$. $\square$

Theorem 14.7. *Let $f \in APW_{N\times N}$ and suppose*

$$\Omega(f) \subset \{-\nu\} \cup [\lambda-\nu, \lambda) \quad \textit{where} \quad -\nu < 0 < \lambda - \nu.$$

Then $G_f^{(\lambda)}$ has a canonical APW factorization if and only if $a_0 := M_{-\nu}(f)$ and $a_1 := M_{\lambda-\nu}(f)$ are invertible. In that case $\tilde{\mathbf{d}}(G_f^{(\lambda)})$ is of the form

$$\tilde{\mathbf{d}}(G_f^{(\lambda)}) = \begin{pmatrix} -a_1^{-1}a_0 & 0 \\ * & -a_1a_0^{-1} \end{pmatrix}.$$

Proof. Suppose a_0 and a_1 are invertible. Then $G_f^{(\lambda)}$ has a canonical *APW* factorization by virtue of Theorem 14.5.

To compute $\tilde{\mathbf{d}}(G_f^{(\lambda)})$, we proceed as in the proof of Theorem 14.5. The sole invertibility of a_0 gives $g_1, g_2 \in APW^+_{N\times N}$ satisfying (14.8) such that

$$g_2 = a_0^{-1} + qe_\lambda + \sum_{\eta>\lambda} q_\eta e_\eta, \; q = -a_0^{-1}a_1a_0^{-1}.$$

Put $f_1 := g_2e_{-\lambda}$. Since q is also invertible, we obtain from Theorem 14.2 that

$$\tilde{\mathbf{d}}(G^{(\nu)}_{P_{(-\nu,\nu)}f_1}) = \begin{pmatrix} 0 & -q^{-1} \\ q & * \end{pmatrix}.$$

As $M_{-\nu}(f_1) = 0$, Proposition 13.4 yields

$$\tilde{\mathbf{d}}(G^{(\nu)}_{f_1}) = \begin{pmatrix} I & 0 \\ * & I \end{pmatrix}\begin{pmatrix} 0 & -q^{-1} \\ q & * \end{pmatrix}\begin{pmatrix} I & 0 \\ 0 & I \end{pmatrix} = \begin{pmatrix} 0 & -q^{-1} \\ q & * \end{pmatrix}.$$

Finally, we have

$$\begin{pmatrix} -M(e_\nu f) & 0 \\ M(g_1) & M(g_2) \end{pmatrix}^{-1} = \begin{pmatrix} -a_0 & 0 \\ * & a_0^{-1} \end{pmatrix}^{-1} = \begin{pmatrix} -a_0^{-1} & 0 \\ * & a_0 \end{pmatrix},$$

and (13.33) therefore gives

$$\begin{aligned} \tilde{\mathbf{d}}(G_f^{(\lambda)}) &= \begin{pmatrix} -a_0^{-1} & 0 \\ * & a_0 \end{pmatrix}\begin{pmatrix} 0 & -q^{-1} \\ q & * \end{pmatrix}\begin{pmatrix} 0 & I \\ I & 0 \end{pmatrix} \\ &= \begin{pmatrix} a_0^{-1}q^{-1} & 0 \\ * & a_0q \end{pmatrix} = \begin{pmatrix} -a_1^{-1}a_0 & 0 \\ * & -a_1a_0^{-1} \end{pmatrix}. \end{aligned}$$

Now assume $G_f^{(\lambda)}$ has a canonical *APW* factorization. Then, by Lemma 14.6, a_0 is invertible. Construct $f_1 := g_2e_{-\lambda}$ as in the first part of the proof. From Section 13.3 we know that $G^{(\nu)}_{P_{(-\nu,\nu)}f_1}$ also has a canonical *APW* factorization. Theorem 14.2 therefore implies that $q = M(P_{(-\nu,\nu)}f_1)$ is invertible. Hence, the matrix a_1 is also invertible. □

In the case where f is a scalar periodic polynomial, Theorem 14.7 can be made more precise.

Theorem 14.8. *Let* $f = a_0 e_{-\nu} + \sum_{k=1}^m a_k e_{\gamma_k - \nu}$ *where*

$$-\nu < 0 < \lambda - \nu = \gamma_1 - \nu < \gamma_2 - \nu < \ldots < \gamma_m - \nu < \lambda. \tag{14.12}$$

Then $G_f^{(\lambda)}$ *has a canonical APW factorization if and only if* $a_0 \neq 0$ *and* $a_1 \neq 0$, *in which case*

$$\tilde{\mathbf{d}}(G_f^{(\lambda)}) = \begin{pmatrix} -a_1^{-1} a_0 & 0 \\ a_0 s & -a_0^{-1} a_1 \end{pmatrix} \tag{14.13}$$

where

$$s := \sum_l \frac{(l_2 + \ldots + l_m)!}{l_2! \ldots l_m!} \Big(-\frac{a_2}{a_1} \Big)^{l_2} \ldots \Big(-\frac{a_m}{a_1} \Big)^{l_m}, \tag{14.14}$$

the sum taken over all $l = (l_2, \ldots, l_m) \in \mathbf{Z}_+^{m-1}$ *such that*

$$l_2(\gamma_2 - \lambda) + \ldots + l_m(\gamma_m - \lambda) = \nu.$$

Proof. In view of Theorem 14.7, it suffices to verify formula (14.13). We first solve the corona problem

$$g_1 e_{\lambda+\nu} + g_2 e_\nu f = 1 \tag{14.15}$$

using Theorem 13.7. Thus, put $b_k := -a_0^{-1} a_k$ $(k = 1, \ldots, m)$, for $n \in \mathbf{Z}_+^m$ define c_n by (13.41), set $\gamma := (\gamma_1, \ldots, \gamma_m)$, and put

$$\begin{aligned}
g_2 &:= \sum_{(n,\gamma)<\lambda+\nu} a_0^{-1} c_n e_{(n,\gamma)}, \\
f_1 &:= g_2 e_{-\lambda} = \sum_{(n,\gamma)-\lambda<\nu} a_0^{-1} c_n e_{(n,\gamma)-\lambda}, \\
g_1 &:= \sum_{k=1}^m b_k \sum_{\lambda+\nu-\gamma_k \leq (n,\gamma) < \lambda+\nu} c_n e_{(n,\gamma)+\gamma_k-\lambda-\nu}.
\end{aligned}$$

If $n \in \mathbf{Z}_+^m$ has at least two positive components, then, by (14.12),

$$(n,\gamma) = n_1\gamma_1 + \ldots + n_m\gamma_m \geq \gamma_1 + \gamma_2 > 2\lambda > \lambda + \nu.$$

This implies that

$$\begin{aligned} g_2 &= a_0^{-1} + a_0^{-1}\sum_{k=1}^{m} b_k e_{\gamma_k}, \\ f_1 &= a_0^{-1} e_{-\lambda} + a_0^{-1}\sum_{k=1}^{m} b_k e_{\gamma_k-\lambda}, \\ g_1 &= \sum_{k=1}^{m} b_k \sum_{\lambda+\nu-\gamma_k\le\gamma_j} b_j e_{\gamma_j+\gamma_k-\lambda-\nu}. \end{aligned} \tag{14.16}$$

By Theorem 14.3,

$$\tilde{\mathbf{d}}(G^{(\nu)}_{P_{(-\nu,\nu)}f_1}) = \begin{pmatrix} 0 & -b_1^{-1}a_0 \\ a_0^{-1}b_1 & s \end{pmatrix} \tag{14.17}$$

where $s := \sum_{(l,\delta)=\nu} \tilde{c}_l$, $l = (l_2, \dots, l_m) \in \mathbf{Z}_+^{m-1}$, $\delta := (\gamma_2 - \lambda, \dots, \gamma_m - \lambda)$,

$$\tilde{c}_l := \frac{(l_2 + \dots + l_m)!}{l_2! \dots l_m!}\Big(-\frac{b_2}{b_1}\Big)^{l_2} \dots \Big(-\frac{b_m}{b_1}\Big)^{l_m}.$$

Since $M_{-\nu}(f_1) = M_\nu(f_1) = 0$ due to (14.12), we obtain from Proposition 13.4 that

$$\tilde{\mathbf{d}}(G^{(\nu)}_{f_1}) = \tilde{\mathbf{d}}(G^{(\nu)}_{P_{(-\nu,\nu)}f_1}). \tag{14.18}$$

From (13.33) we get

$$\tilde{\mathbf{d}}(G^{(\lambda)}_{f}) = \begin{pmatrix} -M(e_\nu f) & 0 \\ M(g_1) & M(g_2) \end{pmatrix}^{-1} \tilde{\mathbf{d}}(G^{(\nu)}_{f_1}) \begin{pmatrix} 0 & 1 \\ 1 & 0 \end{pmatrix}.$$

Clearly, $M(e_\nu f) = a_0$ and $M(g_2) = a_0^{-1}$. Because $\gamma_j + \gamma_k \ge 2\lambda > \lambda + \nu$ for all j, k, we deduce from (14.16) that $M(g_1) = 0$. Thus,

$$\tilde{\mathbf{d}}(G^{(\lambda)}_{f}) = \begin{pmatrix} -a_0 & 0 \\ 0 & a_0^{-1} \end{pmatrix} \tilde{\mathbf{d}}(G^{(\nu)}_{f_1}) \begin{pmatrix} 0 & 1 \\ 1 & 0 \end{pmatrix}. \tag{14.19}$$

Combining (14.17) to (14.19) and taking into account that $b_1 = -a_0^{-1}a_1$, we arrive at (14.13), (14.14). □

14.3 Scalar Binomials

Our next concern is the 2×2 matrix function

$$G^{(\lambda)}_{f} = \begin{pmatrix} e_\lambda & 0 \\ f & e_{-\lambda} \end{pmatrix} \quad \text{with } f = be_{-\nu} - ce_\mu. \tag{14.20}$$

Thus, f is a scalar binomial.

Theorem 14.9. *The matrix function* (14.20) *always possesses an APW factorization. If* $-\lambda < -\nu < \mu < \lambda$, *then the AP indices are*

$$\begin{aligned}
&\pm\lambda \ \text{ if } b = c = 0,\\
&\pm\nu \ \text{ if } b \neq 0,\ c = 0 \ \text{ or } \nu \leq 0,\ b \neq 0,\\
&\pm\mu \ \text{ if } b = 0,\ c \neq 0 \ \text{ or } \mu \leq 0,\ c \neq 0,\\
&\pm \min\left\{\nu, \mu, \min_{l \in \mathbf{N}} |l(\mu+\nu) - \lambda|\right\} \ \text{ if } \mu > 0,\ \nu > 0,\ bc \neq 0.
\end{aligned}$$

Proof. If one of the coefficients b, c is zero, we have the one-sided case and the assertion follows from Theorem 14.1. So suppose $bc \neq 0$. If $-\nu, \mu \leq 0$ or $-\nu, \mu \geq 0$, we are again in the one-sided case and can employ Theorem 14.1. Thus, we are left with the situation in which

$$-\lambda < -\nu < 0 < \mu < \lambda$$

(also recall Proposition 13.4). Let us further assume that $\nu \leq \mu$, since otherwise we can replace f by $\overline{f}$ and can have recourse to Proposition 13.2 (also see (13.12) and (13.13)).

We now apply the Portuguese transformation. This leads us to the corona problem

$$g_1 e_{\lambda+\nu} + g_2 e_\nu f = 1 \ \text{ with } \ e_\nu f = b - c e_{\mu+\nu}. \tag{14.21}$$

Let $\Sigma \subset [0, \infty)$ be the smallest semigroup which contains $0, \mu+\nu, \lambda+\nu$. The data of problem (14.21) belong to APW_Σ. From Theorem 13.8 we deduce that (14.21) has a solution $g_1, g_2 \in APW^+$ such that $P_{[0,\lambda+\nu)} g_2 \in APW_\Sigma$. The Portuguese transformation of $G_f^{(\lambda)}$ is

$$G^{(\nu)}_{P_{(-\nu,\nu)}}(g_2 e_{-\lambda}). \tag{14.22}$$

Let k be the largest natural number such that $k(\mu+\nu) \leq \lambda + \nu$. Since $P_{[0,\lambda+\nu)} g_2$ is in APW_Σ, we have

$$\Omega(g_2) \subset \{0, \mu+\nu, 2(\mu+\nu), \dots, k(\mu+\nu)\} \cup [\lambda+\nu, \infty),$$

whence

$$\Omega(g_2 e_{-\lambda}) \subset \{-\lambda, \mu+\nu-\lambda, \dots, k(\mu+\nu) - \lambda\} \cup [\nu, \infty).$$

Because $\nu \leq \mu$, we see that $\Omega(g_2 e_{-\lambda}) \cap (-\nu, \nu)$ contains at most one point. Thus, we arrived again at the one-sided case, and Theorem 14.1 implies that (14.22) has an AP factorization whose AP indices are $\pm\nu$ if $\Omega(g_2 e_{-\lambda}) \cap (-\nu, \nu)$ is empty and are $\pm(l(\mu+\nu) - \lambda)$ if

$$\Omega(g_2 e_{-\lambda}) \cap (-\nu, \nu) = \{l(\mu+\nu) - \lambda\}. \qquad \square$$

Theorem 14.10. *Let* $-\lambda < -\nu < \mu < \lambda$, *put* $k := \lambda/(\mu+\nu)$, *and let*

$$\omega := \begin{cases} 1 & \text{if } k \in \mathbf{N} \text{ and } bc \neq 0, \\ 0 & \text{if } k \notin \mathbf{N} \text{ or } bc = 0. \end{cases}$$

There are exactly three cases in which the matrix function (14.20) *has a canonical APW factorization, and in these cases* $\tilde{\mathbf{d}}(G_f)$ *has the form indicated below:*
(a) $\nu = 0$ *and* $b \neq 0$,

$$\tilde{\mathbf{d}}(G_f) = \begin{pmatrix} 0 & -1/b \\ b & (c/b)^k \omega \end{pmatrix};$$

(b) $\mu = 0$ *and* $c \neq 0$,

$$\tilde{\mathbf{d}}(G_f) = \begin{pmatrix} (b/c)^k \omega & 1/c \\ -c & 0 \end{pmatrix};$$

(c) $\mu > 0, \nu > 0, bc \neq 0,$ *and* $k \in \mathbf{N}$,

$$\tilde{\mathbf{d}}(G_f) = \begin{pmatrix} (b/c)^k & 0 \\ 0 & (c/b)^k \end{pmatrix}. \tag{14.23}$$

Proof. Theorem 14.9 shows that G_f has a canonical *APW* factorization if and only if (a), (b), or (c) holds. We are left with the computation of $\tilde{\mathbf{d}}(G_f)$.
(a) If $c = 0$, then

$$G_f = \begin{pmatrix} e_\lambda & 0 \\ b & e_{-\lambda} \end{pmatrix} \quad \text{and hence} \quad \tilde{\mathbf{d}}(G_f) = \begin{pmatrix} 0 & -1/b \\ b & 0 \end{pmatrix}$$

by Theorem 14.3. Let $c \neq 0$. Then $f = b - ce_\mu$ and Theorem 14.3 gives

$$\tilde{\mathbf{d}}(G_f) = \begin{pmatrix} 0 & -b^{-1} \\ b & x \end{pmatrix}$$

with

$$x = \sum_{n(\mu+\nu)=\lambda} (c/b)^n = \begin{cases} (c/b)^k & \text{if } k \in \mathbf{N}, \\ 0 & \text{if } k \notin \mathbf{N}. \end{cases}$$

(b) This case is similar to case (a).
(c) In this case we must explicitly solve problem (14.21). Assume first that $\nu \leq \mu$. From Theorem 13.7 we infer that

$$g_2 := \sum_{n(\mu+\nu)<\lambda+\nu} b^{-1}(c/b)^n e_{n(\mu+\nu)},$$

$$g_1 := \sum_{\lambda-\mu \leq n(\mu+\nu)<\lambda+\nu} (c/b)^{n+1} e_{n(\mu+\nu)+\mu-\lambda}$$

is a solution of (14.21). Let $f_1 = g_2 e_{-\lambda}$. Proposition 13.4 yields

$$\tilde{\mathbf{d}}(G_{f_1}^{(\nu)}) = \begin{pmatrix} 1 & 0 \\ M_\nu(f_1) & 1 \end{pmatrix} \tilde{\mathbf{d}}(G_{P_{(-\nu,\nu)}f_1}^{(\nu)}) \begin{pmatrix} 1 & 0 \\ M_{-\nu}(f_1) & 1 \end{pmatrix}.$$

We have $M_\nu(f_1) \neq 0$ if and only if there is an $n \in \mathbf{N}$ such that $n(\mu+\nu) = \lambda + \nu$, which is impossible because $k(\mu+\nu) = \lambda$ and thus $(k+1)(\mu+\nu) > \lambda + \nu$. Analogously, $M_{-\nu}(f_1) = 0$. Hence

$$\tilde{\mathbf{d}}(G_{f_1}^{(\nu)}) = \tilde{\mathbf{d}}(G_{P_{(-\nu,\nu)}f_1}^{(\nu)}). \tag{14.24}$$

The assumption $\nu \leq \mu$ implies that $\Omega(f_1) \cap (-\nu, \nu) = \{0\}$. Therefore

$$P_{(-\nu,\nu)}f_1 = M(f_1)e_0 = b^{-1}(c/b)^k e_0.$$

Consequently, by Theorem 14.3,

$$\tilde{\mathbf{d}}(G_{P_{(-\nu,\nu)}f_1}^{(\nu)}) = \begin{pmatrix} 0 & -b(b/c)^k \\ b^{-1}(c/b)^k & 0 \end{pmatrix}. \tag{14.25}$$

Since $k(\mu+\nu) = \lambda$ and $(k-1)(\mu+\nu) < \lambda - \mu$, it follows that $M(g_1) = 0$. As $M(g_2) = 1/b$, we finally obtain from (13.33), (14.24), (14.25) that

$$\tilde{\mathbf{d}}(G_f) = \begin{pmatrix} -b & 0 \\ 0 & 1/b \end{pmatrix}^{-1} \begin{pmatrix} 0 & -b(b/c)^k \\ b^{-1}(c/b)^k & 0 \end{pmatrix} \begin{pmatrix} 0 & 1 \\ 1 & 0 \end{pmatrix},$$

which is (14.23).

If $\nu > \mu$, we can apply the above reasoning to $\overline{f} = -\overline{c}e_{-\mu} + \overline{b}e_\nu$. It results that

$$\tilde{\mathbf{d}}(G_f) = \begin{pmatrix} (\overline{c}/\overline{b})^k & 0 \\ 0 & (\overline{b}/\overline{c})^k \end{pmatrix},$$

and Proposition 13.2 along with (13.12) then gives (14.23). □

The following theorem provides explicit formulas in the general case. The theorem can be verified by direct computations. Recall that $[x]$ stands for the integer part of x.

Theorem 14.11. *Let $-\lambda < -\nu < \mu < \lambda$ and $|b|+|c| \neq 0$. Put $s = \left[\frac{\lambda+\nu}{\mu+\nu}\right]$, $r = \left[\frac{\lambda+\mu}{\nu+\mu}\right]$. Then an APW factorization $G_f^{(\lambda)} = G_+ \Lambda G_-$ of (14.20) is given by*

$$\begin{aligned} G_+ &= \begin{pmatrix} -\sum_{j=0}^{s-1}(c^{-1}b)^j c^{-1} e_{\lambda-\mu-j(\mu+\nu)} & 1 \\ 1 & 0 \end{pmatrix}, \quad \Lambda = \begin{pmatrix} e_\mu & 0 \\ 0 & e_{-\mu} \end{pmatrix}, \\ G_- &= \begin{pmatrix} be_{-\mu-\nu} - c & e_{-\lambda-\mu} \\ (c^{-1}b)^s e_{\lambda+\mu-s(\mu+\nu)} & \sum_{j=0}^{s-1}(c^{-1}b)^j c^{-1} e_{-j(\mu+\nu)} \end{pmatrix} \end{aligned}$$

if $c \neq 0$ *and either* $\lambda + \mu \leq s(\mu + \nu)$ *or* $b = 0$*;*

$$G_+ = \begin{pmatrix} e_{\lambda+\nu} & -b^{-1}\sum_{j=0}^{r-1}(cb^{-1})^j e_{j(\mu+\nu)} \\ b - ce_{\mu+\nu} & (cb^{-1})^r e_{-\lambda-\nu+r(\mu+\nu)} \end{pmatrix},$$

$$\Lambda = \begin{pmatrix} e_{-\nu} & 0 \\ 0 & e_{\nu} \end{pmatrix}, \quad G_- = \begin{pmatrix} 1 & b^{-1}\sum_{j=0}^{r-1}(cb^{-1})^j e_{-\lambda+\nu+j(\mu+\nu)} \\ 0 & 1 \end{pmatrix}$$

if $b \neq 0$ *and either* $\lambda + \nu \leq r(\mu + \nu)$ *or* $c = 0$*;*

$$G_+ = \begin{pmatrix} 1 + \sum_{j=1}^{s}(b^{-1}c)^j e_{j(\mu+\nu)} & -\sum_{j=0}^{s-1}(c^{-1}b)^j c^{-1} e_{\lambda-\mu-j(\mu+\nu)} \\ -c(b^{-1}c)^s e_{-\lambda+\mu+s(\mu+\nu)} & 1 \end{pmatrix},$$

$$\Lambda = \begin{pmatrix} e_{\lambda-s(\mu+\nu)} & 0 \\ 0 & e_{-\lambda+s(\mu+\nu)} \end{pmatrix},$$

$$G_- = \begin{pmatrix} (c^{-1}b)^s & \sum_{j=1}^{s}(c^{-1}b)^{s-j}c^{-1}e_{-\lambda-\mu+j(\mu+\nu)} \\ be_{\lambda-\nu-s(\mu+\nu)} & e_{-s(\mu+\nu)} + \sum_{j=0}^{s-1}(cb^{-1})^{s-j}e_{-j(\mu+\nu)} \end{pmatrix}$$

if $\lambda + \mu > s(\mu + \nu)$, $\lambda + \nu > r(\mu + \nu)$, *and* $bc \neq 0$.

14.4 Commensurable Distances

We now show by repeated application of the Portuguese transformation that the 2×2 matrix function

$$G_f^{(\lambda)} = \begin{pmatrix} e_\lambda & 0 \\ f & e_{-\lambda} \end{pmatrix}, \quad f \in APW,$$

has an *APW* factorization in case the distances between the points of $\Omega(f) \cap (-\lambda, \lambda)$ are commensurable. Denoting by $-\nu$ the leftmost point of $\Omega(f) \cap (-\lambda, \lambda)$, our hypothesis is equivalent to requiring that

$$\Omega(f) \cap (-\lambda, \lambda) \subset -\nu + h\mathbf{Z}_+ \tag{14.26}$$

for some $h \in (0, \infty)$.

Lemma 14.12. *Let* $f \in APW \setminus \{0\}$ *and suppose there are* $-\nu \in (-\lambda, 0)$ *and* $h \in (0, \infty)$ *such that* (14.26) *holds. Then there exist a number* $\lambda_1 \in (0, \lambda)$ *and a function* $f_2 \in APW$ *such that* $G_f^{(\lambda)}$ *is APW factorable if and only if* $G_{f_2}^{(\lambda_1)}$ *is APW factorable and such that either* $f_2 = 0$ *or*

$$\emptyset \neq \Omega(f_2) \subset (-\lambda_1, \lambda_1) \cap (-\nu + h\mathbf{Z}_+) \quad \text{and} \quad \lambda_1 \leq \lambda - h.$$

Proof. We will twice apply the Portuguese transformation. If (14.26) holds for some $-\nu$, then it is true with $-\nu$ being the leftmost point of $\Omega(f) \cap (-\lambda, \lambda)$. By Theorem 13.8 with $\Sigma := h\mathbf{Z}_+$, the problem

$$g_1 e_{\lambda+\nu} + g_2 e_\nu f = 1 \quad \text{where} \quad \{0\} \subset \Omega(e_\nu f) \subset h\mathbf{Z}_+$$

has a solution $g_1, g_2 \in APW^+$ such that $\Omega(P_{[0,\lambda+\nu)} g_2) \subset h\mathbf{Z}_+$. Put

$$f_1 := P_{(-\nu,\nu)}(g_2 e_{-\lambda}).$$

Clearly, $\Omega(f_1) \subset (-\lambda + h\mathbf{Z}_+) \cap (-\nu, \nu)$. From Section 13.3 we know that $G_f^{(\lambda)}$ and $G_{f_1}^{(\nu)}$ are APW factorable only simultaneously. If $f_1 = 0$, we get the assertion of the lemma with $\lambda_1 := \nu$ and $f_2 := f_1 = 0$. So let $f_1 \neq 0$ and denote by $-\lambda_1$ the leftmost point of $\Omega(f_1)$. If $-\lambda_1 \geq 0$, we are in the one-sided case, and Theorem 14.1 implies that $G_{f_1}^{(\lambda_1)}$ has an APW factorization with the AP indices $\pm\lambda_1$. Letting $f_2 = 0$ in this case, we obtain the assertion of the lemma again. Thus, assume $-\lambda_1 < 0$. Obviously, $-\lambda_1 > -\nu$ and $-\lambda_1 \geq -\lambda + h$.

We now use the Portuguese transformation a second time to reduce $G_{f_1}^{(\nu)}$ to $G_{f_2}^{(\lambda_1)}$. Once more by Theorem 13.8 with $\Sigma := h\mathbf{Z}_+$, the problem

$$g_3 e_{\nu+\lambda_1} + g_4 e_{\lambda_1} f_1 = 1 \quad \text{where} \quad \{0\} \subset \Omega(e_{\lambda_1} f_1) \subset h\mathbf{Z}_+$$

possesses a solution $g_3, g_4 \in APW^+$ such that $\Omega(P_{[0,\nu+\lambda_1)} g_4) \subset h\mathbf{Z}_+$. Let

$$f_2 := P_{(-\lambda_1,\lambda_1)}(g_4 e_{-\nu}).$$

It is obvious that f_2 is as required in the lemma, and from Section 13.3 we infer that $G_{f_2}^{(\lambda_1)}$ and $G_{f_1}^{(\nu)}$ (and thus $G_f^{(\lambda)}$) are APW factorable only simultaneously. □

Theorem 14.13. *If $f \in APW$ and* (14.26) *holds for some $\nu, h \in (0, \infty)$, then $G_f^{(\lambda)}$ has an APW factorization.*

Proof. Repeated application of Lemma 14.12 either gives a diagonal matrix $G_0^{(\lambda_k)}$, which is obviously APW factorable, or terminates at

$$G_{f_{2n}}^{(\lambda_n)} = \begin{pmatrix} e_{\lambda_n} & 0 \\ f_{2n} & e_{-\lambda_n} \end{pmatrix}$$

where n is the largest integer such that $\lambda_n := \lambda - nh > 0$ and

$$\emptyset \neq \Omega(f_{2n}) \subset (-\lambda_n, \lambda_n) \cap (-\nu + h\mathbf{Z}_+).$$

Since $\lambda - (n+1)h < 0$, we get $2\lambda_n < 2h$, which shows that $-\nu + h\mathbf{Z}_+ \cap (-\lambda_n, \lambda_n)$ consists of at most two points whose distance is $h > \lambda_n$. Thus, we can have recourse to Theorem 14.5 (big gap case) or to Theorem 14.9 (binomials) to deduce that $G_{f_{2n}}^{(\lambda_n)}$ is APW factorable. □

The above construction gives the AP indices and $\tilde{\mathbf{d}}(G_f)$ via multiple application of the Portuguese transformation. This works well in case G_f is a concrete matrix function, but it is clear that this recursive procedure is very inconvenient whenever G_f contains parameters. However, from the proof of Lemma 14.12 it follows that the AP indices of the matrix G_f^λ are of the form $\pm(\nu - mh)$ or $\pm(\lambda - mh)$ with some $m \in \mathbf{Z}_+$. Hence, a canonical factorization can exist only if ν/h or λ/h is integer. Moreover, the following result holds.

Theorem 14.14. *Let $\Omega(f) \subset \mathbf{M} = -\nu + h\mathbf{Z}$. Put $c_j = M_{\tau+jh}(f)$, where τ is the smallest non-negative element of $\mathbf{M}$, and let*

$$T_n = T_n(c) = (c_{j-i})_{i,j=1}^n = \begin{pmatrix} c_0 & c_1 & \dots & c_{n-1} \\ c_{-1} & c_0 & \dots & c_{n-2} \\ \vdots & \vdots & \ddots & \vdots \\ c_{1-n} & c_{2-n} & \dots & c_0 \end{pmatrix},$$

$$\Delta_n = \Delta_n(c) = (c_{j-i-1})_{i,j=1}^n = \begin{pmatrix} c_{-1} & c_0 & \dots & c_{n-2} \\ c_{-2} & c_{-1} & \dots & c_{n-3} \\ \vdots & \vdots & \ddots & \vdots \\ c_{-n} & c_{1-n} & \dots & c_{-1} \end{pmatrix}.$$

Finally, let N be the integer part of λ/h. The matrix $G_{\lambda,f}$ admits a canonical AP factorization if and only if one of the following three conditions holds:

(a) $\dfrac{\nu}{h}, \dfrac{\lambda}{h} \in \mathbf{Z}$, *and* $\det T_N \neq 0$;

(b) $\dfrac{\nu}{h} \in \mathbf{Z}$, $\dfrac{\lambda}{h} \notin \mathbf{Z}$, *and* $\det T_N \det T_{N+1} \neq 0$;

(c) $\dfrac{\nu}{h} \notin \mathbf{Z}$, $\dfrac{\lambda}{h} \in \mathbf{Z}$, *and* $\det T_N \det \Delta_N \neq 0$.

If these conditions are satisfied, then, respectively:

$$\mathbf{d}(G) = \begin{pmatrix} (\det T_N)^{-1} A_{1,N+1} & -(\det T_N)^{-1} \det T_{N-1} \\ (\det T_N)^{-1} \det T_{N+1} & (\det T_N)^{-1} A_{N+1,1} \end{pmatrix},$$

$$\mathbf{d}(G) = \begin{pmatrix} 0 & -(\det T_{N+1})^{-1} \det T_N \\ (\det T_N)^{-1} \det T_{N+1} & 0 \end{pmatrix},$$

and

$$\mathbf{d}(G) = \begin{pmatrix} (\det T_N)^{-1} \tilde{A}_{1,N+1} & 0 \\ 0 & \det T_N (\tilde{A}_{1,N+1})^{-1} \end{pmatrix}.$$

Here $\tilde{A}_{1,N+1}$ is the cofactor of the upper-right entry of the matrix T_{N+1}, and $A_{1,N+1}$, $A_{N+1,1}$ are the cofactors of the upper-right and lower-left entries of T_{N+1} in which the entries $c_{\pm N}$ are substituted by zeros.

14.5 The Division Algorithm

For $f \in AP$, put $\operatorname{ord} f = \sup\{\lambda \colon \lambda \in \Omega(f)\}$. We say that $\operatorname{ord} f$ is the *order* of the function f. According to this definition, $\operatorname{ord} f = -\infty$ if and only if $f = 0$. On the other hand, there are many AP functions f with $\operatorname{ord} f = +\infty$. Of course, $\operatorname{ord} f$ is finite for all nonzero AP polynomials. In this section we suppose that $f \in AP^0$.

Due to Proposition 13.4, the factorization of the matrix function $G_f^{(\lambda)}$ can be reduced to the factorization of

$$\begin{pmatrix} e_\lambda & 0 \\ r_0 & e_{-\lambda} \end{pmatrix}, \tag{14.27}$$

where $r_0 = P_{(-\lambda,\lambda)} f$. Let $r_{-1} = e_\lambda$. We want to construct a recursive sequence of functions $r_j \in AP$, with $\operatorname{ord} r_j = \mu_j$, such that

$$r_{j-2} = q_j^+ r_{j-1} + r_j, \quad q_j^+ \in AP^+, \quad \mu_j < \mu_{j-1} \quad (j = 1, 2, \dots). \tag{14.28}$$

The functions q_j^+ and r_j are defined by (14.28) uniquely. Indeed, if $r_{j-2} = q_j' r_{j-1} + r_j'$, where $q_j' \in AP^+$, $q_j' \neq q_j^+$, and $\operatorname{ord} r_j' < \mu_{j-1}$, then

$$\operatorname{ord}((q_j^+ - q_j')r_{j-1}) \geq \operatorname{ord} r_{j-1} > \operatorname{ord}(r_j - r_j'),$$

which is a contradiction to the equality $(q_j^+ - q_j')r_{j-1} = r_j' - r_j$.

Let us now show that such functions q_j^+ and r_j exist, and that, moreover, these functions are AP polynomials. To that end it suffices to construct AP polynomials q_j^+ and r_j provided we are given AP polynomials r_{j-2} and r_{j-1} for which $\mu_{j-2} > \mu_{j-1}$. Thus, suppose that

$$r_{j-2} = \sum_{s=1}^{t} a_s e_{\nu_s}, \quad r_{j-1} = b_0 e_{\mu_{j-1}} \left(1 - \sum_{l=1}^{m} b_l e_{\gamma_l}\right), \tag{14.29}$$

where $\mu_{j-2} = \nu_1 > \dots > \nu_{s_0} \geq \mu_{j-1} > \nu_{s_0+1} > \dots > \nu_t$ for a certain number $s_0 \in \{2, \dots, t\}$, $0 > \gamma_1 > \dots > \gamma_m$, and $a_1 a_2 \cdots a_t b_0 \neq 0$. We put

$$\begin{aligned}
&\gamma = (\gamma_1, \dots, \gamma_m), \quad n = (n_1, \dots, n_m), \quad \langle n, \gamma \rangle = \sum_{l=1}^{m} n_l \gamma_l, \\
&N(\nu, r_{j-1}) = \{n = (n_1, \dots, n_m) \colon n_1, \dots, n_m \in \mathbf{Z}_+; \nu + \langle n, \gamma \rangle - \mu_{j-1} \geq 0\}, \\
&N_l(\nu, r_{j-1}) = \{n \in N(\nu, r_{j-1}) \colon n_l = f(n_1, \dots, n_{l-1}, n_{l+1}, \dots, n_m)\},
\end{aligned}$$

where $l = 1, \dots, m$ and

$$f(n_1, \dots, n_{l-1}, n_{l+1}, \dots, n_m) = \sup\{n_l \colon \nu + \langle n, \gamma \rangle - \mu_{j-1} \geq 0\}.$$

Consider the AP polynomials

$$q_j^+ = \sum_{s=1}^{s_0} a_s b_0^{-1} e_{\nu_s-\mu_{j-1}} \sum_{n\in N(\nu_s,r_{j-1})} c_n e_{\langle n,\gamma\rangle}, \tag{14.30}$$

$$r_j = \sum_{s=s_0+1}^{t} a_s e_{\nu_s} + \sum_{s=1}^{s_0} a_s e_{\nu_s} \sum_{l=1}^{m} \sum_{n\in N_l(\nu_s,r_{j-1})} c_n b_l e_{\langle n,\gamma\rangle+\gamma_l} \tag{14.31}$$

with

$$c_n = \frac{(n_1+n_2+\cdots+n_m)!}{n_1!n_2!\cdots n_m!} b_1^{n_1} b_2^{n_2}\cdots b_m^{n_m} \tag{14.32}$$

(as usual, a sum with an empty set of summation indices is assumed to be equal zero). Obviously, $q_j^+ \in AP^+$ and $\mu_j < \mu_{j-1}$. Further, for $n \in N(\nu, r_{j-1}) \setminus \{0\}$ we have the recursive relation

$$c_n = \sum_{l=1}^{m} c_{n^{(l)}} b_l, \tag{14.33}$$

where $n^{(l)} = (n_1, \dots, n_{l-1}, n_l - 1, n_{l+1}, \dots, n_m)$ for $n_l > 0$, and $c_{n^{(l)}} = 0$ for $n_l = 0$. Let $\tilde{N}_l(\nu) = \{n \in N(\nu, r_{j-1})\colon n_l \geq 1\}$, $N_l'(\nu) = N(\nu, r_{j-1}) \setminus N_l(\nu, r_{j-1})$. According to (14.33),

$$\begin{aligned}\sum_{l=1}^{m} \sum_{n\in N_l'(\nu)} c_n b_l e_{\langle n,\gamma\rangle+\gamma_l} &= \sum_{l=1}^{m} \sum_{n\in \tilde{N}_l(\nu)} c_{n^{(l)}} b_l e_{\langle n,\gamma\rangle} \\ &= \sum_{n\in N(\nu,r_{j-1})\setminus\{0\}} \sum_{l=1}^{m} c_{n^{(l)}} b_l e_{\langle n,\gamma\rangle} = \sum_{n\in N(\nu,r_{j-1})\setminus\{0\}} c_n e_{\langle n,\gamma\rangle}.\end{aligned}$$

Since $c_0 = 1$, it follows that

$$\begin{aligned}&r_{j-2} - q_j^+ r_{j-1} - r_j = \\ &\sum_{s=1}^{s_0} a_s e_{\nu_s} \left(1 - \sum_{n\in N(\nu_s,r_{j-1})} c_n e_{\langle n,\gamma\rangle} + \sum_{l=1}^{m} \sum_{n\in N_l'(\nu_s)} c_n b_l e_{\langle n,\gamma\rangle+\gamma_l}\right) = 0.\end{aligned}$$

Therefore, the recursive relations (14.28) define the AP polynomials q_j^+ and r_j uniquely, and these polynomials are delivered by formulas (14.30) and (14.31) if r_{j-2} and r_{j-1} are of the form (14.29).

If at a certain step of this recursion we arrive at $r_k = 0$, then the corresponding value of μ_k is $-\infty$, and the process terminates. Otherwise we can continue, and the resulting sequence $\{\mu_j\}_{j=0}^{\infty}$, being decreasing, has a limit $\mu_\infty \geq -\infty$. Let us denote by K the set of indices of the sequence $\{\mu_j\}$, that is, the segment $\{0, \ldots, k\}$ in the first case and $\mathbf{Z}_+$ in the second.

Relations (14.28) are equivalent to the representation of the function r_{-1}/r_0 in the form of a continued fraction,

$$\frac{r_{-1}}{r_0} = q_1^+ + \cfrac{1}{q_2^+ + \cfrac{1}{\ddots\, q_k^+ + \cfrac{r_k}{r_{k-1}}}}.$$

If $\mu_k = -\infty$, the latter equality means that

$$\frac{r_{-1}}{r_0} = q_1^+ + \cfrac{1}{q_2^+ + \cfrac{1}{\ddots\, q_k^+}} \quad (=: [q_1^+, \ldots, q_k^+]).$$

If $K = \mathbf{Z}_+$, we will associate with the function r_{-1}/r_0 a formal infinite continued fraction,

$$\frac{r_{-1}}{r_0} \sim q_1^+ + \cfrac{1}{q_2^+ + \cfrac{1}{\ddots\, q_k^+ + \cdots}} \quad (=: [q_1^+, \ldots, q_k^+, \ldots]),$$

irrespective to the convergence of the latter. Denote by p_j^+ the numerator of jth convergent $[q_1^+, \ldots, q_j^+]$ of the continued fraction $[q_1^+, \ldots, q_j^+, \ldots]$. These numerators satisfy the recursive relations

$$p_j^+ = q_j^+ p_{j-1}^+ + p_{j-2}^+ \ (j = 1, 2, \ldots), \quad p_{-1}^+ = 0, \quad p_0^+ = 1, \tag{14.34}$$

and are therefore AP polynomials lying in AP^+ together with q_j^+. In addition,

$$\operatorname{ord} p_j^+ = \operatorname{ord}(q_j^+ \cdots q_1^+) = \lambda - \mu_{j-1}.$$

Finally, put

$$Q_j^+ = \begin{pmatrix} q_j^+ & 1 \\ 1 & 0 \end{pmatrix}.$$

Lemma 14.15. *For all $k \in K$ the matrix function* (14.27) *admits a representation*

$$G = G_k^+ \Lambda_k G_k^-, \tag{14.35}$$

where

$$G_k^+ = Q_1^+ \cdots Q_k^+, \quad \Lambda_k = \operatorname{diag}(e_{\mu_{k-1}}, e_{-\mu_{k-1}}),$$
$$G_k^- = \begin{pmatrix} r_{k-1}e_{-\mu_{k-1}} & (-1)^{k-1}p_{k-1}^+ e_{-(\mu_{k-1}+\lambda)} \\ r_k e_{\mu_{k-1}} & (-1)^k p_k^+ e_{\mu_{k-1}-\lambda} \end{pmatrix}. \tag{14.36}$$

Proof. For $k = 0$, equality (14.35) takes the form

$$G = \begin{pmatrix} e_\lambda & 0 \\ 0 & e_{-\lambda} \end{pmatrix} \begin{pmatrix} 1 & 0 \\ r_0 e_\lambda & 1 \end{pmatrix}$$

and is therefore valid. With the use of (14.28) and (14.34) it can be verified that $G_j^- = (Q_j^+\Lambda_j)^{-1}\Lambda_{j-1}G_{j-1}^-$. Hence, if (14.35) holds for $k = j-1$, then

$$G_j^+\Lambda_j G_j^- = G_{j-1}^+ Q_j^+ \Lambda_j (Q_j^+\Lambda_j)^{-1}\Lambda_{j-1}G_{j-1}^- = G_{j-1}^+\Lambda_{j-1}G_{j-1}^- = G,$$

that is, (14.35) is also true for $k = j$. □

In (14.35), the factor G_k^+ is an AP polynomial with $\Omega(G_k^+) \subset [0, \lambda - \mu_{k-1}]$ and $\det G_k^+ = (-1)^k$. Hence, $(G_k^+)^{\pm 1} \in AP^+$. The factor G_k^- has the same determinant $(-1)^k$ as G_k^+ since $\det G_k^+ G_k^- = \det G / \det \Lambda_k = 1$. Further, the entries of G_k^- are AP polynomials, whose orders equal $\mu_k + \mu_{k-1}$, 0, and $-(\mu_{k-1} + \mu_{k-2})$. Therefore, (14.35) delivers an AP-factorization of the matrix (14.27) if and only if

$$\mu_{k-2} + \mu_{k-1} \geq 0 \geq \mu_{k-1} + \mu_k. \tag{14.37}$$

This implies the following.

Theorem 14.16. *Let r_0 in* (14.27) *be an AP polynomial whose spectrum $\Omega(r_0)$ is located in the interval $(-\lambda, \lambda)$ and let $\{r_j\}$ be the sequence of AP polynomials of orders $\operatorname{ord} r_j = \mu_j$ that is constructed according to* (14.29), (14.30), (14.31). *Suppose that* (14.37) *is satisfied for some k. Then the matrix* (14.27) *has an AP factorization, its partial AP-indices equal $\pm\mu_{k-1}$, and the factors $G_\pm (= G_k^\pm)$ can be constructed by formulas* (14.36). □

Obviously, the desired value of k exists if and only if the sequence $\{\mu_j\}$ terminates or its limit μ_∞ is nonpositive. If, in addition, the sequence $\mu_{j-1} + \mu_j$ skips 0, then such a value of k is unique, and both inequalities in (14.37) are strict. On the other hand, if there exists a j_0 such that $\mu_{j_0-1} + \mu_{j_0} = 0$ (which is unique, due to the strict monotonicity of μ_j), then both $k = j_0$ and $k = j_0 + 1$ are acceptable for our purposes and the right and left inequality in (14.37) turns into equality, respectively. From the computational point of view, it is more convenient to choose $k = j_0$, that is, the minimal possible k such that $\mu_{k-1} + \mu_k \leq 0$.

Using the explicit formulas for the factors of the AP-factorization (14.35), it is now possible to find a necessary and sufficient condition for the partial indices of

G to be equal to 0 and to derive also a formula for $\mathbf{d}(G)$. Denote by a_j the leading coefficient of q_j^+, that is, $a_j = M(q_j^+ e_{\mu_{j-1}-\mu_{j-2}})$, and put $c_j = M(q_j^+)$.

Corollary 14.17. *Under the conditions of Theorem* 14.16, *the partial AP-indices of the matrix function* (14.27) *equal zero if and only if the sequence* μ_j *assumes the value* 0, *that is, if and only if there exists a* $k \in \mathbf{Z}_+$ *such that* $\mu_{k-1} = 0$. *If this condition is satisfied, then*

$$\tilde{\mathbf{d}}(G) = \begin{pmatrix} c_1 & 1 \\ 1 & 0 \end{pmatrix} \cdots \begin{pmatrix} c_k & 1 \\ 1 & 0 \end{pmatrix} \begin{pmatrix} a_k^{-1} \cdots a_1^{-1} & 0 \\ 0 & (-1)^k a_k \cdots a_1 \end{pmatrix}. \tag{14.38}$$

Observe that $a_k^{-1} \cdots a_1^{-1}$ is nothing but the leading coefficient of r_{k-1}.

Notes

Historically, the division method of Section 14.5 was developed before the techniques of Sections 14.1 to 14.4. The division algorithm was established in [129], [130]; see also [131], [134]. In the periodic case this method boils down to the well known Chebotarev factorization algorithm [56], for a description of which see also [155, Theorem 4.9] and for a modified version of which we refer to [76]. The first step of this algorithm can be carried out (with minor changes in the formulas) even in the matrix case, provided that the leftmost or rightmost Fourier coefficient of f is invertible.

Many results of the preceding Sections 14.1-14.4 were originally obtained with the help of the division algorithm. The scalar case of Theorem 14.3 was established in [130], [132], and the matrix case was disposed of in [135, Theorem 2.1 and Corollary 2.2]. The results of Sections 13.2, 13.3 and 13.6 allow us to construct explicit APW factorizations of the matrix functions G_f with $f \in APW^{\pm}_{N\times N}$. Theorem 13.13, in its turn, leads to explicit APW factorization of matrix functions satisfying the conditions of Theorems 14.5 and 14.7. Theorem 14.8 without formula (14.13) for the geometric mean was obtained in [86]. The results of Section 14.3 were announced in [131] and published with proofs in [130], [132]. We remark that the proofs of [130], [132] made use of the division algorithm although this was not explicitly mentioned there. Theorem 14.13 is from [130].

The approach using the Portuguese transformation was systematically realized in [177], where unified proofs of Theorem 14.3, Theorems 14.5, 14.7, Lemma 14.6 in their present form were given and formula (14.13) for AP polynomials was established. The passage from AP polynomials to APW functions with infinite Fourier spectrum (Theorems 14.1 and 14.2) is in [184]. We believe that the approach based on the Portuguese transform is simpler and more natural than the one using the division algorithm, and therefore in our exposition we followed the Portuguese approach. At the same time, we want the reader to be aware of alternative possibilities, because so far some matrices can be factored only by using the division algorithm; see [86], [130], [134].

The results of Section 14.3, though formulated in the scalar case only, allow matrix generalizations [130]. As it happens, in this case an *AP* factorization also always exists. This was proved and the factorization was explicitly constructed in [130] with the help of canonical forms of linear pencils of matrices [82]. Part of these results, corresponding to canonical *AP* factorization, appeared as Theorem 2.3 in [132].

Theorem 14.14 was proved in [134] with the use of Theorem 8.27 for the case of *AP* polynomials and represents a joint effort of two of the authors with Tishin [123], [220]. The canonical factorability criterion and the formulas for the geometric mean carry over to the matrix case: see [138], where they were proved by yet another method. The formulas for the periodic situation (situation (a) in Theorem 14.14) were, of course, known earlier; see [77] for the matrix case and [76] for simpler formulas in the scalar case.

Some applications of the results of this chapter to sampling and interpolating problems can be found in [156].

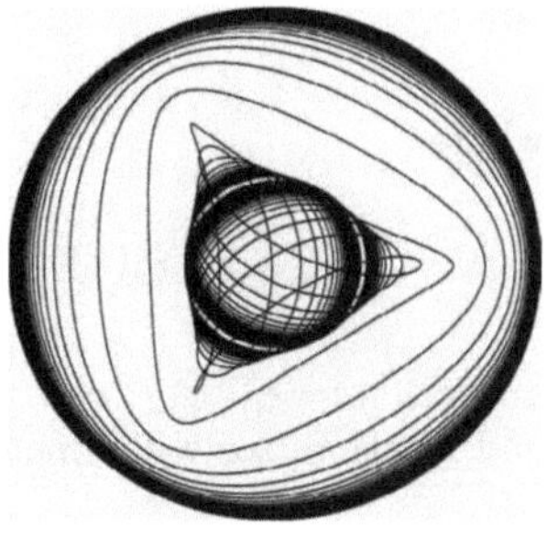

Chapter 15

Scalar Trinomials

In this chapter we continue the investigations of the preceding chapter and consider triangular matrix functions in which $f(x)$ is an almost periodic trinomial. This case is more difficult by at least one order than the case of binomials, because, as known from Section 8.5, the matrix functions studied here need not at all have an AP factorization.

15.1 $\nu + \delta = \lambda$, $\mu = 0$

We now turn to 2×2 matrix functions of the form

$$G(x) := G_f^{(\lambda)}(x) := \begin{pmatrix} e^{i\lambda x} & 0 \\ c_{-1}e^{-i\nu x} - c_0 + c_1 e^{i\delta x} & e^{-i\lambda x} \end{pmatrix}, \quad \lambda > 0, \tag{15.1}$$

that is, to the case where $f(x) = c_{-1}e^{-i\nu x} - c_0 + c_1 e^{i\delta x}$ is a trinomial. We assume that $c_{-1}c_0c_1 \neq 0$, since otherwise f is at most a binomial, and this case was disposed of in Section 14.3. We also suppose that $\nu > 0$ and $\delta > 0$: if ν and δ have different signs, we have the one-sided case (Section 14.1). Finally, if ν/δ is rational, we are in the case of commensurable distances, which was considered in Section 14.4; we therefore assume that ν/δ is irrational.

In Section 15.3 and 15.4 we will study trinomials in which $-c_0$ is replaced by $-c_0e^{i\mu x}$. In this and the following section we take $\mu = 0$. Furthermore, in the present section we restrict ourselves to the case where $\nu + \delta = \lambda$.

We have

$$G(x) = \begin{pmatrix} 1 & 0 \\ 0 & c_0 \end{pmatrix} G_0(x) \begin{pmatrix} 1 & 0 \\ 0 & c_0^{-1} \end{pmatrix} \tag{15.2}$$

where

$$G_0(x) = \begin{pmatrix} e^{i\lambda x} & 0 \\ c_{-1,0}e^{-i\nu x} - 1 + c_{1,0}e^{i\delta x} & e^{-i\lambda x} \end{pmatrix}$$

and $c_{-1,0} := c_0^{-1}c_{-1}$, $c_{1,0} := c_0^{-1}c_1$. Using the expansion of $\beta := \nu/\delta$ into a continued fraction, we will construct a sequence G_k of matrix functions

$$G_k(x) = \begin{pmatrix} e^{i\lambda_k x} & 0 \\ c_{-1,k}e^{-i\nu_k x} - 1 + c_{1,k}e^{i\delta_k x} & e^{-i\lambda_k x} \end{pmatrix} \tag{15.3}$$

such that G and G_k are APW factorable only simultaneously.

Thus, let

$$\beta = n_1 + \cfrac{1}{n_2 + \cfrac{1}{n_3 + \dots}} \tag{15.4}$$

be the expansion of the irrational number $\beta = \nu/\delta$ into an infinite continued fraction. Here $n_1 \in \mathbf{Z}_+$ and $n_j \in \mathbf{N}$ for $j \geq 2$. The kth convergent of the continued fraction (15.4) is

$$\frac{p_k}{q_k} = n_1 + \cfrac{1}{n_2 + \cfrac{1}{\dots + \cfrac{1}{n_k}}}. \tag{15.5}$$

It is well known that (15.5) holds if p_k, q_k are given by the recursion formulas

$$p_k = n_k p_{k-1} + p_{k-2}, \quad p_0 = 1, \quad p_{-1} = 0, \tag{15.6}$$

$$q_k = n_k q_{k-1} + q_{k-2}, \quad q_0 = 0, \quad q_{-1} = 1. \tag{15.7}$$

Moreover, one can show that

$$\frac{1}{q_{k+1} + q_k} < |p_k - \beta q_k| \leq \frac{1}{q_{k+1}}. \tag{15.8}$$

We define the matrix functions $G_k(x)$ by (15.3) where

$$\nu_0 = \nu, \quad \delta_0 = \delta, \quad \lambda_0 = \lambda,$$

$$\nu_k = \begin{cases} \nu_{k-1} - n_k\delta_{k-1} \\ \nu_{k-1}, \end{cases} \quad \delta_k = \begin{cases} \delta_{k-1} \\ \delta_{k-1} - n_k\nu_{k-1}, \end{cases} \quad \lambda_k = \begin{cases} \lambda_{k-1} - n_k\delta_{k-1} \\ \lambda_{k-1} - n_k\nu_{k-1}, \end{cases}$$

$$c_{-1,k} = \begin{cases} c_{1,k-1}^{n_k} c_{-1,k-1} \\ c_{-1,k-1}, \end{cases} \quad c_{1,k} = \begin{cases} c_{1,k-1} \\ c_{1,k-1}c_{-1,k-1}^{n_k} \end{cases} \quad \text{for} \begin{cases} k \text{ odd} \\ k \text{ even.} \end{cases} \tag{15.9}$$

Using (15.9), it is easy to show by induction on k that

$$n_{k+1}+\cfrac{1}{n_{k+2}+\cfrac{1}{n_{k+3}+\dots}}=\begin{cases}\delta_k/\nu_k & \text{if } k \text{ is odd,}\\ \nu_k/\delta_k & \text{if } k \text{ is even.}\end{cases}$$

This implies (again by induction) that

$$\begin{aligned}\nu_{k-1} &> n_k\delta_{k-1} \quad \text{if } k \text{ is odd,} &(15.10)\\ \delta_{k-1} &> n_k\nu_{k-1} \quad \text{if } k \text{ is even,} &(15.11)\end{aligned}$$

whence

$$\nu_k>0, \quad \delta_k>0 \quad \text{for all} \quad k. \tag{15.12}$$

Further, the assumption $\nu+\delta=\lambda$ yields the equalities

$$\nu_k+\delta_k=\lambda_k \ \text{ for all } \ k. \tag{15.13}$$

Finally, put

$$a_0=1,\ a_k=\begin{cases}a_{k-1}c_{1,k-1}^{-n_k} & \text{for} \quad k \quad \text{odd,}\\ a_{k-1}c_{-1,k-1}^{n_k} & \text{for} \quad k \quad \text{even.}\end{cases} \tag{15.14}$$

Taking into account (15.6), (15.7), (15.9), one can show by induction on k that

$$c_{(-1)^k,k+1}=c_{(-1)^k,k}=c_{1,0}^{p_k}c_{-1,0}^{q_k} \ \text{ for all } \ k, \tag{15.15}$$

$$a_k=c_{1,0}^{-1}c_{-1,0}\big(c_{1,0}^{p_k-p_{k-1}}c_{-1,0}^{q_k-q_{k-1}}\big)^{(-1)^k} \quad \text{for all } \ k. \tag{15.16}$$

Lemma 15.1. *If $\nu>0$, $\delta>0$, $\nu+\delta=\lambda$, and $\beta=\nu/\delta$ is irrational, then*

$$G_k=B_k^+G_{k-1}B_k^- \ \text{ for all } \ k\in\mathbf{N}, \tag{15.17}$$

where $B_k^\pm\in GAPW_{2\times2}^\pm$ are almost periodic polynomials with nonzero constant determinants and

$$M(B_k^+)=\begin{pmatrix}a_k^{-1}a_{k-1} & 0\\ 0 & 1\end{pmatrix}, \quad M(B_k^-)=\begin{pmatrix}1 & 0\\ 0 & a_{k-1}^{-1}a_k\end{pmatrix}. \tag{15.18}$$

Proof. Put

$$\psi_k(x)=\sum_{j=0}^{n_k-1}(c_{1,k-1}e^{i\delta_{k-1}x})^j, \quad \varphi_k(x)=\sum_{j=0}^{n_k-1}(c_{-1,k-1}e^{-i\nu_{k-1}x})^j$$

and define

$$B_k^+(x) = \left(\begin{array}{c|c} c_{1,k-1}^{n_k} & -c_{1,k-1}^{n_k} e^{i\lambda_{k-1}x} \sum_{j=1}^{n_k} (c_{1,k-1} e^{i\delta_{k-1}x})^{-j} \\ \hline 0 & 1 \end{array}\right),$$

$$B_k^-(x) =$$

$$\left(\begin{array}{c|c} 1 + \psi_k(x) c_{-1,k-1} e^{-i\nu_{k-1}x} & \psi_k(x) c_{1,k-1}^{-n_k} e^{-i\lambda_{k-1}x} \\ \hline -c_{-1,k-1}\psi_k(x) c_{-1,k-1} e^{i(\lambda_{k-1}-2\nu_{k-1})x} & c_{1,k-1}^{-n_k} - c_{-1,k-1}\psi_k(x) c_{1,k-1}^{-n_k} e^{-i\nu_{k-1}x} \end{array}\right)$$

for k odd, and

$$B_k^+(x) = \left(\begin{array}{c|c} c_{-1,k-1}^{-n_k} - c_{-1,k-1}^{-n_k}\varphi_k(x) c_{1,k-1} e^{i\delta_{k-1}x} & c_{-1,k-1}^{-n_k} e^{i\lambda_{k-1}x}\varphi_k(x) \\ \hline -c_{1,k-1} e^{i(-\lambda_{k-1}+2\delta_{k-1})x}\varphi_k(x) c_{1,k-1} & 1 + c_{1,k-1} e^{i\delta_{k-1}x}\varphi_k(x) \end{array}\right),$$

$$B_k^-(x) = \left(\begin{array}{c|c} 1 & -\sum_{j=1}^{n_k} (c_{-1,k-1} e^{-i\nu_{k-1}x})^{-j} c_{-1,k-1}^{n_k} e^{-i\lambda_{k-1}x} \\ \hline 0 & c_{-1,k-1}^{n_k} \end{array}\right)$$

for k even. Equality (15.17) can be verified straightforwardly. It is obvious that $\det B_k^+(x)$ is a nonzero constant for k odd and that $\det B_k^-(x)$ is a nonzero constant for k even. From (15.17) we infer that $\det B_k^+(x) \det B_k^-(x) = 1$ for all k, which implies that $\det B_k^+(x)$ and $\det B_k^-(x)$ are nonzero constants also for k even and k odd, respectively.

Let k be odd. From (15.13) and (15.10) we get

$$\lambda_{k-1} - n_k\delta_{k-1} > \nu_{k-1} - n_k\delta_{k-1} > 0. \tag{15.19}$$

This shows that $B_k^+ \in APW_{2\times 2}^+$ and that

$$M(B_k^+) = \begin{pmatrix} c_{1,k-1}^{n_k} & 0 \\ 0 & 1 \end{pmatrix} = \begin{pmatrix} a_k^{-1} a_{k-1} & 0 \\ 0 & 1 \end{pmatrix},$$

the latter equality resulting from (15.14). Again by (15.10), (15.12), (15.13),

$$\begin{aligned} (n_k - 1)\delta_{k-1} - \nu_{k-1} &= (n_k\delta_{k-1} - \nu_{k-1}) - \delta_{k-1} < 0, \\ (n_k - 1)\delta_{k-1} - \lambda_{k-1} &= (n_k\delta_{k-1} - \nu_{k-1}) - 2\delta_{k-1} < 0, \qquad (15.20) \\ (n_k - 1)\delta_{k-1} + \lambda_{k-1} - 2\nu_{k-1} &= (n_k - 1)\delta_{k-1} + \delta_{k-1} + \nu_{k-1} - 2\nu_{k-1} \\ &= n_k\delta_{k-1} - \nu_{k-1} < 0, \qquad (15.21) \end{aligned}$$

whence $B_k^- \in APW_{2\times 2}^-$ and

$$M(B_k^-) = \begin{pmatrix} 1 & 0 \\ 0 & c_{1,k-1}^{-n_k} \end{pmatrix} = \begin{pmatrix} 1 & 0 \\ 0 & a_{k-1}^{-1} a_k \end{pmatrix}.$$

Analogously one can show that $B_k^\pm \in APW_{2\times 2}^\pm$ for k even and that in this case (15.18) holds as well. □

Fix any argument $\arg c_{1,0}$ of $c_{1,0}$ and define

$$c_{1,0}^{\alpha}=|c_{1,0}|^{\alpha}e^{i\alpha\arg c_{1,0}}$$

for $\alpha\in\mathbf{R}$. Put

$$z=c_{1,0}^{\beta}c_{-1,0}. \tag{15.22}$$

Since $c_{1,0}^{p_k}=(c_{1,0}^{\beta})^{q_k}c_{1,0}^{p_k-\beta q_k}$, we obtain from (15.15) and (15.16) that

$$c_{(-1)^k,k+1}=c_{(-1)^k,k}=z^{q_k}c_{1,0}^{p_k-\beta q_k}, \tag{15.23}$$

$$a_k=c_{1,0}^{-1}c_{-1,0}\left(z^{q_k-q_{k-1}}c_{1,0}^{\eta_k}\right)^{(-1)^k}, \tag{15.24}$$

where $\eta_k:=p_k-p_{k-1}-\beta(q_k-q_{k-1})$. Taking into consideration (15.8) and the inequalities $q_k-q_{k-1}\geq q_{k-2}$ (which follow from (15.7)), we see that the exponents in (15.23) and (15.24) have the following behavior as $k\to\infty$:

$$p_k-\beta q_k\to 0,\quad \eta_k\to 0,\quad q_k\to\infty,\quad q_k-q_{k-1}\to\infty. \tag{15.25}$$

Lemma 15.2. *If $|z|<1$, then the matrix function G given by* (15.1) *has a canonical APW factorization and*

$$\tilde{\mathbf{d}}(G)=\begin{pmatrix}0 & c_0^{-1}\\ -c_0 & 0\end{pmatrix}.$$

Proof. From(15.23) and (15.25) we see that

$$c_{\pm1,k}\to 0 \text{ as } k\to\infty. \tag{15.26}$$

Rewrite (15.3) in the form

$$G_k(x)=\begin{pmatrix}e^{i\lambda_k x} & 1\\ -1 & 0\end{pmatrix}F_k(x)\begin{pmatrix}1 & -e^{-i\lambda_k x}\\ 0 & 1\end{pmatrix}$$

with

$$F_k(x)=\begin{pmatrix}1-f_k(x) & -e^{-i\lambda_k x}f_k(x)\\ e^{i\lambda_k x}f_k(x) & 1+f_k(x)\end{pmatrix}$$

and $f_k(x):=c_{-1,k}e^{-i\nu_k x}+c_{1,k}e^{i\delta_k x}$. By (15.26), F_k converges to I in $APW_{2\times2}$. Applying Proposition 8.25(a) to JF_k^TJ (recall Proposition 13.1), we conclude that F_k has a canonical APW factorization whenever k is sufficiently large. Since

$$\Omega(F_k-I)=\{-\nu_k+l\lambda_k : l=-1,0,1,2\}$$

is linearly independent over $\mathbf{Z}_+$, Proposition 8.25(c) and Proposition 13.1 yield

$$\tilde{\mathbf{d}}(F_k)=J\mathbf{d}(JF_k^TJ)J=JM(JF_k^TJ)J=JIJ=I$$

and thus

$$\tilde{\mathbf{d}}(G_k) = \begin{pmatrix} 0 & 1 \\ -1 & 0 \end{pmatrix} \tilde{\mathbf{d}}(F_k) \begin{pmatrix} 1 & 0 \\ 0 & 1 \end{pmatrix} = \begin{pmatrix} 0 & 1 \\ -1 & 0 \end{pmatrix}$$

for all k large enough.

From Lemma 15.1 we now deduce that G has a canonical APW factorization. Combining (15.2) and (15.17) we get

$$G = \begin{pmatrix} 1 & 0 \\ 0 & c_0 \end{pmatrix} (B_1^+)^{-1} \dots (B_k^+)^{-1} G_k (B_k^-)^{-1} \dots (B_1^-)^{-1} \begin{pmatrix} 1 & 0 \\ 0 & c_0^{-1} \end{pmatrix}, \tag{15.27}$$

and taking into account (15.18) we finally arrive at the equality

$$\begin{aligned}
\tilde{\mathbf{d}}(G) &= \begin{pmatrix} 1 & 0 \\ 0 & c_0 \end{pmatrix} \begin{pmatrix} a_1 a_0^{-1} & 0 \\ 0 & 1 \end{pmatrix} \cdots \begin{pmatrix} a_k a_{k-1}^{-1} & 0 \\ 0 & 1 \end{pmatrix} \begin{pmatrix} 0 & 1 \\ -1 & 0 \end{pmatrix} \times \\
&\quad \times \begin{pmatrix} 1 & 0 \\ 0 & a_{k-1} a_k^{-1} \end{pmatrix} \cdots \begin{pmatrix} 1 & 0 \\ 0 & a_0 a_1^{-1} \end{pmatrix} \begin{pmatrix} 1 & 0 \\ 0 & c_0^{-1} \end{pmatrix} \\
&= \begin{pmatrix} 1 & 0 \\ 0 & c_0 \end{pmatrix} \begin{pmatrix} a_k & 0 \\ 0 & 1 \end{pmatrix} \begin{pmatrix} 0 & 1 \\ -1 & 0 \end{pmatrix} \begin{pmatrix} 1 & 0 \\ 0 & a_k^{-1} \end{pmatrix} \begin{pmatrix} 1 & 0 \\ 0 & c_0^{-1} \end{pmatrix} \\
&= \begin{pmatrix} 0 & c_0^{-1} \\ -c_0 & 0 \end{pmatrix}.
\end{aligned}$$

□

Lemma 15.3. *If* $|z| > 1$, *then the matrix function* G *defined by* (15.1) *has a canonical* APW *factorization and*

$$\tilde{\mathbf{d}}(G) = \begin{pmatrix} -c_1^{-1} c_{-1} & 0 \\ -2c_0 & -c_1 c_{-1}^{-1} \end{pmatrix}. \tag{15.28}$$

Proof. We now write $G_k = A_k^+ F_k A_k^-$ where

$$\begin{aligned}
F_k(x) &= \begin{pmatrix} 1 & 0 \\ 0 & 1 \end{pmatrix} - \begin{pmatrix} c_{1,k}^{-1} & 0 \\ 0 & c_{1,k}^{-1} \end{pmatrix} \begin{pmatrix} -e^{i\nu_k x} \\ 1 - c_{-1,k}^{-1} c_{1,k} e^{i\lambda_k x} \end{pmatrix} \times \\
&\quad \times (e^{-i\lambda_k x} - c_{-1,k}^{-1} c_{1,k},\ e^{-i\delta_k x}), \\
A_k^+(x) &= \begin{pmatrix} 1 & 0 \\ 0 & c_{1,k} \end{pmatrix} \begin{pmatrix} 1 - c_{-1,k}^{-1} c_{1,k} e^{i\lambda_k x} & e^{i\nu_k x} \\ -c_{-1,k}^{-1} c_{1,k} e^{i\delta_k x} & 1 \end{pmatrix}, \\
A_k^-(x) &= \begin{pmatrix} -c_{1,k}^{-1} c_{-1,k} & -e^{-i\delta_k x} \\ c_{1,k}^{-1} c_{-1,k} e^{-i\nu_k x} & e^{-i\lambda_k x} - c_{-1,k}^{-1} c_{1,k} \end{pmatrix} \begin{pmatrix} 1 & 0 \\ 0 & c_{1,k}^{-1} \end{pmatrix}.
\end{aligned}$$

It can be directly verified that $A_k^\pm \in APW_{2\times 2}^\pm$ have nonzero determinants. From (15.23) we obtain that

$$c_{\pm 1,k}^{-1} \to 0 \text{ as } k \to \infty. \tag{15.29}$$

Equality (15.23) shows that if k is odd, then

$$c_{-1,k}^{-1}c_{1,k}=z^{-(q_k-q_{k-1})}c_{1,0}^{-\eta_k},$$

whence

$$c_{-1,2j-1}^{-1}c_{1,2j-1}\to 0 \text{ as } j\to\infty. \tag{15.30}$$

Consequently, F_{2j-1} converges to I in $APW_{2\times 2}$ and therefore has a canonical APW factorization for all sufficiently large j by virtue of Proposition 8.25 (a). Lemma 15.1 now implies that G also has a canonical APW factorization.

We are left with computing $\tilde{\mathbf{d}}(G)$. Let k be a sufficiently large odd number. Since $\Omega(F_k)$ contains $\pm\lambda_k$, Proposition 8.25 (c) is unfortunately not applicable to the case at hand. Instead, we make use of formula (8.30), which in the case of left canonical factorization reads

$$\tilde{\mathbf{d}}(a)=\Big(\sum_{j=0}^{\infty}M(R_c^jI)\Big)^{-1},\quad R_cx=\tilde{P}(xc).$$

In our situation, we have

$$\tilde{\mathbf{d}}(F_k)=\Big(\sum_{j=0}^{\infty}M(X_j)\Big)^{-1},\quad X_j:=R_{I-F_k}^jI,\quad R_{I-F_k}X=\tilde{P}\big(X(I-F_k)\big).$$

Induction on j shows that

$$X_j=\begin{pmatrix}\sum_{l=0}^{j}a_{jl}e^{i(-j\delta_k+l\lambda_k)x} & \sum_{l=1}^{j}b_{jl}e^{i(-(j+1)\delta_k+l\lambda_k)x}\\ \sum_{l=-1}^{j}c_{jl}e^{i(-(j-1)\delta_k+l\lambda_k)x} & \sum_{l=0}^{j}d_{jl}e^{i(-j\delta_k+l\lambda_k)x}\end{pmatrix}.$$

Thus,

$$\Omega(X_j)\subset\Big\{-j\delta_k+l\lambda_k+\varepsilon(\lambda_k-\delta_k):\varepsilon\in\{-1,0,1\},\ l\in\{0,\dots,j+1\}\Big\}.$$

Since λ_k/δ_k is irrational (which can also be proved by induction on k), the origin cannot belong to $\Omega(X_j)$ for $j>1$. This gives

$$\begin{aligned}\tilde{\mathbf{d}}(F_k)&=(I-M(X_1))^{-1}=\Big(I+M\big(\tilde{P}(I-F_k)\big)\Big)^{-1}\\ &=(I+I-M(F_k))^{-1}=\begin{pmatrix}1&0\\-2c_{-1,k}^{-1}&1\end{pmatrix}^{-1}=\begin{pmatrix}1&0\\2c_{-1,k}^{-1}&1\end{pmatrix}.\end{aligned}$$

So

$$\tilde{\mathbf{d}}(G_k)=M(A_k^+)\tilde{\mathbf{d}}(F_k)M(A_k^-)=\begin{pmatrix}-c_{1,k}^{-1}c_{-1,k}&0\\-2&-c_{1,k}c_{-1,k}^{-1}\end{pmatrix},$$

and from (15.27) we finally obtain

$$\begin{aligned}\tilde{\mathbf{d}}(G) &= \begin{pmatrix} 1 & 0 \\ 0 & c_0 \end{pmatrix}\begin{pmatrix} a_k & 0 \\ 0 & 1 \end{pmatrix}\tilde{\mathbf{d}}(G_k)\begin{pmatrix} 1 & 0 \\ 0 & a_k^{-1} \end{pmatrix}\begin{pmatrix} 1 & 0 \\ 0 & c_0^{-1} \end{pmatrix} \\ &= \begin{pmatrix} -a_k c_{1,k}^{-1} c_{-1,k} & 0 \\ -2c_0 & -a_k^{-1} c_{1,k} c_{-1,k}^{-1} \end{pmatrix}.\end{aligned}$$

Using (15.9) and (15.14), one can show by induction on k that

$$a_k c_{1,k}^{-1} c_{-1,k} = c_{1,0}^{-1} c_{-1,0} \quad (= c_1^{-1} c_{-1}),$$

and hence we arrive at (15.28). □

Lemma 15.4. *If* $|z| = 1$, *then the matrix function* G *given by* (15.1) *has no AP factorization.*

Proof. Contrary to what we want, let us assume that G has an AP factorization. Then, by Lemma 15.1, all G_k have an AP factorization with the same AP indices as G. By Proposition 8.26, the AP indices are included between $-\lambda_k$ and λ_k.

We claim that $\lambda_k \to 0$ as $k \to \infty$. Indeed, from (15.9) and (15.12) we infer that $\{\nu_k\}$, $\{\delta_k\}$, $\{\lambda_k\}$ are monotoneously decreasing sequences of positive numbers. Hence, all three sequences are convergent. Denote the limits by ν_∞, δ_∞, λ_∞. By (15.9),

$$\begin{aligned} n_{2j+1}\delta_{2j} &= \nu_{2j} - \nu_{2j+1} \to \nu_\infty - \nu_\infty = 0, \\ n_{2j}\nu_{2j-1} &= \delta_{2j-1} - \delta_{2j} \to \delta_\infty - \delta_\infty = 0, \end{aligned}$$

whence $\delta_\infty = \nu_\infty = 0$. As $\lambda_k = \nu_k + \delta_k$ for all k by (15.13), it follows that $\lambda_\infty = 0$, which proves our claim.

From what was said in the preceding two paragraphs, we conclude that the AP indices of G are zero. Thus, G has a canonical AP factorization. Corollary 10.7 now implies that G has even a canonical APW factorization. For $\theta \in \mathbf{R}$, put

$$G^{(\theta)} := \begin{pmatrix} e^{i\lambda x} & 0 \\ (1+\theta)c_{-1}e^{-i\nu x} - c_0 + c_1 e^{i\delta x} & e^{-i\lambda x} \end{pmatrix}.$$

By Proposition 8.25(a),(b), the matrix functions $G^{(\theta)}$ have a canonical APW factorization for all θ close enough to zero, and $\tilde{\mathbf{d}}(G^{(\theta)})$ is a continuous function of θ. In accordance with (15.22), let

$$z_\theta := (c_0^{-1}c_1)^\beta (c_0^{-1}(1+\theta)c_{-1}) = (1+\theta)z.$$

If $\theta \in (-1, 0)$, then Lemma 15.2 shows that

$$\tilde{\mathbf{d}}(G^{(\theta)}) = \begin{pmatrix} 0 & c_0^{-1} \\ -c_0 & 0 \end{pmatrix},$$

while for $\theta\in(0,\infty)$ we deduce from Lemma 15.3 that

$$\tilde{\mathbf{d}}(G^{(\theta)})=\begin{pmatrix} -c_1^{-1}c_{-1}(1+\theta) & 0 \\ -2c_0 & -c_1c_{-1}^{-1}(1+\theta)^{-1}\end{pmatrix}.$$

Consequently,

$$\lim_{\theta\to 0-0}\tilde{\mathbf{d}}(G^{(\theta)})=\begin{pmatrix} 0 & c_0^{-1} \\ -c_0 & 0\end{pmatrix}\neq\begin{pmatrix} -c_1^{-1}c_{-1} & 0 \\ -2c_0 & -c_1c_{-1}^{-1}\end{pmatrix}=\lim_{\theta\to 0+0}\tilde{\mathbf{d}}(G^{(\theta)}).$$

We see that $\tilde{\mathbf{d}}(G^{(\theta)})$ is discontinuous at $\theta=0$. This contradiction tells us that G cannot have an AP factorization. □

We are now in a position to establish the main result of this section.

Theorem 15.5. *Let G be the matrix function* (15.1), *suppose $\nu>0$, $\delta>0$, $\nu+\delta=\lambda$, $c_{-1}c_0c_1\neq 0$, and ν/δ is irrational. Then G has an AP factorization if and only if*

$$|c_1|^{\nu}|c_{-1}|^{\delta}\neq|c_0|^{\lambda}. \tag{15.31}$$

If (15.31) *is satisfied, then G has a canonical APW factorization and*

$$\tilde{\mathbf{d}}(G)=\begin{pmatrix} -\omega c_1^{-1}c_{-1} & (1-\omega)c_0^{-1} \\ -(1+\omega)c_0 & -\omega c_1c_{-1}^{-1}\end{pmatrix}$$

where

$$\omega:=\begin{cases} 0 & \text{if}\ \ |c_1|^{\nu}|c_{-1}|^{\delta}<|c_0|^{\lambda},\\ 1 & \text{if}\ \ |c_1|^{\nu}|c_{-1}|^{\delta}>|c_0|^{\lambda}.\end{cases}$$

Proof. From Lemmas 15.2 to 15.4 we know that G is AP factorable if and only if

$$\begin{aligned}1\neq|z|&=|c_{1,0}|^{\beta}|c_{-1,0}|=|c_0^{-1}c_1|^{\beta}|c_0^{-1}c_{-1}|\\ &=|c_0|^{-(\beta+1)}|c_1|^{\beta}|c_{-1}|=\left(|c_0|^{-\lambda}|c_1|^{\nu}|c_{-1}|^{\delta}\right)^{1/\delta},\end{aligned}$$

which is equivalent to (15.31). The remaining assertions are immediate from Lemmas 15.2 and 15.3. □

Taking $c_{-1}=c_0=c_1=1$, $\nu=(\sqrt{5}-1)/2$, $\delta=1$, $\lambda=(\sqrt{5}+1)/2$, we get the matrix function considered in Theorem 8.17. We there proved its non-factorability by a different argument.

15.2 $\nu+\delta>\lambda,\ \mu=0$

Let G be given by (15.1), where $c_{-1}c_0c_1\neq 0$, $\nu>0$, $\delta>0$, and $\beta:=\nu/\delta$ is irrational. Throughout this section we suppose that $\lambda<\nu+\delta$.

Define $n_k, \nu_k, \delta_k, \lambda_k, c_{-1,k}, c_{1,k}, a_k$ as in the previous section. In addition, put

$$\gamma_{-1} = \nu, \qquad \gamma_k = \begin{cases} \nu_k & \text{if } k \text{ is odd,} \\ \delta_k & \text{if } k \text{ is even,} \end{cases}$$

$$\varrho_0 = \nu, \qquad \varrho_k = \gamma_{k-1} + \sum_{l=1}^{k} n_l \gamma_{l-1}.$$

By induction on k, it is easy to show that

$$\gamma_k = \gamma_{k-2} - n_k \gamma_{k-1}, \tag{15.32}$$

$$\gamma_{k-1} + \gamma_k = \nu_k + \delta_k, \tag{15.33}$$

$$\lambda_k = \lambda + \gamma_{k-1} - \varrho_k, \tag{15.34}$$

$$\delta_k + \nu_k = \delta + \nu + \gamma_{k-1} - \varrho_k. \tag{15.35}$$

Lemma 15.6. *If $\lambda < \nu + \delta$, then there is a number $m \in \mathbf{Z}_+$ such that $\lambda \leq \varrho_m$.*

Proof. We first show that there is an $m \in \mathbf{Z}_+$ such that $\lambda_m \leq \gamma_{m-1}$. Assume the contrary, i.e., let $\lambda_m > \gamma_{m-1}$ for all $m \in \mathbf{Z}_+$. Then, by (15.34),

$$\nu + \delta > \lambda = \lambda_m - \gamma_{m-1} + \varrho_m > \varrho_m \tag{15.36}$$

for all $m \in \mathbf{Z}_+$. From (15.35) and (15.33) we see that the difference of the outer terms in (15.36) is

$$\nu + \delta - \varrho_m = \delta_m + \nu_m - \gamma_{m-1} = \gamma_m,$$

and in the proof of Lemma 15.4 we observed that $\gamma_m \to 0$ as $m \to \infty$ (this piece of the proof did not make use of the equality $\nu + \delta = \lambda$), which contradicts the inequality $\nu + \delta > \lambda$ in (15.36). Thus, there is an $m \in \mathbf{Z}_+$ such that $\lambda_m \leq \gamma_{m-1}$. From (15.34) we now obtain that $\lambda = \lambda_m - \gamma_{m-1} + \varrho_m \leq \varrho_m$. □

Theorem 15.7. *Let G be the matrix function* (15.1), *let $\nu > 0$, $\delta > 0$, $\nu + \delta > \lambda$, $c_{-1}c_0c_1 \neq 0$, suppose ν/δ is irrational. Then*

(a) *the matrix function G has an APW factorization;*

(b) *the AP indices of G are zero if and only if*

$$\gamma_m = 0 \textit{ and } c_{(-1)^m,m} \neq 1 \tag{15.37}$$

or

$$\gamma_m > 0, \tag{15.38}$$

where $m \in \mathbf{Z}_+$ is the smallest number for which $\lambda \leq \varrho_m$.

(c) *If* (15.37) *is satisfied, then*

$$\tilde{\mathbf{d}}(G) = \begin{pmatrix} 0 & (1 - c_{(-1)^m,m})^{-1} c_0^{-1} \\ -c_0(1 - c_{(-1)^m,m}) & 0 \end{pmatrix}, \tag{15.39}$$

and in case (15.38) *holds, we have*

$$\tilde{\mathbf{d}}(G)=\begin{pmatrix} a_m c_{-1,m}^{\varkappa_m}\omega(\varkappa_m) & c_0^{-1} \\ c_0 c_{-1,m}^{\varkappa_m} c_{1,m}\omega(\varkappa_m)\omega(\varkappa_m')-c_0 & a_m^{-1}\omega(\varkappa_m')c_{1,m} \end{pmatrix} \tag{15.40}$$

if m is odd and

$$\tilde{\mathbf{d}}(G)=\begin{pmatrix} a_m c_{-1,m}\omega(\varkappa_m') & c_0^{-1} \\ c_0 c_{1,m}^{\varkappa_m} c_{-1,m}\omega(\varkappa_m)\omega(\varkappa_m')-c_0 & a_m^{-1}\omega(\varkappa_m)c_{1,m}^{\varkappa_m} \end{pmatrix} \tag{15.41}$$

if m is even, where

$$\varkappa_m:=\frac{\lambda_m}{\gamma_m},\quad \varkappa_m':=\frac{\lambda_m}{\gamma_{m-1}},\quad \omega(\varkappa):=\begin{cases} 1 & \textit{for} \quad \varkappa\in\mathbf{N}, \\ 0 & \textit{for} \quad \varkappa\notin\mathbf{N}. \end{cases}$$

Proof. Define G_k by (15.2) and (15.3). Let $m>0$. We first show that Lemma 15.1 is also true under the assumption $\nu+\delta>\lambda$ provided $k\le m$. Except for the inequalities (15.19), (15.20) and (15.21), all arguments of the proof of Lemma 15.1 can be literally applied to the case at hand. Thus, we are left with (15.19),(15.20) and (15.21). Let k be odd. We have $\lambda_0=\lambda<\delta+\nu=\gamma_0+\gamma_{-1}$. If $\lambda_{k-1}\le\gamma_{k-1}+\gamma_{k-2}$ then by (15.9) and (15.32),

$$\lambda_k=\lambda_{k-1}-n_k\gamma_{k-1}\le\gamma_{k-1}+\gamma_{k-2}-n_k\gamma_{k-1}=\gamma_k+\gamma_{k-1},$$

which shows that $\lambda_k\le\gamma_k+\gamma_{k-1}$ for all k. From (15.33) we therefore get $\nu_k+\delta_k\ge\lambda_k$ for all k. The latter inequality in conjunction with (15.10) proves (15.21):

$$\begin{aligned}(n_k-1)\delta_{k-1}+\lambda_{k-1}-2\nu_{k-1}&=n_k\delta_{k-1}-\delta_{k-1}+\lambda_{k-1}-\nu_{k-1}-\nu_{k-1}\\ &\le n_k\delta_{k-1}-\nu_{k-1}<0.\end{aligned}$$

If $k\le m$, then $\lambda>\varrho_{k-1}$. From (15.34) and (15.32) we get

$$\lambda_{k-1}=\lambda+\gamma_{k-2}-\varrho_{k-1}>\gamma_{k-2}=\gamma_k+n_k\gamma_{k-1},$$

and since $\gamma_{k-1}=\delta_{k-1}$, it follows that $\lambda_{k-1}-n_k\delta_{k-1}>\gamma_k\ge 0$, which is (15.19). Finally, from (15.19) and (15.12) we obtain (15.20):

$$(n_k-1)\delta_{k-1}-\lambda_{k-1}=(n_k\delta_{k-1}-\lambda_{k-1})-\delta_{k-1}<0.$$

The case where k is even can be treated analogously.

Thus, we can employ Lemma 15.1 for $k\le m$ also if $\lambda<\nu+\delta$ and $m>0$. It results that G is APW factorable if and only if so is

$$G_m(x):=\begin{pmatrix} e^{i\lambda_m x} & 0 \\ c_{-1,m}e^{-i\nu_m x}-1+c_{1,m}e^{i\delta_m x} & e^{-i\lambda_m x} \end{pmatrix},$$

that G and G_m have the same AP indices, and that

$$\tilde{\mathbf{d}}(G) = \begin{pmatrix} 1 & 0 \\ 0 & c_0 \end{pmatrix} \begin{pmatrix} a_m & 0 \\ 0 & 1 \end{pmatrix} \tilde{\mathbf{d}}(G_m) \begin{pmatrix} 1 & 0 \\ 0 & a_m^{-1} \end{pmatrix} \begin{pmatrix} 1 & 0 \\ 0 & c_0^{-1} \end{pmatrix} \tag{15.42}$$

(see the proof of Lemma 15.2 for the last equality). For $m = 0$, the conclusions of this paragraph are trivial.

(a) Taking into account that $\nu_m = \gamma_m$, $\delta_m = \gamma_{m-1}$ if m is odd and $\nu_m = \gamma_{m-1}$, $\delta_m = \gamma_m$ if m is even, one can straightforwardly verify that

$$G_m = C_m^+ H_m C_m^- \tag{15.43}$$

where

$$\begin{aligned} C_m^+(x) &= \begin{pmatrix} 1 & 0 \\ \frac{1-(-1)^m}{2} c_{1,m} e^{i(\gamma_{m-1}-\lambda_m)x} & 1 \end{pmatrix}, \\ C_m^-(x) &= \begin{pmatrix} 1 & 0 \\ \frac{1+(-1)^m}{2} c_{-1,m} e^{i(\lambda_m-\gamma_{m-1})x} & 1 \end{pmatrix}, \\ H_m(x) &= \begin{pmatrix} e^{i\lambda_m x} & 0 \\ g_m(x) & e^{-i\lambda_m x} \end{pmatrix}, \ g_m(x) = \begin{cases} -1 + c_{-1,m} e^{-i\gamma_m x} & \text{for } m \text{ odd}, \\ -1 + c_{1,m} e^{i\gamma_m x} & \text{for } m \text{ even}. \end{cases} \end{aligned}$$

By (15.34) and the definition of m,

$$\lambda_m = \lambda + \gamma_{m-1} - \varrho_m \le \gamma_{m-1}, \tag{15.44}$$

which shows that $C_m^\pm \in GAPW_{2\times 2}^\pm$. As g_m is a binomial, we deduce from Theorem 14.9 that H_m has an APW factorization. Consequently, G_m and thus G are also APW factorable.

(b) The AP indices of G are zero if and only if H_m has zero AP indices. If $\gamma_m = 0$, then

$$H_m(x) = \begin{pmatrix} e^{i\lambda_m x} & 0 \\ c_{(-1)^m,m} - 1 & e^{-i\lambda_m x} \end{pmatrix}.$$

The AP indices of H_m are $\pm\lambda_m$ in case $b := c_{(-1)^m,m} - 1 = 0$. If $b \ne 0$, a canonical APW factorization is

$$H_m(x) = \begin{pmatrix} e^{i\lambda_m x} & 1 \\ b & 0 \end{pmatrix} \begin{pmatrix} 1 & b^{-1} e^{-i\lambda_m x} \\ 0 & -b^{-1} \end{pmatrix},$$

whence

$$\tilde{\mathbf{d}}(H_m) = \begin{pmatrix} 0 & 1 \\ b & 0 \end{pmatrix} \begin{pmatrix} 1 & 0 \\ 0 & -b^{-1} \end{pmatrix} = \begin{pmatrix} 0 & -b^{-1} \\ b & 0 \end{pmatrix}. \tag{15.45}$$

If $\gamma_m > 0$, we are in the cases (a) and (b) of Theorem 14.10, and hence H_m has zero AP indices. This completes the proof of (b).

(c) Suppose (15.37) holds. Then, by (15.35) and (15.33),

$$\nu+\delta=\nu_m+\delta_m-\gamma_{m-1}+\varrho_m=\gamma_m+\varrho_m=\varrho_m,$$

and therefore (15.34) implies that

$$\lambda<\nu+\delta=\varrho_m=\lambda-\lambda_m+\gamma_{m-1},$$

which gives $\lambda_m<\gamma_{m-1}$. Thus

$$M(C_m^+)=M(C_m^-)=\begin{pmatrix}1&0\\0&1\end{pmatrix},$$

and from (15.43) and (15.45) we obtain

$$\tilde{\mathbf{d}}(G_m)=\tilde{\mathbf{d}}(H_m)=\begin{pmatrix}0&-b^{-1}\\b&0\end{pmatrix}.$$

Inserting this into (15.42) yields

$$\tilde{\mathbf{d}}(G)=\begin{pmatrix}a_m&0\\0&c_0\end{pmatrix}\begin{pmatrix}0&-b^{-1}\\b&0\end{pmatrix}\begin{pmatrix}1&0\\0&a_m^{-1}c_0^{-1}\end{pmatrix}=\begin{pmatrix}0&-b^{-1}c_0^{-1}\\c_0b&0\end{pmatrix},$$

which is (15.39).

Now suppose (15.38) is satisfied. Let m be odd. From case (b) of Theorem 14.10 we infer that

$$\tilde{\mathbf{d}}(H_m)=\begin{pmatrix}c_{-1,m}^{\varkappa_m}\omega(\varkappa_m)&1\\-1&0\end{pmatrix}$$

where $\varkappa_m:=\lambda_m/\gamma_m$. Because $1+(-1)^m=0$, we have $M(C_m^-)=I$. As

$$M(C_m^+)=\begin{pmatrix}1&0\\0&1\end{pmatrix}\ \text{ if }\gamma_{m-1}>\lambda_m,\quad M(C_m^+)=\begin{pmatrix}1&0\\c_{1,m}&1\end{pmatrix}\ \text{ if }\gamma_{m-1}=\lambda_m,$$

we can write

$$M(C_m^+)=\begin{pmatrix}1&0\\c_{1,m}\omega(\varkappa_m')&1\end{pmatrix}$$

where $\varkappa_m':=\lambda_m/\gamma_{m-1}$. This gives

$$\tilde{\mathbf{d}}(G_m)=M(C_m^+)\tilde{\mathbf{d}}(H_m)=\begin{pmatrix}c_{-1,m}^{\varkappa_m}\omega(\varkappa_m)&1\\c_{-1,m}^{\varkappa_m}c_{1,m}\omega(\varkappa_m)\omega(\varkappa_m')-1&\omega(\varkappa_m')c_{1,m}\end{pmatrix}.$$

Substituting this matrix into (15.42) we arrive at (15.40). Formula (15.41) can be shown analogously. □

15.3 $\nu + \delta = \lambda,\ \mu \neq 0$

We now consider functions of the form

$$G(x) := G_f^{(\lambda)}(x) := \begin{pmatrix} e^{i\lambda x} & 0 \\ c_{-1}e^{-i\nu x} - c_0 e^{i\mu x} + c_1 e^{i\delta x} & e^{-i\lambda x} \end{pmatrix}, \quad \lambda > 0, \tag{15.46}$$

under the assumption that $c_{-1}c_0c_1 \neq 0$, $\nu > 0$, $\delta > 0$, $\nu + \delta = \lambda$, $\mu \neq 0$, $-\nu < \mu < \delta$. We also assume that $\mu > 0$, since for $\mu < 0$ we can pass to the matrix function $G_{\overline{f}}^{(\lambda)} = JG^*J$ with

$$\overline{f}(x) = \overline{c}_1 e^{-i\delta x} - \overline{c}_0 e^{i|\mu|x} + \overline{c}_1 e^{i\nu x}$$

(recall Proposition 13.2). We also suppose that

$$\beta := \frac{\delta - \mu}{\nu + \mu}$$

is an irrational number, as otherwise we are in the case of commensurable distances (Section 14.4).

Theorem 15.8. *Under the above assumptions, G has an APW factorization.*

Proof. We apply the Portuguese transformation to G in order to reduce the problem to an already understood case. Thus, we replace $G = G_f^{(\lambda)}$ by $G_{f_1}^{(\nu)}$ where $f_1 = P_{(-\nu,\nu)}(g_2 e_{-\lambda})$ and g_2 is given by (13.42): writing

$$(a =)\ e_\nu f = c_{-1} - c_0 e_{\nu+\mu} + c_1 e_{\nu+\delta},$$

we have

$$g_2 = \sum_{(n,\gamma)<\lambda+\nu} c_{-1}^{-1} d_n e_{(n,\gamma)}, \tag{15.47}$$

$$\gamma = (\nu + \mu, \nu + \delta) = (\nu + \mu, \lambda), \quad d_{(n_1,n_2)} = \frac{(n_1+n_2)!}{n_1!n_2!}\left(\frac{c_0}{c_{-1}}\right)^{n_1}\left(-\frac{c_1}{c_{-1}}\right)^{n_2}.$$

Clearly,

$$\Omega(f_1) \subset \{n_1(\nu+\mu) + n_2\lambda - \lambda : (n_1, n_2) \in \mathbf{Z}_+^2\} \cap (-\nu, \nu).$$

If $n_2 \geq 2$, then $n_2\lambda - \lambda \geq \lambda > \nu \notin \Omega(f_1)$. For $n_2 = 1$ and $n_1 \geq 1$, we have $n_1(\nu + \mu) \geq \nu + \mu > \nu \notin \Omega(f_1)$. The choice $n_2 = 1$, $n_1 = 0$ gives the point 0 of $\Omega(f_1)$. So let $n_2 = 0$ and let N be the smallest integer such that $N(\nu+\mu) - \lambda > -\nu$. Obviously,

$$N = 1 + \left[\frac{\lambda - \nu}{\nu + \mu}\right]. \tag{15.48}$$

Evidently, $(N+2)(\nu+\mu)-\lambda > -\nu+2(\nu+\mu)=\nu+2\mu>\nu$. Thus,

$$\Omega(f_1)\subset\big\{N(\nu+\mu)-\lambda,\ 0,\ (N+1)(\nu+\mu)-\lambda\big\}.$$

This shows that f_1 is at worst a trinomial. The distance between $N(\nu+\mu)-\lambda$ and $(N+1)(\nu+\mu)-\lambda$ is $\nu+\mu>\nu$. Hence, if $N(\nu+\mu)-\lambda\geq 0$ or $(N+1)(\nu+\mu)-\lambda\leq 0$, then f_1 is a binomial, and Theorem 14.9 guarantees the *APW* factorability of $G_{f_1}^{(\nu)}$ and thus of $G_f^{(\lambda)}$. If

$$N(\nu+\mu)-\lambda<0<(N+1)(\nu+\mu)-\lambda, \tag{15.49}$$

we are in the case covered by Theorem 15.7: note that

$$-\big(N(\nu+\mu)-\lambda\big)+\big((N+1)(\nu+\mu)-\lambda\big)=\nu+\mu>\nu.$$

Thus, again we obtain that $G_{f_1}^{(\nu)}$ and therefore $G_f^{(\lambda)}$ is *APW* factorable. □

To single out the case of zero *AP* indices, suppose N is given by (15.48) and (15.49) holds. Since

$$\frac{\lambda}{\nu+\mu}=\frac{\nu+\delta}{\nu+\mu}=1+\frac{\delta-\mu}{\nu+\mu},$$

we see that $\lambda/(\nu+\mu)$ is irrational together with $\beta:=(\delta-\mu)/(\nu+\mu)$. Obviously, the irrationality of $\lambda/(\nu+\mu)$ implies that

$$\big(\lambda-N(\nu+\mu)\big)/\big((N+1)(\nu+\mu)-\lambda\big)$$

is also irrational. Therefore, we have an infinite continued fraction

$$\frac{\lambda-N(\nu+\mu)}{(N+1)(\nu+\mu)-\lambda}=n_1+\cfrac{1}{n_2+\cfrac{1}{n_3+\dots}},\quad n_1\in\mathbf{Z},\quad n_k\in\mathbf{N}\ \text{ for } k\geq 2.$$

Define γ_k by

$$\gamma_{-1}=\lambda-N(\nu+\mu),\quad \gamma_0=(N+1)(\nu+\mu)-\lambda,\quad \gamma_k=\gamma_{k-2}-n_k\gamma_{k-1}.$$

Lemma 15.6 shows that if (15.49) is satisfied, then there is a number $k\in\mathbf{Z}_+$ such that

$$\nu\leq\gamma_{k-1}+\sum_{l=1}^{k}n_l\gamma_{l-1}.$$

Let m be the smallest of these numbers k. Since $N(\nu+\mu)-\lambda>-\nu$, we have $\nu>\gamma_{-1}$ and thus $m>0$. Finally, put

$$c_{-1,0}=\frac{c_{-1}}{c_1}\left(\frac{c_0}{c_{-1}}\right)^N,\quad c_{1,0}=\frac{c_{-1}}{c_1}\left(\frac{c_0}{c_{-1}}\right)^{N+1}$$

and define $c_{\pm1,k}$ by (15.9).

Theorem 15.9. *Under the assumptions made in the beginning of this section, the AP indices of G are all zero if and only if one of the following three conditions is satisfied:*

$$(a)\ N(\nu+\mu) \geq \lambda - \mu, \qquad (b)\ \gamma_m = 0 \text{ and } c_{(-1)^m,m} \neq 1, \quad (c)\ \gamma_m > 0.$$

Proof. If $N(\nu+\mu) \geq \lambda-\mu$, then $(N+1)(\nu+\mu)-\lambda \geq \nu \notin \Omega(f_1)$, so that f_1 is a binomial and $\Omega(f_1) \subset \{N(\nu+\mu)-\lambda, 0\}$. Theorem 14.10 therefore shows that the AP indices of $G_{f_1}^{(\nu)}$ and thus those of $G_f^{(\lambda)}$ are zero.
So assume $N(\nu+\mu) < \lambda-\mu$. Then $N(\nu+\mu)-\lambda < 0$, and as $N(\nu+\mu)-\lambda > -\nu$ by the construction of N, we have

$$(N+1)(\nu+\mu)-\lambda > -\nu+\nu+\mu = \mu > 0.$$

Moreover, $(N+1)(\nu+\mu)-\lambda < -\mu+\nu+\mu = \nu$. Thus, (15.49) holds. From (15.47) we get

$$\begin{aligned} f_1 &= P_{(-\nu,\nu)}(g_2 e_{-\lambda}) \\ &= c_{-1}^{-1} d_{(N,0)} e_{N(\nu+\mu)-\lambda} + c_{-1}^{-1} d_{(0,1)} e_0 + c_{-1}^{-1} d_{(N+1,0)} e_{(N+1)(\nu+\mu)-\lambda} \\ &= \frac{1}{c_{-1}}\left(\frac{c_0}{c_{-1}}\right)^N e_{N(\nu+\mu)-\lambda} - \frac{c_1}{c_{-1}^2} e_0 + \frac{1}{c_{-1}}\left(\frac{c_0}{c_{-1}}\right)^{N+1} e_{(N+1)(\nu+\mu)-\lambda}. \end{aligned}$$

Theorem 15.7 now shows that the AP indices of $G_{f_1}^{(\nu)}$ and thus of $G_f^{(\lambda)}$ equal zero if and only if (b) and (c) hold. □

15.4 $\nu+\delta > \lambda$, $\mu \neq 0$

Suppose G is given by (15.46) with $c_{-1}c_0c_1 \neq 0$,

$$-\lambda < -\nu < 0, \quad 0 < \mu < \lambda, \quad 0 < \delta < \lambda, \quad \lambda < \nu + \delta.$$

We now do not require that $\mu < \delta$ and we also do not demand that $(\delta-\mu)/(\nu+\mu)$ be irrational. Put

$$\begin{aligned} \lambda_0 &:= \lambda, \quad \nu_0 := \nu, \quad \mu_0 := \mu, \quad \delta_0 := \delta, \\ N_k &:= 1 + \left[\frac{\lambda_k - \nu_k}{\nu_k + \mu_k}\right], \\ \lambda_{k+1} &:= \nu_k, \\ \nu_{k+1} &:= -N_k(\nu_k+\mu_k) + \lambda_k, \\ \mu_{k+1} &:= \nu_k + \delta_k - \lambda_k, \\ \delta_{k+1} &:= (N_k+1)(\nu_k+\mu_k) - \lambda_k. \end{aligned}$$

Theorem 15.10. *Let the above assumptions be satisfied. Then G has an AP factorization. Moreover, there exist natural numbers k such that $\delta_k \geq \lambda_k$, and if m is the smallest of these numbers, then the AP indices of G are*

$$\pm\min\left\{\nu_m, \mu_m, \min_{j\in\mathbf{N}} |j(\nu_m+\mu_m)-\lambda_m|\right\} \quad \text{if } -\nu_m<0,$$
$$\pm\min\{-\nu_m, \mu_m\} \quad \text{if } -\nu_m \geq 0.$$

In particular, the AP indices are zero if and only if $\nu_m = 0$ or if $-\nu_m<0$ and $\lambda_m/(\nu_m+\mu_m)\in\mathbf{N}$.

Proof. Suppose for a moment that the existence of m is proved. We show that if $k\leq m$, then after k Portuguese transformations we get a matrix function

$$G_k(x)=\begin{pmatrix} e^{i\lambda_k x} & 0 \\ c_{-1}^{(k)}e^{-i\nu_k x}-c_0^{(k)}e^{i\mu_k x}+c_1^{(k)}e^{i\delta_k x} & e^{-i\lambda_k x}\end{pmatrix}$$

where $-\lambda_k<-\nu_k<0$, $0<\mu_k<\lambda_k$, $0<\delta_k<\lambda_k$, $\lambda_k<\nu_k+\delta_k$ for $k\leq m-1$ and $\delta_k\geq\lambda_k$ for $k=m$. As for $k=m$ we are in the binomial case, G_k and therefore G itself is *APW* factorable with *AP* indices computable by Theorem 14.9.

By assumption, $G_0=G$ is of the asserted form. Assuming that $G_k (k\leq m-1)$ is as above, we apply one Portuguese transformation to G_k in order to obtain G_{k+1} of the desired form. We have $f_{k+1}:=P_{(-\nu_k,\nu_k)}(g_2^{(k)}e_{-\lambda_k})$ where $g_2^{(k)}$ is given by a formula of the type (15.47). Since $\gamma_k=(\nu_k+\mu_k,\,\nu_k+\delta_k)$, it follows that

$$\Omega(f_{k+1})\subset\left\{n_1(\nu_k+\mu_k)+n_2(\nu_k+\delta_k)-\lambda_k:(n_1,n_2)\in\mathbf{Z}_+^2\right\}\cap(-\nu_k,\nu_k).$$

If $n_2\geq 2$, then $n_2(\nu_k+\delta_k)-\lambda_k>2\lambda_k-\lambda_k=\lambda_k>\nu_k\notin\Omega(f_{k+1})$. If $n_1\geq 1$ and $n_2=1$, we have

$$\begin{aligned} & n_1(\nu_k+\mu_k)+n_2(\nu_k+\delta_k)-\lambda_k \\ & \quad \geq \nu_k+\mu_k+\nu_k+\delta_k-\lambda_k>\nu_k+\mu_k>\nu_k\notin\Omega(f_{k+1}).\end{aligned}$$

The smallest integer N_k such that $N_k(\nu_k+\mu_k)-\lambda_k>-\nu_k$ is

$$N_k=1+\left[\frac{\lambda_k-\nu_k}{\nu_k+\mu_k}\right].$$

If $n_1\geq N_k+2$, then

$$n_1(\nu_k+\mu_k)-\lambda_k>-\nu_k+2(\nu_k+\mu_k)=\nu_k+2\mu_k>\nu_k\notin\Omega(f_{k+1}).$$

So the only $(n_1,n_2)\in\mathbf{Z}_+^2$ which can contribute to $\Omega(f_{k+1})$ are $(N_k,0)$, $(0,1)$, $(N_k+1,0)$, and at worst we have

$$\begin{aligned}\Omega(f_{k+1}) &= \left\{N_k(\nu_k+\mu_k)-\lambda_k,\ \nu_k+\delta_k-\lambda_k,\ (N_k+1)(\nu_k+\mu_k)-\lambda_k\right\} \\ &= \{-\nu_{k+1},\ \mu_{k+1},\ \delta_{k+1}\}.\end{aligned}$$

By assumption, $\mu_{k+1} = \nu_k + \delta_k - \lambda_k > 0$ and $\mu_{k+1} = \nu_k + \delta_k - \lambda_k < \nu_k = \lambda_{k+1}$. Also by assumption,

$$\delta_{k+1} = N_k(\nu_k + \mu_k) - \lambda_k + \nu_k + \mu_k > -\nu_k + \nu_k + \mu_k = \mu_k > 0.$$

Further,

$$\begin{aligned}\nu_{k+1} + \delta_{k+1} &= -N_k(\nu_k + \mu_k) + \lambda_k + (N_k + 1)(\nu_k + \mu_k) - \lambda_k \\ &= \nu_k + \mu_k > \nu_k = \lambda_{k+1}\end{aligned}$$

and

$$\nu_{k+1} = -N_k(\nu_k + \mu_k) + \lambda_k < \nu_k = \lambda_{k+1}.$$

If $k \le m - 2$, then $\delta_{k+1} < \lambda_{k+1}$ by the definition of m. Moreover, for $k \le m - 2$ we also have $-\nu_{k+1} < 0$, because otherwise

$$\begin{aligned}\delta_{k+1} &= N_k(\nu_k + \mu_k) - \lambda_k + \nu_k + \mu_k \\ &= -\nu_{k+1} + \nu_k + \mu_k \ge \nu_k + \mu_k > \nu_k = \lambda_{k+1},\end{aligned}$$

which is impossible for $k + 1 < m$. Finally, for $k = m - 1$ we get $\delta_{k+1} \ge \lambda_{k+1}$.

We now show the existence of the number m. Contrary to what we want, let us assume that $\delta_k < \lambda_k$ for all $k \in \mathbf{N}$. Then, by what was proved in the preceding paragraph,

$$-\lambda_k < -\nu_k < 0, \quad 0 < \mu_k < \lambda_k, \quad 0 < \delta_k < \lambda_k, \quad \lambda_k < \nu_k + \delta_k$$

for all $k \in \mathbf{N}$. Suppose there is an $l \in \mathbf{N}$ such that $\nu_l + \mu_l \ge \lambda_l$. Then $N_l = 1$, because $\nu_l + \mu_l - \lambda_l \ge 0 > -\nu_l$. Consequently,

$$\delta_{l+1} = (N_l + 1)(\nu_l + \mu_l) - \lambda_l = 2(\nu_l + \mu_l) - \lambda_l \ge \nu_l + \mu_l > \nu_l = \lambda_{l+1},$$

which contradicts our assumption. Thus, $\nu_l + \mu_l < \lambda_l$ for all $l \in \mathbf{N}$. It follows that $\lambda_{l+1} + \mu_l < \lambda_l$ for all $l \in \mathbf{N}$. Adding the two inequalities $\lambda_{l+2} < \lambda_{l+1} - \mu_{l+1}$, $\lambda_{l+1} < \lambda_l - \mu_l$ we get $\lambda_{l+2} < \lambda_l - \mu_l - \mu_{l+1}$. As

$$\begin{aligned}\mu_{k+2} &= \nu_{k+1} + \delta_{k+1} - \lambda_{k+1} \\ &= -N_k(\nu_k + \mu_k) + \lambda_k + (N_k + 1)(\nu_k + \mu_k) - \lambda_k - \nu_k = \mu_k\end{aligned}$$

for all k, we conclude that μ_l or μ_{l+1} equals $\mu_0 = \mu$. Thus,

$$\lambda_{l+2} < \lambda_l - \mu_l - \mu_{l-1} < \lambda_l - \mu \ \text{ for all } \ l.$$

It results that $\lambda_k < 0$ for sufficiently large k, which is impossible. This contradiction proves that there is an $m \in \mathbf{N}$ such that $\delta_m \ge \lambda_m$.

Thus, after the mth Portuguese transformation we obtain $\delta_m \geq \lambda_m$. We have

$$\begin{aligned} \mu_m &= \nu_{m-1} + \delta_{m-1} - \lambda_{m-1} > 0, \\ \mu_m &= \nu_{m-1} + \delta_{m-1} - \lambda_{m-1} < \nu_{m-1} = \lambda_m. \end{aligned}$$

Hence, if $-\nu_m \geq 0$, we are in the one-sided case, and G_m has an APW factorization with the AP indices $\pm \min\{-\nu_m, \mu_m\}$ due to Theorem 14.1. If $-\nu_m < 0$, we can apply Theorem 14.9 to deduce that G_m is factorable and that the AP indices are as asserted. The last assertion of Theorem 15.10 is obvious. □

Notes

The results of Sections 15.1 and 15.2 were announced in [131]; full proofs are contained in [130] and [135]. In [130] and [135], we considered the case of invertible matrix coefficients, and for $\nu + \delta = \lambda$, on top of that, a commutativity condition was imposed. Thus, in the general case of non-commuting matrix coefficients the problem is still open. The invertibility condition for canonical factorization was lifted in [24] (for $\lambda = \nu + \delta$) and [138] (for $\lambda < \nu + \delta$). Some results on non-canonical factorization in the case of singular coefficients can be found in [216]. As was noticed there, the general case of commuting (not necessarily invertible) coefficients covers the case of arbitrary (non-commuting) coefficients, and so there is not much hope in solving the former while the latter remains unsolved. Some cases, reducing after Portuguese transformation to the trinomials covered by the results of Sections 15.1 and 15.2, were pointed out in [24].

The results of Sections 15.3 and 15.4 were obtained in [177]. Their matrix generalizations are in [138].

We want to emphasize that, even for scalar coefficients, the factorization problem for the matrix $G_f^{(\lambda)}$ with $\lambda > \nu + \delta$ remains open. This is probably the biggest mystery of the AP factorization theory.

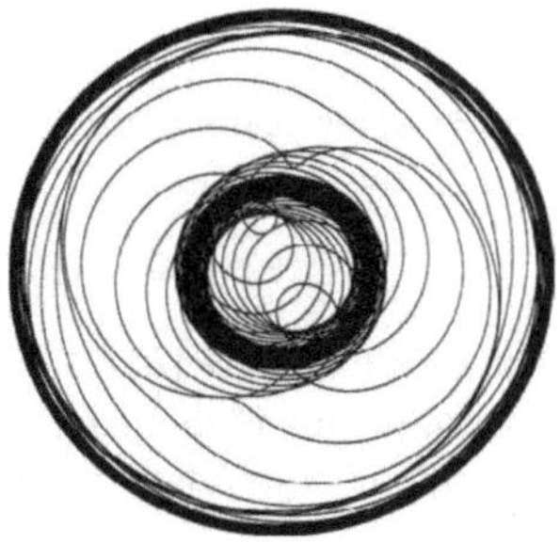

Chapter 16

Toeplitz Operators

We now leave Hilbert space and turn to operators on Banach spaces. Natural generalizations of the Hilbert spaces $L^2(\mathbf{R}_+)$ and H^2_+ are the weighted Lebesgue spaces $L^p(\mathbf{R}_+, w)$ and the weighted Hardy spaces $H^p(\mathbf{R}, w)$, respectively. Accordingly, the theory of Wiener-Hopf operators we have studied in Chapters 2 and 10 bifurcates into the theory of Wiener-Hopf operators on $L^p(\mathbf{R}_+, w)$ and the theory of Toeplitz operators on $H^p(\mathbf{R}, w)$. Although the final results of these two theories almost coincide, the theories are not equivalent and are based on different techniques. In the following, we first consider Toeplitz operators and turn to the (slightly more difficult) theory of Wiener-Hopf operators in the forthcoming chapters.

We already made acquaintance with Toeplitz operators in Chapter 11. There we worked with operators on $H^2(\mathbf{T})$ and noted some simple properties of these operators. We now consider Toeplitz operators on H^p with Muckenhoupt weights w. This class of weights is the natural terrain for operators with piecewise continuous symbols. It turns out that the exponent p and the weight w cause a metamorphosis of the line segments prevailing in the H^2 case to so-called horns. The resulting theory is very beautiful.

16.1 Muckenhoupt Weights

For $a \in L^\infty(\mathbf{R})$, the Wiener-Hopf operator $W(a)$ is defined on $L^2(\mathbf{R}_+)$ by the formula

$$W(a)f = \chi_+ F^{-1} a F f, \quad f \in L^2(\mathbf{R}_+).$$

The Fourier transform $F : L^2(\mathbf{R}_+) \to H^2_+$ is an isomorphism, and $FW(a)F^{-1}$ acts on H^2_+ by the rule

$$FW(a)F^{-1}g = F\chi_+F^{-1}(ag) = P(ag), \quad g \in H^2_+.$$

The operator $FW(a)F^{-1}$ is frequently called the *Toeplitz operator* on H^2_+ with the *symbol* a and is denoted by $T(a)$. Thus, the theory of Wiener-Hopf operators on $L^2(\mathbf{R}_+)$ is equivalent to the theory of Toeplitz operators on H^2_+.

We begin with a few definitions. Let E be a connected subset of $\mathbf{R}$ (e.g., $E = \mathbf{R}_+$ or $E = \mathbf{R}$). A (Lebesgue) measurable function $w : E \to [0, \infty]$ is called a weight if $w^{-1}(\{0, \infty\})$ has (Lebesgue) measure zero. For $1 \le p < \infty$, we denote by $L^p(E, w)$ the usual Lebesgue space with the norm

$$\|f\|_{p,w} := \left(\int_E |f(x)|^p w(x)^p \, dx\right)^{1/p}.$$

Notice that w enters the norm in the pth power.

The operator $P := F\chi_+F^{-1}$ is the orthogonal projection of $L^2(\mathbf{R})$ onto H^2_+. We denote by $A_p(\mathbf{R})$ the set of all weights w on $\mathbf{R}$ for which $L^2(\mathbf{R}) \cap L^p(\mathbf{R}, w)$ is dense in $L^p(\mathbf{R}, w)$ and the operator P satisfies the inequality

$$\|Pf\|_{p,w} \le C_{p,w}\|f\|_{p,w} \text{ for all } f \in L^2(\mathbf{R}) \cap L^p(\mathbf{R}, w)$$

with some constant $C_{p,w} < \infty$ independent of f. Since

$$P = F\chi_+F^{-1} = F^{-1}\chi_-F = W^0(\chi_-)$$

and $W^0(-\operatorname{sgn} x)$ is the Cauchy singular integral operator S, we have

$$P = \frac{1}{2}(I + S).$$

Thus, $w \in A_p(\mathbf{R})$ if and only if $L^2(\mathbf{R}) \cap L^p(\mathbf{R}, w)$ is dense in $L^p(\mathbf{R}, w)$ and S extends from $L^2(\mathbf{R}) \cap L^p(\mathbf{R}, w)$ to a bounded operator on $L^p(\mathbf{R}, w)$.

Clearly, if $w \in A_p(\mathbf{R})$ then P extends (in a unique way) to a bounded projection on $L^p(\mathbf{R}, w)$. The image of $L^p(\mathbf{R}, w)$ under P is then a closed subspace of $L^p(\mathbf{R}, w)$. This subspace is denoted by $H^p(\mathbf{R}, w)$ and called the *Hardy space* of $L^p(\mathbf{R}, w)$. Thus, a weighted Hardy space $H^p(\mathbf{R}, w)$ is a space of the form

$$H^p(\mathbf{R}, w) := PL^p(\mathbf{R}, w)$$

with $w \in A_p(\mathbf{R})$. We emphasize that in general the set $H^p(\mathbf{R}, w)$ is different from the set $FL^p(\mathbf{R}_+, w)$. The following theorem characterizes the weights in $A_p(w)$.

Theorem 16.1 (Hunt, Muckenhoupt, Wheeden). *Let $1 < p < \infty$. A weight w on $\mathbf{R}$ belongs to $A_p(\mathbf{R})$ if and only if*

$$\sup_I \left(\frac{1}{|I|}\int_I w(x)^p \, dx\right)^{1/p} \left(\frac{1}{|I|}\int_I w(x)^{-q} \, dx\right)^{1/q} < \infty,$$

where $1/p+1/q=1$ and I ranges over all bounded intervals $I \subset \mathbf{R}$; here $|I|$ stands for the length of I.

Frequently, it is convenient to pass from the line $\mathbf{R}$ to the unit circle $\mathbf{T}$. Given a weight $\varrho : \mathbf{T} \to [0,\infty]$, we denote by $L^p(\mathbf{T},\varrho)$ $(1 \le p < \infty)$ the Lebesgue space with the norm

$$\|f\|_{p,\varrho} := \left(\int_{\mathbf{T}} |f(t)|^p \varrho(t)^p \, |dt|\right)^{1/p},$$

where $|dt|$ is Lebesgue measure on $\mathbf{T}$. Let P_0 be the orthogonal projection of $L^2(\mathbf{T})$ onto $H^2_+(\mathbf{T})$. We let $A_p(\mathbf{T})$ stand for the set of all weights ϱ on $\mathbf{T}$ such that $L^2(\mathbf{T}) \cap L^p(\mathbf{T},\varrho)$ is dense in $L^p(\mathbf{T},\varrho)$ and

$$\|P_0 g\|_{p,\varrho} \le C_{p,\varrho}\|g\|_{p,\varrho} \quad \text{for all} \quad g \in L^2(\mathbf{T}) \cap L^p(\mathbf{T},\varrho)$$

with some constant $C_{p,\varrho} < \infty$ which does not depend on g.

We know from Section 6.1 that

$$P_0 = \frac{1}{2}(I + S_0)$$

where S_0 is the Cauchy singular integral operator on $\mathbf{T}$. Thus, $\varrho \in A_p(\mathbf{T})$ if and only if S_0 is a well-defined and bounded operator on $L^p(\mathbf{T},\varrho)$. If $\varrho \in A_p(\mathbf{T})$, then P_0 extends to a bounded projection on $L^p(\mathbf{T},\varrho)$. The weighted Hardy space $H^p(\mathbf{T},\varrho)$ is then the closed subspace of $L^p(\mathbf{T},\varrho)$ defined by

$$H^p(\mathbf{T},\varrho) := P_0 L^p(\mathbf{T},\varrho).$$

Here is the circle analogue of Theorem 16.1.

Theorem 16.2 (Hunt, Muckenhoupt, Wheeden). *Let $1 < p < \infty$. Then $\varrho \in A_p(\mathbf{T})$ if and only if*

$$\sup_I \left(\frac{1}{|I|}\int_I \varrho(t)^p \, |dt|\right)^{1/p} \left(\frac{1}{|I|}\int_I \varrho(t)^{-q} \, |dt|\right)^{1/q} < \infty,$$

where $1/p+1/q=1$, I ranges over all subarcs of $\mathbf{T}$, and $|I|$ is the length of I.

Weights in $A_p(\mathbf{R})$ and $A_p(\mathbf{T})$ are usually referred to as *Muckenhoupt weights.*

Given a weight w on $\mathbf{R}$, define a weight ϱ on $\mathbf{T}$ by

$$\varrho(t) := w\left(i\frac{1+t}{1-t}\right)|1-t|^{1-2/p}, \quad t \in \mathbf{T}. \tag{16.1}$$

It can be checked directly (also recall Section 6.1) that the operator $B : L^p(\mathbf{T},\varrho) \to L^p(\mathbf{R},w)$ defined by

$$(Bg)(x) := \frac{2^{1-1/p}}{x+i} g\left(\frac{x-i}{x+i}\right), \quad x \in \mathbf{R}, \quad g \in L^p(\mathbf{T},\varrho) \tag{16.2}$$

is an isometric isomorphism, the inverse operator being

$$(B^{-1}f)(t) = \frac{i\,2^{1/p}}{1-t} f\left(i\frac{1+t}{1-t}\right), \quad t \in \mathbf{T}, \quad f \in L^p(\mathbf{R}, w).$$

We have

$$BS_0B^{-1} = S \text{ and thus: } w \in A_p(\mathbf{R}) \Longleftrightarrow \varrho \in A_p(\mathbf{T}). \tag{16.3}$$

Weights of the form

$$w(x) = |x+i|^{\mu_0} \prod_{j=1}^{n} |x-\beta_j|^{\mu_j}, \quad x \in \mathbf{R},$$

where $\beta_j \in \mathbf{R}$ and $\mu_0, \mu_j \in \mathbf{R}$, are called *power weights on* **R**. One can show that if $1 < p < \infty$, then

$$w \in A_p(\mathbf{R}) \Longleftrightarrow \mu := \mu_0 + \mu_1 + \ldots + \mu_n, \mu_1, \ldots, \mu_n \in \left(-\frac{1}{p}, \frac{1}{q}\right) \tag{16.4}$$

where $1/p + 1/q = 1$. Analogously, a *power weight on* **T** is a weight of the form

$$\varrho(t) = \prod_{j=1}^{n} |t-\zeta_j|^{\mu_j}, \quad t \in \mathbf{T},$$

where $\zeta_j \in \mathbf{T}$ and $\mu_j \in \mathbf{R}$. We have

$$\varrho \in A_p(\mathbf{T}) \Longleftrightarrow \mu_1, \ldots, \mu_n \in \left(-\frac{1}{p}, \frac{1}{q}\right), \tag{16.5}$$

where again $1 < p < \infty$ and $1/p + 1/q = 1$.

16.2 Simonenko's Theorems

Throughout what follows we assume that $1 < p < \infty$ and $w \in A_p(\mathbf{R})$. Also, q is always given by $1/p + 1/q = 1$.

For $a \in L^\infty(\mathbf{R})$, the Toeplitz operator $T(a) : f \mapsto P(af)$ is clearly bounded on $H^p(\mathbf{R}, w)$. Some of the properties of Wiener-Hopf operators in $L^2(\mathbf{R}_+)$ we cited in Sections 2.2 and 5.1 have analogues for Toeplitz operators on $H^p(\mathbf{R}, w)$.

Theorem 16.3 (Simonenko). (a) *If $a \in L^\infty(\mathbf{R}) \setminus \{0\}$ then $T(a)$ has a trivial kernel or a dense image in $H^p(\mathbf{R}, w)$.*

(b) *If $a \in L^\infty_{N\times N}(\mathbf{R})$ and $T(a)$ is semi-Fredholm on $H^p_N(\mathbf{R}, w)$, then a is necessarily in $GL^\infty_{N\times N}(\mathbf{R})$.*

(c) *Let $a \in L^\infty_{N\times N}(\mathbf{R})$ and suppose for each $x_0 \in \dot{\mathbf{R}}$ there is an $a_{x_0} \in L^\infty_{N\times N}(\mathbf{R})$ such that $T(a_{x_0})$ is Fredholm (left-Fredholm, right-Fredholm) on $H^p(\mathbf{R}, w)$. Then $T(a)$ itself is Fredholm (left-Fredholm, right-Fredholm) on $H^p(\mathbf{R}, w)$.*

For Toeplitz operators on $L^p(\mathbf{T}, \varrho)$ proofs are in [37] and [46], for example. The passage from the circle case to the line case can be carried out with the help of the isomorphism (16.2): from (16.3) we infer that $B^{-1}PB = P_0$, which shows that B is an isometric isomorphism of $H^p(\mathbf{T}, \varrho)$ onto $H^p(\mathbf{R}, w)$ and

$$B^{-1}T(a)B = T_0(B_0 a) \tag{16.6}$$

where $T_0(b)$ acts on $H^p(\mathbf{T}, \varrho)$ by the rule $g \mapsto P_0(bg)$ and

$$(B_0 a)(t) := a\left(i\frac{1+t}{1-t}\right), \quad t \in \mathbf{T} \setminus \{1\}.$$

We now define the analogue of Wiener-Hopf factorization for weighted Hardy spaces. We begin with the circle case, which is a little simpler. Put

$$H^p_-(\mathbf{T}, \varrho) := \mathbf{C} + Q_0 L^p(\mathbf{T}, \varrho),$$

where $\mathbf{C}$ stands for the constant functions and $Q_0 := I - P_0$. A matrix function $b \in GL^\infty_{N\times N}(\mathbf{T})$ is said to have a *right WH factorization in* $L^p(\mathbf{T}, \varrho)$ if it can be written in the form

$$b(t) = b_-(t)v(t)b_+(t) \text{ for almost all } t \in \mathbf{T}$$

where

$$\begin{array}{ll} (i) & v(t) = \operatorname{diag}(t^{\varkappa_1}, \dots, t^{\varkappa_N}) \text{ with } \varkappa_1, \dots, \varkappa_N \in \mathbf{Z}, \\ (ii) & b_- \in \left[H^p_-(\mathbf{T}, \varrho)\right]_{N\times N}, \quad b_-^{-1} \in \left[H^q_-(\mathbf{T}, \varrho^{-1})\right]_{N\times N}, \\ & b_+ \in \left[H^q_+(\mathbf{T}, \varrho^{-1})\right]_{N\times N}, \quad b_+^{-1} \in \left[H^p_+(\mathbf{T}, \varrho)\right]_{N\times N}, \\ (iii) & \text{the operator } b_- S_0 b_-^{-1} I \text{ is bounded on } L^p_N(\mathbf{T}, \varrho). \end{array}$$

Note that in the scalar case ($N = 1$), condition (iii) is equivalent to the requirement that $|b_-|\varrho \in A_p(\mathbf{T})$. Also notice that, by Theorem 16.2, $\varrho \in A_p(\mathbf{T})$ if and only if $\varrho^{-1} \in A_q(\mathbf{T})$.

Theorem 16.4 (Simonenko). *Let $b \in GL^\infty_{N\times N}(\mathbf{T})$. The operator $T_0(b)$ is Fredholm of index $-\varkappa$ on $H^p_N(\mathbf{T}, \varrho)$ if and only if b has a right WH factorization in $L^p(\mathbf{T}, \varrho)$ such that $\varkappa = \varkappa_1 + \dots + \varkappa_N$. The operator $T_0(b)$ is invertible if and only if it is Fredholm and $\varkappa_1 = \dots = \varkappa_N = 0$.*

Let $H^p_-(\mathbf{R}, w) := QL^p(\mathbf{R}, w)$. One says that a matrix function $a \in GL^\infty_{N\times N}(\mathbf{R})$ has a *right WH factorization* in $L^p(\mathbf{R}, w)$ if

$$a(x) = a_-(x)\, d(x)\, a_+(x) \text{ for almost all } x \in \mathbf{R}$$

where

$$(i)\quad d(x)=\operatorname{diag}\left[\left(\frac{x-i}{x+i}\right)^{\varkappa_1},\dots,\left(\frac{x-i}{x+i}\right)^{\varkappa_N}\right]\quad\text{with }\varkappa_1,\dots,\varkappa_N\in\mathbf{Z},$$

$$(ii)\quad \frac{a_-(x)}{x-i}\in\left[H^p_-(\mathbf{R},w)\right]_{N\times N},\quad \frac{a_-^{-1}(x)}{x-i}\in\left[H^q_-(\mathbf{R},w^{-1})\right]_{N\times N},$$

$$\frac{a_+(x)}{x+i}\in\left[H^q(\mathbf{R},w^{-1})\right]_{N\times N},\quad \frac{a_+^{-1}(x)}{x+i}\in\left[H^p(\mathbf{R},w)\right]_{N\times N},$$

(iii) the operator $a_-Sa_-^{-1}I$ is bounded on $L^p_N(\mathbf{R},w)$.

In the scalar case, that is, for $N=1$, condition (iii) can again be replaced by the inclusion $|a_-|w\in A_p(\mathbf{R})$.

Theorem 16.5 (Simonenko). *Let $a\in GL^\infty_{N\times N}(\mathbf{R})$. Then $T(a)$ is Fredholm of index $-\varkappa$ on $H^p_N(\mathbf{R},w)$ if and only if a admits a right WH factorization in $L^p(\mathbf{R},w)$ with $\varkappa=\varkappa_1+\ldots+\varkappa_N$. The operator $T(a)$ is invertible if and only if it is Fredholm and $\varkappa_1=\ldots=\varkappa_N=0$.*

Theorems 5.6 and 5.7 also extend to Toeplitz operators on $H^p_N(\mathbf{R},w)$. We confine ourselves to formulating the generalization of Theorem 5.7.

Theorem 16.6. *Let $a\in[C(\dot{\mathbf{R}})+H^\infty_+]_{N\times N}$.*

(a) *If $a\notin GL^\infty_{N\times N}(\mathbf{R})$, then $T(a)$ is not semi-Fredholm on $H^p_N(\mathbf{R},w)$.*

(b) *If $a\in GL^\infty_{N\times N}(\mathbf{R})$ but $a\notin G[C(\dot{\mathbf{R}})+H^\infty_+]_{N\times N}$, then $T(a)$ is properly n-normal on $H^p_N(\mathbf{R},w)$ and $T(a^{-1})$ is a left regularizer of $T(a)$.*

(c) *If $a\in G[C(\dot{\mathbf{R}})+H^\infty_+]_{N\times N}$, then $T(a)$ is Fredholm on $H^p_N(\mathbf{R},w)$, $T(a^{-1})$ is a regularizer of $T(a)$, and*

$$\operatorname{Ind}T(a)=-\operatorname{wind}\det a.$$

The Toeplitz analogue of Proposition 2.17 and Corollary 2.12 is as follows.

Proposition 16.7. *If $b\in L^\infty_{N\times N}(\mathbf{R})$ and $a_\pm\in[C(\dot{\mathbf{R}})+H^\infty_\pm]_{N\times N}$, then the operator*

$$T(a_-)T(b)T(a_+)-T(a_-ba_+)$$

is compact on $H^p_N(\mathbf{R},w)$.

This proposition implies in particular part (b) of Theorem 16.6.

16.3 Emergence of Horns

Things become surprisingly beautiful for Toeplitz operators on $H^p(\mathbf{R},w)$ with piecewise continuous symbols: the line segments we had to fill in between the endpoints of the jumps in the H_+^2 case metamorphose into so-called horns, that is, into closed sets bounded by two circular arcs, in the $H^p(\mathbf{R}, w)$ case.

Let $1 < p < \infty$ and $\varrho \in A_p(\mathbf{T})$. For each $t \in \mathbf{T}$, we define the set $I_t(p, \varrho)$ by

$$I_t(p, \varrho) := \{\lambda \in \mathbf{R} : |\tau - t|^\lambda \varrho(\tau) \in A_p(\mathbf{T})\};$$

of course, $|\tau - t|^\lambda \varrho(\tau) \in A_p(\mathbf{T})$ means that the weight given by $\tau \mapsto |\tau - t|^\lambda \varrho(\tau)$ is a weight in $A_p(\mathbf{T})$.

Proposition 16.8. *The set $I_t(p, \varrho)$ is an open interval of length at most 1 which contains the origin.*

Proof. We make use of two general properties of Muckenhoupt weights:
(a) if $\varrho \in A_p(\mathbf{T})$, then $\varrho^{1+\varepsilon} \in A_p(\mathbf{T})$ for all $\varepsilon \in \mathbf{R}$ sufficiently close to zero;
(b) if $\varrho_1, \varrho_2 \in A_p(\mathbf{T})$, then $\varrho_1^s \varrho_2^{1-s} \in A_p(\mathbf{T})$ for all $s \in [0, 1]$.
First of all, $0 \in I_t(p, \varrho)$ since $\varrho \in A_p(\mathbf{T})$. Now suppose $\lambda, \mu \in I_t(p, \varrho)$. Then, by property (b),

$$|\tau - t|^{\lambda s} \varrho(\tau)^s |\tau - t|^{\mu(1-s)} \varrho(\tau)^{1-s} \in A_p(\mathbf{T})$$

for all $s \in [0, 1]$, which implies that $\lambda s + \mu(1 - s) \in I_t(p, \varrho)$ for all $s \in [0, 1]$. Consequently, $I_t(p, \varrho)$ is connected. To show that $I_t(p, \varrho)$ is open, pick $\lambda \in I_t(p, \varrho)$. By (a),

$$|\tau - t|^{\lambda(1+\varepsilon)} \varrho(\tau)^{1+\varepsilon} \in A_p(\mathbf{T})$$

for all sufficiently small $\varepsilon > 0$. As $|\tau - t|^\mu \in A_p(\mathbf{T})$ if and only if $-1/p < \mu < 1/q$, we see that $|\tau - t|^\mu \in A_p(\mathbf{T})$ whenever $|\mu|$ is sufficiently small. Hence, by (b),

$$|\tau - t|^{\lambda(1+\varepsilon)s + \mu(1-s)} \varrho(\tau)^{(1+\varepsilon)s} \in A_p(\mathbf{T})$$

for all sufficiently small $\varepsilon > 0$, $|\mu|$ and all $s \in [0, 1]$. Choosing $\varepsilon > 0$ sufficiently small and letting $s := 1/(1 + \varepsilon)$, we obtain that

$$|\tau - t|^{\lambda + \mu\varepsilon/(1+\varepsilon)} \varrho(\tau) \in A_p(\mathbf{T}).$$

Thus, $\lambda + \mu\varepsilon/(1 + \varepsilon) \in I_t(p, \varrho)$ for all $|\mu|$ small enough, which proves that $I_t(p, \varrho)$ is open. Finally, if the length of $I_t(p, \varrho)$ were greater than 1, there were λ and $\lambda + 1$ in $I_t(p, \varrho)$. Theorem 16.2 would imply that

$$|\tau - t|^\lambda \varrho(\tau) \in L^p(\mathbf{T}), \ |\tau - t|^{-\lambda-1} \varrho(\tau)^{-1} \in L^q(\mathbf{T}),$$

whence $|\tau - t|^{-1} \in L^1(\mathbf{T})$, which is not true. □

Example 16.9. Let $\varrho(\tau) = |\tau - t|^\mu$ with $-1/p < \mu < 1/q$. Then $\varrho \in A_p(\mathbf{T})$ and

$$I_t(p, \varrho) = \{\lambda \in \mathbf{R} : -1/p < \lambda + \mu < 1/q\} = (-1/p - \mu, \, 1/q - \mu).$$

Thus, $I_t(p, \varrho)$ is an open interval of length 1 containing the origin. Conversely, if $\alpha < 0 < \beta$ and $\beta - \alpha = 1$, we can easily find $p \in (1, \infty)$ such that

$$I_t(p, 1) = (-1/p, 1/q) = (\alpha, \beta). \qquad \square$$

The following result provides us with a sufficient supply of nontrivial Muckenhoupt weights.

Theorem 16.10. *Let $1 < p < \infty$ and $t \in \mathbf{T}$. Let further f be any function in $C^2(\mathbf{R})$ such that f, f', f'' are bounded on $\mathbf{R}$ and define the weight ϱ on $\mathbf{T}$ by*

$$\varrho(\tau) = \exp\left(f\left(\log\left(-\log\frac{r}{2}\right)\right)\log\frac{r}{2}\right) \tag{16.7}$$

for $0 < r := |\tau - t| < 2$. Then $\varrho \in A_p(\mathbf{T})$ if and only if

$$-1/p < \liminf_{x\to+\infty}\left(f(x) + f'(x)\right) \le \limsup_{x\to+\infty}\left(f(x) + f'(x)\right) < 1/q.$$

For a proof we refer to [37, Section 2.7 and Theorem 2.36].

Example 16.11. If $f(x) = \mu$ for all $x \in \mathbf{R}$, Theorem 16.10 is equivalent to saying that $\varrho(\tau) = |\tau - t|^\mu/2^\mu$ belongs to $A_p(\tau)$ if and only if $-1/p < \mu < 1/q$. Choosing

$$f(x) = \mu + \varepsilon \sin x, \tag{16.8}$$

we see that the weight given by (16.7) is in $A_p(\mathbf{T})$ if and only if

$$-1/p < \mu - |\varepsilon|\sqrt{2} \le \mu + |\varepsilon|\sqrt{2} < 1/q. \tag{16.9}$$

$\square$

Proposition 16.12. *Let $1 < p < \infty$ and $t \in \mathbf{T}$. If α, β are any real numbers such that $\alpha < 0 < \beta$ and $\beta - \alpha < 1$, then there is a $\varrho \in A_p(\mathbf{T})$ such that $I_t(p, \varrho) = (\alpha, \beta)$.*

Proof. Put

$$\mu := 0, \quad \frac{1}{q} := \frac{\alpha + \beta + 1}{2}, \quad |\varepsilon|\sqrt{2} := \frac{1 - (\beta - \alpha)}{2} \quad (> 0).$$

Then $1 < q < \infty$ and (16.9) holds. Thus, by Theorem 16.10 and Example 16.11, the weight ϱ given by (16.7), (16.8) belongs to $A_p(\mathbf{T})$ and

$$\begin{aligned} I_t(p, \varrho) &= \{\lambda \in \mathbf{R} : -1/p < \lambda - |\varepsilon|\sqrt{2} \le \lambda + |\varepsilon|\sqrt{2} < 1/q\} \\ &= \left(-1 + 1/q + |\varepsilon|\sqrt{2},\ 1/q - |\varepsilon|\sqrt{2}\right) = (\alpha, \beta). \end{aligned} \qquad \square$$

From Proposition 16.8 we infer that $I_t(p, \varrho)$ can be written in the form

$$I_t(p, \varrho) = \left(-\nu_t^-(p, \varrho),\ 1 - \nu_t^+(p, \varrho)\right) \text{ with } 0 < \nu_t^-(p, \varrho) \le \nu_t^+(p, \varrho) < 1. \tag{16.10}$$

If ϱ is the power weight $\varrho(t) = |\tau - t|^\mu$, we know from Example 16.9 that

$$\nu_t^-(p,\varrho) = \nu_t^+(p,\varrho) = 1/p + \mu.$$

Therefore, we call the two numbers $\nu_t^\pm(p,\varrho) - 1/p$ the *indices of powerlikeness* of the weight ϱ at the point t.

Let $\nu \in (0,1)$. The set $\{e^{2\pi(x+i\nu)} : x \in \mathbf{R}\}$ is a ray starting at the origin and making the angle $2\pi\nu \in (0, 2\pi)$ with the positive real half-line. For $z_1, z_2 \in \mathbf{C}$, the Möbius transform

$$M_{z_1,z_2}(\zeta) := \frac{z_2\zeta - z_1}{\zeta - 1}$$

maps 0 to z_1 and ∞ to z_2. Thus,

$$\mathcal{A}(z_1, z_2; \nu) := \left\{M_{z_1,z_2}(e^{2\pi(x+i\nu)}) : x \in \mathbf{R}\right\} \cup \{z_1, z_2\}$$

is a circular arc between z_1 and z_2 (which contains its endpoints z_1, z_2). Clearly, $\mathcal{A}(z, z; \nu)$ degenerates to the point $\{z\}$. Also notice that $\mathcal{A}(z_1, z_2; 1/2)$ is the line segment between z_1 and z_2. Finally, given $0 < \nu_1 \le \nu_2 < 1$, we put

$$\mathcal{H}(z_1, z_2; \nu_1, \nu_2) := \bigcup_{\nu \in [\nu_1, \nu_2]} \mathcal{A}(z_1, z_2; \nu)$$

and refer to $\mathcal{H}(z_1, z_2; \nu_1, \nu_2)$ as the *horn* between z_1, z_2 determined by ν_1, ν_2. See Figure 16.1.

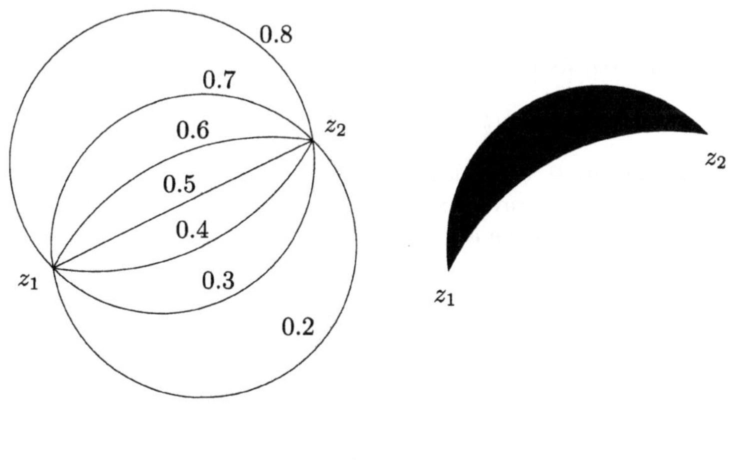

Figure 16.1: The circular arcs $\mathcal{A}(z_1, z_2; \nu)$ for several values of ν (left) and the horn $\mathcal{H}(z_1, z_2; 0.6, 0.7)$ (right).

Now let $b \in PC(\mathbf{T})$. We denote by $b_{\#}^{p,\varrho}(\mathbf{T})$ the set resulting from the essential range of b by filling in the horn

$$\mathcal{H}\big(b(t-0),\, b(t+0);\; \nu_t^-(p,\varrho),\, \nu_t^+(p,\varrho)\big)$$

whenever $b(t-0) \neq b(t+0)$, where $\nu_t^\pm(p,\varrho)$ are given by (16.10). Further, let $b_0^{p,\varrho}(\mathbf{T})$ be the set which is obtained from the essential range of b by joining $b(t-0)$ to $b(t+0)$ by the circular arc

$$\mathcal{A}\left(b(t-0),\, b(t+0);\; \frac{1}{2}\big(\nu_t^-(p,\varrho) + \nu_t^+(p,\varrho)\big)\right)$$

in case $b(t-0) \neq b(t+0)$. We may think of $b_0^{p,\varrho}(\mathbf{T})$ as a closed continuous curve which is completely contained in $b_{\#}^{p,\varrho}(\mathbf{T})$. The counter-clockwise orientation of $\mathbf{T}$ induces an orientation of $b_0^{p,\varrho}(\mathbf{T})$ in a natural way. Thus, if $0 \notin b_{\#}^{p,\varrho}(\mathbf{T})$ then the winding number of $b_0^{p,\varrho}(\mathbf{T})$ about the origin is well-defined. We denote this number by $\operatorname{wind}^{p,\varrho} b$.

Theorem 16.13. *Let $b \in PC(\mathbf{T}) \setminus \{0\}$.*

(a) *If $0 \in b_{\#}^{p,\varrho}(\mathbf{T})$, then $T_0(b)$ is not normally solvable on $H^p(\mathbf{T},\varrho)$.*

(b) *If $0 \notin b_{\#}^{p,\varrho}(\mathbf{T})$, then $T_0(b)$ is Fredholm on $H^p(\mathbf{T},\varrho)$ and*

$$\operatorname{Ind} T_0(b) = -\operatorname{wind}^{p,\varrho} b. \tag{16.11}$$

This theorem was established in [213]. A full proof is also in [37]. We therefore confine ourselves to the steps of the proof which show where the horns come from. Figure 16.2 illustrates the theorem.

Let $b \in GPC(\mathbf{T})$ and fix $t \in \mathbf{T}$. We have

$$b(t+0)/b(t-0) = e^{-2\pi i \gamma_t}$$

with some number $\gamma_t \in \mathbf{C}$ (which is uniquely determined up to a summand in $2\pi\mathbf{Z}$). Let $\Lambda_t \subset \mathbf{C}$ be the ray starting at the origin and passing through t. Then let $\arg z$ be any branch of the argument which is continuous in $\mathbf{C} \setminus \Lambda_t$. Define

$$z^{\gamma_t} := |z|^{\gamma_t} e^{i\gamma_t \arg z}, \quad z \in \mathbf{C} \setminus \Lambda_t.$$

Clearly, z^{γ_t} is analytic in $\mathbf{C} \setminus \Lambda_t$. The restriction of z^{γ_t} to $\mathbf{T} \setminus \{t\}$ will be denoted by g_{t,γ_t}. Obviously, $g_{t,\gamma_t} \in GPC(\mathbf{T})$. Since

$$g_{t,\gamma_t}(t+0)/g_{t,\gamma_t}(t-0) = e^{-2\pi i \gamma_t},$$

it results that b is locally equivalent to $c_t g_{t,\gamma_t}$ with some $c_t \in \mathbf{C} \setminus \{0\}$ at the point t. The T_0 analogue of Theorem 16.3(c) therefore shows that $T_0(b)$ is Fredholm on

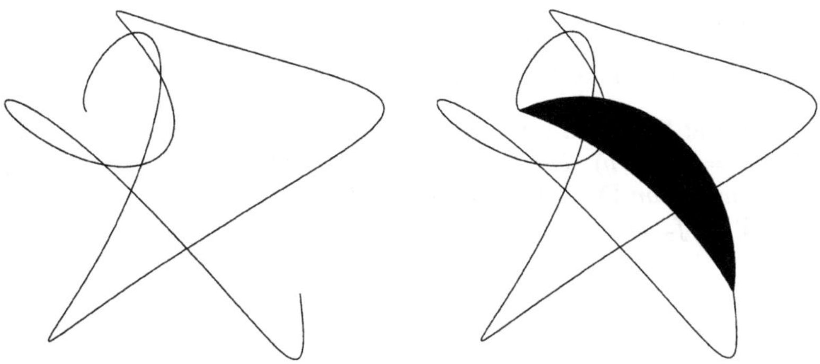

Figure 16.2: The essential range of a piecewise continuous function b with a single jump (left) and the essential spectrum of $T_0(b)$, which results by filling in a horn (right).

$H^p(\mathbf{T}, \varrho)$ if and only if $T_0(g_{t,\gamma_t})$ is Fredholm on $H^p(\mathbf{T}, \varrho)$ for every $t \in \mathbf{T}$. Thus, we are left with the study of the "local representatives" $T_0(g_{t,\gamma_t})$. In what follows we abbreviate γ_t to γ and g_{t,γ_t} to g.

We must show that $T_0(g)$ is Fredholm if and only if $0 \notin g_{\#}^{p,\varrho}(\mathbf{T})$. A moment's thought reveals that

$$0 \notin \mathcal{A}(z_1, z_2; \nu) \Longleftrightarrow \nu + \frac{1}{2\pi} \arg \frac{z_2}{z_1} \notin \mathbf{Z},$$

whence

$$0 \notin \mathcal{H}(z_1, z_2; \nu_1, \nu_2) \Longleftrightarrow \nu + \frac{1}{2\pi} \arg \frac{z_2}{z_1} \notin \mathbf{Z} \text{ for all } \nu \in [\nu_1, \nu_2].$$

Since

$$\arg \frac{g(t+0)}{g(t-0)} = \arg e^{-2\pi i \gamma} \in -2\pi \mathrm{Re}\, \gamma + 2\pi \mathbf{Z},$$

we arrive at the conclusion that

$$0 \notin g_{\#}^{p,\varrho}(\mathbf{T}) \Longleftrightarrow \varkappa_t(\nu) := \nu - \mathrm{Re}\, \nu \notin \mathbf{Z} \text{ for all } \nu \in [\nu_-, \nu_+]$$

where $\nu_\pm := \nu_t^\pm(p, \varrho)$. Thus, our aim is to prove the following.

Proposition 16.14. *The operator $T_0(g)$ is Fredholm on $H^p(\mathbf{T}, \varrho)$ if and only if*

$$\varkappa_t(\nu) := \nu - \mathrm{Re}\, \gamma \notin \mathbf{Z} \textit{ for all } \nu \in [\nu_-, \nu_+]. \tag{16.12}$$

In that case the integral part $[\varkappa_t(\nu)]$ *of* $\varkappa_t(\nu)$ *is independent of* $\nu \in [\nu_-, \nu_+]$ *and* $\operatorname{Ind} T_0(g) = [\varkappa_t(\nu)]$.

The proof of Proposition 16.14 is based on two lemmas.

Lemma 16.15. *Put* $\mathbf{D}_+ := \{z \in \mathbf{C} : |z| < 1\}$, $\mathbf{D}_- := \{z \in \mathbf{C} : |z| > 1\} \cup \{\infty\}$. *Also set* $H^p_+(\mathbf{T}, \varrho) := H^p(\mathbf{T}, \varrho)$. *Let* $f_\pm$ *be functions in* $\mathbf{D}_\pm \cup \mathbf{T}$ *which are analytic in* $\mathbf{D}_\pm$ *and continuous on* $\mathbf{D}_\pm \cup \mathbf{T}$ *with the possible exception of finitely many points* $t_1, \dots, t_n \in \mathbf{T}$. *If* $f_\pm \in L^p(\mathbf{T}, \varrho)$ *and*

$$|f_\pm(z)| \le M|z - t_k|^{-\mu_k} \quad (k = 1, \dots, n)$$

with $M < 0$ *and* $\mu_k < \infty$ *for all* $z \in \mathbf{D}_\pm$ *sufficiently close to* t_k, *then* $f_\pm|\mathbf{T} \in H^p_\pm(\mathbf{T}, \varrho)$.

For a proof see [37, Lemma 6.10].

In what follows, we make use of the factorization

$$g(\tau) = \left(1 - \frac{t}{\tau}\right)^{\varkappa - \gamma} \tau^{\varkappa} (\tau - t)^{\gamma - \varkappa}, \quad \tau \in \mathbf{T} \setminus \{t\}. \tag{16.13}$$

The functions occurring in (16.13) are defined as follows. Choose a continuous branch of $\arg z$ for $z \in \mathbf{C} \setminus \Lambda_t$, take any continuous branch of $\arg(z - t)$ for $z \in \mathbf{C} \setminus \Lambda_t^\infty$ where $\Lambda_t^\infty := \Lambda_t \cap \{z \in \mathbf{C} : |z| \ge 1\}$, and put

$$\arg\left(1 - \frac{t}{z}\right) = \arg \frac{z - t}{z} := \arg(z - t) - \arg z \tag{16.14}$$

for $z \in \mathbf{C} \setminus \Lambda_t$. It is easily seen that the function (16.14) is actually continuous on $\mathbf{C} \setminus \Lambda_t^0$ where $\Lambda_t^0 = \Lambda_t \cap \{z \in \mathbf{C} : |z| \le 1\}$. Hence, for each $\alpha \in \mathbf{C}$ the functions

$$(z - t)^\alpha := |z - t|^\alpha e^{i\alpha \arg(z - t)}, \quad \left(1 - \frac{t}{z}\right)^\alpha := \left|1 - \frac{t}{z}\right|^\alpha e^{i\alpha \arg(1 - t/z)}$$

are analytic and nonzero in $\mathbf{C} \setminus \Lambda_t^\infty$ and $\mathbf{C} \setminus \Lambda_t^0$, respectively. Note that $\mathbf{D}_+ \subset \mathbf{C} \setminus \Lambda_t^\infty$ and $\mathbf{D}_- \subset \mathbf{C} \setminus \Lambda_t^0$ and that $(z - t)^\alpha$ and $(1 - t/z)^\alpha$ are continuous on $\mathbf{D}_+ \cup (\mathbf{T} \setminus \{t\})$ and $\mathbf{D}_- \cup (\mathbf{T} \setminus \{t\})$, respectively.

Lemma 16.16. *If* $-\nu_- < \operatorname{Re} \alpha < 1 - \nu_+$ *then*

$$(\tau - t)^\alpha \in H^p_+(\mathbf{T}, \varrho), \quad (\tau - t)^{-\alpha} \in H^q_+(\mathbf{T}, \varrho^{-1}),$$
$$\left(1 - \frac{t}{\tau}\right)^\alpha \in H^p_-(\mathbf{T}, \varrho), \quad \left(1 - \frac{t}{\tau}\right)^{-\alpha} \in H^q_-(\mathbf{T}, \varrho^{-1}).$$

Proof. If $\alpha \in \mathbf{C}$, then

$$|(z - t)^\alpha| \simeq |z - t|^{\operatorname{Re} \alpha} \text{ for } z \in \mathbf{D}_+ \cup (\mathbf{T} \setminus \{t\}), \tag{16.15}$$
$$|(1 - t/z)^\alpha| \simeq |1 - t/z|^{\operatorname{Re} \alpha} \text{ for } z \in \mathbf{D}_- \cup (\mathbf{T} \setminus \{t\}), \tag{16.16}$$

where $f_1(z) \simeq f_2(z)$ means that $f_1(z)/f_2(z)$ is bounded away from zero and from infinity. Hence, by the definition of $I_t(p, \varrho)$ and by (16.10),

$$|(\tau - t)^\alpha|\varrho(\tau) \in A_p(\mathbf{T}) \Longleftrightarrow -\nu_- < \operatorname{Re}\alpha < 1 - \nu_+,$$
$$|(1 - t/\tau)^\alpha|\varrho(\tau) \in A_p(\mathbf{T}) \Longleftrightarrow -\nu_- < \operatorname{Re}\alpha < 1 - \nu_+. \qquad (16.17)$$

But if $|f|\varrho \in A_p(\mathbf{T})$, then $f \in L^p(\mathbf{T}, \varrho)$ and $f^{-1} \in L^q(\mathbf{T}, \varrho^{-1})$ by Theorem 16.2. From (16.15) and (16.16) in conjunction with Lemma 16.15 we therefore obtain the assertion. □

Proof of Proposition 16.14. Suppose (16.12) holds. Then the integer $[\varkappa_t(\nu)]$ is obviously independent of $\nu \in [\nu_-, \nu_+]$. Put $\varkappa := -[\varkappa_t(\nu)]$ and consider the factorization (16.13). By the definition of $\varkappa$,

$$-\varkappa < \nu_- - \operatorname{Re}\gamma \le \nu_+ - \operatorname{Re}\gamma < -\varkappa + 1,$$

whence $-\nu_- < -\operatorname{Re}\gamma + \varkappa < 1 - \nu_+$. Consequently, by Lemma 16.16,

$$\Big(1 - \frac{t}{\tau}\Big)^{\varkappa-\gamma} \in H^p_-(\mathbf{T}, \varrho), \quad \Big(1 - \frac{t}{\tau}\Big)^{\gamma-\varkappa} \in H^q_-(\mathbf{T}, \varrho^{-1}),$$
$$(\tau - t)^{\gamma-\varkappa} \in H^q_+(\mathbf{T}, \varrho^{-1}), \quad (\tau - t)^{\varkappa-\gamma} \in H^p_+(\mathbf{T}, \varrho),$$

and from (16.17) we also infer that

$$|(1 - t/\tau)^{\varkappa-\gamma}|\varrho(\tau) \in A_p(\mathbf{T}).$$

This shows that (16.13) is a right WH factorization of g in $L^p(\mathbf{T}, \varrho)$, and Theorem 16.4 therefore implies that $T_0(g)$ is Fredholm of index $-\varkappa = [\varkappa_t(\nu)]$.

Now suppose that $T_0(g)$ is Fredholm of index $\varkappa$ on $H^p(\mathbf{T}, \varrho)$. Then, by Theorem 16.4, g admits a right WH factorization

$$g(\tau) = b_-(\tau)\tau^\varkappa b_+(\tau), \quad \tau \in \mathbf{T} \setminus \{t\} \qquad (16.18)$$

in $L^p(\mathbf{T}, \varrho)$. Contrary to what we want, let us assume that (16.12) is violated. Then $-\operatorname{Re}\gamma$ belongs to the set $[-\nu_+, -\nu_-] + \mathbf{Z}$ and hence, there is an integer k such that

$$-\nu_+ \le -\operatorname{Re}\gamma + k \le -\nu_-. \qquad (16.19)$$

We have

$$g(\tau) = \Big(1 - \frac{t}{\tau}\Big)^{k-\gamma} \tau^k (\tau - t)^{\gamma-k}, \quad \tau \in \mathbf{T} \setminus \{t\}. \qquad (16.20)$$

Combining (16.18) and (16.20) we get

$$b_+^{-1}(\tau)(\tau - t)^{\gamma-k} = \tau^{\varkappa-k} b_-(\tau)\Big(1 - \frac{t}{\tau}\Big)^{\gamma-k}, \quad \tau \in \mathbf{T} \setminus \{t\}. \qquad (16.21)$$

By virtue of (16.19), there is a number $s > 0$ such that $-\nu_- < -\mathrm{Re}\,\gamma + k + s < 1 - \nu_+$. Therefore, by Lemma 16.16,

$$(\tau - t)^{\gamma-k-s} \in H^q_+(\mathbf{T}, \varrho^{-1}), \quad \Big(1 - \frac{t}{\tau}\Big)^{\gamma-k-s} \in H^q_-(\mathbf{T}, \varrho^{-1}).$$

The positivity of s implies that $(\tau - t)^s$ and $(1 - t/\tau)^s$ are in $H^\infty_+(\mathbf{T})$ and $H^\infty_-(\mathbf{T})$, respectively, which gives

$$(\tau - t)^{\gamma-k} = (\tau - t)^{\gamma-k-s}(\tau - t)^s \in H^q_+(\mathbf{T}, \varrho^{-1}),$$
$$\Big(1 - \frac{t}{\tau}\Big)^{\gamma-k} = \Big(1 - \frac{t}{\tau}\Big)^{\gamma-k-s}\Big(1 - \frac{t}{\tau}\Big)^s \in H^q_-(\mathbf{T}, \varrho^{-1}).$$

By assumption, $b_+^{-1} \in H^p_+(\mathbf{T}, \varrho)$ and $b_- \in H^p_-(\mathbf{T}, \varrho)$. It follows that the left-hand side of (16.21) lies in $H^1_+(\mathbf{T})$ while its right-hand side is in $\tau^{\varkappa-k}H^1_-(\mathbf{T})$. Thus, if $\varkappa < k$, the both sides of (16.21) must be zero, which is impossible because $b_+^{-1}(\tau)$ and $(\tau - t)^{\gamma-k}$ vanish almost nowhere on $\mathbf{T}$. Hence $\varkappa \geq k$.

Suppose $\varkappa = k$. Then both sides of (16.21) are equal to some nonzero constant, and as $|b_+^{-1}|\varrho = |g^{-1}|\,|b_-\varrho| \in A_p(\mathbf{T})$ by the definition of WH factorization, it follows that $|(\tau - t)^{k-\gamma}|\varrho(\tau) \in A_p(\mathbf{T})$ and hence $-\nu_- < -\mathrm{Re}\,\gamma + k < 1 - \nu_+$, which contradicts (16.19). Thus, $\varkappa \geq k + 1 =: l$. Now consider the factorization

$$g(\tau) = \Big(1 - \frac{t}{\tau}\Big)^{l-\gamma}\tau^l(\tau - t)^{\gamma-l}, \quad \tau \in \mathbf{T} \setminus \{t\}. \tag{16.22}$$

On comparing (16.18) and (16.22) we obtain

$$b_+(\tau)(\tau - t)^{l-\gamma} = \tau^{l-\varkappa}b_-^{-1}(\tau)\Big(1 - \frac{t}{\tau}\Big)^{l-\gamma}, \quad \tau \in \mathbf{T} \setminus \{t\}. \tag{16.23}$$

From (16.19) we see that

$$1 - \nu_+ \leq -\mathrm{Re}\,\gamma + l \leq 1 - \nu_-, \tag{16.24}$$

and hence there is an $s > 0$ such that $-\nu_- < -\mathrm{Re}\,\gamma + l - s < 1 - \nu_+$. Proceeding as above we conclude that

$$(\tau - t)^{l-\gamma} \in H^p_+(\mathbf{T}, \varrho), \quad \Big(1 - \frac{t}{\tau}\Big)^{l-\gamma} \in H^p_-(\mathbf{T}, \varrho)$$

and then that the left-hand side of (16.23) is in $H^1_+(\mathbf{T})$ and the right-hand side of (16.23) belongs to $\tau^{l-\varkappa}H^1_-(\mathbf{T})$. This is impossible for $l < \varkappa$, whence $\varkappa = l$. Consequently, both sides of (16.23) represent a nonzero constant. It follows that $|(\tau - t)^{l-\gamma}|\varrho(\tau)$ belongs to $A_p(\mathbf{T})$ together with $|b_+^{-1}|\varrho$. Therefore $-\nu_- < -\mathrm{Re}\,\gamma + l < 1 - \nu_+$, which contradicts (16.24). This contradiction completes the proof. □

16.4 Toeplitz Operators with PC Symbols

Let $1 < p < \infty$ and $w \in A_p(\mathbf{R})$.

Theorem 16.17. (a) *Each of the sets*

$$\begin{aligned} I_x(p,w) &:= \left\{\lambda \in \mathbf{R} : \left|\frac{\xi - x}{\xi + i}\right|^\lambda w(\xi) \in A_p(\mathbf{R})\right\} \quad (x \in \mathbf{R}), \\ I_\infty(p,w) &:= \left\{\lambda \in \mathbf{R} : |\xi + i|^{-\lambda} w(\xi) \in A_p(\mathbf{R})\right\} \end{aligned}$$

is an open interval of length not greater than 1 *which contains the origin:*

$$I_x(p,w) = \big(-\nu_x^-(p,w),\, 1 - \nu_x^+(p,w)\big) \quad (x \in \dot{\mathbf{R}})$$

with $0 < \nu_x^-(p,w) \le \nu_x^+(p,w) < 1$.
(b) *If* $a \in PC$, *then the essential spectrum of* $T(a)$ *on* $H^p(\mathbf{R}, w)$ *is equal to the set*

$$\begin{aligned} a_{\#}^{p,w}(\dot{\mathbf{R}}) &:= \Big(\bigcup_{x \in \mathbf{R}} \mathcal{H}\big(a(x-0), a(x+0);\, \nu_x^-(p,w), \nu_x^+(p,w)\big)\Big) \\ &\quad \cup \mathcal{H}\big(a(+\infty), a(-\infty);\, \nu_\infty^-(p,w), \nu_\infty^+(p,w)\big). \end{aligned}$$

Proof. We reduce the case of the line to the circle case by means of the isomorphism B given by (16.2). Thus, $T(a) - \mu I$ is Fredholm on $H^p(\mathbf{R}, w)$ if and only if the operator $T_0(B_0 a) - \mu I$ is Fredholm on $H^p(\mathbf{T}, \varrho)$, where ϱ is the weight (16.1) and $B_0 a$ is defined by (6.5):

$$(B_0 a)(t) := a\left(i\frac{1+t}{1-t}\right), \quad t \in \mathbf{T} \setminus \{1\}.$$

From Theorem 16.13 we therefore deduce that

$$\operatorname{sp_{ess}} T(a) = \bigcup_{t \in \mathbf{T}} \mathcal{H}\big((B_0 a)(t-0), (B_0 a)(t+0);\, \nu_t^-(p,\varrho), \nu_t^+(p,\varrho)\big), \tag{16.25}$$

where $(-\nu_t^-(p,\varrho), 1 - \nu_t^+(p,\varrho)) = I_t(p,\varrho)$. If $t \in \mathbf{T} \setminus \{1\}$, then

$$(B_0 a)(t \pm 0) = a(x \pm 0) \text{ where } x = i(1+t)/(1-t), \tag{16.26}$$

and λ belongs to $I_t(p,\varrho)$ if and only if

$$|\tau - t|^\lambda \varrho(\tau) = |\tau - t|^\lambda w\left(i\frac{1+\tau}{1-\tau}\right) |1 - \tau|^{1-2/p} \in A_p(\mathbf{T}),$$

which, by (16.3), is equivalent to requiring that

$$\left|\frac{\xi - i}{\xi + i} - \frac{x - i}{x + i}\right|^\lambda \varrho\Big(\frac{\xi - i}{\xi + i}\Big) \left|\frac{\xi + i}{2}\right|^{1-2/p} = \frac{2^\lambda}{|x + i|^\lambda} \left|\frac{\xi - x}{\xi + i}\right|^\lambda w(\xi) \in A_p(\mathbf{R}).$$

As $2^\lambda/|x+i|^\lambda$ is a nonzero constant, it follows that

$$I_t(p,\varrho) = I_x(p,w). \tag{16.27}$$

If $t=1$, then $x := i(1+t)/(1-t) = \infty$ and

$$(B_0 a)(t\pm 0) = a(\mp\infty). \tag{16.28}$$

Furthermore, $\lambda \in I_1(p,\varrho)$ exactly if

$$|\tau-1|^\lambda \varrho(\tau) = |\tau-1|^\lambda w\left(i\frac{1+\tau}{1-\tau}\right)|1-\tau|^{1-2/p} \in A_p(\mathbf{T}),$$

and by (16.3) this happens if and only if

$$\left|\frac{\xi-i}{\xi+i}-1\right|^\lambda \varrho\Big(\frac{\xi-i}{\xi+i}\Big)\left|\frac{\xi+i}{2}\right|^{1-2/p} = 2^\lambda|\xi+i|^{-\lambda}w(\xi) \in A_p(\mathbf{R}),$$

whence

$$I_1(p,\varrho) = I_\infty(p,w). \tag{16.29}$$

Combining (16.27) and (16.29) with Proposition 16.8, we get part (a), while (16.25) to (16.29) in conjunction with Theorem 16.13 give part (b). □

Clearly, the set $a_\#^{p,w}(\dot{\mathbf{R}})$ results from the essential range of a by filling in certain horns between the endpoints of the jumps. The set

$$a_0^{p,w}(\dot{\mathbf{R}}) := \Big(\bigcup_{x\in\mathbf{R}} \mathcal{A}(a(x-0),a(x+0);\nu_x^0(p,w))\Big) \cup \mathcal{A}\big(a(+\infty),a(-\infty);\nu_\infty^0(p,w)\big),$$

where $\nu_x^0 := (\nu_x^- + \nu_x^+)/2$, is a closed continuous and naturally oriented curve which is completely contained in $a_\#^{p,w}(\dot{\mathbf{R}})$. Thus, if $0\notin a_\#^{p,w}(\dot{\mathbf{R}})$ then

$$\operatorname{wind}^{p,w} a := \operatorname{wind} a_0^{p,w}$$

is a well-defined integer.

Theorem 16.18. *If $a\in PC$ and $0\notin a_\#^{p,w}(\dot{\mathbf{R}})$, then the index of $T(a)$ on $H^p(\mathbf{R},w)$ is given by*

$$\operatorname{Ind} T(a) = -\operatorname{wind}^{p,w} a. \tag{16.30}$$

Proof. This is immediate from Theorems 16.13 and 16.17 and the equality $\operatorname{Ind} T(a) = \operatorname{Ind} T_0(B_0 a)$. □

Suppose $a \in PC$ has only finitely many jumps, at the points $\Lambda_a \subset \mathbf{R}$ and possibly at ∞. If $0 \notin a_{\#}^{p,w}(\mathbf{R})$, we define the Cauchy index $\operatorname{ind}^{p,w} a$ of a with respect to p, w by

$$\operatorname{ind}^{p,w} a := \sum_l \operatorname{ind}_l a + \sum_{x\in\Lambda_a} \left(-\nu_x^0 + \left\{\nu_x^0 + \frac{1}{2\pi} \arg \frac{a(x+0)}{a(x-0)}\right\}\right),$$

where l ranges over the connected components of $\mathbf{R} \setminus \Lambda_a$. One can show that

$$\operatorname{wind}^{p,w} a = \operatorname{ind}_{p,w} a + \left(-\nu_\infty^0 + \left\{\nu_\infty^0 + \frac{1}{2\pi} \arg \frac{a(-\infty)}{a(+\infty)}\right\}\right).$$

Thus, we can write (16.30) also in the form

$$\operatorname{Ind} T(a) = -\sum_l \operatorname{ind}_l a + \sum_{x\in\Lambda_a\cup\{\infty\}} \left(\nu_x^0 - \left\{\nu_x^0 + \frac{1}{2\pi} \arg \frac{a(x+0)}{a(x-0)}\right\}\right)$$

where, as usual, $a(\infty \pm 0) := a(\mp\infty)$.

Example 16.19. Consider $T(-\operatorname{sgn} x)$ on $H^p(\mathbf{R}, |x|^\mu)$ where $-1/p < \mu < 1/q$. The function $-\operatorname{sgn} x$ has exactly two jumps: at 0 and at ∞. We have

$$\begin{aligned}
I_0(p, |x|^\mu) &= \left\{\lambda \in \mathbf{R} : \left|\frac{\xi}{\xi+i}\right|^\lambda |\xi|^\mu \in A_p(\mathbf{R})\right\} \\
&= \{\lambda \in \mathbf{R} : -1/p < \lambda + \mu < 1/q\} = (-1/p - \mu, 1/q - \mu), \\
I_\infty(p, |x|^\mu) &= \{\lambda \in \mathbf{R} : |\xi + i|^{-\lambda} |\xi|^\mu \in A_p(\mathbf{R})\} \\
&= \{\lambda \in \mathbf{R} : -1/p < \lambda + \mu < 1/q\} = (-1/q + \mu, 1/p + \mu)
\end{aligned}$$

whence $\nu_0^- = \nu_0^+ = 1/p + \mu$ and $\nu_\infty^- = \nu_\infty^+ = 1/q - \mu$. Thus, by Theorem 16.17,

$$\begin{aligned}
&\operatorname{sp}_{\mathrm{ess}} T(-\operatorname{sgn} x) = \mathcal{H}(1, -1; \nu_0^-, \nu_0^+) \cup \mathcal{H}(-1, 1; \nu_\infty^-, \nu_\infty^+) \\
&= \mathcal{A}(1, -1; 1/p + \mu) \cup \mathcal{A}(-1, 1; 1/q - \mu).
\end{aligned}$$

Since $\mathcal{A}(1, -1; 1/p + \mu) = \mathcal{A}(-1, 1; 1/q - \mu)$, we get

$$\operatorname{sp}_{\mathrm{ess}} T(-\operatorname{sgn} x) = \mathcal{A}(-1, 1; 1/q - \mu), \tag{16.31}$$

and Theorems 16.3(a) and 16.18 now imply that

$$\operatorname{sp} T(-\operatorname{sgn} x) = \mathcal{A}(-1, 1; 1/q - \mu). \tag{16.32}$$

□

We now turn to the $H^p(\mathbf{R}, w)$ analogue of Theorem 2.21. Let $\operatorname{alg}_{p,w} T(PC)$ be the smallest closed subalgebra of $\mathcal{B}(H^p(\mathbf{R}, w))$ which contains the set $\{T(a) : a \in PC\}$. One can show that $\operatorname{alg}_{p,w} T(PC)$ contains the set $\mathcal{K}(H^p(\mathbf{R}, w))$ of all compact operators on $H^p(\mathbf{R}, w)$. For $A \in \mathcal{B}(H^p(\mathbf{R}, w))$, let A^π stand for the coset $A + \mathcal{K}(H^p(\mathbf{R}, w))$.

Theorem 16.20. *The Banach algebra*

$$\operatorname{alg}_{p,w} T^\pi(PC) := \operatorname{alg}_{p,w} T(PC)/\mathcal{K}(H^p(\mathbf{R}, w))$$

is inverse closed in $\mathcal{B}(H^p(\mathbf{R}, w))/\mathcal{K}(H^p(\mathbf{R}, w))$ and commutative. The maximal ideal space can be identified with the "horn bundle"

$$\mathcal{M}_{p,w} := \bigcup_{x \in \dot{\mathbf{R}}} \left(\{x\} \times \mathcal{H}(0, 1; \nu_x^-(p, w), \nu_x^+(p, w)\right)$$

in the following sense: the Gelfand map

$$\Gamma : \operatorname{alg}_{p,w} T^\pi(PC) \to C(\mathcal{M}_{p,w})$$

is for $a \in PC$ and $(x, \mu) \in \mathcal{M}_{p,w}$ given by

$$\big(\Gamma T^\pi(a)\big)(x, \mu) = (1 - \mu)a(x - 0) + \mu a(x + 0).$$

The theorem can be reduced to the circle case as above, and in this case a full proof is in [34] and [37, Chapter 8].

We finally consider matrix symbols. Let $\mathcal{M}_{p,w}$ be as in Theorem 16.20 and put

$$\mathcal{M}_{p,w}^0 = \bigcup_{x \in \dot{\mathbf{R}}} \Big(\{x\} \times \mathcal{A}(0, 1; \nu_x^0(p, w))\Big), \quad \nu_x^0(p, w) := \frac{1}{2}\big(\nu_x^-(p, w) + \nu_x^+(p, w)\big).$$

Given $a \in PC_{N \times N}$, we define

$$a^{p,w} : \mathcal{M}_{p,w} \to \mathbf{C}^{N \times N}, \quad \text{and} \quad a_0^{p,w} : \mathcal{M}_{p,w}^0 \to \mathbf{C}^{N \times N}$$

by

$$a^{p,w}(x, \mu) := (1 - \mu)a(x - 0) + \mu a(x + 0),$$

$$a_0^{p,w}(x, \mu) := (1 - \mu)a(x - 0) + \mu a(x + 0)$$

(on their respective domains). One can think of the range of $\det a_0^{p,w}$ on $\mathcal{M}_{p,w}^0$ as a closed continuous and naturally oriented curve which results from the essential range of $\det a$ by joining $\det a(x - 0)$ to $\det a(x + 0)$ by the curve

$$\Big\{\det\big((1 - \mu)a(x - 0) + \mu a(x + 0)\big) : \mu \in \mathcal{A}(0, 1; \nu_x^0(p, w))\Big\}$$

whenever $a(x - 0) \neq a(x + 0)$. Thus, if $\det a_0^{p,w}(x, \mu) \neq 0$ for all $(x, \mu) \in \mathcal{M}_{p,w}^0$, then the winding number $\operatorname{wind}(\det a_0^{p,w})$ is a well-defined integer.

Theorem 16.21. *Let $a \in PC_{N\times N}$.*
(a) *If $\det a^{p,w}(x_0,\mu_0) = 0$ for some $(x_0,\mu_0) \in \mathcal{M}_{p,w}$, then $T(a)$ is not semi-Fredholm on $H^p_N(\mathbf{R},w)$.*
(b) *If $\det a^{p,w}(x,\mu) \neq 0$ for all $(x,\mu) \in \mathcal{M}_{p,w}$, then $T(a)$ is Fredholm on $H^p_N(\mathbf{R},w)$ and*

$$\operatorname{Ind} T(a) = -\operatorname{wind}(\det a_0^{p,w}).$$

This can be reduced to the scalar case as in the H^2_+ case by having recourse to the factorization $a = b\varphi c$ with $b, c \in GC_{N\times N}(\dot{\mathbf{R}})$ and an upper-triangular matrix function $\varphi \in GPC_{N\times N}$. Of course, the Fredholm criterion is also a direct consequence of Theorems 5.5(a) and 16.20.
We remark that if $a \in PC_{N\times N}$ has at most finitely many discontinuities and $\det a^{p,w}(x,\mu) \neq 0$ for all $(x,\mu) \in \mathcal{M}_{p,w}$, then

$$\begin{aligned}
\operatorname{wind}(\det a^{p,w}) &= \operatorname{ind}(\det a_0^{p,w}) + \sum_{k=1}^{N}\Big(-\nu^0_\infty + \big\{\nu^0_\infty + \frac{1}{2\pi}\arg\xi_k(\infty)\big\}\Big),\\
\operatorname{ind}(\det a_0^{p,w}) &= \sum_l \operatorname{ind}_l(\det a) + \sum_{x\in\Lambda_a}\Big(-\nu^0_x + \big\{\nu^0_x + \frac{1}{2\pi}\arg\xi_k(x)\big\}\Big),
\end{aligned}$$

where $\xi_1(x),\dots,\xi_N(x)$ are the eigenvalues of $a^{-1}(x-0)a(x+0)$.
Finally, for symbols in $C_{N\times N}(\overline{\mathbf{R}})$, the Fredholm criterion contained in Theorem 16.21 can be restated as follows.

Theorem 16.22. *Let $a \in C_{N\times N}(\overline{\mathbf{R}})$ and put $\nu^\pm_\infty := \nu^\pm_\infty(p,w)$. Then the following are equivalent:*
(i) *$T(a)$ is Fredholm on $H^p_N(\mathbf{R},w)$;*
(ii) *$\det a(x) \neq 0$ for all $x \in \overline{\mathbf{R}}$ and*

$$\nu + \frac{1}{2\pi}\arg\xi_j\big(a^{-1}(+\infty)a(-\infty)\big) \notin \mathbf{Z} \tag{16.33}$$

for all $\nu \in [\nu^-_\infty,\nu^+_\infty]$ and all eigenvalues $\xi_j = \xi_j(a^{-1}(+\infty)a(-\infty))$ of the matrix $a^{-1}(+\infty)\,a(-\infty)$.
(iii) *$\det a(x) \neq 0$ for all $x \in \overline{\mathbf{R}}$ and*

$$\operatorname{sp}\big(a^{-1}(+\infty)a(-\infty)\big) \cap \Sigma[1-\nu^+_\infty, 1-\nu^-_\infty] = \emptyset \tag{16.34}$$

where $\Sigma[\alpha,\beta]$ is the closed sector

$$\Sigma[\alpha,\beta] := \big\{re^{2\pi i\varphi} : r\in[0,\infty],\ \varphi\in[\alpha,\beta]\big\}. \tag{16.35}$$

Proof. Put $\mathcal{H}(z_1,z_2) := \mathcal{H}(z_1,z_2;\nu_\infty^-,\nu_\infty^+)$. If $\det a(\pm\infty)\neq 0$, we have

$$\begin{aligned}
&\det\big((1-\mu)a(+\infty)+\mu a(-\infty)\big)\neq 0 \quad \forall\,\mu\in\mathcal{H}(0,1)\\
&\Longleftrightarrow \det\big((1-\mu)I+\mu a^{-1}(+\infty)a(-\infty)\big)\neq 0 \quad \forall\,\mu\in\mathcal{H}(0,1)\\
&\Longleftrightarrow 1-\mu+\mu\xi_j\neq 0 \quad \forall\, j \quad \forall\,\mu\in\mathcal{H}(0,1)\\
&\Longleftrightarrow 1+\mu(\xi_j-1)\neq 0 \quad \forall\, j \quad \forall\,\mu\in\mathcal{H}(0,1)\\
&\Longleftrightarrow 0\notin\mathcal{H}(1,\xi_j) \quad \forall\, j\\
&\Longleftrightarrow \nu+\frac{1}{2\pi}\arg\xi_j\notin\mathbf{Z} \quad \forall\, j \quad \forall\,\nu\in[\nu_\infty^-,\nu_\infty^+].
\end{aligned}$$

From Theorem 16.21 we therefore get the equivalence (i)$\Longleftrightarrow$(ii). Further,

$$\begin{aligned}
&\nu+\frac{1}{2\pi}\arg\xi_j\notin\mathbf{Z} \quad \forall\,\nu\in[\nu_\infty^-,\nu_\infty^+]\\
&\Longleftrightarrow 1-\mu+\mu\xi_j\neq 0 \quad \forall\,\mu\in\mathcal{H}(0,1)\\
&\Longleftrightarrow \xi_j\neq\frac{\mu-1}{\mu} \quad \forall\,\mu\in\mathcal{H}(0,1)\\
&\Longleftrightarrow \xi_j\neq\frac{\zeta/(\zeta-1)-1}{\zeta/(\zeta-1)} \quad \forall\,\zeta\in\big\{\varrho e^{2\pi i\nu}:\varrho\in[0,\infty],\,\nu\in[\nu_\infty^-,\nu_\infty^+]\big\}\\
&\Longleftrightarrow \xi_j\neq\frac{1}{\zeta} \quad \forall\,\zeta\in\big\{\varrho e^{2\pi i\nu}:\varrho\in[0,\infty],\,\nu\in[\nu_\infty^-,\nu_\infty^+]\big\}\\
&\Longleftrightarrow \xi_j\notin\big\{re^{2\pi i(1-\nu)}:r\in[0,\infty],\,\nu\in[\nu_\infty^-,\nu_\infty^+]\big\}\\
&\Longleftrightarrow \xi_j\notin\Sigma[1-\nu_\infty^+,1-\nu_\infty^-],
\end{aligned}$$

which proves the equivalence (ii)$\Longleftrightarrow$(iii). □

Taking the eigenvalues η_j of $a^{-1}(-\infty)a(+\infty)$, we must replace (16.33), (16.34) by

$$\begin{gathered}
\nu-\frac{1}{2\pi}\arg\eta_j\big(a^{-1}(-\infty)a(+\infty)\big)\notin\mathbf{Z},\\
\operatorname{sp}\big(a^{-1}(-\infty)a(+\infty)\big)\cap\Sigma[\nu_\infty^-,\nu_\infty^+]=\emptyset,
\end{gathered}$$

respectively. If $T(a)$ is Fredholm, the index can also be computed by the formulas

$$\begin{aligned}
\operatorname{Ind}T(a) &= -\operatorname{ind}(\det a)+N\nu_\infty^0-\sum_{j=1}^N\Big\{\nu_\infty^0+\frac{1}{2\pi}\arg\xi_j\Big\}\\
&= -\operatorname{ind}(\det a)+N\nu_\infty^0-\sum_{j=1}^N\Big\{\nu_\infty^0-\frac{1}{2\pi}\arg\eta_j\Big\}. \qquad (16.36)
\end{aligned}$$

Notes

A_p weights on $\mathbf{R}$ were introduced by Muckenhoupt in [158], Theorems 16.1 and 16.2 were proved by Hunt, Muckenhoupt, and Wheeden in [113]. Many analysis books contain a detailed discussion of A_p weights and their properties; see, for example, [84] and [142]. Generalizations to weights defined on Carleson curves in $\mathbf{C}$ different from $\mathbf{R}$ or $\mathbf{T}$ can be found, e.g., in [37] and [83]. We also remark that the transition to the matrix equivalent of the A_p condition is by no means easy: see [222] for $p = 2$ and [162], [223] for general p.

The results of Section 16.2 are from Simonenko's papers [204] and [207].

The Fredholm theory of Toeplitz operators with PC symbols on L^p spaces with power weights has a long history: see, e.g., [46], [99], and [114]. Singular integral operators with PC coefficients on the spaces $L^p(\mathbf{R})$ and thus Toeplitz operators with PC symbols on H^p were studied by Widom [226]. In the case of piecewise Lyapunov curves, a complete solution of the problem was given by Gohberg and Krupnik [96]. They also disposed of the matrix case [97]. Their paper [98] is devoted to Banach algebras generated by singular integral operators with PC coefficients, thus covering the algebras generated by Toeplitz operators with PC symbols. For individual operators (both with scalar and matrix coefficients), the passage to general A_p weights is in [213]. Toeplitz operators on weighted Hardy spaces on the real line were subsequently considered in [48].

For algebras generated by singular integral operators with PC symbols, the transition to general weights was made in [100] (including an index formula) and [79] (Fredholm criterion). The curves considered in [100] are not the most general ones, but this class of curves is larger than the class of Lyapunov curves. The most general case, that is, the case of Carleson curves and Muckenhoupt weights on them, was disposed of in [26], [34], [35], [37], [38]. A different approach to singular integral operators with slowly oscillating data, which makes use of the theory of Mellin pseudodifferential operators with slowly oscillating double symbols and limit operators techniques, was developed in [39], [40], [41], [42], [179], [180], [181]. An exposition of the theory of boundary value problems and singular integral operators with discontinuous coefficients on weighted L^p spaces is also given in the recent publication [140].

For $w = 1$, the index formula (16.30) was established by Widom [226] in terms of the jumps of the argument of the symbol. In the form cited here, formula (16.30) was obtained by Gohberg and Krupnik [98], [99] in the case of power weights and in [48] for general Muckenhoupt weights. In the case of operators on Lyapunov curves, see also Karapetiants and Samko's works [117] and [118] for index formulas.

The indices of powerlikeness for Muckenhoupt weights were (implicitly) introduced in [213]. Explicit formulas for these indices, along with examples of their computation, are in [36], [38]; see also [37]. Lemma 16.15 for $\mu_k \in (0,1)$ is in [99, Chapter 2, Theorem 4.8], the general case is due to Grudsky [101].

Chapter 17

Zero-Order Pseudodifferential Operators

Our next objective is Wiener-Hopf operators on $L^p(\mathbf{R}_+, w)$. One of the technical differences between Wiener-Hopf and Toeplitz operators is the emergence of a multiplier problem in the Wiener-Hopf case. This problem is studied in the first section of this chapter, where we in particular prove the important Stechkin inequality. Another point is that Wiener-Hopf factorization is in fact a technique for Toeplitz operators that does not work for Wiener-Hopf operators on L^p spaces. This obstacle can be overcome by appropriately switching between different spaces and naturally leads to the consideration of pseudodifferential operators with piecewise continuous symbols.

17.1 Fourier Multipliers

Throughout what follows, we assume that $1 < p < \infty$ and $w \in A_p(\mathbf{R})$. By $L^p(\mathbf{R}_+, w)$ we understand the space $L^p(\mathbf{R}_+, w|\mathbf{R}_+)$.

A function $a \in L^\infty(\mathbf{R})$ is called a *Fourier multiplier* on $L^p(\mathbf{R}, w)$ if the map $f \mapsto F^{-1}aFf$ maps $L^2(\mathbf{R}) \cap L^p(\mathbf{R}, w)$ into itself and extends to a bounded operator on $L^p(\mathbf{R}, w)$ (notice that $L^2(\mathbf{R}) \cap L^p(\mathbf{R}, w)$ is dense in $L^p(\mathbf{R}, w)$ whenever w is in $A_p(\mathbf{R})$). The latter operator is then denoted by $W^0(a)$. We let $M_{p,w}$ stand for the set of all Fourier multipliers on $L^p(\mathbf{R}, w)$. One can show that $M_{p,w}$ is a Banach algebra under the norm

$$\|a\|_{M_{p,w}} := \|W^0(a)\|_{\mathcal{B}(L^p(\mathbf{R},w))}.$$

If $w(x) = 1$ for all $x \in \mathbf{R}$, we write M_p for $M_{p,w}$. The algebra M_2 equals $L^\infty(\mathbf{R})$. The following theorem gives us an important subset of $M_{p,w}$.

Theorem 17.1 (Stechkin's inequality). *If $a \in PC$ and a has finite total variation $V_1(a)$, then $a \in M_{p,w}$ and*

$$\|a\|_{M_{p,w}} \le \|S\|_{\mathcal{B}(L^p(\mathbf{R},w))}\big(\|a\|_\infty + V_1(a)\big). \tag{17.1}$$

Proof. If $a(x) = \chi_{(c,\infty)}(x) = (1/2)(1 + \operatorname{sgn}(x - c))$, then

$$W^0(a) = \frac{1}{2}(I - e_{-c}Se_cI)$$

due to Example 1.18. Consequently, $\|W^0(a)\| \le (1 + \|S\|)/2$. Because $S^2 = I$, it follows that $1 \le \|S\|$, whence $\|W^0(a)\| \le \|S\|$. In case $a \in PC$ is a piecewise constant function, we can write

$$a(x) = a_0 + \sum_{k=1}^{n}(a_k - a_{k-1})\chi_{(c_k,\infty)}.$$

From what was already proved we deduce that

$$\begin{aligned}\|W^0(a)\| &\le |a_0| + \sum_{k=1}^{n} |a_k - a_{k-1}|\,\|S\| \\ &\le \Big(|a_0| + \sum_{k=1}^{n} |a_k - a_{k-1}|\Big)\|S\| \le \big(\|a\|_\infty + V_1(a)\big)\|S\|.\end{aligned}$$

Thus, we have the desired inequality for all piecewise constant functions. The extension of the inequality to all $a \in PC$ with $V_1(a) < \infty$ is based on interpolation and approximation.

Let $a \in PC$ be piecewise constant. For the sake of definiteness, assume $p \ge 2$ (otherwise we can pass to adjoints). If $w \in A_p(\mathbf{R})$, then $w^{1+\delta} \in A_{p(1+\varepsilon)}(\mathbf{R})$ whenever $|\varepsilon|$ and $|\delta|$ are sufficiently small (see, e.g., [84, Corollary VI.6.10] or [37, Theorem 2.31]). We know that

$$W^0(a) \in \mathcal{B}\big(L^{p(1+\varepsilon)}(\mathbf{R}, w^{1+\delta})\big), \quad W^0(a) \in \mathcal{B}\big(L^2(\mathbf{R})\big),$$

and there are sufficiently small $\varepsilon > 0$, $\delta > 0$ and a number $t \in (0,1)$ such that

$$\frac{1}{p} = \frac{1-t}{2} + \frac{t}{p(1+\varepsilon)}, \qquad w = 1^{1-t}w^{(1+\delta)t}.$$

Consequently, by the Stein-Weiss interpolation theorem [218],

$$\begin{aligned}\|W^0(a)\|_{\mathcal{B}(L^p(\mathbf{R},w))} &\le \|W^0(a)\|^{1-t}_{\mathcal{B}(L^2(\mathbf{R}))}\|W^0(a)\|^t_{\mathcal{B}(L^{p(1+\varepsilon)}(\mathbf{R},w^{1+\delta}))} \\ &= \|a\|^{1-t}_\infty\|W^0(a)\|^t_{\mathcal{B}(L^{p(1+\varepsilon)}(\mathbf{R},w^{1+\delta}))}.\end{aligned}$$

Given $a \in PC$ with $V_1(a) < \infty$, we can find piecewise constant functions a_n in PC such that

$$\|a - a_n\|_\infty \to 0 \text{ as } n \to \infty, \quad V_1(a_n) \le V_1(a) \text{ for all } n$$

(see, e.g., [70, Lemmas 2.9 and 2.10]). Since

$$\begin{aligned}\|W^0(a_n)-W^0(a_m)\|_{\mathcal{B}(L^p(\mathbf{R},w))} &\le \|a_n-a_m\|_\infty^{1-t}\|W^0(a_n-a_m)\|^t_{\mathcal{B}(L^{p(1+\varepsilon)}(\mathbf{R},w^{1+\delta}))}\\ &\le \|S\|^t\|a_n - a_m\|_\infty^{1-t}\big(\|a_n - a_m\|_\infty + V_1(a_n - a_m)\big)^t\\ &\le \|S\|^t\|a_n - a_m\|_\infty^{1-t}\big(\|a_n - a_m\|_\infty + V_1(a_n) + V_1(a_m)\big)^t,\end{aligned}$$

it results that $\{W^0(a_n)\}$ is a Cauchy sequence and therefore has a limit A in $\mathcal{B}(L^p(\mathbf{R}, w))$ (in the last two inequalities $\|S\|$ is understood as the norm of S on the space $L^{p(1+\varepsilon)}(\mathbf{R}, w^{1+\delta})$). As

$$\|W^0(a_n) - W^0(a)\|_{\mathcal{B}(L^2(\mathbf{R}))} = \|a_n - a\|_\infty = o(1),$$

we see that $W^0(a)f = Af$ for all $f \in L^2(\mathbf{R}) \cap L^p(\mathbf{R}, w)$, whence $A = W^0(a)$. Finally, because

$$\begin{aligned}\|W^0(a)\| &\le \|W^0(a_n)\| + \|W^0(a) - W^0(a_n)\|\\ &\le \|S\|\big(\|a_n\|_\infty + V_1(a_n)\big) + \|W^0(a) - W^0(a_n)\|\\ &\le \|S\|\,\big(\|a_n\|_\infty + V_1(a)\big) + \|W^0(a) - W^0(a_n)\|,\end{aligned}$$

passage to the limit $n \to \infty$ gives the assertion. □

For spaces without weight (= spaces with the weight $w(x) = 1$), one can characterize the set M_p also as follows. An operator $A \in \mathcal{B}(L^p(\mathbf{R}))$ is said to be *translation invariant* if $U_hAU_h^{-1} = A$ for all $h \in \mathbf{R}$ where $(U_hf)(t) := f(t - h)$. The following result is well known.

Theorem 17.2 (Hörmander). *An operator $A \in \mathcal{B}(L^p(\mathbf{R}))$ is translation invariant if and only if $A = W^0(a)$ with $a \in M_p$.*

We also remark that $M_p = M_q$ $(1/p+1/q = 1)$ and that if $a \in M_p$, then $a \in M_r$ for all r between p and q. Furthermore, M_p is continuously embedded in $M_2 = L^\infty(\mathbf{R})$ for all $p \in (1, \infty)$. Moreover,

$$\|a\|_\infty \le \|a\|_{M_p} = \|a\|_{M_q}. \tag{17.2}$$

Also notice that the norms $\|U_h\|$ are all equal to 1 on $L^p(\mathbf{R})$, whereas spaces with weights cause problems with the uniform boundedness of the norms $\|U_h\|$.

We denote by $C_{p,w}(\dot{\mathbf{R}})$ (resp. $PC_{p,w}$) the closure of the set of all $a \in C(\dot{\mathbf{R}})$ (resp. $a \in PC$) with finite total variation in $M_{p,w}$. Clearly, $C_{p,w}(\dot{\mathbf{R}}) \subset PC_{p,w}$. One can show (see, e.g., [183, Proposition 12.2] for power weights) that $PC_{p,w}$ is continuously embedded in $L^\infty(\mathbf{R})$. This implies that

$$C_{p,w}(\dot{\mathbf{R}}) \subset C(\dot{\mathbf{R}}), \quad PC_{p,w} \subset PC.$$

Note, however, that $C_{p,w}$ and $PC_{p,w}$ are in general properly contained in the intersections $M_{p,w} \cap C(\dot{\mathbf{R}})$ and $M_{p,w} \cap PC$, respectively.

Proposition 17.3. (a) *The algebras $C_{p,w}(\dot{\mathbf{R}})$ and $PC_{p,w}$ are inverse closed in $L^\infty(\mathbf{R})$.*

(b) *If $a \in PC_{p,w}$, then both the spectrum and the essential spectrum of the operator $W^0(a) \in \mathcal{B}(L^p(\mathbf{R}, w))$ are equal to the essential range $\mathcal{R}(a) = a(\overline{\mathbf{R}})$ of a.*

For power weights, this result is in [70] and [183]. The proofs given there can be extended to general $w \in A_p(\mathbf{R})$.

17.2 Wiener-Hopf Operators with PC Symbols

The Wiener-Hopf operator $W(a)$ generated by a function $a \in M_{p,w}$ (its so-called symbol) is the compression of $W^0(a)$ to the half-line $\mathbf{R}_+ = (0,\infty)$, i.e., $W(a)$ is the bounded operator on $L^p(\mathbf{R}_+, w)$ $(= L^p(\mathbf{R}_+, w|\mathbf{R}_+))$ acting by the rule $f \mapsto (W^0(a)f)|\mathbf{R}_+$. We may identify $L^p(\mathbf{R}_+, w)$ with $\chi_+ L^p(\mathbf{R}, w)$ and may therefore also think of $W(a)$ as the operator $\chi_+ W^0(a)|\mathrm{Im}\,\chi_+ I$.

Theorem 17.4. *If $a \in PC_{p,w}$, then for $W(a)$ to be invertible on $L^p(\mathbf{R}_+, w)$, it is necessary and sufficient that $W(a)$ be Fredholm of index zero on $L^p(\mathbf{R}_+, w)$.*

Theorem 17.5. *Let $a \in C_{p,w}(\dot{\mathbf{R}})$. The operator $W(a)$ is Fredholm on $L^p(\mathbf{R}_+, w)$ if and only if $a(x) \neq 0$ for all $x \in \dot{\mathbf{R}}$. In that case* $\mathrm{Ind}\, W(a) = -\,\mathrm{wind}\, a$.

The proofs of these two theorems given in [70] and [183] for power weights also work in the case of general weights.

Example 17.6. The operator $S_+ = W(-\,\mathrm{sgn}\, x)$ is the simplest example of a Wiener-Hopf operator with a symbol in $PC_{p,w}$. This operator can be studied on $L^p(\mathbf{R}_+, t^\mu)$ $(-1/p < \mu < 1/q)$ with the help of the trick already employed in Example 1.24. The map

$$\Phi : L^p(\mathbf{R}_+, t^\mu) \to L^p(\mathbf{R}),\ (\Phi f)(t) := e^{-(1/p+\mu)t} f(e^{-t})$$

is an isometric isomorphism. Clearly,

$$(\Phi^{-1} g)(s) = s^{-(1/p+\mu)} g(-\log s), \quad s > 0,$$

and therefore

$$\begin{aligned}(\Phi S_+\Phi^{-1}f)(t) &= \frac{1}{\pi i}\int_{\mathbf{R}_+}\frac{e^{(-1/p+\mu)t}s^{-(1/p+\mu)}f(-\log s)}{s-e^{-t}}\,ds\\ &= \frac{1}{\pi i}\int_{\mathbf{R}}\frac{e^{-(1/p+\mu)(t-y)}f(y)}{1-e^{-(t-y)}}\,dy.\end{aligned}$$

Thus, $\Phi S_+\Phi^{-1}$ is convolution by the kernel

$$k(t)=\frac{1}{\pi i}\frac{e^{-(1/p+\mu)t}}{1-e^{-t}}$$

on $L^p(\mathbf{R})$. The Fourier transform of k is

$$\begin{aligned}&\int_{-\infty}^{\infty}k(t)e^{ixt}\,dt=\frac{1}{\pi i}\int_0^{\infty}\left(\frac{e^{-(1/p+\mu)t}}{1-e^{-t}}e^{ixt}-\frac{e^{-(1/q-\mu)t}}{1-e^{-t}}e^{-ixt}\right)dt\\ &=\frac{1}{\pi i}\int_0^{\infty}\frac{\sinh\left((\frac{1}{2}-\frac{1}{p}-\mu+ix)t\right)}{\sinh\frac{t}{2}}\,dt=\tanh\left(\pi\left(x-i\left(\frac{1}{2}-\frac{1}{\pi}-\mu\right)\right)\right)\\ &=\coth\left(\pi\left(x+i\left(\frac{1}{p}+\mu\right)\right)\right)=M_{-1,1}\left(e^{2\pi(x+i(1/p+\mu))}\right)\end{aligned}$$

where $M_{-1,1}(\zeta):=(\zeta+1)/(\zeta-1)$. From what was said in Section 16.3, we infer that $(Fk)(\overline{\mathbf{R}})$ is the circular arc $\mathcal{A}(-1,1;1/p+\mu)$. Thus, by Proposition 17.3(b),

$$\operatorname{sp}(\Phi S_+\Phi^{-1})=\operatorname{sp}_{\rm ess}(\Phi S_+\Phi^{-1})=\operatorname{sp}W^0(Fk)=\mathcal{A}(-1,1;1/p+\mu).$$

In summary, both the spectrum and the essential spectrum of the operator $S_+=W(-\operatorname{sgn} n)$ on $L^p(\mathbf{R}_+,t^\mu)$ $(-1/p<\mu<1/q)$ are equal to the arc $\mathcal{A}(-1,1;1/p+\mu)$:

$$\operatorname{sp}W(-\operatorname{sgn}x)=\operatorname{sp}_{\rm ess}W(-\operatorname{sgn}x)=\mathcal{A}(-1,1;1/p+\mu). \tag{17.3}$$

Comparing (17.3) and (16.31), (16.32), we see that there is a difference between the Wiener-Hopf and Toeplitz cases. For example, if $\mu=0$ and $2<p<\infty$, then (17.3) is the circular arc in the lower half-plane at the points of which the segment $[-1,1]$ is seen at the angle $2\pi/p$, while (16.31), (16.32) is the circular arc in the upper half-plane determined by the same angle $2\pi/p$ (see Figure 17.1). Nevertheless, the surprise is that in both the Wiener-Hopf and Toeplitz cases we arrive at circular arcs, although (17.3) and (16.31), (16.32) were established by completely different techniques. □

To tackle Wiener-Hopf operators with $PC_{p,w}$ symbols on L^p with general Muckenhoupt weights, the argument of Example 17.6 cannot be used. We proceed in another way.

The link between Toeplitz operators on $H^p(\mathbf{R},w)$ and Wiener-Hopf operators on $L^p(\mathbf{R}_+,w)$ is singular integral operators on $L^p(\mathbf{R},w)$, that is, operators of the form

$$bI+cS=bI+cW^0(-\operatorname{sgn}x)$$

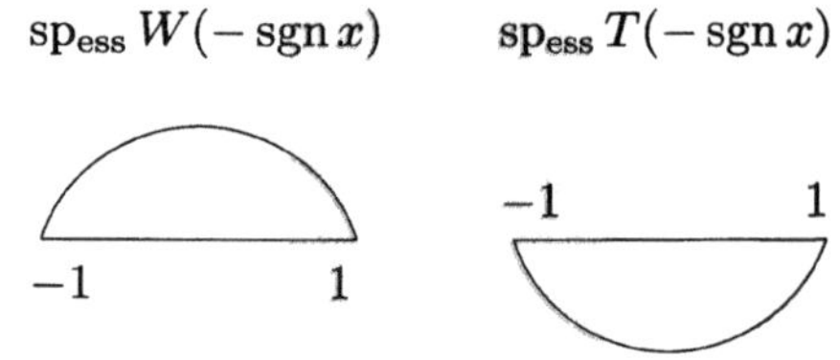

Figure 17.1: The essential spectrum of $W(-\operatorname{sgn} x)$ on $L^p(\mathbf{R}_+)$ (left) and the essential spectrum of $T(-\operatorname{sgn} x)$ on $H^p(\mathbf{R})$ (right).

or, slightly more generally, operators of the form

$$e_{-\eta}(bI + cS)e_\eta I = bI + cW^0\big(-\operatorname{sgn}(x-\eta)\big),$$

where $e_\eta(x) := e^{i\eta x}$. We have

$$\big(I + \chi_+ W^0(a)\chi_- I\big)\big(\chi_- I + \chi_+ W^0(a)\chi_+ I\big) = \chi_- I + \chi_+ W^0(a),$$

and since $I + \chi_+ W^0(a)\chi_- I$ has the inverse $I - \chi_+ W^0(a)\chi_- I$, it follows that the Wiener-Hopf operator $W(a) = \chi_+ W^0(a)|\operatorname{Im}\chi_+ I$ is strongly Φ-equivalent to the operator

$$\chi_- I + \chi_+ W^0(a) \ \text{ on } \ L^p(\mathbf{R}, w). \tag{17.4}$$

Using standard localization techniques, the study of the operator (17.4) can be reduced to the investigation of the operators

$$\chi_- I + \chi_+ W^0\big(a(\eta-0)\chi_{(-\infty,\eta)} + a(\eta+0)\chi_{(\eta,\infty)}\big), \quad \eta \in \mathbf{R} \tag{17.5}$$

and

$$\chi_- I + \chi_+ W^0\big(a(-\infty)\chi_{(-\infty,0)} + a(+\infty)\chi_{(0,\infty)}\big). \tag{17.6}$$

The operators (17.5) and (17.6) are the "local representatives" of the operator (17.4) at $\eta \in \mathbf{R}$ and at $\eta = \infty$, respectively. But if $\alpha, \beta \in \mathbf{C}$, then

$$\begin{aligned}
&\chi_- I + \chi_+ W^0\big(\alpha\chi_{(-\infty,\eta)} + \beta\chi_{(\eta,\infty)}\big)\\
&= \chi_- I + \chi_+ W^0\left(\alpha\frac{1-\operatorname{sgn}(x-\eta)}{2} + \beta\frac{1+\operatorname{sgn}(x-\eta)}{2}\right)\\
&= \left(\chi_- + \frac{\alpha+\beta}{2}\chi_+\right) I + \chi_+\frac{\alpha-\beta}{2} W^0\big(-\operatorname{sgn}(x-\eta)\big)\\
&=: bI + cW^0\big(-\operatorname{sgn}(x-\eta)\big) = e_{-\eta}(bI + cS)e_\eta I.
\end{aligned}$$

As $e_{\pm\eta} I$ are isometric isomorphisms, we arrive at the operators

$$\begin{aligned}
bI + cS &= b(P+Q) + c(P-Q) = (b+c)P + (b-c)Q\\
&= (b-c)\left(\frac{b+c}{b-c}P + Q\right) =: (b-c)(dP+Q).
\end{aligned}$$

Since $dP + Q = (PdP + Q)(I + QdP)$ and $I + QdP$ has the inverse $I - QdP$, we are finally led to operators of the form $Pd|\mathrm{Im}\, P$ with $d \in PC$. But the latter operators are just the Toeplitz operators $T(d)$ on $H^p(\mathbf{R}, w)$ we have considered in Section 16.4. In this way, one can derive the following theorem from Theorems 16.17 and 16.18 (see [48] for the details).

Theorem 17.7. *Let $\nu_x^{\pm}(p, w)$ be as in Theorem 16.17(a). If $a \in PC_{p,w}$, then the essential spectrum of $W(a)$ on $L^p(\mathbf{R}_+, w)$ is the set*

$$\begin{aligned} a^{\#}_{p,w}(\dot{\mathbf{R}}) \;:=\; & \Big(\bigcup_{x\in\mathbf{R}} \mathcal{H}\big(a(x-0), a(x+0);\, \nu_\infty^-(p,w), \nu_\infty^+(p,w)\big)\Big) \\ & \cup \mathcal{H}\big(a(+\infty), a(-\infty);\, \nu_0^-(p,w), \nu_0^+(p,w)\big). \end{aligned}$$

If $0 \notin a^{\#}_{p,w}(\dot{\mathbf{R}})$, then the index of $W(a)$ on $L^p(\mathbf{R}_+, w)$ equals

$$\mathrm{Ind}\, W(a) = -\,\mathrm{wind}_{p,w}\, a,$$

where $\mathrm{wind}_{p,w}\, a$ is the winding number about the origin of the naturally oriented curve

$$\begin{aligned} a^{0}_{p,w}(\dot{\mathbf{R}}) \;:=\; & \Big(\bigcup_{x\in\mathbf{R}} \mathcal{A}\Big(a(x-0), a(x+0);\, \big(\nu_\infty^-(p,w) + \nu_\infty^+(p,w)\big)/2\Big)\Big) \\ & \cup \mathcal{A}\Big(a(+\infty), a(-\infty);\, \big(\nu_0^-(p,w) + \nu_0^+(p,w)\big)/2\Big). \end{aligned}$$

Note that the horns we have to fill in between the endpoints of the jumps at points on $\mathbf{R}$ are all exclusively determined by the behavior of the weight at infinity, while the horn between the endpoints of the possible jump at ∞ depends only on the behavior of the weight at the origin. Figure 17.2 illustrates Theorems 16.17 and 17.7.

Example 17.8. Let $w(t) = t^\mu$ with $-1/p < \mu < 1/q$. By Example 16.19,

$$\nu_0^- = \nu_0^+ = 1/p + \mu, \quad \nu_\infty^- = \nu_\infty^+ = 1/q - \mu.$$

From Theorem 17.7 we therefore deduce that the essential spectrum of $W(-\operatorname{sgn} x)$ on $L^p(\mathbf{R}_+, t^\mu)$ is

$$\mathcal{A}(1, -1; 1/q - \mu) \cup \mathcal{A}(-1, 1; 1/p + \mu) = \mathcal{A}(-1, 1; 1/p + \mu).$$

This in conjunction with Theorems 17.4 and 17.7 shows that $\mathrm{sp}\, W(-\operatorname{sgn} x)$ equals $\mathcal{A}(-1, 1;\ 1/p + \mu)$. Of course, the result of this example is in accordance with Example 17.6. □

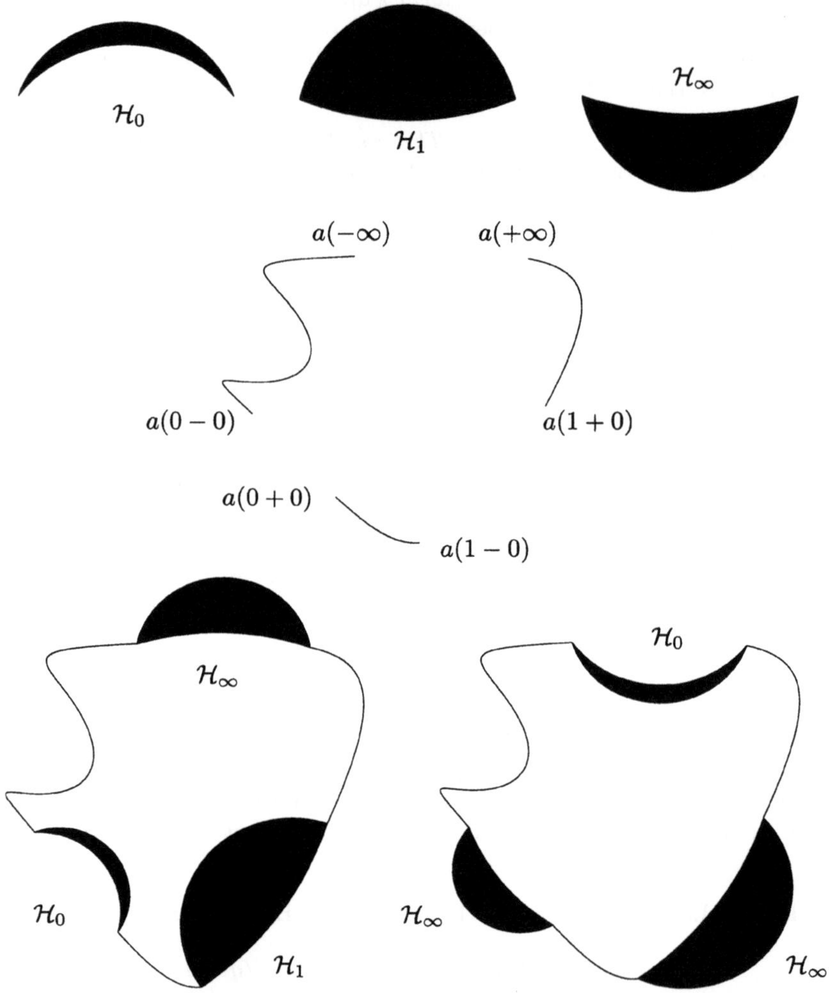

Figure 17.2: In the center we see the range of a function a with three jumps, at 0, 1, and ∞. The three top pictures show the horns $\mathcal{H}(0,1;\nu_0^-(p,w),\nu_0^+(p,w))$, $\mathcal{H}(0,1;\nu_1^-(p,w),\nu_1^+(p,w))$, and $\mathcal{H}(0,1;\nu_\infty^-(p,w),\nu_\infty^+(p,w))$ for given p and given $w \in A_p(\mathbf{R})$. The left set in the bottom is the essential spectrum of $T(a)$ on $H^p(\mathbf{R},w)$, while the right set in the bottom shows the essential spectrum of $W(a)$ on $L^p(\mathbf{R}_+,w)$.

We denote by $\mathrm{alg}_{p,w} W(PC)$ the smallest closed subalgebra of $\mathcal{B}(L^p(\mathbf{R}_+, w))$ that contains all the operators $W(a)$ with $a \in PC_{p,w}$. The set $\mathcal{K}(L^p(\mathbf{R}_+, w))$ of all compact operators can be shown to be a subset of $\mathrm{alg}_{p,w} W(PC)$. As usual, we let A^π denote the coset $A + \mathcal{K}(L^p(\mathbf{R}_+, w))$.

Theorem 17.9. *The Banach algebra*

$$\mathrm{alg}_{p,w} W^\pi(PC) := \mathrm{alg}_{p,w} W(PC)/\mathcal{K}(L^p(\mathbf{R}_+, w))$$

is an inverse closed and commutative subalgebra of the Calkin algebra, that is, of the algebra $\mathcal{B}(L^p(\mathbf{R}_+, w))/\mathcal{K}(L^p(\mathbf{R}_+, w))$. The maximal ideal space can be identified with the "horn bundle"

$$\begin{aligned} \mathcal{N}_{p,w} \quad := \quad & \Big(\bigcup_{x \in \mathbf{R}} (\{x\} \times \mathcal{H}(0,1; \nu_\infty^-(p,w), \nu_\infty^+(p,w)) \Big) \\ & \cup (\{\infty\} \times \mathcal{H}(0,1; \nu_0^-(p,w), \nu_0^+(p,w)) \end{aligned}$$

in the following sense: the Gelfand map

$$\Gamma : \mathrm{alg}_{p,w} W^\pi(PC) \to C(\mathcal{N}_{p,w})$$

is for $a \in PC_{p,w}$ and $(x, \mu) \in \mathcal{N}_{p,w}$ given by

$$\big(\Gamma W^\pi(a)\big)(x,\mu) = (1-\mu) a(x-0) + \mu a(x+0),$$

again with the convention $a(\infty \pm 0) = a(\mp\infty)$.

The analogue of Theorem 16.21 for Wiener-Hopf operators is true. We confine ourselves to stating the analogue of Theorem 16.22. We put

$$C_{p,w}(\overline{\mathbf{R}}) := \big\{ a \in PC_{p,w} : a|\mathbf{R} \text{ is continuous} \big\}. \tag{17.7}$$

Theorem 17.10. *Let $a \in [C_{p,w}(\overline{\mathbf{R}})]_{N \times N}$ and $\nu_0^\pm := \nu_0^\pm(p,w)$. Then the following are equivalent:*

(i) *$W(a)$ is Fredholm on $L^p_N(\mathbf{R}_+, w)$;*

(ii) *$\det a(x) \neq 0$ for all $x \in \overline{\mathbf{R}}$ and*

$$\nu + \frac{1}{2\pi} \arg \xi_j \big(a^{-1}(+\infty) a(-\infty)\big) \notin \mathbf{Z} \tag{17.8}$$

for all $\nu \in [\nu_0^-, \nu_0^+]$ and all eigenvalues $\xi_j = \xi_j(a^{-1}(+\infty)a(-\infty))$ of the matrix $a^{-1}(+\infty)\, a(-\infty)$.

(iii) *$\det a(x) \neq 0$ for all $x \in \overline{\mathbf{R}}$ and*

$$\mathrm{sp}\ \big(a^{-1}(+\infty) a(-\infty)\big) \cap \Sigma[1-\nu_0^+, 1-\nu_0^-] = \emptyset \tag{17.9}$$

where $\Sigma[1-\nu_0^+, 1-\nu_0^-]$ is defined by (16.35).

The proof is similar to the $H^p(\mathbf{R}, w)$ case. Notice that in the language of the eigenvalues η_j of $a^{-1}(-\infty)a(+\infty)$ conditions (17.8), (17.9) read

$$\nu - \frac{1}{2\pi}\arg\eta_j\big(a^{-1}(-\infty)a(+\infty)\big) \notin \mathbf{Z},$$
$$\mathrm{sp}\,\big(a^{-1}(-\infty)a(+\infty)\big) \cap \Sigma[\nu_0^-, \nu_0^+] = \emptyset,$$

respectively. In case $W(a)$ is Fredholm on $L^p(\mathbf{R}_+, w)$ the index is given by

$$\begin{aligned} \operatorname{Ind} W(a) &= -\operatorname{ind}(\det a) + N\nu_0^0 - \sum_{j=1}^{N}\Big\{\nu_0^0 + \frac{1}{2\pi}\arg\xi_j\Big\} \\ &= -\operatorname{ind}(\det a) + N\nu_0^0 - \sum_{j=1}^{N}\Big\{\nu_0^0 - \frac{1}{2\pi}\arg\eta_j\Big\} \end{aligned}$$

where $\nu_0^0 := (\nu_0^- + \nu_0^+)/2$.

17.3 The Symbol Calculus

The *zero-order pseudodifferential operator* Ψ_σ on the real line $\mathbf{R}$ which is generated by a function $\sigma(t, x)$ of two real variables t and x is formally defined by

$$(\Psi_\sigma f)(t) := \frac{1}{2\pi}\int_{\mathbf{R}} e^{-ixt}\sigma(t,x)\Big(\int_{\mathbf{R}} e^{ixs} f(s)\,ds\Big)\,dx, \quad t \in \mathbf{R}.$$

This is frequently also written in the very suggestive form $\Psi_\sigma = F^{-1}_{x\to t}\sigma(t,x)F_{t\to x}$. Let $1 < p < \infty$ and $w \in A_p(\mathbf{R})$. Suppose σ is a finite sum

$$\sigma(t,x) = \sum_{j=1}^{n} a_j(t) b_j(x), \quad t \in \mathbf{R},\ x \in \mathbf{R},$$

where $a_j \in PC$ and $b_j \in PC_{p,w}$. In that case we can write

$$\Psi_\sigma = \sum_{j=1}^{n} a_j W^0(b_j), \tag{17.10}$$

and from the inclusion $PC_{p,w} \subset M_{p,w}$ we see that Ψ_σ is bounded on $L^p(\mathbf{R}, w)$. Let $OP^c_{p,w}$ (resp. $OP_{p,w}$) be the smallest closed subalgebra of $\mathcal{B}(L^p(\mathbf{R}, w))$ that contains all operators of the form (17.10) with $a_j \in C(\dot{\mathbf{R}})$ and $b_j \in C_{p,w}(\dot{\mathbf{R}})$ (resp. $a_j \in PC$ and $b_j \in PC_{p,w}$).

A classical result on pseudodifferential operators says that if $a_j \in C(\dot{\mathbf{R}})$ and $b_j \in C_{p,w}(\dot{\mathbf{R}})$, then the operator (17.10) is Fredholm on $L^p(\mathbf{R}, w)$ if and only if

$$\begin{aligned}\sigma(t,\infty) &:= \sum a_j(t)b_j(\infty) \neq 0,\\ \sigma(\infty,x) &:= \sum a_j(\infty)b_j(x) \neq 0,\\ \sigma(\infty,\infty) &:= \sum a_j(\infty)b_j(\infty) \neq 0\end{aligned}$$

for all $t \in \mathbf{R}$ and all $x \in \mathbf{R}$. This can be rephrased and generalized as follows. Put

$$M := (\dot{\mathbf{R}} \times \dot{\mathbf{R}}) \setminus (\mathbf{R} \times \mathbf{R}) = (\mathbf{R} \times \{\infty\}) \cup (\{\infty\} \times \mathbf{R}) \cup \{(\infty,\infty)\} \tag{17.11}$$

and equip M with the discrete topology.

Theorem 17.11. *There exists a (unique) map*

$$\mathrm{Sym} : OP^c_{p,w} \to BC(M)$$

with the following properties:
(a) *if* $a \in C(\dot{\mathbf{R}})$ *and* $b \in C_{p,w}(\dot{\mathbf{R}})$ *then*

$$(\mathrm{Sym}\, aI)(t,x) = a(t), \ (\mathrm{Sym}\, W^0(b))(t,x) = b(x) \ \textit{for} \ (t,x) \in M;$$

(b) *if* K *is compact then* $(\mathrm{Sym}\, K)(t,x) = 0$ *for all* $(t,x) \in M$*;*
(c) Sym *is an algebra homomorphism;*
(d) *for each* $(t,x) \in M$ *the function*

$$OP^c_{p,w} \to \mathbf{C}, \ A \mapsto (\mathrm{Sym}\, A)(t,x)$$

is continuous;
(e) *an operator* $A \in OP^c_{p,w}$ *is Fredholm on the space* $L^p(\mathbf{R}, w)$ *if and only if* $(\mathrm{Sym}\, A)(t,x) \neq 0$ *for all* $(t,x) \in M$*;*
(f) *if* $A \in OP^c_{p,w}$ *is Fredholm on* $L^p(\mathbf{R}, w)$*, then there is a* $B \in OP^c_{p,w}$ *such that* $AB - I$ *and* $BA - I$ *are compact on* $L^p(\mathbf{R}, w)$*.*

We now formulate the analogue of Theorem 17.11 for the algebra $OP_{p,w}$. Given $x \in \mathbf{R}_+$, define $\nu_x^\pm := \nu_x^\pm(p,w)$ as in Theorem 16.17(a). Let M be the set (17.11). We associate a horn $\mathcal{H}_m = \mathcal{H}_{(t,x)}$ with each $m = (t,x) \in M \setminus \{(\infty,\infty)\}$ as follows:

$$\begin{aligned}\mathcal{H}_{(t,\infty)} &:= \mathcal{H}(0,1;\nu_t^-,\nu_t^+),\\ \mathcal{H}_{(\infty,x)} &:= \mathcal{H}(0,1;1-\nu_\infty^+,1-\nu_\infty^-) = \mathcal{H}(1,0;\nu_\infty^-,\nu_\infty^+).\end{aligned}$$

Also put

$$\mathcal{H}_{(\infty,\infty)} := \{0,1\} \ \text{(two “satellites”)}.$$

Let

$$\mathcal{M} := \bigcup_{m \in M} (\{m\} \times \mathcal{H}_m)$$

and provide $\mathcal{M}$ with the discrete topology. For $a \in PC$ and $b \in PC_{p,w}$, we now define two functions

$$\operatorname{Sym} aI : \mathcal{M} \to \mathbf{C}^{2\times 2}, \quad \operatorname{Sym} W^0(b) : \mathcal{M} \to \mathbf{C}^{2\times 2}.$$

A point of $\mathcal{M}$ is of the form (m, μ) with $m \in M$ and $\mu \in \mathcal{H}_m$. For m we have the possibilities

$$m = (t, \infty) \ (t \in \mathbf{R}), \quad m = (\infty, x) \ (x \in \mathbf{R}), \quad m = (\infty, \infty),$$

and hence we have to define the above functions at points of the three types

$$(t, \infty, \mu), \quad (\infty, x, \mu), \quad (\infty, \infty, j) \quad (j = 0, 1).$$

For $\alpha, \beta, \mu \in \mathbf{C}$, let

$$E(\alpha, \beta, \mu) := \begin{pmatrix} \alpha\mu + \beta(1-\mu) & (\beta - \alpha)\sqrt{\mu(1-\mu)} \\ (\beta - \alpha)\sqrt{\mu(1-\mu)} & \beta\mu + \alpha(1-\mu) \end{pmatrix}$$

where $\sqrt{\mu(1-\mu)}$ denotes any function $\varrho : \mathbf{C} \to \mathbf{C}$ such that $(\varrho(\mu))^2 = \mu(1-\mu)$ for all $\mu \in \mathbf{C}$. Put

$$\begin{aligned}
(\operatorname{Sym} aI)(t, \infty, \mu) &= \operatorname{diag}\big(a(t+0),\, a(t-0)\big), \\
(\operatorname{Sym} aI)(\infty, x, \mu) &= \operatorname{diag}\big(a(+\infty),\, a(-\infty)\big), \\
(\operatorname{Sym} aI)(\infty, \infty, 0) &= \operatorname{diag}\big(a(+\infty),\, a(-\infty)\big), \\
(\operatorname{Sym} aI)(\infty, \infty, 1) &= \operatorname{diag}\big(a(+\infty),\, a(-\infty)\big), \\
\big(\operatorname{Sym} W^0(b)\big)(t, \infty, \mu) &= E\big(b(-\infty),\, b(+\infty), \mu\big), \\
\big(\operatorname{Sym} W^0(b)\big)(\infty, x, \mu) &= E\big(b(x-0),\, b(x+0), \mu\big), \\
\big(\operatorname{Sym} W^0(b)\big)(\infty, \infty, 0) &= \operatorname{diag}\big(b(-\infty),\, b(+\infty)\big), \\
\big(\operatorname{Sym} W^0(b)\big)(\infty, \infty, 1) &= \operatorname{diag}\big(b(+\infty),\, b(-\infty)\big).
\end{aligned}$$

Theorem 17.12. *There exists a (unique) map*

$$\operatorname{Sym} : OP_{p,w} \to BC(\mathcal{M}, \mathbf{C}^{2\times 2})$$

with the following properties:

(a) *for* $a \in PC$ *and* $b \in PC_{p,w}$, *the matrix functions* $\operatorname{Sym} aI$ *and* $\operatorname{Sym} W^0(b)$ *are as above;*

(b) $\operatorname{Sym} K$ *is identically zero if* K *is compact;*

(c) Sym *is an algebra homomorphism;*

(d) *for each* $(m, \mu) \in \mathcal{M}$ *the matrix function*

$$OP_{p,w} \to \mathbf{C}^{2\times 2}, \quad A \mapsto (\operatorname{Sym} A)(m, \mu)$$

is continuous;

(e) *an operator* $A \in OP_{p,w}$ *is Fredholm on the space* $L^p(\mathbf{R}, w)$ *if and only if* $\det(\operatorname{Sym} A)(m, \mu) \neq 0$ *for all* $(m, \mu) \in \mathcal{M}$*;*

(f) *if* $A \in OP_{p,w}$ *is Fredholm on* $L^p(\mathbf{R}, w)$*, then there is an operator* $B \in OP_{p,w}$ *such that* $AB - I$ *and* $BA - I$ *are compact on* $L^p(\mathbf{R}, w)$*.*

A full proof of Theorem 17.12 is in [49]. We remark that a key ingredient of the proof is the following result. We denote by BV the set of all functions of bounded total variation on $\mathbf{R}$.

Lemma 17.13 (Duduchava). *Let* $1 < p < \infty$*.*

(a) *If* $a \in PC$*,* $b \in BV$*,* $a(\pm\infty) = b(\pm\infty) = 0$*, then the operators* $aW^0(b)$ *and* $W^0(b)aI$ *are compact on* $L^p(\mathbf{R}, w)$*.*

(b) *If either* $a \in C, b \in BV$ *or* $a \in PC, b \in BV \cap C$*, then* $aW^0(b) - W^0(b)aI$ *is compact on* $L^p(\mathbf{R}, w)$*.*

(c) *If* $a \in C(\overline{\mathbf{R}})$ *and* $b \in BV \cap C(\overline{\mathbf{R}})$*, then* $aW^0(b) - W^0(b)aI$ *is compact on* $L^p(\mathbf{R}, w)$*.*

Example 17.14. Let $u \in PC$, suppose $u(-\infty) = 0$, $u(+\infty) = 1$, and let χ be the characteristic function of some finite interval $(\alpha, \beta) \subset \mathbf{R}$. We want to determine $\mathrm{sp}_{\mathrm{ess}}\,(uW^0(\chi))$. By Theorem 17.12, $\mathrm{sp}_{\mathrm{ess}}\,(uW^0(\chi))$ is the set of all $\lambda \in \mathbf{C}$ for which $\det(\operatorname{Sym}(\lambda I - uW^0(\chi)))$ has a zero on $\mathcal{M}$. Put $A := \lambda I - uW^0(\chi)$. We have

$$(\operatorname{Sym} A)(t, \infty, \mu) = \lambda \begin{pmatrix} 1 & 0 \\ 0 & 1 \end{pmatrix} - \begin{pmatrix} u(t+0) & 0 \\ 0 & u(t-0) \end{pmatrix} E(0, 0, \mu) = \begin{pmatrix} \lambda & 0 \\ 0 & \lambda \end{pmatrix},$$

$$(\operatorname{Sym} A)(\infty, x, \mu) = \lambda \begin{pmatrix} 1 & 0 \\ 0 & 1 \end{pmatrix} - \begin{pmatrix} 1 & 0 \\ 0 & 0 \end{pmatrix} E(\chi(x-0), \chi(x+0), \mu)$$

$$= \begin{pmatrix} \lambda - \chi(x-0)\mu - \chi(x+0)(1-\mu) & -(\chi(x+0) - \chi(x-0))\sqrt{\mu(1-\mu)} \\ 0 & \lambda \end{pmatrix},$$

$$(\operatorname{Sym} A)(\infty, \infty, 0) = \lambda \begin{pmatrix} 1 & 0 \\ 0 & 1 \end{pmatrix} - \begin{pmatrix} 1 & 0 \\ 0 & 0 \end{pmatrix} \begin{pmatrix} 0 & 0 \\ 0 & 0 \end{pmatrix} = \begin{pmatrix} \lambda & 0 \\ 0 & \lambda \end{pmatrix},$$

$$(\operatorname{Sym} A)(\infty, \infty, 1) = \lambda \begin{pmatrix} 1 & 0 \\ 0 & 1 \end{pmatrix} - \begin{pmatrix} 1 & 0 \\ 0 & 0 \end{pmatrix} \begin{pmatrix} 0 & 0 \\ 0 & 0 \end{pmatrix} = \begin{pmatrix} \lambda & 0 \\ 0 & \lambda \end{pmatrix}.$$

Thus, $\mathrm{sp}_{\mathrm{ess}}\,(uW^0(\chi))$ is the union of $\{0\}$ and

$$\bigcup_{x \in \mathbf{R}} \{\chi(x-0)\mu + \chi(x+0)(1-\mu) : \mu \in \mathcal{H}(1, 0; \nu_\infty^-, \nu_\infty^+)\}. \tag{17.12}$$

If χ is continuous at x, then the corresponding set in (17.12) is $\{\chi(x)\} \subset \{0,1\}$. The two jumps of χ make the contribution

$$\begin{aligned}&\{1-\mu : \mu \in \mathcal{H}(1,0;\nu_\infty^-,\nu_\infty^+)\} \cup \{\mu : \mu \in \mathcal{H}(1,0;\nu_\infty^-,\nu_\infty^+)\}\\&= \mathcal{H}(0,1;\nu_\infty^-,\nu_\infty^+) \cup \mathcal{H}(1,0;\nu_\infty^-,\nu_\infty^+).\end{aligned}$$

Since $\{0,1\}$ is contained in the last two horns, we finally get

$$\mathrm{sp}_{\mathrm{ess}}\left(uW^0(\chi)\right) = \mathcal{H}(0,1;\nu_\infty^-,\nu_\infty^+) \cup \mathcal{H}(1,0;\nu_\infty^-,\nu_\infty^+). \qquad \square$$

Notes

The classical theory of Fourier multipliers on unweighted L^p spaces can be found in [111]. For power weights, see [70] and [183]. Stechkin's inequality goes back to [217], for power weights it was established by Duduchava [70], and for general A_p it was stated in [48].

The Fredholm theory of Wiener-Hopf operators with PC coefficients on unweighted L^p spaces was pioneered by Duduchava in a series of papers, starting with [68]. A complete exposition is in Duduchava's works [69] and [70]. The passage to power weights is in Schneider's paper [198], and general A_p weights were considered in [48]. The investigation of algebras generated by Wiener-Hopf operators with PC symbols was also started by Duduchava in the unweighted case, see [71]. The case of general weights is studied in [49]. The paper [49] contains the symbol calculus presented in Section 17.3 and also an index formula for some operators from the algebra. These results were generalized to a bigger algebra in [127], with index computation for all its elements. For the proof of Lemma 17.13 see Lemmas 7.1-7.4 in [70]; though given there for the unweighted case, they carry over to our setting with no changes.

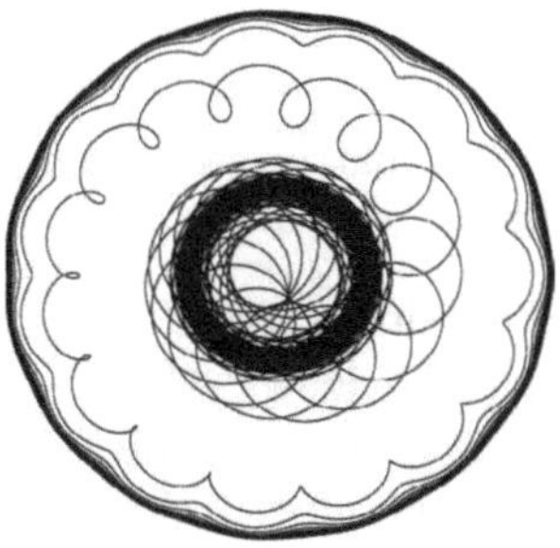

Chapter 18

Toeplitz Operators with SAP Symbols on Hardy Spaces

In this chapter we establish a Fredholm criterion and an index formula for Toeplitz operators with symbols in $SAP_{N\times N}$ on $H^p_N(\mathbf{R},w)$, where w is a power weight. This is a quite deep result and the proof is laborious.

The main steps of the proof are as follows. In Section 18.1 we settle the Fredholm theory of operators with symbols in $APW_{N\times N}$ on $H^p_N(\mathbf{R})$, and in Section 18.4 we prove that if $T(a)$ is Fredholm on $H^p_N(\mathbf{R})$, then the two operators $T(a_l)$ and $T(a_r)$ generated by the almost periodic representatives a_l and a_r are invertible. Section 18.5 is devoted to the definition of the geometric mean for matrix functions in $SAP_{N\times N}$. These results give the necessity portion of the main theorem. The proof of the sufficiency part is more difficult and based on the following strategy: we approximate $a \in SAP_{N\times N}$ by matrix functions $a^{(n)} \in SAP_{N\times N}$ for which $a_l^{(n)}$ and $a_r^{(n)}$ belong to $AP^0_{N\times N}$, we then prove the main theorem for the operators $T(a^{(n)})$ and, in addition, we show that the operators $T(a^{(n)})$ have regularizers R_n satisfying $\limsup_{n\to\infty} \|R_n\|_{\mathrm{ess}} < \infty$. Clearly, the Fredholmness of $T(a^{(n)})$ in conjunction with this estimate implies that $T(a)$ is Fredholm. To prove that the operators $T(a^{(n)})$ are Fredholm and to obtain the above estimate we use a local principle, which reduces things to simpler auxiliary operators. These are studied in Sections 18.6 and 18.7.

18.1 Matrix-Valued *APW* Symbols

For scalar Toeplitz operators with AP symbols, there are no differences between the H^2_+ and $H^p(\mathbf{R},w)$ cases: Theorem 2.28 and its proof remain literally true for the operator $T(a)$ on $H^p(\mathbf{R},w)$ where $1 < p < \infty$ and $w \in A_p(\mathbf{R})$.

In the matrix case, things are more complicated. We begin with the following simple observation.

Proposition 18.1. *Let* $1 < p < \infty$, $w \in A_p(\mathbf{R})$, *and suppose* $a \in AP_{N\times N}$ *has a right* AP *factorization.*

(a) *If* a *has both positive and negative right* AP *indices, then* $T(a)$ *is not semi-Fredholm on* $H^p_N(\mathbf{R}, w)$.

(b) *The operator* $T(a)$ *is n-normal (resp. d-normal) on* $H^p_N(\mathbf{R}, w)$ *if and only if all right* AP *indices are nonnegative (resp. nonpositive), in which case* $T(a)$ *is left-invertible (resp. right-invertible).*

(c) *The operator* $T(a)$ *is Fredholm on* $H^p_N(\mathbf{R}, w)$ *if and only if all right* AP *indices are zero, in which case* $T(a)$ *is invertible.*

Proof. If $a = a_- d a_+$ is a right AP factorization, then $T(a)$ is strongly Φ-equivalent to $T(d)$. This easily implies all assertions. □

We now consider operators with $APW_{N\times N}$ symbols on $H^p_N(\mathbf{R})$, that is, on H^p spaces without weights. Our aim is to show that $a \in APW_{N\times N}$ necessarily has a canonical right APW factorization if $T(a)$ is invertible on $H^p_N(\mathbf{R})$.

Let $\mathcal{S}^W_{N\times N}$ be the set of all operators $A \in \mathcal{B}(L^p_N(\mathbf{R}))$ which can be written in the form

$$A = \sum_{\lambda\in\mathbf{R}} W^0(a_\lambda)e_\lambda I, \quad a_\lambda \in [M_p]_{N\times N}, \quad \sum_{\lambda\in\mathbf{R}} \|a_\lambda\|_{[M_p]_{N\times N}} < \infty, \tag{18.1}$$

where $a_\lambda \neq 0$ for at most countably many $\lambda \in \mathbf{R}$. Recall that $(U_\lambda f)(t) := f(t-\lambda)$. It is easily seen that

$$e_\lambda W^0(a) = W^0(U_{-\lambda} a)e_\lambda I. \tag{18.2}$$

This implies that $U_{-\lambda}a \in [M_p]_{N\times N}$ whenever $a \in [M_p]_{N\times N}$ and also shows that $\mathcal{S}^W_{N\times N}$ is an algebra. Clearly, $\mathcal{S}^W_{N\times N}$ is a Banach algebra with the norm

$$\|A\| := \sum_{\lambda\in\mathbf{R}} \|W^0(a_\lambda)\|_{\mathcal{B}(L^p_N(\mathbf{R}))} = \sum_{\lambda\in\mathbf{R}} \|a_\lambda\|_{[M_p]_{N\times N}}.$$

Theorem 18.2. *The algebra* $\mathcal{S}^W_{N\times N}$ *is inverse closed in* $\mathcal{B}(L^p_N(\mathbf{R}))$, *i.e., if an operator in* $\mathcal{S}^W_{N\times N}$ *is invertible on* $L^p_N(\mathbf{R})$, *then the inverse operator also belongs to* $\mathcal{S}^W_{N\times N}$.

Proof. We proceed as in the proof of Theorem 9.6. Let A be of the form (18.1), suppose A is invertible, put $E := \mathcal{B}(L^p_N(\mathbf{R}))$, and define

$$a : \mathbf{R}_B \to E, \; a(\chi) := \sum_\lambda \chi(\lambda) W^0(a_\lambda)e_\lambda I.$$

From (18.1) we infer that $a \in W(\mathbf{R}_B, E)$. Let $e_h(\lambda) := e^{i\lambda h}$. Clearly,

$$U_h^{-1} e_\lambda U_h = e^{i\lambda h} e_\lambda I. \tag{18.3}$$

The operators $W^0(a_\lambda)$ are translation invariant, that is,

$$U_h W^0(a_\lambda) = W^0(a_\lambda) U_h. \tag{18.4}$$

Using (18.3) and (18.4) we get

$$\begin{aligned} a(e_h) &= \sum_\lambda e_h(\lambda) W^0(a_\lambda) e_\lambda I = \sum_\lambda W^0(a_\lambda) e^{i\lambda h} e_\lambda I \\ &= \sum_\lambda W^0(a_\lambda) U_h^{-1} e_\lambda U_h = U_h^{-1}\Big(\sum_\lambda W^0(a_\lambda) e_\lambda I\Big) U_h = U_h^{-1} A U_h. \end{aligned}$$

It follows that

$$\|a^{-1}(e_h)\| = \|U_h^{-1} A^{-1} U_h\| = \|A^{-1}\| \quad \text{for all} \quad h \in \mathbf{R},$$

and since $\mathbf{R}$ is dense in $\mathbf{R}_B$, we can conclude that a is invertible in $C(\mathbf{R}_B, E)$. The Bochner-Phillips theorem (Theorem 9.5) therefore implies that a^{-1} belongs to $W(\mathbf{R}_B, E)$:

$$a^{-1}(\chi) = \sum_\lambda \chi(\lambda) C_\lambda, \quad C_\lambda \in E, \quad \sum_\lambda \|C_\lambda\| < \infty,$$

and $C_\lambda \neq 0$ for at most countably many $\lambda \in \mathbf{R}$. In particular,

$$A^{-1} = a^{-1}(e_0) = \sum_\lambda e_0(\lambda) C_\lambda = \sum_\lambda C_\lambda.$$

It remains to show that there are $d_\lambda \in [M_p]_{N\times N}$ such that $C_\lambda = W^0(d_\lambda) e_\lambda I$ for all λ. By (18.3) and (18.4),

$$\begin{aligned} U_h^{-1} a(e_s) U_h &= \sum_\lambda U_h^{-1} e^{i\lambda s} W^0(a_\lambda) e_\lambda U_h = \sum_\lambda e^{i\lambda s} W^0(a_\lambda) U_h^{-1} e_\lambda U_h \\ &= \sum_\lambda e^{i\lambda s} W^0(a_\lambda) e^{i\lambda h} e_\lambda I = a(e_{s+h}), \end{aligned}$$

whence $U_h^{-1} a^{-1}(e_s) U_h = a^{-1}(e_{s+h})$. Because $a^{-1}(e_\nu) = \sum_\lambda e^{i\lambda\nu} C_\lambda$, we obtain

$$U_h^{-1}\Big(\sum_\lambda e^{i\lambda s} C_\lambda\Big) U_h = \sum_\lambda e^{i\lambda s} e^{i\lambda h} C_\lambda,$$

and comparison of the coefficients of $e^{i\lambda s}$ gives $U_h^{-1} C_\lambda U_h = e^{i\lambda h} C_\lambda$ and thus

$$U_h^{-1} C_\lambda U_h e_{-\lambda} I = e^{i\lambda h} C_\lambda e_{-\lambda} I.$$

Again by (18.3), $U_h e_{-\lambda} I = e^{i\lambda h} e_{-\lambda} U_h$, which implies that

$$U_h^{-1} C_\lambda e_{-\lambda} U_h = C_\lambda e_{-\lambda} I.$$

Consequently, $C_\lambda e_{-\lambda} I$ is translation invariant. Theorem 17.2 now shows that $C_\lambda e_{-\lambda} I = W^0(d_\lambda)$ with $d_\lambda \in [M_p]_{N\times N}$. □

Corollary 18.3. *Let $a \in APW_{N\times N}$ and $1 < p < \infty$. If $T(a)$ is invertible on $H_N^p(\mathbf{R})$, then $T(a)$ is invertible on $H_N^2(\mathbf{R})$.*

Proof. The invertibility of $T(a)$ on $H_N^p(\mathbf{R})$ is equivalent to the invertibility of the operator $PaI + Q$ on $L_N^p(\mathbf{R})$. We have

$$a = \sum_\lambda a_\lambda e_\lambda, \quad a_\lambda \in \mathbf{C}^{N\times N}, \quad \sum \|a_\lambda\| < \infty,$$
$$P = F^{-1}\chi_- F = W^0(\chi_-), \; Q = F^{-1}\chi_+ F = W^0(\chi_+),$$

and hence,

$$\begin{aligned} PaI + Q &= W^0(\chi_-)\Big(\sum_\lambda a_\lambda e_\lambda I\Big) + W^0(\chi_+) \\ &= \sum_\lambda W^0(a_\lambda \chi_-) e_\lambda I + W^0(\chi_+) e_0 I \in \mathcal{S}^W_{N\times N}. \end{aligned}$$

Theorem 18.2 implies that $(PaI + Q)^{-1} \in \mathcal{S}^W_{N\times N}$. Therefore,

$$(PaI + Q)^{-1} = \sum_\lambda W^0(b_\lambda) e_\lambda I, \quad b_\lambda \in [M_p]_{N\times N}, \quad \sum \|b_\lambda\|_{[M_p]_{N\times N}} < \infty.$$

From (17.2) we now obtain that

$$\sum_\lambda \|W^0(b_\lambda)\|_{\mathcal{B}(L_N^2(\mathbf{R}))} = \sum_\lambda \|b_\lambda\|_{L^\infty_{N\times N}(\mathbf{R})} < \infty.$$

Thus, $(PaI + Q)^{-1}$ is bounded on $L_N^2(\mathbf{R})$. This yields the invertibility of $PaI + Q$ on $L_N^2(\mathbf{R})$ and thus the invertibility of $T(a)$ on $H_N^2(\mathbf{R})$. □

Corollary 18.4. *For $a \in APW_{N\times N}$ and $1 < p < \infty$, the following are equivalent:*

(i) *$T(a)$ is invertible on $H_N^p(\mathbf{R})$;*

(ii) *a has a canonical right APW factorization.*

Proof. In Section 16.1 we observed that

$$T(a) \in G\mathcal{B}\big(H_N^2(\mathbf{R})\big) \Longleftrightarrow W(a) \in G\mathcal{B}\big(L_N^2(\mathbf{R}_+)\big).$$

The implication (i)⇒(ii) is therefore immediate from Corollaries 18.3 and 9.8. The implication (ii)⇒(i) is trivial. □

We will show later that the two conditions of Corollary 18.4 are also equivalent to the Fredholmness of $T(a)$ on $H_N^p(\mathbf{R})$.

18.2 Necessary Conditions for Semi-Fredholmness

While for Hilbert space operators left-Fredholmness is equivalent to n-normality, this is in general not true for Banach space operators: left-Fredholmness implies n-normality, but an n-normal operator is only left-Fredholm if its range is a complementable subspace. The following result provides us with a very useful criterion for n-normality.

Lemma 18.5. *Let X be a Banach space and $A \in \mathcal{B}(X)$. Then the following are equivalent:*

(i) *A is n-normal;*

(ii) *there is a compact operator K and a $\delta > 0$ such that*

$$\|Ax\| + \|Kx\| \geq \delta\|x\| \text{ for all } x \in X;$$

(iii) *there are finitely many compact operators $K_1, \ldots, K_n$ and a $\delta > 0$ such that*

$$\|Ax\| + \sum_{j=1}^{n} \|K_j x\| \geq \delta\|x\| \text{ for all } x \in X.$$

For a proof we refer to [144] and [176, Section 1.3.2].

Let X be a Banach space. We put $\mathcal{B} := \mathcal{B}(X)$ and we denote by $\mathcal{K} := \mathcal{K}(X)$ the ideal of all compact operators on X. In what follows the notation $A \simeq B$ will mean that $A - B \in \mathcal{K}$. Fix an operator $V \in \mathcal{B}$, put

$$V_l := \frac{1}{2}(I - V), \quad V_r := \frac{1}{2}(I + V)$$

and consider the operator

$$A = A_l V_l + A_r V_r + A_0$$

where A_l, A_r, A_0 are arbitrary operators in $\mathcal{B}$.

Lemma 18.6. (a) *Suppose there is an $R \in \mathcal{B}$ such that $V_r R \not\simeq 0$ and*

$$BR \simeq 0 \text{ for } B \in \{A_r, A_r V - V A_r, I - V^2, A_0, A_0 V\}.$$

Then A is not n-normal.

(b) *If there is an $L \in \mathcal{B}$ such that $LV_r \not\simeq 0$ and*

$$LB \simeq 0 \text{ for } B \in \big\{A_r, A_r V - V A_r, I - V^2, A_0, V A_0, A_l V - V A_l, V(A_l V - V A_l)\big\},$$

then A is not d-normal.

Proof. We only prove (a), since (b) can be shown analogously. Observe that

$$\begin{aligned} A_l V_l V_r &= \frac{1}{4} A_l (I - V^2), \\ A_r V_r V_r &= \frac{1}{4} A_r (I + 2V + V^2) \\ &= \frac{1}{2} A_r + \frac{1}{2} V A_r + \frac{1}{2} (A_r V - V A_r) - \frac{1}{4} A_r + \frac{1}{4} A_r V^2 \\ &= V_r A_r + \frac{1}{2} (A_r V - V A_r) - \frac{1}{4} A_r (I - V^2), \\ A_0 V_r &= \frac{1}{2} A_0 + \frac{1}{2} A_0 V. \end{aligned}$$

Adding these three equalities and multiplying the result from the right by R, we get

$$\begin{aligned} A V_r R &= V_r A_r R + \frac{1}{2} (A_r V - V A_r) R + \frac{1}{4} (A_l - A_r)(I - V^2) R \\ &\quad + \frac{1}{2} A_0 R + \frac{1}{2} A_0 V R. \end{aligned}$$

By assumption, all operators on the right are compact, whence $AV_rR \simeq 0$. Contrary to what we want, let us assume that A is n-normal. Then, by Lemma 18.5, $\|Ax\| + \|Kx\| \ge \delta\|x\|$ with some $K \in \mathcal{K}$ and some $\delta > 0$. Letting $x = V_rRy$, we obtain

$$\|AV_rRy\| + \|KV_rRy\| \ge \delta \|V_rRy\| \text{ for all } y \in X. \tag{18.5}$$

Let $\{y_\tau\}_{\tau\in T}$ be a bounded net in X. Since AV_rR and KV_rR are compact, there is a subnet $\{y_\tau\}_{\tau\in T_0}$ such that $\{AV_rRy_\tau\}_{\tau\in T_0}$ and $\{KV_rRy_\tau\}_{\tau\in T_0}$ are convergent. From (18.5) we see that $\{V_rRy_\tau\}_{\tau\in T_0}$ is also convergent. This implies that V_rR must be compact, which contradicts the hypothesis of the lemma. □

Now let $1 < p < \infty$ and $w \in A_p(\mathbf{R})$. Put $v(x) := \sqrt{(x-i)/(x+i)}$ where the branch of the root is chosen so that $v(\pm\infty) = \pm 1$. By Theorem 1.21, every matrix function $a \in SAP_{N\times N}$ can be written in the form

$$a = a_l \frac{1-v}{2} + a_r \frac{1+v}{2} + a_0$$

with $a_l, a_r \in AP_{N\times N}$ and $a_0 \in [C_0(\dot{\mathbf{R}})]_{N\times N}$. We know that $T(a) \in \mathcal{B}(H^p_N(\mathbf{R}, w))$ is strongly Φ-equivalent to $aP + Q \in \mathcal{B}(L^p_N(\mathbf{R}, w))$. We have

$$\begin{aligned} a_l \frac{1-v}{2} P &= a_l P \frac{1-v}{2} I + \frac{1}{2} a_l (PvI - vP), \\ a_r \frac{1+v}{2} P &= a_r P \frac{1+v}{2} I - \frac{1}{2} a_r (PvI - vP), \\ Q + a_0 P &= Q \frac{1-v}{2} I + Q \frac{1+v}{2} I + a_0 P. \end{aligned}$$

Addition of these identities gives

$$aP + Q = A_l V_l + A_r V_r + A_0 \tag{18.6}$$

where

$$\begin{aligned} A_l &:= a_l P + Q, \quad V_l := \frac{1}{2}(I - V), \\ A_r &:= a_r P + Q, \quad V_r := \frac{1}{2}(I + V), \\ A_0 &:= a_0 P + \frac{1}{2}(a_l - a_r)(PV - VP), \quad V := vI. \end{aligned} \tag{18.7}$$

Theorem 18.7. *Let $a \in SAP_{N\times N}$ and suppose the almost periodic representatives $a_l, a_r \in AP_{N\times N}$ have right AP factorizations. If $T(a)$ is n-normal (resp. d-normal) on $H^p_N(\mathbf{R}, w)$, then $T(a_l)$ and $T(a_r)$ are left-invertible (resp. right-invertible) on $H^p_N(\mathbf{R}, w)$.*

Proof. Suppose $T(a)$ is n-normal. Then so also is the operator (18.6). Let us assume that $T(a_r)$ is not left-invertible. Let $a_r = a_- d a_+$ be a right AP factorization of a_r. By Proposition 18.1, at least one right AP index of a_r must be negative. Without loss of generality, we can assume that

$$d(x) = \operatorname{diag}(e^{-i\lambda x}, e^{i\lambda_2 x}, \dots, e^{i\lambda_N x})$$

with $\lambda > 0$. Choose numbers ν, μ so that $0 < \nu < \mu < \lambda$ and put

$$\begin{aligned} R &:= (a_+^{-1} I - a_- e_{-\lambda} I) D, \\ D &:= (e_\nu P e_{-\nu} I - e_\mu P e_{-\mu} I) \operatorname{diag}(1, 0, \dots, 0) I. \end{aligned}$$

We will show that R has all the properties of Lemma 18.6(a).

Clearly, $Qa_+^{-1}D = 0$, and $Qa_- e_{-\lambda} D$ equals

$$\begin{aligned} &Qa_- e_{-\lambda}(-e_\nu Q e_{-\nu} I + e_\mu Q e_{-\mu} I) \operatorname{diag}(1, 0, \dots, 0) I \\ &= a_- e_{-\lambda}(-e_\nu Q e_{-\nu} I + e_\mu Q e_{-\mu} I) \operatorname{diag}(1, 0, \dots, 0) I \\ &= a_- e_{-\lambda}(e_\nu P e_{-\nu} I - e_\mu P e_{-\mu} I) \operatorname{diag}(1, 0, \dots, 0) I \\ &= a_- e_{-\lambda} D = a_- d D. \end{aligned}$$

Thus,

$$QR = -a_- d D. \tag{18.8}$$

Analogously we get $Pa_+^{-1} D = a_+^{-1} D$ and $Pa_- e_{-\lambda} D = 0$, whence

$$PR = a_+^{-1} D. \tag{18.9}$$

Combining (18.8) and (18.9) we obtain

$$A_r R = (a_r P + Q)R = a_r a_+^{-1} D - a_- dD = 0.$$

Let u be an arbitrary function in $C(\overline{\mathbf{R}})$. By Lemma 3.5, $ue_\alpha \in C(\dot{\mathbf{R}}) + H_+^\infty$ for $\alpha > 0$ and $ue_{-\alpha} \in C(\dot{\mathbf{R}}) + H_-^\infty$ for $\alpha < 0$. Since $PcQ \simeq 0$ and $QcP \simeq 0$ whenever $c \in C(\dot{\mathbf{R}})$ (recall the proof of Proposition 2.11), the arguments leading to (18.8) and (18.9) also give

$$QuR \simeq -ua_- dD, \quad PuR \simeq ua_+^{-1} D. \tag{18.10}$$

Now assume $V_r R = v_r R \simeq 0$, where $v_r = (1+v)/2$. Then, by (18.10), $a_+^{-1} v_r D = v_r a_+^{-1} D \simeq P v_r R \simeq 0$ and thus $v_r D \simeq 0$. We have

$$\begin{aligned} e_\nu P e_{-\nu} I &= F^{-1} F e_\nu F^{-1} \chi_- F e_\nu F^{-1} F \\ &= F^{-1} U_{-\nu} \chi_- U_\nu F = W^0(\chi_{(-\infty,-\nu)}) \end{aligned}$$

and hence,

$$e_\nu P e_{-\nu} I - e_\mu P e_{-\mu} I = W^0(\chi_{(-\mu,-\nu)}). \tag{18.11}$$

Therefore the relation $v_r D \simeq 0$ is equivalent to the compactness of $v_r W^0(\chi_{(-\mu,-\nu)})$ on $L^2(\mathbf{R}, w)$. However, from Example 17.14 we know that the essential spectrum of this operator always contains the point 1. This proves that $V_r R \not\simeq 0$.

The operator $I - V^2$ is multiplication by $1 - v^2 \in C_0(\dot{\mathbf{R}})$. By (18.11),

$$(I - V^2)R = (1 - v^2)(a_+^{-1} - a_- e_{-\lambda}) W^0(\chi_{(-\mu,-\nu)}) \operatorname{diag}(1, 0, \dots, 0) I,$$

and since $(1 - v^2)(a_+^{-1} - a_- e_{-\lambda}) \in [C_0(\dot{\mathbf{R}})]_{N\times N}$, we can use Lemma 17.13 to prove that $(I - V^2)R \simeq 0$. From (18.9) and (18.10) we get

$$(PV - VP)R = PvR - vPR \simeq va_+^{-1} D - va_+^{-1} D = 0.$$

Taking into account (18.7), we therefore have $A_0 R \simeq a_0 PR$. By (18.9), $a_0 PR = a_0 a_+^{-1} D$. As $a_0 a_+^{-1} \in [C_0(\dot{\mathbf{R}})]_{N\times N}$, we obtain from (18.11) with the help of Lemma 17.13 that $A_0 R \simeq 0$. We now consider $A_0 VR$. By (18.10), $a_0 PVR \simeq a_0 va_+^{-1} D$, and as $a_0 va_+^{-1} \in [C_0(\dot{\mathbf{R}})]_{N\times N}$, we see as above that $a_0 va_+^{-1} D \simeq 0$. Also by (18.10),

$$(PV - VP)VR \simeq v^2 a_+^{-1} D - vva_+^{-1} D = 0.$$

Thus, $A_0 VR \simeq 0$ due to (18.7). Finally, once again by (18.10) and (18.9), (18.8),

$$\begin{aligned} &(A_r V - VA_r) = (a_r P + Q)VR - V(a_r P + Q)R \\ &\simeq a_r va_+^{-1} D - va_- dD - va_r a_+^{-1} D + va_- dD = 0. \end{aligned}$$

In summary, R satisfies the hypotheses of Lemma 18.6(a). It follows that the operator (18.6) cannot be n-normal. This contradiction shows that $T(a_r)$ must be left-invertible. Since Lemma 18.6(a) remains true with r replaced by l, we obtain that $T(a_l)$ is also left-invertible.

The case of right-Fredholmness and invertibility from the right can now be disposed of by passage to adjoints. □

18.3 Saginashvili's Theorem

Here is the $H^p(\mathbf{R}, w)$ analogue of Theorems 3.9 and 3.14. For power weights, this result was established by Saginashvili. Figure 18.1 shows an example.

Theorem 18.8. *Let* $1 < p < \infty$, $w \in A_p(\mathbf{R})$, $a \in SAP \setminus \{0\}$.

(a) *If* $a \notin GSAP$, *then* $T(a)$ *is not normally solvable on* $H^p(\mathbf{R}, w)$.

(b) *If* $a \in GSAP$ *and* $\kappa(a_l)\kappa(a_r) < 0$, *then* $T(a)$ *is not normally solvable on* $H^p(\mathbf{R}, w)$.

(c) *If* $a \in GSAP$, $\kappa(a_l)\kappa(a_r) \geq 0$, *and* $\kappa(a_l) + \kappa(a_r) > 0$, *then* $T(a)$ *is properly* n*-normal and has trivial kernel on* $H^p(\mathbf{R}, w)$.

(d) *If* $a \in GSAP$, $\kappa(a_l)\kappa(a_r) \geq 0$, *and* $\kappa(a_l) + \kappa(a_r) < 0$, *then* $T(a)$ *is properly* d*-normal and its range is all of* $H^p(\mathbf{R}, w)$.

(e) *If* $a \in GSAP$, $\kappa(a_l) = \kappa(a_r) = 0$, *and*

$$0 \notin \mathcal{H}\big(\mathbf{d}(a_r), \mathbf{d}(a_l);\, \nu_\infty^-(p, w), \nu_\infty^+(p, w)\big), \tag{18.12}$$

then $T(a)$ *is Fredholm on* $H^p(\mathbf{R}, w)$ *and*

$$\operatorname{Ind} T(a) = -\operatorname{ind} a + \nu_\infty^0(p, w) - \Big\{\nu_\infty^0(p, w) + \frac{1}{2\pi} \arg \frac{\mathbf{d}(a_l)}{\mathbf{d}(a_r)}\Big\}.$$

(f) *If* $a \in GSAP$, $\kappa(a_l) = \kappa(a_r) = 0$, *and* (18.12) *is violated, then* $T(a)$ *is not normally solvable on* $H^p(\mathbf{R}, w)$.

Proof. Part (a) follows from Theorem 16.3(a),(b). If $T(a)$ is semi-Fredholm, we can find a $d \in SAPW$ such that $T(d)$ is also semi-Fredholm and

$$\operatorname{Ind} T(d) = \operatorname{Ind} T(a), \quad \kappa(d_l) = \kappa(a_l), \quad \kappa(d_r) = \kappa(a_r)$$

(Theorem 2.1 and Proposition 8.18). Notice that $a \in GSAP$ and $d \in GSAPW$ due to (a), which implies that all mean motions $\kappa(\cdot)$ are well-defined. Since d_l, d_r belong to $GAPW$ and are therefore APW factorable (Theorem 8.11), we can employ Theorem 18.7 and Proposition 18.1 to obtain the following:

$$T(a) \text{ is Fredholm} \implies \kappa(a_l) = \kappa(a_r) = 0, \tag{18.13}$$

$$T(a) \text{ is } n\text{-normal} \implies \kappa(a_l) \geq 0,\ \kappa(a_r) \geq 0, \tag{18.14}$$

$$T(a) \text{ is } d\text{-normal} \implies \kappa(a_l) \leq 0,\ \kappa(a_r) \leq 0. \tag{18.15}$$

Hence, if $\kappa(a_l)\kappa(a_r) < 0$, then $T(a)$ is not semi-Fredholm and therefore not normally solvable (Theorem 16.3(a),(b)), which proves (b).

Suppose $a \in GSAP$. Proceeding as in the proof of Theorem 3.9, one can show that $T(a)$ is weakly Φ-equivalent to $T(b)$ where

$$b(x) := \big(1 - u(x)\big)\mathbf{d}(a_l)e^{i\kappa(a_l)x} + u(x)\mathbf{d}(a_r)e^{i\kappa(a_r)x} + b_0(x)$$

and $b_0 \in C_0(\dot{\mathbf{R}})$ is such that $b \in GSAP$.

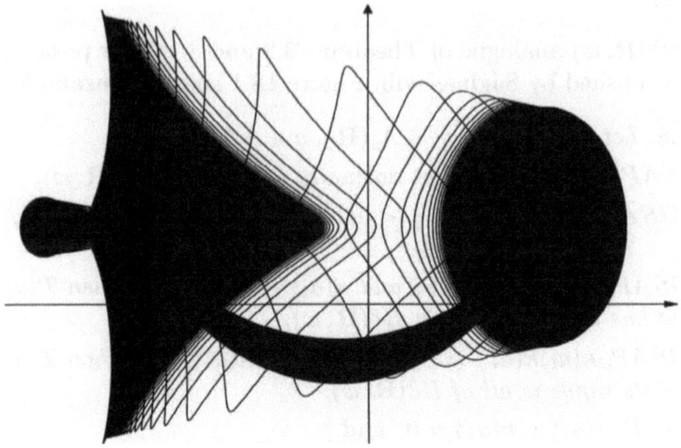

Figure 18.1: The picture shows the range of a function $a \in SAP$, the points at which $\kappa(a_l) \neq 0$ or $\kappa(a_r) \neq 0$, and the horn $\mathcal{H}(\mathbf{d}(a_r), \mathbf{d}(a_l); \nu_\infty^-(p,w), \nu_\infty^+(p,w))$. As this set of "forbidden" points does not contain the origin, the operator $W(a)$ is Fredholm. Notice that the set seen in the picture is *not* the essential spectrum of $W(a)$ (see Chapter 4).

If $\kappa(a_l) = \kappa(a_r) = 0$, then $b \in C(\overline{\mathbf{R}})$ and Theorems 16.13(a), 16.17, 16.18 yield (e) (without the index formula) and (f). To calculate $\operatorname{Ind} T(a)$ in the case (e) we can proceed as in the proof of Theorem 3.14, the only difference being that Theorem 2.20 must be replaced by Theorem 16.18. If $\kappa(a_l) > 0$ and $\kappa(a_r) > 0$, then $b \in C(\dot{\mathbf{R}}) + H_+^\infty$ due to Lemma 3.5, which, by Theorem 16.6, shows that $T(b)$ is left-Fredholm and thus n-normal. Analogously, $T(b)$ is d-normal provided $\kappa(a_l) < 0$ and $\kappa(a_r) < 0$. Parts (c) and (d) will therefore follow from (18.13) to (18.15) and Theorem 16.3(a) once we have shown that $T(b)$ is n-normal (resp. d-normal) if one of the numbers $\kappa(a_l)$, $\kappa(a_r)$ is zero and the other one is positive (resp. negative).

For the sake of definiteness, we show that $T(b)$ is n-normal if $\kappa(a_l) = 0$ and $\kappa(a_r) > 0$. To do this, we use the argument of the proof of Lemma 3.8. Put

$$s_1(x) := \frac{1}{\mathbf{d}(a_l)}\big(1 - u(x)\big) + \frac{4}{\mathbf{d}(a_r)}u(x)e^{-i\kappa(a_r)x}$$

and choose $s_0 \in C_0(\dot{\mathbf{R}})$ so that $s := s_1 + s_0 \in GSAP$. As in the proof of Lemma 3.8, we can use Proposition 16.7 to get

$$T(s)T(b) \simeq \big(T(1-u)\big)^2 + T(4u^2 + \delta_0)$$

with $\delta_0 \in C_0(\dot{\mathbf{R}})$. The operator $T(s)T(b)$ belongs to $\mathrm{alg}_{p,w}T(PC)$, and we can therefore have recourse to Theorem 16.20. Obviously,

$$\Big(\Gamma\big(T(s)T(b)\big)^\pi\Big)(x,\mu) = s(x)b(x) \neq 0 \text{ for } x \in \mathbf{R}.$$

Furthermore,

$$\Big(\Gamma\big(T(s)T(b)\big)^\pi\Big)(\infty,\mu) = \mu^2 + 4(1-\mu) = (\mu-2)^2,$$

and since $2 \notin \mathcal{H}(0,1;\nu_\infty^-,\nu_\infty^+)$, it follows that $T(s)T(b)$ is Fredholm, which implies that $T(b)$ is left-Fredholm and thus n-normal. □

18.4 Invertibility of the Almost Periodic Representatives

Let $\mathcal{A}$ and $\mathcal{F}$ denote the smallest closed subalgebras of $\mathcal{B}(L^p_N(\mathbf{R}))$ that contain the operator P and the sets $\{aI : a \in AP_{N\times N}\}$ and $\{aI : a \in SAP_{N\times N}\}$, respectively:

$$\mathcal{A} := \mathrm{alg}_p(P, AP_{N\times N}), \quad \mathcal{F} := \mathrm{alg}_p(P, SAP_{N\times N}).$$

One can show that $\mathcal{F}$ contains the set $\mathcal{K} := \mathcal{K}(L^p_N(\mathbf{R}))$ of all compact operators on $L^p_N(\mathbf{R})$. Recall that for $h \in \mathbf{R}$ the operator U_h is defined by $(U_h f)(x) := f(x-h)$.

Lemma 18.9. *If $K \in \mathcal{K}(L^p(\mathbf{R}))$ and $h_n \to +\infty$, then $U_{-h_n}KU_{h_n}$ converges strongly to zero on $L^p(\mathbf{R})$.*

Proof. Since $\|U_h\| = 1$ for all $h \in \mathbf{R}$, it suffices to show that U_{h_n} converges weakly to zero on $L^p(\mathbf{R})$, i.e., that

$$(U_{h_n}f, g) := \int_{\mathbf{R}} (U_{h_n}f)(x)\overline{g(x)}\,dx \to 0 \text{ as } n \to \infty$$

for every $f \in L^p(\mathbf{R})$ and every $g \in (L^p(\mathbf{R}))^* = L^q(\mathbf{R})$. Since functions with finite support are dense in $L^p(\mathbf{R})$ and $L^q(\mathbf{R})$, we may assume that f and g have finite support. But then

$$(U_{h_n}f, g) = \int_{\Delta_n} f(x-h_n)\overline{g(x)}\,dx, \tag{18.16}$$

where $\Delta_n := (h_n + \operatorname{supp} f) \cap \operatorname{supp} g$, and because $\Delta_n = \emptyset$ for all sufficiently large n, it follows that (18.16) is zero whenever n is large enough. □

Theorem 18.10. *The maps*

$$\mu_r : P \mapsto P,\ aI \mapsto a_r I \quad (a \in SAP_{N\times N}),$$
$$\mu_l : P \mapsto P,\ aI \mapsto a_l I \quad (a \in SAP_{N\times N})$$

extend to (continuous) Banach algebra homomorphisms of $\mathcal{F}$ onto $\mathcal{A}$ such that

$$\|\mu_r(A)\| \le \|A\|_{\rm ess}, \quad \|\mu_l(A)\| \le \|A\|_{\rm ess}$$

for all $A \in \mathcal{F}$.

Proof. Once Lemmas 18.9 and 10.2 are available, this can be shown by nearly exactly the same arguments as in the proof of Theorem 10.3, the difference being only that instead of (10.8) we now must prove that

$$\text{s-}\lim_{n\to\infty} U_{-h_n} A U_{h_n} = \mu_r(A).$$

For $A = P$, this is trivial because $U_{-h_n} P U_{h_n} = P$. If $A = aI$ with $a \in SAP$, we have

$$\|U_{-h_n} A U_{h_n} f - a_r f\|_p^p = \int_{\mathbf{R}} |a(x+h_n) - a_r(x)|^p |f(x)|^p \, dx, \tag{18.17}$$

and since f can be assumed to have compact support, one can show that (18.17) converges to zero as in the proof of Theorem 10.3. □

Corollary 18.11. *If $A \in \mathcal{F}$ is Fredholm on $L_N^p(\mathbf{R})$, then $\mu_l(A)$ and $\mu_r(A)$ are invertible on $L_N^p(\mathbf{R})$.*

Proof. We cannot use the simple C^*-algebra argument of the proof of the corresponding Corollary 10.4. Therefore we proceed in another way.

Let $m := \operatorname{Ind} A$. Put $A_0 := A(dP + Q)$ where

$$d(x) := \operatorname{diag}\left(\left(\frac{x-i}{x+i}\right)^m, 1, \dots, 1\right).$$

Since $\operatorname{Ind}(dP+Q) = \operatorname{Ind} T(d) = -m$, it follows that $\operatorname{Ind} A_0 = 0$. Consequently, $A_0 = V + K$ where V is invertible and K is compact (see, e.g., [99, Vol. 1, p. 167]). As $A_0 \in \mathcal{F}$ and $K \in \mathcal{K} \subset \mathcal{F}$, we see that V also belongs to $\mathcal{F}$. From the proof of Theorem 18.10 (and Theorem 10.3) we know that there is a sequence $h_n \to +\infty$ such that

$$\mu_r(A_0) = \lim_{n\to\infty} U_{-h_n} V U_{h_n}, \quad (\mu_r(A_0))^* = \lim_{n\to\infty} U_{-h_n} V^* U_{h_n},$$

where $(\mu_r(A_0))^*$ and V^* are the adjoint operators acting on $L_N^q(\mathbf{R})$. The operator $U_{-h_n} V U_{h_n}$ and $U_{-h_n} V^* U_{h_n}$ are invertible, and the norms of their inverses are uniformly bounded:

$$\|(U_{-h_n} V U_{h_n})^{-1}\|_p = \|U_{-h_n} V^{-1} U_{h_n}\|_p = \|V^{-1}\|_p,$$
$$\|(U_{-h_n} V^* U_{h_n})^{-1}\|_q = \|U_{-h_n} (V^*)^{-1} U_{h_n}\|_q = \|(V^*)^{-1}\|_q = \|V^{-1}\|_p.$$

Consequently,

$$\|f\|_p \le \|V^{-1}\|_p \|U_{-h_n} V U_{h_n} f\|_p, \quad \|g\|_q \le \|V^{-1}\|_p \|U_{-h_n} V^* U_{h_n} g\|_q,$$

whence

$$\|f\|_p \le \|V^{-1}\|_p \|\mu_r(A_0) f\|_p, \quad \|g\|_q \le \|V^{-1}\|_p \|\big(\mu_r(A_0)\big)^* g\|_q$$

for all $f \in L^p_N(\mathbf{R})$, $g \in L^q_N(\mathbf{R})$. These two inequalities imply that $\mu_r(A_0)$ is invertible. Because

$$\mu_r(A_0) = \mu_r(A)\mu_r(dP + Q) = \mu_r(A)\big(d(+\infty)P + Q\big) = \mu_r(A),$$

we finally arrive at the conclusion that $\mu_r(A)$ is invertible. The proof for $\mu_l(A)$ is analogous. □

18.5 Existence and Continuity of the Geometric Mean

Throughout the rest of this chapter we abbreviate $L^p_N(\mathbf{R})$ to L^p_N, $H^p_\pm(\mathbf{R})$ to $H^p_\pm$ and H^p_+ to H^p. In what follows we also abbreviate "canonical right *APW* factorization" to "canonical *APW* factorization", and write $(\mathbf{d}(a))^{-1} =: \mathbf{d}^{-1}(a)$, $(M(a))^{-1} =: M^{-1}(a)$, etc.

Recall from Chapter 8 (Definition 8.5) that if $a \in AP_{N\times N}$ has a canonical *AP* factorization $a = a_- a_+$, then the geometric mean $\mathbf{d}(a)$ of a is defined by

$$\mathbf{d}(a) = M(a_-)M(a_+).$$

The geometric mean does not depend on the choice of a factorization (Proposition 8.4). Due to Corollary 9.8, the geometric mean is well-defined for all a in $APW_{N\times N}$ for which $T(a)$ is invertible on H^2_N (and thus on all H^p_N with $1 < p < \infty$). The following theorem allows us to extend the notion of the geometric mean from $APW_{N\times N}$ to $AP_{N\times N}$.

Theorem 18.12. *Let $1 < p < \infty$ and put*

$$\mathcal{G} = \{a \in AP_{N\times N} : T(a) \text{ is invertible on } H^p_N\}.$$

If $a \in \mathcal{G}$ and $\{a_n\} \subset AP^0_{N\times N}$ is any sequence of almost periodic matrix polynomials which converges uniformly to a, then the matrix polynomials a_n have canonical APW factorizations for all sufficiently large n and the limit

$$\mathbf{d}(a) := \lim_{n\to\infty} \mathbf{d}(a_n) \tag{18.18}$$

exists and is a matrix in $G\mathbf{C}^{N\times N}$. The map

$$\mathcal{G} \to G\mathbf{C}^{N\times N}, \quad a \mapsto \mathbf{d}(a) \tag{18.19}$$

is continuous.

Proof. Fix a sequence $\{a_n\}$ of almost periodic matrix polynomials a_n which converges uniformly to a. Since $T(a)$ is invertible, the operators $T(a_n)$ are invertible for all sufficiently large n and $\|T^{-1}(a_n) - T^{-1}(a)\| \to 0$. Without loss of generality assume that $T(a_n)$ is invertible for all n. From Corollary 18.4 we deduce that a_n has a canonical APW factorization $a_n = a_n^- a_n^+$. It is easily seen that $T^{-1}(a_n) = (a_n^+)^{-1} P (a_n^-)^{-1} I$.

Put $\chi(x) = \tanh(\pi x)$. By Lemma 3.5,

$$\chi (a_n^{\pm})^{-1} = \chi M^{-1}(a_n^{\pm}) + f_n^{\pm} \tag{18.20}$$

where $f_n^{\pm} \in [C + H_{\pm}^{\infty}]_{N\times N}$. Since Pf^-Q and Qf^+P are compact for $f^{\pm} \in C + H_{\pm}^{\infty}$, we obtain from (18.20) that

$$\begin{aligned} Q\chi T^{-1}(a_n)\chi Q &= Q\chi (a_n^+)^{-1} P (a_n^-)^{-1} \chi Q \\ &\simeq Q\chi M^{-1}(a_n^+) P M^{-1}(a_n^-)\chi Q = \mathbf{d}^{-1}(a_n) Q\chi P\chi Q, \end{aligned}$$

where, as before, $A \simeq B$ means that $A - B$ is compact. Hence

$$Q\chi(T^{-1}(a_n) - T^{-1}(a_m))\chi Q \simeq (\mathbf{d}^{-1}(a_n) - \mathbf{d}^{-1}(a_m)) Q\chi P\chi Q,$$

and passing to the jk entry on each side of this equality we get

$$\begin{aligned} &\left| \left(\mathbf{d}^{-1}(a_n) - \mathbf{d}^{-1}(a_m)\right)_{jk} \right| \|Q\chi P\chi Q\|_{\mathrm{ess}} \\ &= \left\| \left(Q\chi(T^{-1}(a_n) - T^{-1}(a_m))\chi Q\right)_{jk} \right\|_{\mathrm{ess}} \\ &\le \|Q\| \, \|T^{-1}(a_n) - T^{-1}(a_m)\| \, \|Q\|. \end{aligned}$$

Theorem 17.12(b) shows that the operator $Q\chi P\chi Q$ is not compact. Therefore $c_0 := \|Q\chi P\chi Q\|_{\mathrm{ess}} > 0$, and it follows that

$$\left| \left(\mathbf{d}^{-1}(a_n) - \mathbf{d}^{-1}(a_m)\right)_{jk} \right| \le c_0^{-1} \|Q\|^2 \|T^{-1}(a_n) - T^{-1}(a_m)\|. \tag{18.21}$$

Consequently, the limit $b := \lim \mathbf{d}^{-1}(a_n)$ exists. To prove that $b \in G\mathbf{C}^{N\times N}$, it suffices to verify that there exists an $M \in (0, \infty)$ such that for all j, k

$$\limsup_{n\to\infty} |(\mathbf{d}(a_n))_{jk}| \le M. \tag{18.22}$$

Indeed, (18.22) gives $|\det \mathbf{d}(a_n)| \le N! M^N$, whence

$$|\det b| = \lim_{n\to\infty} |\det \mathbf{d}^{-1}(a_n)| \ge 1/(N! M^N) > 0.$$

For a measurable subset E of $\mathbf{R}$, let χ_E be the characteristic function of E. Put $\chi_+ := \chi_{[0,\infty)}$ and $\chi_- := \chi_{(-\infty,0]}$. We know that $P = W^0(\chi_-)$ and $Q = W^0(\chi_+)$.

We now prove (18.22). Without loss of generality assume that a_n^- and a_n^+ belong to $AP^0_{N\times N}$. Fix a number n. Then there is an $\alpha = \alpha_n > 0$ such that

$$\Omega(a_n^-)\cap(-\alpha,0)=\emptyset,\quad \Omega(a_n^+)\cap(0,\alpha)=\emptyset. \tag{18.23}$$

Put

$$W_0 := W^0(\chi_{[-\alpha,0]}),\quad W_- := W^0(\chi_{(-\infty,-\alpha]}) \tag{18.24}$$

and notice that $W_- = e_\alpha P e_{-\alpha} I$. The invertibility of $T(a_n) = Pa_nP$ on H^p_N implies that the compression of the operator

$$W_-a_nW_- = e_\alpha Pe_{-\alpha}a_ne_\alpha Pe_{-\alpha}I = e_\alpha Pa_nPe_{-\alpha}I$$

to $W_-H^p_N$ is invertible. Consider the operator

$$B_n := W_0a_nW_0 - W_0a_nW_-(W_-a_nW_-)^{-1}W_-a_nW_0. \tag{18.25}$$

Since $W_- = (I+e_\alpha Se_{-\alpha}I)/2$ and $W_0 = (S - e_\alpha Se_{-\alpha}I)/2$, we have

$$\|W_-\|\le(1+\|S\|)/2\le\|S\|,\quad \|W_0\|\le\|S\|, \tag{18.26}$$

and thus $\|B_n\|\le\|S\|^2\|a_nI\| + \|S\|^4\|a_nI\|^2\|T^{-1}(a_n)\|$. This gives

$$M := \limsup_{n\to\infty}\|B_n\| < \infty. \tag{18.27}$$

As $a_n = a_n^-a_n^+$, we get $W_-a_nW_- = W_-a_n^-W_-a_n^+W_-$ and hence

$$(W_-a_nW_-)^{-1}W_- = (a_n^+)^{-1}W_-(a_n^-)^{-1}W_-. \tag{18.28}$$

Furthermore, the equality $a_n = a_n^-a_n^+$ implies that

$$W_0a_nW_0 = W_0a_n^-W^0(\chi_-)a_n^+W_0. \tag{18.29}$$

Inserting (18.28) and (18.29) in (18.25) we arrive at the equality

$$\begin{aligned} B_n &= W_0a_nW_0 - W_0a_n^-a_n^+W_-(a_n^+)^{-1}W_-(a_n^-)^{-1}W_-a_n^-a_n^+W_0\\ &= W_0a_n^-W^0(\chi_-)a_n^+W_0 - W_0a_n^-W_-a_n^+W_0 = W_0a_n^-W_0a_n^+W_0. \end{aligned} \tag{18.30}$$

By (18.23), $a_n^+ = M(a_n^+) + e_\alpha b_n^+$ and $a_n^- = M(a_n^-) + e_{-\alpha}b_n^-$ with $b_n^\pm \in APW^\pm_{N\times N}$. Because

$$\begin{aligned} W_0e_\alpha b_n^+P &= (P - e_\alpha Pe_{-\alpha}I)e_\alpha b_n^+P = 0,\\ Pe_{-\alpha}b_n^-W_0 &= Pe_{-\alpha}b_n^-(P - e_\alpha Pe_{-\alpha}I) = 0, \end{aligned}$$

it results that

$$\begin{aligned} W_0a_n^-W_0a_n^+W_0 &= W_0Pa_n^-W_0a_n^+PW_0\\ &= W_0PM(a_n^-)W_0M(a_n^+)PW_0\\ &= W_0M(a_n^-)W_0M(a_n^+)W_0 = \mathbf{d}(a_n)W_0. \end{aligned} \tag{18.31}$$

Putting (18.30) and (18.31) together we see that $\mathbf{d}(a_n)W_0 = B_n$, and hence (18.27) yields the estimate

$$\limsup_{n\to\infty} \left(|(\mathbf{d}(a_n))_{jk}|\,\|W_0\|\right) \le M < \infty.$$

As $\|W_0\| \ge 1$, we arrive at the desired estimate (18.22).

We now prove the continuity of the map (18.19). Fix $a \in \mathcal{G}$ and $\varepsilon > 0$. Then there is a $\delta > 0$ such that $b \in \mathcal{G}$ and $\|T^{-1}(a) - T^{-1}(b)\| < \varepsilon$ whenever $b \in AP_{N\times N}$ and $\|b - a\| < \delta$. Fix such a b and choose a_n, b_n in $AP^0_{N\times N}$ so that $a_n \to a$ and $b_n \to b$ uniformly. In the same way we proved (18.21) we get

$$\|\mathbf{d}^{-1}(a_n) - \mathbf{d}^{-1}(b_n)\| \le C\|T^{-1}(a_n) - T^{-1}(b_n)\|$$

with some constant $C < \infty$, and (18.18) allows us to pass to the limit $n \to \infty$ to obtain

$$\|\mathbf{d}^{-1}(a) - \mathbf{d}^{-1}(b)\| \le C\|T^{-1}(a) - T^{-1}(b)\| < C\varepsilon.$$

This shows that the map $a \mapsto \mathbf{d}^{-1}(a)$ is continuous, which implies that the map $a \mapsto \mathbf{d}(a)$ is also continuous. □

18.6 The First Auxiliary Operator

In this and the next section we establish some technical results that will be needed in the proof of the main result of this chapter.

Let $a \in GAPW_{N\times N}$ and suppose a has a canonical APW factorization $a = a_-a_+$ with $a_\pm \in AP^0_{N\times N}$ (which, of course, implies that a itself is in $AP^0_{N\times N}$). There is a $\lambda > 0$ such that

$$\Omega(a_-) \cap (-2\lambda, 0) = \emptyset, \quad \Omega(a_+) \cap (0, 2\lambda) = \emptyset. \tag{18.32}$$

Let $\hat{\chi}_- \in C(\overline{\mathbf{R}})$ be any real-valued and monotonically decreasing function such that $\hat{\chi}_-(t) = 0$ for $t \ge 0$ and $\hat{\chi}_-(t) = 1$ for $t \le -\lambda$. Put $\hat{\chi}_+ = 1 - \hat{\chi}_-$. In addition to the operators $P = W^0(\chi_-)$ and $Q = W^0(\chi_+)$, we consider the operators

$$\hat{P} = W^0(\hat{\chi}_-), \quad \hat{Q} = W^0(\hat{\chi}_+).$$

Put

$$A = aP + Q, \quad \hat{A} = a\hat{P} + \hat{Q} + W^0(f)$$

where f is any function in $[BV \cap C]_{N\times N}$ such that $\operatorname{supp} f \subset [-\lambda, 0]$. Let I_N be the $N \times N$ identity matrix.

Lemma 18.13. *The operator $\hat{A}$ is invertible on L^p_N $(1 < p < \infty)$ if and only if*

$$\det\left(\mathbf{d}(a)\hat{\chi}_-(t) + I_N\hat{\chi}_+(t) + f(t)\right) \ne 0 \quad \textit{for all} \quad t \in [-\lambda, 0]. \tag{18.33}$$

Proof. Since $\hat{P}P = \hat{P}$, $\hat{P}Q = 0$, $\hat{Q}Q = Q$, the matrix representation of $\hat{A}$ corresponding to the direct sum decomposition $L^p_N = [H^p_-]_N \oplus [H^p_+]_N$ is

$$\begin{pmatrix} Q & Qa\hat{P} \\ 0 & Pa\hat{P} + P\hat{Q} + W^0(f) \end{pmatrix} =: \begin{pmatrix} Q & Qa\hat{P} \\ 0 & C \end{pmatrix}. \tag{18.34}$$

Consequently, $\hat{A}$ is invertible on L^p_N if and only if C is invertible on H^p_N, in which case

$$\hat{A}^{-1} = Q + (P - Qa\hat{P})C^{-1}P. \tag{18.35}$$

As in Section 18.5, let

$$W_0 := W^0(\chi_{[-\lambda,0]}), \quad W_- := W^0(\chi_{(-\infty,-\lambda]}).$$

Then $H^p_N = W_0 H^p_N \oplus W_- H^p_N$, and the corresponding matrix decomposition of C is

$$\begin{pmatrix} W_0 C W_0 & W_0 a W_- \\ W_- a W_0 W^0(\hat{\chi}_-) & W_- a W_- \end{pmatrix}. \tag{18.36}$$

The compression of $W_- a W_-$ to $W_- H^p_N$ is invertible, the inverse operator being $a_+^{-1} W_- a_-^{-1} W_-$. Hence, the matrix (18.36) equals

$$\begin{pmatrix} W_0 & W_0 a W_- \\ 0 & W_- a W_- \end{pmatrix} \times$$
$$\times \begin{pmatrix} W_0 C W_0 - W_0 a W_- (W_- a W_-)^{-1} W_- a W_0 W^0(\hat{\chi}_-) & 0 \\ (W_- a W_-)^{-1} W_- a W_0 W^0(\hat{\chi}_-) & W_- \end{pmatrix}. \tag{18.37}$$

It follows that C is invertible on H^p_N if and only if the compression of

$$B := W_0 C W_0 - W_0 a W_- (W_- a W_-)^{-1} W_- a W_0 W^0(\hat{\chi}_-)$$

to $W_0 H^p_N$ is invertible. Since

$$W_0 C W_0 = W_0 a W_0 W^0(\hat{\chi}_-) + W_0 W^0(\hat{\chi}_+) + W^0(f),$$

we see that B equals

$$\left(W_0 a W_0 - W_0 a W_- a_+^{-1} W_- a_-^{-1} W_- a W_0\right) W^0(\hat{\chi}_-) + W_0 W^0(\hat{\chi}_+) + W^0(f)$$
$$= W_0 a_- W_0 a_+ W_0 W^0(\hat{\chi}_-) + W_0 W^0(\hat{\chi}_+) + W^0(f).$$

In analogy to (18.31), $W_0 a_- W_0 a_+ W_0 = \mathbf{d}(a) W_0$, whence

$$B = W^0\Big((\mathbf{d}(a)\hat{\chi}_- + I_N \hat{\chi}_+ + f)\chi_{[-\lambda,0]}\Big). \tag{18.38}$$

As $\mathbf{d}(a)\hat{\chi}_- + I_N\hat{\chi}_+ + f \in [BV \cap C(\overline{\mathbf{R}})]_{N\times N}$, the compression of B to $W_0H^p_N$ is invertible if and only if (18.33) holds. □

Lemma 18.14. *If $\hat{A}$ is invertible, then*

$$\|\hat{A}^{-1}\| \le \|S\|\Big(1+\|S\|(1+2\|QaP\|)\|C^{-1}\|\Big) \tag{18.39}$$

where

$$\|C^{-1}\| \le \|S\|\Big(\|T^{-1}(a)\| + \|B^{-1}\|(1+\|S\|^2\kappa(T(a)))\Big)\Big(1+3\|S\|^2\kappa(T(a))\Big) \tag{18.40}$$

with $\kappa(T(a)) := \|T(a)\|\,\|T^{-1}(a)\|$ and

$$\|B^{-1}\| = \|W^0((\mathbf{d}(a)\hat{\chi}_- + I_N\hat{\chi}_+ + f)^{-1}\chi_{[-\lambda,0]})\|. \tag{18.41}$$

Proof. Equality (18.41) is immediate from (18.38) and Lemma 18.13. Since C is given by (18.37), the operator C^{-1} has the matrix

$$\begin{pmatrix} B^{-1} & -B^{-1}W_0aW_-(W_-aW_-)^{-1} \\ -(W_-aW_-)^{-1}W_-aW_0W^0(\hat{\chi}_-)B^{-1} & (C^{-1})_{22} \end{pmatrix} \tag{18.42}$$

where

$$(C^{-1})_{22} = (W_-aW_-)^{-1}(I + W_-aW_0W^0(\hat{\chi}_-)B^{-1}W_0aW_-(W_-aW_-)^{-1}).$$

This shows that $\|C^{-1}\|$ does not exceed

$$\begin{aligned}
&\|B^{-1}\|\,\|W_0\| + \|B^{-1}\|\,\|W_0aW_-\|\,\|(W_-aW_-)^{-1}\|\,\|W_-\| \\
&+\|(W_-aW_-)^{-1}\|\,\|W_-aW_0W^0(\hat{\chi}_-)\|\,\|B^{-1}\|\,\|W_0\| \\
&+\|(W_-aW_-)^{-1}\|\,\|W_-\| \\
&+\|(W_-aW_-)^{-1}\|\,\|W_-aW_0W^0(\hat{\chi}_-)\|\,\|B^{-1}\|\,\|W_0aW_-\| \times \\
&\quad \times\|(W_-aW_-)^{-1}\|\,\|W_-\|.
\end{aligned}$$

Clearly, $\|(W_-aW_-)^{-1}\| = \|T^{-1}(a)\|$, and from (17.1) and (18.26) we get

$$\begin{aligned}
&\|W_0aW_-\| \le \|S\|^2\|T(a)\|, \\
&\|W_-aW_0W^0(\hat{\chi}_-)\| \le \|W_-\|\,\|T(a)\|\,\|W^0(\chi_{[-\lambda,0]}\hat{\chi}_-)\| \le 3\|S\|^2\|T(a)\|.
\end{aligned}$$

This gives (18.40). Finally, since $\|\hat{P}\| = \|W^0(\hat{\chi}_-)\| \le 2\|S\|$ due to (17.1), we obtain from (18.35) and the equality $\hat{P} = P\hat{P}$ that

$$\begin{aligned}
\|\hat{A}^{-1}\| &\le \|Q\| + \|P\|^2\|C^{-1}\| + \|QaP\|\,\|\hat{P}\|\,\|C^{-1}\|\,\|P\| \\
&\le \|S\| + \|S\|^2\|C^{-1}\| + \|QaP\|\,(2\|S\|)\,\|C^{-1}\|\,\|S\|,
\end{aligned}$$

which is (18.39). □

18.7 The Second Auxiliary Operator

We denote by $\Lambda \subset \mathcal{B}(L_N^p)$ the Banach subalgebra of all operators of local type. Thus, $B \in \Lambda$ if and only if $cB - BcI \in \mathcal{K}(L_N^p)$ for all $c \in C := C(\dot{\mathbf{R}})$. Obviously, $\mathcal{K} := \mathcal{K}(L_N^p)$ is a closed ideal of Λ. An operator $B \in \Lambda$ is Fredholm if and only if $B + \mathcal{K}$ is invertible in $\Lambda/\mathcal{K}$. In particular, every regularizer of a Fredholm operator $B \in \Lambda$ belongs to Λ.

The set $Z := \{cI + K : c \in C, K \in \mathcal{K}\}$ is a closed subalgebra of Λ. The quotient algebra $Z/\mathcal{K}$ is a central subalgebra of $\Lambda/\mathcal{K}$ and the maximal ideal space of $Z/\mathcal{K}$ can be identified with $\dot{\mathbf{R}} = \mathbf{R} \cup \{\infty\}$. For $t \in \dot{\mathbf{R}}$, we denote by J_t the smallest closed ideal of Λ which contains the set $\{cI + K : c \in C, c(t) = 0, K \in \mathcal{K}\}$.

Theorem 18.15 (Simonenko/Allan/Douglas). *Let $B \in \Lambda$. The operator B is Fredholm if and only if $B + J_t$ is invertible in Λ/J_t for every $t \in \dot{\mathbf{R}}$. In that case*

$$\|B^{(-1)}\|_{\text{ess}} = \sup_{t \in \dot{\mathbf{R}}} \|(B + J_t)^{-1}\| \tag{18.43}$$

for every regularizer $B^{(-1)}$ of B.

A proof of the Fredholm criterion of Theorem 18.15 can be found in [46, Theorem 1.34] and [102, Theorem 1.5], for example. Equality (18.43) is proved in [45] (also see [102, Section 1.7]).

Let $a \in GSAP_{N\times N}$ and write a in the form

$$a = (1-u)a_l + ua_r + a_0 \tag{18.44}$$

where $u \in C(\overline{\mathbf{R}})$ is any fixed function such that $0 \le u \le 1$, $u(-\infty) = 0$, $u(+\infty) = 1$, a_l and a_r belong to $AP_{N\times N}$, and a_0 is in $[C_0]_{N\times N}$. Suppose that a_l and a_r are almost periodic matrix polynomials and that a_l and a_r have canonical APW factorizations $a_l = a_l^- a_l^+$ and $a_r = a_r^- a_r^+$ with $a_l^\pm$, $a_r^\pm$ in $AP^0_{N\times N}$. Choose $\lambda > 0$ so that

$$\Omega(a_l^-) \cap (-2\lambda, 0) = \emptyset, \quad \Omega(a_l^+) \cap (0, 2\lambda) = \emptyset, \tag{18.45}$$
$$\Omega(a_r^-) \cap (-2\lambda, 0) = \emptyset, \quad \Omega(a_r^+) \cap (0, 2\lambda) = \emptyset, \tag{18.46}$$

let $f_l, f_r \in [BV \cap C]_{N\times N}$ be any matrix functions such that $\operatorname{supp} f_l \subset [-\lambda, 0]$, $\operatorname{supp} f_r \subset [-\lambda, 0]$, and

$$\zeta_l := \mathbf{d}(a_l)\hat{\chi}_- + I_N\hat{\chi}_+ + f_l \in G[C(\overline{\mathbf{R}})]_{N\times N}, \tag{18.47}$$
$$\zeta_r := \mathbf{d}(a_r)\hat{\chi}_- + I_N\hat{\chi}_+ + f_r \in G[C(\overline{\mathbf{R}})]_{N\times N}. \tag{18.48}$$

We consider the operator

$$R := a\hat{P} + \hat{Q} + (1-u)W^0(f_l) + uW^0(f_r).$$

The operator R can be represented in the form

$$R = (1-u)\hat{A}_l + u\hat{A}_r + a_0\hat{P}$$

with

$$\hat{A}_l := a_l\hat{P} + \hat{Q} + W^0(f_l), \quad \hat{A}_r := a_r\hat{P} + \hat{Q} + W^0(f_r).$$

The operators $\hat{A}_l$ and $\hat{A}_r$ are invertible by virtue of (18.47), (18.48) and Lemma 18.13.

Lemma 18.16. *The operator R is Fredholm on L^p_N.*

Proof. Part (b) of Lemma 17.13 shows that $R \in \Lambda$. From Theorem 18.15 we therefore deduce that R is Fredholm if and only if $R + J_t$ is invertible in Λ/J_t for every $t \in \dot{\mathbf{R}}$.

Fix $t \in \mathbf{R}$ and choose $c_t \in C$ so that $c_t(\infty) = 1$ and $c_t(t) = 0$. If $b \in BV_{N\times N}$ and $b(\pm\infty) = 0$, then, by part (a) of Lemma 17.13, $W^0(b) \simeq c_t W^0(b) \in J_t$. Consequently, the operators

$$W^0(f_l), \quad W^0(f_r), \quad \hat{P} - P = W^0(\hat{\chi}_- - \chi_-), \quad \hat{Q} - Q = W^0(\hat{\chi}_+ - \chi_+)$$

all belong to J_t. This implies that $R + J_t = a(t)P + Q + J_t$. Since $a \in GSAP_{N\times N}$, it follows that

$$(R + J_t)^{-1} = a^{-1}(t)P + Q + J_t. \tag{18.49}$$

Let now $t = \infty$. We have $R + J_\infty = (1-u)\hat{A}_l + u\hat{A}_r + J_\infty$. The operators $\hat{A}_l^{-1}$ and $\hat{A}_r^{-1}$ belong to Λ due to the inverse closedness of Λ in $\mathcal{B}(L^p_N)$. We claim that

$$(R + J_\infty)^{-1} = (1-u)\hat{A}_l^{-1} + u\hat{A}_r^{-1} + J_\infty. \tag{18.50}$$

Indeed, from part (c) of Lemma 17.13 we deduce that

$$u\hat{P} - \hat{P}uI \in \mathcal{K}, \quad u\hat{Q} - \hat{Q}uI \in \mathcal{K},$$

$$uW^0(f_l) - W^0(f_l)uI \in \mathcal{K}, \quad uW^0(f_r) - W^0(f_r)uI \in \mathcal{K},$$

which shows that

$$u\hat{A}_l - \hat{A}_l uI \in \mathcal{K} \subset J_\infty, \quad u\hat{A}_r - \hat{A}_r uI \in \mathcal{K} \subset J_\infty. \tag{18.51}$$

Thus,

$$\begin{aligned}
&(R + J_\infty)((1-u)\hat{A}_l^{-1} + u\hat{A}_r^{-1} + J_\infty) \\
&= (1-u)^2 I + u(1-u)\hat{A}_r\hat{A}_l^{-1} + (1-u)u\hat{A}_l\hat{A}_r^{-1} + u^2 I + J_\infty \\
&= I + u(1-u)(-2I + \hat{A}_r\hat{A}_l^{-1} + \hat{A}_l\hat{A}_r^{-1}) + J_\infty = I + J_\infty,
\end{aligned}$$

the last equality resulting from the inclusion $u(1-u)I \in J_\infty$. Similarly, taking into account that (18.51) implies

$$u\hat{A}_l^{-1} - \hat{A}_l^{-1}uI \in \mathcal{K} \subset J_\infty, \quad u\hat{A}_r^{-1} - \hat{A}_r^{-1}uI \in \mathcal{K} \subset J_\infty, \tag{18.52}$$

one can verify that

$$((1-u)\hat{A}_l^{-1} + u\hat{A}_r^{-1} + J_\infty)(R + J_\infty) = I + J_\infty.$$

This completes the proof of (18.50).

From (18.49), (18.50), and Theorem 18.15 we obtain that R is Fredholm. □

Lemma 18.17. *If $R^{(-1)}$ is any regularizer of R, then*

$$\|R^{(-1)}\|_{\text{ess}} \le \max\Big\{(\|a^{-1}I\|+1)\|S\|, \|\hat{A}_l^{-1}\|, \|\hat{A}_r^{-1}\|\Big\}. \tag{18.53}$$

Proof. From (18.49) we see that if $t \in \mathbf{R}$, then

$$\|(R+J_t)^{-1}\| \le \|a^{-1}(t)P + Q\| \le (\|a^{-1}I\|+1)\|S\|. \tag{18.54}$$

Let $u_l, u_r \in C(\overline{\mathbf{R}})$ be any functions such that $0 \le u_l \le 1$, $0 \le u_r \le 1$, $\operatorname{supp} u_l = (-\infty, 0]$, $\operatorname{supp} u_r = [0,\infty)$, $u_l(-\infty) = 1$, $u_r(+\infty) = 1$. Then, by (18.52),

$$(1-u)\hat{A}_l^{-1} + u\hat{A}_r^{-1} + J_\infty = u_l\hat{A}_l^{-1}u_lI + u_r\hat{A}_r^{-1}u_rI + J_\infty$$

and hence (18.50) gives

$$\begin{aligned} \|(R+J_\infty)^{-1}\| &= \|u_l\hat{A}_l^{-1}u_lI + u_r\hat{A}_r^{-1}u_rI + J_\infty\| \\ &\le \max\{\|u_l\hat{A}_l^{-1}u_lI\|, \|u_r\hat{A}_r^{-1}u_rI\|\} \\ &\le \max\{\|\hat{A}_l^{-1}\|, \|\hat{A}_r^{-1}\|\}. \end{aligned} \tag{18.55}$$

Theorem 18.15 in conjunction with (18.54) and (18.55) yields (18.53). □

18.8 Matrix-Valued SAP Symbols

Here is the main result of this chapter.

Theorem 18.18. *Let $a \in SAP_{N\times N}$ and $1 < p < \infty$. The operator $T(a)$ is Fredholm on H_N^p if and only if the following three conditions are satisfied:*

(i) *$a \in GSAP_{N\times N}$;*

(ii) *$T(a_l)$ and $T(a_r)$ are invertible on H_N^p;*

(iii) *$\operatorname{sp}(\mathbf{d}^{-1}(a_r)\mathbf{d}(a_l)) \cap e^{2\pi i/p}\mathbf{R}_+ = \emptyset$, where $\operatorname{sp} A$ denotes the set of the eigenvalues of the matrix A and $\mathbf{R}_+ := [0,\infty)$.*

If $T(a)$ is Fredholm, then

$$\operatorname{Ind} T(a) = -\operatorname{ind}(\det a) - \sum_{k=1}^{N} \left(\frac{1}{p} - \left\{ \frac{1}{p} - \frac{1}{2\pi} \arg \xi_k \right\} \right) \tag{18.56}$$

where

$$\operatorname{ind}(\det a) := \frac{1}{2\pi} \lim_{T\to\infty} \frac{1}{T} \int_0^T \Big((\arg\det a)(x) - (\arg\det a)(-x) \Big)\, dx,$$

$\arg\det a$ *is any argument of the function* $\det a$ *which is continuous on the line* $\mathbf{R}$*,* $\xi_1, \dots, \xi_N \in \mathbf{C} \setminus e^{2\pi i/p}\mathbf{R}_+$ *are the eigenvalues of* $\mathbf{d}^{-1}(a_r)\mathbf{d}(a_l)$*, and* $\{y\} \in [0,1)$ *denotes the fractional part of* $y \in \mathbf{R}$*.*

The rest of this section is devoted to the proof of this theorem. The proof of the necessity of conditions (i) to (iii) is short, because the nontrivial part of it, namely, the fact that the Fredholmness of $T(a)$ implies (ii), follows from Corollary 18.11. The proof of the sufficiency portion of Theorem 18.18 is more difficult and based on the following strategy: we approximate $a \in SAP_{N\times N}$ by matrix functions $a^{(n)} \in SAP_{N\times N}$ for which $a_l^{(n)}$ and $a_r^{(n)}$ belong to $AP^0_{N\times N}$, we then prove Theorem 18.18 for the operators $T(a^{(n)})$ and, in addition, we show that the operators $T(a^{(n)})$ have regularizers R_n satisfying

$$\limsup_{n\to\infty} \|R_n\|_{\mathrm{ess}} < \infty; \tag{18.57}$$

clearly, the Fredholmness of $T(a^{(n)})$ in conjunction with (18.57) implies that $T(a)$ is Fredholm. To prove that the operators $T(a^{(n)})$ are Fredholm and to obtain estimate (18.57) we use Theorem 18.15, which reduces things to simpler auxiliary operators. These were studied in Sections 18.6 and 18.7.

Suppose (i), (ii), (iii) hold. We first assume that a is as in Section 18.7, that is, a_l and a_r are almost periodic matrix polynomials which have canonical APW factorizations with factors in $AP^0_{N\times N}$. Choose $\lambda > 0$ so that (18.45) and (18.46) are satisfied. Lemmas 18.16 and 18.17 tell us that the operator R is Fredholm and that $\|R^{(-1)}\|_{\mathrm{ess}}$ can be estimated as in (18.53).

Put $R_a = aP + Q$ and $V = R^{(-1)}R_a$. Since $R_a \in \Lambda$ and $R \in \Lambda$, the operators $R^{(-1)}$ and V also belong to Λ.

Lemma 18.19. *The operator V is Fredholm on L^p_N.*

Proof. If $t \in \mathbf{R}$, then, by condition (i) and by (18.49),

$$\begin{aligned} V + J_t &= (R + J_t)^{-1}(R_a + J_t) \\ &= (a^{-1}(t)P + Q + J_t)(a(t)P + Q + J_t) \\ &= I + J_t, \end{aligned} \tag{18.58}$$

which shows that $V+J_t$ is invertible. So let $t=\infty$. Then, by (18.50) and (18.52),

$$\begin{aligned} V+J_\infty &= (R+J_\infty)^{-1}(R_a+J_\infty) \\ &= [(1-u)\hat{A}_l^{-1}+u\hat{A}_r^{-1}+J_\infty][(1-u)R_{a_l}+uR_{a_r}+J_\infty] \\ &= \hat{A}_l^{-1}(1-u)R_{a_l}+\hat{A}_r^{-1}uR_{a_r}+J_\infty =: D+J_\infty. \end{aligned} \tag{18.59}$$

Define ζ_l and ζ_r by (18.47) and (18.48) and put $\chi_0=\chi_{[-\lambda,0]}$. From (18.35) we get

$$\hat{A}_l^{-1}=Q+(P-Qa_l\hat{P})C_l^{-1}P. \tag{18.60}$$

By virtue of (18.42) and the equality

$$(W_-a_lW_-)^{-1}W_-=(a_l^+)^{-1}W_-(a_l^-)^{-1}W_-,$$

this implies that

$$\begin{aligned} C_l^{-1} &= \left(Pa_l\hat{P}+P\hat{Q}+W^0(f_l)\right)^{-1} \\ &= W^0(\zeta_l^{-1}\chi_0)-(W_-a_lW_-)^{-1}W_-a_lW_0W^0(\hat{\chi}_-)W^0(\zeta_l^{-1}\chi_0) \\ &\quad -W^0(\zeta_l^{-1}\chi_0)W_0a_lW_-(W_-a_lW_-)^{-1}W_-+(W_-a_lW_-)^{-1}W_- \\ &\quad +(W_-a_lW_-)^{-1}W_-a_lW_0W^0(\hat{\chi}_-)W^0(\zeta_l^{-1}\chi_0)W_0a_lW_-(W_-a_lW_-)^{-1}W_- \\ &= (a_l^+)^{-1}W_-(a_l^-)^{-1}I \\ &\quad +[W_0-(a_l^+)^{-1}W_-a_l^+W_0W^0(\hat{\chi}_-)]W^0(\zeta_l^{-1})\times \\ &\quad \times[W_0-W_0a_l^-W_-(a_l^-)^{-1}I]. \end{aligned} \tag{18.61}$$

On the other hand,

$$\begin{aligned} &\hat{A}_l^{-1}(1-u)R_{a_l} \\ &\quad = \hat{A}_l^{-1}\left[\hat{A}_l(1-u)I+(1-u)\left(a_l(P-\hat{P})+(Q-\hat{Q})-W^0(f_l)\right)\right] \\ &\quad = (1-u)I+\hat{A}_l^{-1}(1-u)B_l \end{aligned} \tag{18.62}$$

where

$$\begin{aligned} B_l &:= a_l(P-\hat{P})+(Q-\hat{Q})-W^0(f_l) \\ &= a_lW^0(\chi_--\hat{\chi}_-)+W^0(\chi_+-\hat{\chi}_+)-W^0(f_l). \end{aligned}$$

In the direct sum

$$L_N^p=[H_-^p]_N\oplus W_0[H_+^p]_N\oplus W_-[H_+^p]_N$$

the operator $(1-u)B_l$ is given by the matrix

$$(1-u)B_l=\begin{pmatrix} 0 & Q(1-u)B_l & 0 \\ 0 & W_0(1-u)B_l & 0 \\ 0 & W_-(1-u)B_l & 0 \end{pmatrix}. \tag{18.63}$$

Combining (18.62) and (18.63) we see that the operator D defined by (18.59) has a matrix of the form

$$D = \begin{pmatrix} Q & d_{12} & 0 \\ 0 & W_0 + d_{22} & 0 \\ 0 & d_{32} & W_- \end{pmatrix}, \tag{18.64}$$

and hence $V + J_\infty = D + J_\infty$ is invertible whenever

$$W_0 + d_{22} + W_0 J_\infty W_0 \in G[(W_0 \mathcal{A} W_0)/(W_0 J_\infty W_0)]. \tag{18.65}$$

The definitions of D and B_l together with (18.61) and (18.62) give

$$\begin{aligned} d_{22} &= W_0\Big(\hat{A}_l^{-1}(1-u)B_l + \hat{A}_r^{-1}uB_r\Big)W_0 \\ &= W^0(\zeta_l^{-1})\Big[W_0 - W_0 a_l^- W_-(a_l^-)^{-1}I\Big](1-u)\times \\ &\qquad \times\Big[a_l W^0(\chi_- - \hat{\chi}_-) + W^0(\chi_+ - \hat{\chi}_+) - W^0(f_l)\Big] \\ &\quad + W^0(\zeta_r^{-1})\Big[W_0 - W_0 a_r^- W_-(a_r^-)^{-1}I\Big]u\times \\ &\qquad \times\Big[a_r W^0(\chi_- - \hat{\chi}_-) + W^0(\chi_+ - \hat{\chi}_+) - W^0(f_r)\Big]. \end{aligned} \tag{18.66}$$

Furthermore,

$$\begin{aligned} &\Big[W_0 - W_0 a_l^- W_-(a_l^-)^{-1}I\Big]a_l(1-u)W_0 \\ &= W_0 a_l^- P a_l^+(1-u)W_0 - W_0 a_l^- W_- a_l^+(1-u)W_0 \\ &= W_0 a_l^- W_0 a_l^+(1-u)W_0 \end{aligned} \tag{18.67}$$

and

$$\begin{aligned} &\Big[W_0 - W_0 a_l^- W_-(a_l^-)^{-1}I\Big](1-u)W_0 \\ &= W_0 a_l^- P(a_l^-)^{-1}(1-u)W_0 - W_0 a_l^- W_-(a_l^-)^{-1}(1-u)W_0 \\ &= W_0 a_l^- W_0(a_l^-)^{-1}(1-u)W_0. \end{aligned} \tag{18.68}$$

By (18.45) and (18.46), there is a $\mu > \lambda > 0$ such that

$$\begin{aligned} a_l^- &= M(a_l^-) + e_{-\mu} b_l^-, \quad & a_l^+ &= M(a_l^+) + e_\mu b_l^+, \\ a_r^- &= M(a_r^-) + e_{-\mu} b_r^-, \quad & a_r^+ &= M(a_r^+) + e_\mu b_r^+ \end{aligned}$$

with $b_l^\pm, b_r^\pm \in APW^\pm_{N\times N}$. Obviously, this implies that

$$\begin{aligned} (a_l^-)^{-1} &= M^{-1}(a_l^-) + e_{-\mu} c_l^-, \quad & (a_l^+)^{-1} &= M^{-1}(a_l^+) + e_\mu c_l^+, \\ (a_r^-)^{-1} &= M^{-1}(a_r^-) + e_{-\mu} c_r^-, \quad & (a_r^+)^{-1} &= M^{-1}(a_r^+) + e_\mu c_r^+ \end{aligned}$$

with $c_l^\pm, c_r^\pm \in APW_{N\times N}^\pm$. From Lemma 3.5 we know that $vf \in C + H_\pm^\infty$ whenever $v \in C(\overline{\mathbf{R}})$, $f \in AP^\pm$, and $M(f) = 0$. Hence,

$$a_l^+(1-u) = M(a_l^+)(1-u) + e_\lambda \varphi_l^+,$$
$$(a_l^-)^{-1}(1-u) = M^{-1}(a_l^-)(1-u) + e_{-\lambda}\varphi_l^-$$

where $\varphi_l^+ := e_{\mu-\lambda} b_l^+ \in [C + H_+^\infty]_{N\times N}$ and $\varphi_l^- := e_{-\mu+\lambda} c_l^- \in [C + H_-^\infty]_{N\times N}$. It follows that

$$\begin{aligned} W_0 a_l^+(1-u)W_0 &= W_0 M(a_l^+)(1-u)W_0 + W_0 e_\lambda \varphi_l^+ P W_0 \\ &= M(a_l^+)W_0(1-u)W_0 + (P - e_\lambda P e_{-\lambda} I)e_\lambda \varphi_l^+ P W_0 \\ &\simeq M(a_l^+)W_0(1-u)W_0 \end{aligned} \tag{18.69}$$

and

$$\begin{aligned} W_0(a_l^-)^{-1}(1-u)W_0 &= W_0 M^{-1}(a_l^-)(1-u)W_0 + W_0 P e_{-\lambda}\varphi_l^- W_0 \\ &= M^{-1}(a_l^-)W_0(1-u)W_0 + W_0 P e_{-\lambda}\varphi_l^-(P - e_\lambda P e_{-\lambda} I) \\ &\simeq M^{-1}(a_l^-)W_0(1-u)W_0. \end{aligned} \tag{18.70}$$

Since $W_0 a_l^- W_0 = M(a_l^-)W_0$, we infer from (18.67), (18.69) and (18.68), (18.70) that

$$\begin{aligned} &\Big[W_0 - W_0 a_l^- W_-(a_l^-)^{-1} I\Big](1-u)a_l W_0 \\ &\quad \simeq M(a_l^-)M(a_l^+)W_0(1-u)W_0 = W_0 \mathbf{d}(a_l)(1-u)W_0 \end{aligned} \tag{18.71}$$

and

$$\begin{aligned} &\Big[W_0 - W_0 a_l^- W_-(a_l^-)^{-1} I\Big](1-u)W_0 \\ &\quad \simeq M(a_l^-)M^{-1}(a_l^-)W_0(1-u)W_0 = W_0(1-u)W_0. \end{aligned} \tag{18.72}$$

Inserting (18.71) and (18.72) in (18.66) we get

$$\begin{aligned} d_{22} \simeq\ & W^0(\zeta_l^{-1})W_0(1-u)\Big[\mathbf{d}(a_l)W^0(\chi_- - \hat{\chi}_-) + W^0(\chi_+ - \hat{\chi}_+) - W^0(f_l)\Big] \\ & + W^0(\zeta_r^{-1})W_0 u\Big[\mathbf{d}(a_r)W^0(\chi_- - \hat{\chi}_-) + W^0(\chi_+ - \hat{\chi}_+) - W^0(f_r)\Big], \end{aligned}$$

whence

$$\begin{aligned} & W_0 + d_{22} + W_0 J_\infty W_0 \\ &= W_0\Big[W^0(\zeta_l^{-1})(1-u)W^0(\zeta_l) + W^0(\zeta_r^{-1})uW^0(\zeta_r)\Big] + d_{22} + W_0 J_\infty W_0 \\ &= W_0 W^0(\zeta_l^{-1})(1-u)\Big(\mathbf{d}(a_l)P + Q\Big) + W_0 W^0(\zeta_r^{-1})u\Big(\mathbf{d}(a_r)P + Q\Big) \\ &\quad + W_0 J_\infty W_0 \\ &= W_0\Big[\hat{A}_l^{-1}(1-u)\Big(\mathbf{d}(a_l)P + Q\Big) + \hat{A}_r^{-1}u\Big(\mathbf{d}(a_r)P + Q\Big)\Big] \\ &\quad + W_0 J_\infty W_0. \end{aligned} \tag{18.73}$$

Condition (iii) and the Fredholm criterion for Toeplitz operators with symbols in $[C(\overline{\mathbf{R}})]_{N\times N}$, that is, Theorem 16.22 show that

$$\widetilde{R}_a + J_\infty := \Big(\mathbf{d}(a_l)(1-u) + \mathbf{d}(a_r)u\Big)P + Q + J_\infty \tag{18.74}$$

is invertible in Λ/J_∞. Consequently

$$\begin{aligned}(R+J_\infty)^{-1}&(\widetilde{R}_a + J_\infty)\\ &= \Big[(1-u)\hat{A}_l^{-1} + u\hat{A}_r^{-1} + J_\infty\Big]\Big[(\mathbf{d}(a_l)(1-u) + \mathbf{d}(a_r)u)P + Q + J_\infty\Big]\\ &= \hat{A}_l^{-1}(1-u)(\mathbf{d}(a_l)P + Q) + \hat{A}_r^{-1}u(\mathbf{d}(a_r)P+Q) + J_\infty \end{aligned}\tag{18.75}$$

is also invertible in Λ/J_∞. In analogy to (18.64), the operator in (18.75) can be represented by a matrix of the form

$$\begin{pmatrix} Q & d'_{12} & 0 \\ 0 & W_0 + d'_{22} & 0 \\ 0 & d'_{32} & W_- \end{pmatrix} \tag{18.76}$$

where, by (18.73) and (18.75),

$$W_0 + d'_{22} + W_0 J_\infty W_0 = W_0 + d_{22} + W_0 J_\infty W_0. \tag{18.77}$$

This implies (18.65) and hence proves that $V + J_\infty$ is invertible.

Thus, we have shown that $V + J_t$ is invertible for all $t \in \dot{\mathbf{R}}$. Theorem 18.15 now gives the assertion. □

Lemma 18.20. *If $V^{(-1)}$ is any regularizer of V, then*

$$\|V^{(-1)}\|_{\text{ess}} \le 2\|S\| + \|S\|^3\Big(2\|S\|^2 M_1 + 1\Big)M_2 \tag{18.78}$$

where

$$M_1 = \Big(\|aI\| + 1\Big)\Big(\|\hat{A}_l^{-1}\| + \|\hat{A}_r^{-1}\|\Big), \quad M_2 = \|(\widetilde{R}_a + J_\infty)^{-1}\|\Big(\|\hat{A}_l\| + \|\hat{A}_r\|\Big),$$

and $\widetilde{R}_a$ is given by (18.74).

Proof. From (18.58), (18.59), and Theorem 18.15 we deduce that

$$\|V^{(-1)}\|_{\text{ess}} = \max\{1, \|(V+J_\infty)^{-1}\|\} = \max\{1, \|(D+J_\infty)^{-1}\|\}. \tag{18.79}$$

In view of (18.64)

$$\begin{aligned}(D+J_\infty)^{-1} &= Q + \Big(-Q(D+J_\infty)W_0 + W_0 - W_-(D+J_\infty)W_0\Big) \times \\ &\quad \times (W_0 + d_{22} + W_0 J_\infty W_0)^{-1} W_0 + W_- + J_\infty,\end{aligned}$$

whence

$$\begin{aligned}
&\|(D+J_\infty)^{-1}\| \\
&\le \|Q\| + \Big(\|Q\|\,\|D+J_\infty\|\,\|W_0\| + \|W_0\| + \|W_-\|\,\|D+J_\infty\|\,\|W_0\|\Big) \times \\
&\qquad \times \|(W_0+d_{22}+W_0J_\infty W_0)^{-1}W_0\| + \|W_-\| \\
&\le 2\|S\| + \Big(2\|S\|^2\|V+J_\infty\| + \|S\|\Big) \times \\
&\qquad \times \|(W_0+d_{22}+W_0J_\infty W_0)^{-1}W_0\|. \qquad (18.80)
\end{aligned}$$

We have

$$\begin{aligned}
\|V+J_\infty\| &\le \|(R+J_\infty)^{-1}\|\,\|R_a+J_\infty\| \\
&\le \Big(\|\hat{A}_l^{-1}\| + \|\hat{A}_r^{-1}\|\Big)\Big(\|aI\|+1\Big)\|S\|, \qquad (18.81)
\end{aligned}$$

and from (18.76) and (18.77) we obtain that

$$\begin{aligned}
&\|(W_0+d_{22}+W_0J_\infty W_0)^{-1}W_0\| \\
&= \|W_0(\widetilde{R}_a+J_\infty)^{-1}(R+J_\infty)W_0\| \\
&\le \|S\|^2\|(\widetilde{R}_a+J_\infty)^{-1}\|\,\|R+J_\infty\| \\
&\le \|S\|^2\|(\widetilde{R}_a+J_\infty)^{-1}\|\Big(\|\hat{A}_l\|+\|\hat{A}_r\|\Big). \qquad (18.82)
\end{aligned}$$

Putting (18.79) to (18.82) together we arrive at (18.78). □

Lemma 18.21. *The operator R_a is Fredholm, and if $R_a^{(-1)}$ is any regularizer of R_a, then $\|R_a^{(-1)}\|_{\text{ess}}$ has an upper estimate that contains only constants (depending possibly on p) and the values*

$$\begin{aligned}
&\|aI\|,\ \|W^0(f_l)\|,\ \|W^0(f_r)\|,\ \|a^{-1}I\|,\ \|T^{-1}(a_l)\|,\ \|T^{-1}(a_r)\|, \\
&\|W^0(\zeta_l^{-1})\|,\ \|W^0(\zeta_r^{-1})\|,\ \|(\widetilde{R}_a+J_\infty)^{-1}\|.
\end{aligned}$$

Proof. The operator $R_a^{(-1)} := V^{(-1)}R^{(-1)}$ is a regularizer of R_a. Obviously,

$$\|R_a^{(-1)}\|_{\text{ess}} \le \|V^{(-1)}\|_{\text{ess}}\|R^{(-1)}\|_{\text{ess}}, \qquad (18.83)$$

and the essential norms on the right of (18.83) can be estimated by (18.53) and (18.78). In this way we arrive at an upper estimate for $\|R_a^{(-1)}\|_{\text{ess}}$ that contains only constants and the norms

$$\|aI\|,\ \|(\widetilde{R}_a+J_\infty)^{-1}\|,\ \|a^{-1}I\|,\ \|\hat{A}_l\|,\ \|\hat{A}_r\|,\ \|\hat{A}_l^{-1}\|,\ \|\hat{A}_r^{-1}\|.$$

Lemma 18.14 implies that $\|\hat{A}_l^{-1}\|$ can be estimated from above by constants and the quantities

$$\|Qa_lP\|, \|a_lI\|, \|T^{-1}(a_l)\|, \|W^0(\zeta_l^{-1}\chi_0)\|.$$

Taking into account that

$$\|Qa_lP\| \le \|S\|^2\|a_lI\| \le \|S\|^2\|aI\|,$$
$$\|T(a_l)\| \le \|S\|^2\|a_lI\| \le \|S\|^2\|aI\|,$$
$$\|W^0(\zeta_l^{-1}\chi_0)\| \le \|S\|\,\|W^0(\zeta_l^{-1})\|,$$

we get an upper estimate for $\|\hat{A}_l^{-1}\|$ in terms of constants and the values $\|aI\|$, $\|T^{-1}(a_l)\|$, $\|W^0(\zeta_l^{-1})\|$. Analogously, $\|\hat{A}_r^{-1}\|$ can be estimated from above by constants and the numbers $\|aI\|$, $\|T^{-1}(a_r)\|$, $\|W^0(\zeta_r^{-1})\|$. Finally, because

$$\|\hat{A}_l\| \le \|a_lI\|2\|S\| + 2\|S\| + \|W^0(f_l)\| \le 2(\|aI\|+1)\|S\| + \|W^0(f_l)\|,$$
$$\|\hat{A}_r\| \le 2(\|aI\|+1)\|S\| + \|W^0(f_r)\|,$$

we arrive at the assertion. □

Proof of Theorem 18.18.

Sufficiency. Since R_a is Fredholm on L^p_N if and only if $T(a)$ is Fredholm on H^p_N, Lemma 18.21 proves the sufficiency of conditions (i), (ii), (iii) under the assumptions made at the beginning of this section. We now turn to the case of general a in $GSAP_{N\times N}$. Let

$$a_n = (1-u)a_l^{(n)} + ua_r^{(n)} + a_0 \in GSAP_{N\times N}$$

be matrix functions such that $a_l^{(n)}, a_r^{(n)} \in AP^0_{N\times N}$ and $\|a - a_n\| \to 0$. By (ii), the operators $T(a_l^{(n)})$ and $T(a_r^{(n)})$ are invertible for all $n \ge n_1$. From Corollary 18.4 we deduce the existence of canonical APW factorizations

$$a_l^{(n)} = (a_l^{(n)})^-(a_l^{(n)})^+, \quad a_r^{(n)} = (a_r^{(n)})^-(a_r^{(n)})^+, \tag{18.84}$$

and on approximating the factors uniformly by matrix functions in $AP^0_{N\times N}$, we can a priori assume that the factors are in $AP^0_{N\times N}$. Hence, there exist $\mu_n > \lambda_n > 0$ such that

$$\Omega((a_l^{(n)})^-) \cap (-\mu_n, 0) = \emptyset, \quad \Omega((a_l^{(n)})^+) \cap (0, \mu_n) = \emptyset,$$

$$\Omega((a_r^{(n)})^-) \cap (-\mu_n, 0) = \emptyset, \quad \Omega((a_r^{(n)})^+) \cap (0, \mu_n) = \emptyset.$$

Choose monotonous functions $\hat{\chi}_\pm^{(n)} \in C(\overline{\mathbf{R}})$ so that $0 \le \hat{\chi}_\pm^{(n)} \le 1$, $\hat{\chi}_+^{(n)} + \hat{\chi}_-^{(n)} = 1$, $\hat{\chi}_-^{(n)}(t) = 0$ for $t \ge 0$, $\hat{\chi}_-^{(n)}(t) = 1$ for $t \le -\lambda_n$. Then pick $f_l^{(n)}$, $f_r^{(n)}$ in $[BV \cap C]_{N\times N}$ so that

$$\mathrm{supp} f_l^{(n)} \subset [-\lambda_n, 0], \quad \mathrm{supp} f_r^{(n)} \subset [-\lambda_n, 0] \tag{18.85}$$

and so that, for all $n \ge n_2 \ge n_1$,

$$\zeta_l^{(n)} := \mathbf{d}(a_l^{(n)})\hat{\chi}_-^{(n)} + I_N\hat{\chi}_+^{(n)} + f_l^{(n)} \in G[C(\overline{\mathbf{R}})]_{N\times N}, \tag{18.86}$$

$$\zeta_r^{(n)} := \mathbf{d}(a_r^{(n)})\hat{\chi}_-^{(n)} + I_N\hat{\chi}_+^{(n)} + f_r^{(n)} \in G[C(\overline{\mathbf{R}})]_{N\times N}. \tag{18.87}$$

By virtue of (iii) and the continuity of the map $a \mapsto \mathbf{d}(a)$, the matrix functions a_n satisfy condition (iii) for all $n \geq n_3 \geq n_2$. Clearly, conditions (i) and (ii) are also satisfied for $n \geq n_3$. Thus, from what was already proved we deduce that the operators $T(a_n)$ and R_{a_n} are Fredholm. The Fredholmness of $T(a)$ and R_a will follow once we have shown that

$$\limsup_{n\to\infty} \|R_{a_n}^{(-1)}\|_{\text{ess}} < \infty. \tag{18.88}$$

The norms $\|a_n I\|$, $\|a_n^{-1} I\|$, $\|T^{-1}(a_l^{(n)})\|$, $\|T^{-1}(a_r^{(n)})\|$, $\|(\widetilde{R}_{a_n} + J_\infty)^{-1}\|$ are obviously uniformly bounded from above. Hence, to prove (18.88) we are by virtue of Lemma 18.21 left to prove the uniform boundedness of the norms

$$\|W^0((\zeta_l^{(n)})^{-1})\|,\ \|W^0((\zeta_r^{(n)})^{-1})\|,\ \|W^0(f_l^{(n)})\|,\ \|W^0(f_r^{(n)})\|.$$

Choose complex-valued functions $\chi_l^\pm$, $\chi_r^\pm$ in $BV \cap C(\overline{\mathbf{R}})$ so that

$$\begin{aligned}
&\chi_l^- + \chi_l^+ = 1, \quad \chi_r^- + \chi_r^+ = 1, \quad 0 \leq |\chi_l^\pm| \leq 1, \quad 0 \leq |\chi_r^\pm| \leq 1,\\
&\chi_l^-(t) = \chi_r^-(t) = 0 \text{ for } t \geq 0, \quad \chi_l^-(t) = \chi_r^-(t) = 1 \text{ for } t \leq -1,\\
&\eta_l := \mathbf{d}(a_l)\chi_l^- + I_N\chi_l^+ \in G[C(\overline{\mathbf{R}})]_{N\times N},\\
&\eta_r := \mathbf{d}(a_r)\chi_r^- + I_N\chi_r^+ \in G[C(\overline{\mathbf{R}})]_{N\times N}.
\end{aligned}$$

Then for all $n \geq n_4 \geq n_3$,

$$\eta_l^{(n)} := \mathbf{d}(a_l^{(n)})\chi_l^- + I_N\chi_l^+ \in G[C(\overline{\mathbf{R}})]_{N\times N},$$

$$\eta_r^{(n)} := \mathbf{d}(a_r^{(n)})\chi_r^- + I_N\chi_r^+ \in G[C(\overline{\mathbf{R}})]_{N\times N},$$

and

$$\|W^0(\eta_l^{(n)})\| \leq 2\|W^0(\eta_l)\|, \quad \|W^0(\eta_r^{(n)})\| \leq 2\|W^0(\eta_r)\|, \tag{18.89}$$

$$\|W^0((\eta_l^{(n)})^{-1})\| \leq 2\|W^0(\eta_l^{-1})\|, \quad \|W^0((\eta_r^{(n)})^{-1})| \leq 2\|W^0(\eta_r^{-1})\|. \tag{18.90}$$

We now specify $f_l^{(n)}, f_r^{(n)} \in [BV \cap C]_{N\times N}$ by

$$f_l^{(n)}(t) := \eta_l^{(n)}(\lambda_n^{-1}t) - \Big(\mathbf{d}(a_l^{(n)})\hat{\chi}_-^{(n)}(t) + I_N\hat{\chi}_+^{(n)}(t)\Big), \tag{18.91}$$

$$f_r^{(n)}(t) := \eta_r^{(n)}(\lambda_n^{-1}t) - \Big(\mathbf{d}(a_r^{(n)})\hat{\chi}_-^{(n)}(t) + I_N\hat{\chi}_+^{(n)}(t)\Big). \tag{18.92}$$

It is clear that with this choice of $f_l^{(n)}$, $f_r^{(n)}$ requirements (18.85), (18.86), (18.87) are met. Moreover, we have

$$\zeta_l^{(n)}(t) = \eta_l^{(n)}(\lambda_n^{-1}t), \quad \zeta_r^{(n)}(t) = \eta_r^{(n)}(\lambda_n^{-1}t). \tag{18.93}$$

From (18.89) to (18.93) we get

$$\|W^0((\zeta_l^{(n)})^{-1})\| = \|W^0((\eta_l^{(n)})^{-1})\| \le 2\|W^0(\eta_l^{-1})\|, \tag{18.94}$$

$$\|W^0((\zeta_r^{(n)})^{-1})\| = \|W^0((\eta_r^{(n)})^{-1})\| \le 2\|W^0(\eta_r^{-1})\|, \tag{18.95}$$

$$\|W^0(f_l^{(n)})\| \le \|W^0(\eta_l^{(n)})\| + \|W^0(\mathbf{d}(a_l^{(n)})\hat{\chi}_-^{(n)} + I_N\hat{\chi}_+^{(n)})\|, \tag{18.96}$$

$$\|W^0(f_r^{(n)})\| \le \|W^0(\eta_r^{(n)})\| + \|W^0(\mathbf{d}(a_r^{(n)})\hat{\chi}_-^{(n)} + I_N\hat{\chi}_+^{(n)})\|, \tag{18.97}$$

Finally, we specify the functions $\hat{\chi}_\pm^{(n)}$ by the choice $\hat{\chi}_\pm^{(n)}(t) = \hat{\chi}_\pm(\lambda_n^{-1}t)$ where $\hat{\chi}_-$ is monotonously increasing, $0 \le \hat{\chi}_- \le 1$, $\hat{\chi}_-(t) = 0$ for $t \ge 0$, $\hat{\chi}_-(t) = 1$ for $t \le -1$, $\hat{\chi}_- + \hat{\chi}_+ = 1$. Since, for all $n \ge n_5 \ge n_4$,

$$\|W^0(\mathbf{d}(a_l^{(n)})\hat{\chi}_- + I_N\hat{\chi}_+)\| \le 2\|W^0(\mathbf{d}(a_l)\hat{\chi}_- + I_N\hat{\chi}_+)\|,$$

$$\|W^0(\mathbf{d}(a_r^{(n)})\hat{\chi}_- + I_N\hat{\chi}_+)\| \le 2\|W^0(\mathbf{d}(a_r)\hat{\chi}_- + I_N\hat{\chi}_+)\|,$$

we obtain from (18.89), (18.90), (18.96), (18.97) that for $n \ge n_5$,

$$\|W^0(f_l^{(n)})\| \le 2\|W^0(\eta_l)\| + 2\|W^0(\mathbf{d}(a_l)\hat{\chi}_- + I_N\hat{\chi}_+)\|, \tag{18.98}$$

$$\|W^0(f_r^{(n)})\| \le 2\|W^0(\eta_r)\| + 2\|W^0(\mathbf{d}(a_r)\hat{\chi}_- + I_N\hat{\chi}_+)\|. \tag{18.99}$$

In summary, (18.94), (18.95), (18.98), (18.99) imply (18.88).

Index formula. In the case where a_l and a_r are in $APW_{N\times N}$, we have canonical APW factorizations by virtue of Corollary 18.4. Under the assumption that a_l and a_r possess canonical APW factorizations, the index formula (18.56) can be established by appropriately modifying the proof of Theorem 10.12 and by using (16.36). Since all terms of formula (18.56) depend continuously on a (also recall Theorem 18.12), we see that (18.56) is valid for general $a \in SAP_{N\times N}$.

Necessity. Suppose $T(a)$ is Fredholm. Theorem 16.3(b) implies that then $a \in GL^\infty_{N\times N}$, which gives (i). Corollary 18.11 shows that the Fredholmness of $T(a)$ implies the invertibility of $T(a_l)$ and $T(a_r)$. Thus, we get (ii). Finally, if (iii) were not satisfied, we could perturb a and use the sufficiency of conditions (i), (ii), (iii) and the index formula to produce b and c in a prescribed neighborhood of a such that $\operatorname{Ind} T(b) \ne \operatorname{Ind} T(c)$. As this contradicts the stability of the index, we see that (iii) must be true. □

18.9 Spaces with Power Weights

Using a trick due to Paatashvili we can pass from $H^p(\mathbf{R})$ to $H^p(\mathbf{R}, w)$ where w is the power weight

$$w(x) = |x+i|^{\mu_0} \prod_{j=1}^{n} |x-\beta_j|^{\mu_j}, \quad x \in \mathbf{R}. \tag{18.100}$$

We know from Section 16.1 that $w \in A_p(\mathbf{R})$ if and only if $1 < p < \infty$ and

$$\mu_1, \ldots, \mu_n \in (-1/p, 1/q) \quad \text{and} \quad \mu \in (-1/p, 1/q), \tag{18.101}$$

where $1/p + 1/q = 1$ and $\mu := \mu_0 + \mu_1 + \ldots + \mu_n$. Suppose these conditions are satisfied.

Let w_μ be the power weight

$$w_\mu(x) := (x^2 + 1)^{\mu/2}. \tag{18.102}$$

Evidently, $w_\mu(x) = |x + i|^\mu = |x - i|^\mu$. Further, let

$$\varphi_\mu(x) := \left(\frac{x-i}{x+i}\right)^\mu, \quad x \in \mathbf{R},$$

be the branch of the function for which

$$\varphi_\mu(-\infty) = 1, \quad \varphi_\mu(+\infty) = e^{2\pi i \mu}.$$

Proposition 18.22 (Paatashvili). *For every $a \in L^\infty_{N\times N}(\mathbf{R})$, the Toeplitz operator $T(a) \in \mathcal{B}(H^p_N(\mathbf{R}, w_\mu))$ is strongly Φ-equivalent to the Toeplitz operator $T(\varphi_\mu a) \in \mathcal{B}(H^p_N(\mathbf{R}))$.*

Proof. Let $\xi_\mu^-(x) := (x-i)^\mu$ and $\eta_\mu^+(x) := (x+i)^{-\mu}$ be the branches of the functions which are analytic in the lower and upper half-planes, respectively, and take the values $e^{3\pi i\mu/2}$ and $e^{-\pi i \mu/2}$ at $x = 0$. Then

$$\varphi_\mu(x) = \left(\frac{x-i}{x+i}\right)^\mu = (x-i)^\mu (x+i)^{-\mu} = \xi_\mu^-(x)\eta_\mu^+(x)$$

for $x \in \mathbf{R}$. Passing from the line to the circle and using Lemma 16.15, one can show that the operators

$$T(\xi_\mu^-) : H^p(\mathbf{R}, w_\mu) \to H^p(\mathbf{R}), \quad f \mapsto P(\xi_\mu^- f),$$
$$T(\eta_\mu^+) : H^p(\mathbf{R}) \to H^p(\mathbf{R}, w_\mu), \quad f \mapsto P(\eta_\mu^+ f)$$

are isomorphisms. Since

$$T(\xi_\mu^-)T(a)T(\eta_\mu^+) = T(\xi_\mu^- a \eta_\mu^+) = T(\varphi_\mu a),$$

we get the assertion. □

Corollary 18.23. *Let $a \in SAP_{N\times N}$, $1 < p < \infty$, and let w be the power weight (18.100), (18.101). The operator $T(a)$ is Fredholm on $H^p_N(\mathbf{R}, w)$ if and only if $a \in GSAP_{N\times N}$, $T(a_l)$ and $T(a_r)$ are invertible on $H^p(\mathbf{R})$, and*

$$\operatorname{sp}(\mathbf{d}^{-1}(a_r)\mathbf{d}(a_l)) \cap e^{2\pi i(1/p+\mu)}\mathbf{R}_+ = \emptyset.$$

In that case the index of $T(a)$ *on* $H^p_N(\mathbf{R}, w)$ *is given by* (18.56) *with the twice occurring* $1/p$ *replaced by* $1/p + \mu$.

Proof. Since a is continuous on $\mathbf{R}$, standard localization techniques, e.g., Theorem 16.3(c), show that $T(a)$ is Fredholm on $H^p_N(\mathbf{R}, w)$ if and only if $T(a)$ is Fredholm on $H^p_N(\mathbf{R}, w_\mu)$, which, by Proposition 18.22, is equivalent to the Fredholmness of the operator $T(\varphi_\mu a)$ on $H^p_N(\mathbf{R})$. It remains to apply Theorem 18.18 to the operator $T(\varphi_\mu a)$. □

Corollary 18.24. *Let* w_μ *be the power weight* (18.102). *For* $a \in APW_{N\times N}$, *the following are equivalent*

(i) $T(a)$ *is Fredholm on* $H^p_N(\mathbf{R}, w_\mu)$;

(ii) $T(a)$ *is invertible on* $H^p_N(\mathbf{R}, w_\mu)$;

(iii) a *has a canonical right* APW *factorization.*

Proof. (i)⇒(iii). Suppose $T(a)$ is Fredholm on $H^p_N(\mathbf{R}, w_\mu)$. Then, by Proposition 18.22, $T(\varphi_\mu a)$ is Fredholm on $H^p_N(\mathbf{R})$. Since $\varphi_\mu a \in SAP_{N\times N}$, Corollary 18.11 implies that

$$T\big((\varphi_\mu a)_l\big) = T(a) \text{ and } T\big((\varphi_\mu a)r\big) = e^{2\pi i\mu} T(a)$$

are invertible on $H^p_N(\mathbf{R})$. From Corollary 18.4 we therefore deduce that a has a canonical right APW factorization.

(iii)⇒(ii)⇒(i). Trivial. □

Corollary 18.25. *Suppose* w *is the power weight* (18.100), (18.101). *If* $a \in SAPW_{N\times N}$ *and* $T(a)$ *is Fredholm on* $H^p_N(\mathbf{R}, w)$, *then* a_l *and* a_r *have canonical right* APW *factorizations.*

Proof. Since a is continuous on $\mathbf{R}$, we may without loss of generality suppose that w is of the form (18.102). Then, by Proposition 18.22, $T(\varphi_\mu a)$ is Fredholm on $H^p_N(\mathbf{R})$. From Corollary 18.11 we obtain that the operators

$$T\big((\varphi_\mu a)_l\big) = T(a_l) \text{ and } T\big((\varphi_\mu a)_r\big) = e^{2\pi i\mu} T(a_r)$$

are invertible on $H^p_N(\mathbf{R})$. The assertion is now immediate from Corollary 18.4. □

Corollary 18.26. *Let* w *be the power weight* (18.100), (18.101) *and let* a *be in* $SAPW_{N\times N}$. *Then* $T(a)$ *is Fredholm on* $H^p_N(\mathbf{R}, w)$ *if and only if* a *is invertible in* $SAPW_{N\times N}$, a_l *and* a_r *have right* APW *factorizations with zero partial indices, and*

$$\operatorname{sp}\big(\mathbf{d}^{-1}(a_r)\mathbf{d}(a_l)\big) \cap e^{2\pi i(1/p+\mu)}\mathbf{R}_+ = \emptyset,$$

with $\mu = \mu_0 + \mu_1 + \cdots + \mu_n$.

Proof. Combine Corollaries 18.23 and 18.25. □

Notes

The results of Section 18.1 were announced in [122] (see also [121]). Corollary 18.4 was first stated in [120], [126].

The exposition in Section 18.2 is based on [130]; it uses some abstract operator theory results from [124] and [125]. In particular, our proof of Lemma 18.6 mimics the proof of Theorem 1.1 in [125]. Note that [130] deals with power weights only, but that the proofs in the general setting are no different.

The L^p theory of scalar Toeplitz operators with SAP symbols for power weights was developed by Saginashvili [190], [191], [192]. The Muckenhoupt weights version presented here is taken from [43]. Invertibility of AP representatives (Section 18.4) was investigated in [122], [121]. The exposition in Sections 18.5-18.9 follows [44]. The exact constants in the local principle were subject of attention in [145]; see also [45], [99, Section 5.4], and [146]. The Paatashvili trick (Proposition 18.22) was first used in [167]; see also [140, Chapter 2, Theorem 5.1]. The Fredholm theory of *matrix* Toeplitz operators with SAP symbols and arbitrary (different from power) Muckenhoupt weights remains undeveloped. The reason is that the Paatashvili trick in the situation of general Muckenhoupt weights does not leave the SAP class of symbols invariant.

The paper [16] contains the Fredholm theory for singular integral operators on $L^p_N(\mathbf{R})$ with a Carleman shift and coefficients in alg $(AP_{N\times N}, PC_{n\times N})$ which have almost periodic representatives in $APW_{N\times N}$. Finally, in [121] and [122], one of the authors studied the Fredholm theory of singular integral operators on $L^p_N(\Gamma, w)$ with a non-Carleman shift and coefficients in $SAP_{N\times N}$ and the associated Haseman boundary value problem; the curve Γ is supposed to be piecewise smooth and to have no cusps and the weight w is taken as a power weight.

Notes

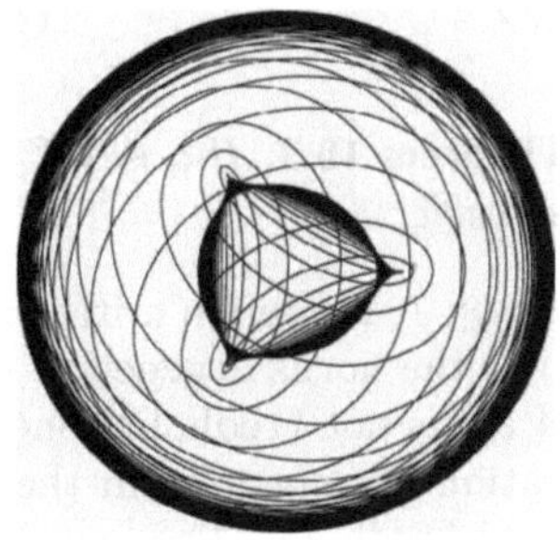

Chapter 19

Wiener-Hopf Operators with SAP Symbols on Lebesgue Spaces

The theory of Wiener-Hopf operators with AP and SAP symbols on $L^p_N(\mathbf{R}_+)$ is more complicated than its Toeplitz counterpart. We have to bound ourselves to symbols that are multipliers, which leads to the introduction of the classes AP_p and SAP_p. The main result of this chapter is a Fredholm criterion and an index formula for Wiener-Hopf operators with symbols in $[SAP_p]_{N\times N}$ on the space $L^p_N(\mathbf{R}_+)$ under the assumption that the almost periodic representatives are known to have AP_p factorizations. This assumption can be removed in the scalar case and in the case where the almost periodic representatives of the symbol at $\pm\infty$ have absolutely convergent Bohr-Fourier series.

19.1 Multipliers from AP and SAP

We henceforth restrict ourselves to operators on spaces without weight, that is, to operators on $L^p(\mathbf{R}_+)$ $(1 < p < \infty)$.

Theorem 19.1. (a) *If $a \in M_p \setminus \{0\}$, then $W(a)$ has a trivial kernel or a dense image in $L^p(\mathbf{R}_+)$.*

(b) *If $a \in M_p \setminus \{0\}$ and $W(a)$ is normally solvable on $L^p(\mathbf{R}_+)$, then $W(a)$ is semi-Fredholm on $L^p(\mathbf{R}_+)$.*

Proofs are in [70], [46], [183].

Theorem 19.2. *If $a \in PC_p \setminus \{0\}$, then $W(a)$ is either Fredholm or not normally solvable.*

Proof. A Fredholm criterion and an index formula are given in Section 17.2. Note that the horns degenerate to circular arcs in the case of spaces without weight. If $W(a)$ is not Fredholm, one can therefore use an approximation and index perturbation argument as in the proof of Theorem 2.20 to show that $W(a)$ cannot be semi-Fredholm. Theorem 19.1(b) so implies the assertion. □

Since $W^0(e_\lambda)$ is the shift operator U_λ (Example 1.10), almost periodic polynomials obviously belong to M_p. We denote by AP_p the closure of the set of all almost periodic polynomials in M_p. Further, let AP_p^+ (resp. AP_p^-) stand for the M_p closure of the set of all almost periodic polynomials $\sum_j a_j e_{\lambda_j}$ with $\lambda_j \geq 0$ (resp. $\lambda_j \leq 0$). Clearly,

$$APW \subset AP_p \subset AP, \quad APW^\pm \subset AP_p^\pm \subset AP^\pm$$

(also recall (17.2)). The algebra $C_p(\overline{\mathbf{R}})$ is given by (17.7). We define SAP_p as the smallest closed subalgebra of M_p which contains $C_p(\overline{\mathbf{R}})$ and AP_p. Let $SAP_pW := SAP_p \cap SAPW$. Obviously,

$$SAP_pW \subset SAP_p \subset SAP.$$

Also notice that $AP_2 = AP$ and $SAP_2 = SAP$. Throughout what follows, let u be a fixed monotonically increasing real-valued function in $C(\overline{\mathbf{R}})$ such that $u(-\infty) = 0$ and $u(+\infty) = 1$. Since $\|u\|_\infty = V_1(u) = 1$, Theorem 17.1 shows that $u \in C_p(\overline{\mathbf{R}})$.

Proposition 19.3. *Every function $a \in SAP_p$ can in a unique way be represented in the form*

$$a = (1-u)a_l + ua_r + a_0 \tag{19.1}$$

where $a_l, a_r \in AP_p$, $a_0 \in C_p(\dot{\mathbf{R}})$, $a_0(\infty) = 0$. The maps $a \mapsto a_l$ and $a \mapsto a_r$ are (continuous) Banach algebra homomorphisms of SAP_p onto AP_p.

Proof. As in the proof of Theorem 1.21, one can show that SAP_p is the M_p closure of the set

$$\{(1-u)a_l + ua_r + a_0 : a_l \in AP_p,\ a_r \in AP_p,\ a_0 \in C(\dot{\mathbf{R}}),\ a_0(\infty) = 0\}.$$

The assertion will therefore follow as soon as we have proved that if a is of the form (19.1), then

$$\|a_l\|_{M_p} \leq \kappa_p \|a\|_{M_p}, \quad \|a_r\|_{M_p} \leq \kappa_p \|a\|_{M_p} \tag{19.2}$$

with some constant $\kappa_p < \infty$ independent of a (actually we have $\kappa_p = \|S\|_p$).

We verify the second inequality of (19.2). Without loss of generality, assume that a_l, a_r are almost periodic polynomials and that $a_0 \in C_p(\dot{\mathbf{R}})$ has compact support. By Lemma 10.2, there is a sequence $h_n \to +\infty$ such that

$$\|a_r - (a_r)_{h_n}\|_\infty \to 0 \text{ as } n \to \infty$$

where $(a_r)_{h_n}(x) := a_r(x + h_n)$. By (17.2), we can also assume that $p \in [2, \infty)$. From (10.3) we infer that

$$e_{h_n} W^0(\chi_{(c,\infty)} a_r) e_{-h_n} I = W^0\big(\chi_{(c-h_n,\infty)}(a_r)_{h_n}\big) \tag{19.3}$$

for every $c \in \mathbf{R}$. Let $\mathcal{S}$ be the Schwartz class of rapidly decreasing C^∞ functions on $\mathbf{R}$. By the Hausdorff-Young inequality,

$$Y_p := \|F^{-1}\|_{\mathcal{B}(L^p(\mathbf{R}), L^q(\mathbf{R}))} < \infty \quad (1/p + 1/q = 1).$$

Thus, if $f \in \mathcal{S}$ then

$$\begin{aligned}
&\|W^0\big(\chi_{(c-h_n,\infty)}(a_r)_{h_n} - a_r\big) f\|_p \\
&\le Y_p \big\| \big(\chi_{(c-h_n,\infty)}(a_r)_{h_n} - a_r\big) F f\big\|_q \\
&\le Y_p \|\chi_{(c-h_n,\infty)}\|_\infty \|(a_r)_{h_n} - a_r\|_\infty \|Ff\|_q \\
&\quad + Y_p \|a_r\|_\infty \|(\chi_{(c-h_n,\infty)} - 1) Ff\|_q,
\end{aligned}$$

and this converges to zero as $n \to \infty$. Because $\mathcal{S}$ is dense in $L^p(\mathbf{R})$ and the operators (19.3) are uniformly bounded on $L^p(\mathbf{R})$, it follows that the operators (19.3) converge strongly to $W^0(a_r)$ on $L^p(\mathbf{R})$. Thus, by the Banach-Steinhaus theorem,

$$\begin{aligned}
\|a_r\|_{M_p} &= \|W^0(a_r)\| \le \liminf_{n\to\infty} \|e_{h_n} W^0(\chi_{(c,\infty)} a_r) e_{-h_n} I\| \\
&= \|W^0(\chi_{(c,\infty)} a_r)\| \\
&= \big\|W^0(\chi_{(c,\infty)} a) - W^0\big(\chi_{(c,\infty)}(a_l - a_r)(1-u)\big) - W^0(\chi_{(c,\infty)} a_0)\big\| \\
&\le \|W^0(\chi_{(c,\infty)} a)\| + \big\|W^0\big(\chi_{(c,\infty)}(a_l - a_r)(1-u)\big)\big\| + \|W^0(\chi_{(c,\infty)} a_0)\| \\
&\le \|\chi_{(c,\infty)}\|_{M_p} \|a\|_{M_p} + \|a_l - a_r\|_{M_p} \|\chi_{(c,\infty)}(1-u)\|_{M_p} + \|\chi_{(c,\infty)} a_0\|_{M_p}.
\end{aligned}$$

Furthermore,

$$\|\chi_{(c,\infty)}\|_{M_p} = \|\chi_+\|_{M_p} = \|W^0(\chi_+)\|_{M_p} = \|Q\|_p \le \|S\|_p.$$

By Stechkin's inequality (Theorem 17.1),

$$\|\chi_{(c,\infty)}(1-u)\|_{M_p} \le 3\|S\|\big(1 - u(c)\big).$$

Obviously, $1 - u(c)$ is smaller than any prescribed $\varepsilon > 0$ if only c is large enough, and since a_0 has compact support, we see that $\chi_{(c,\infty)} a_0 = 0$ for all sufficiently large c. This implies that

$$\|a_r\|_{M_p} \le \|S\|_p \|a\|_{M_p}. \qquad \square$$

Proposition 17.3(a) tells us that $C_p(\dot{\mathbf{R}})$ and $C_p(\overline{\mathbf{R}})$ are inverse closed in $C(\dot{\mathbf{R}})$ and $C(\overline{\mathbf{R}})$, respectively. An analogous result is true for AP_p and SAP_p.

Proposition 19.4. *The algebras AP_p and SAP_p are inverse closed in AP and SAP, respectively, and are therefore inverse closed in $L^\infty(\mathbf{R})$.*

Proof. For AP_p, this was proved by Gohberg and Feldman (see [92, Chapter 7]). We use this result to prove the inverse closedness of SAP_p. By Proposition 19.3, every $a \in SAP_p$ is of the form (19.1). Thus, let

$$a = (1-u)a_l + ua_r + a_0 \in GSAP.$$

Again by Proposition 19.3 for $p = 2$ (or simply by Theorem 1.21), it follows that $a_l, a_r \in GAP$, and as $a_l, a_r \in AP_p$, we conclude that $a_l, a_r \in GAP_p$. Define $b \in SAP_p$ by $b := (1-u)a_l^{-1} + ua_r^{-1}$. Clearly, b is bounded away from zero outside some compact interval $[-x_0, x_0]$. Choose $b_0 \in C^\infty(\mathbf{R})$ so that $\operatorname{supp} b \subset [-x_0, x_0]$ and $b+b_0 \in GL^\infty(\mathbf{R})$. Since $b_0 \in C_p(\dot{\mathbf{R}})$, we have $b+b_0 \in SAP_p$. Put $c := a(b+b_0)$. Obviously, $c \in SAP_p$. As $c_l = c_r = 1$, we see that actually $c \in C_p(\dot{\mathbf{R}})$. Hence,

$$c^{-1} = (b+b_0)^{-1}a^{-1} \in C_p(\dot{\mathbf{R}}),$$

which implies that $a^{-1} = c^{-1}(b+b_0) \in SAP_p$. □

19.2 Invertibility of the Symbol

This section is devoted to the proof of the following result.

Theorem 19.5. *If $a \in [SAP_p]_{N\times N}$ and $W(a)$ is semi-Fredholm on $L^p_N(\mathbf{R}_+)$, then $a \in G[SAP_p]_{N\times N}$.*

Proof. By Proposition 19.4, it suffices to show that $a \in GSAP_{N\times N}$. Assume this is not true. Then

$$\inf_{x\in\mathbf{R}} |\det a(x)| = 0. \tag{19.4}$$

Our first objective is to prove that any M_p neighborhood of a contains a b in $[SAP_p]_{N\times N}$ such that $\det b(x) = 0$ for all x in some open interval of $\mathbf{R}$.

Given any $\delta_1 > 0$, we can find an a_1 such that $a_1 \in [SAP_r]_{N\times N}$ for all $r \in (1,\infty)$ and $\|a-a_1\|_{[M_p]_{N\times N}} < \delta_1$. We denote by $\lambda_1(x), \dots, \lambda_N(x)$ the eigenvalues of $a_1(x)$. Let $\delta_2 > 0$ be any number. From (19.4) we infer that

$$\inf_{j,x} |\lambda_j(x)| < \delta_2 \tag{19.5}$$

if only $\delta_1 > 0$ is small enough. For $\lambda \in \mathbf{C}$, define a_2 by $a_2(x) = a_1(x) - \lambda I$. By (19.5), there are $\lambda \in \mathbf{C}$ and $x_0 \in \mathbf{R}$ such that $|\lambda| < \delta_2$ and $\det a_2(x_0) = 0$. For

$\varepsilon > 0$, let φ_ε be any function $\varphi_\varepsilon : \mathbf{R} \to [0,1]$ such that

$$\varphi_\varepsilon\big|(x_0-\varepsilon, x_0+\varepsilon) = 1, \quad \varphi_\varepsilon\big|(\mathbf{R}\setminus[x_0-2\varepsilon,\, x_0+2\varepsilon]) = 0,$$

and φ_ε is monotonous on each side of x_0. Put

$$b := a_2 + \varphi_\varepsilon\big(a_2(x_0) - a_2\big).$$

Since a_2 and $\varphi_\varepsilon I_N$ are in $[M_r]_{N\times N}$ for all $r \in (1,\infty)$, so is b. If $x \in (x_0-\varepsilon,\, x_0+\varepsilon)$, then $b(x) = a_2(x_0)$ and hence $\det b(x) = 0$. We have

$$b(x) - a_2(x) = \varphi_\varepsilon(x)\big(a_2(x_0) - a_2(x)\big) =: \varphi_\varepsilon(x)\big(c_{jk}(x)\big)_{j,k=1}^N.$$

Now fix $r \in (1, \min\{p,q\})$ and define $t \in (0,1)$ by

$$t := r|p-2|/\big(p|r-2|\big).$$

As in the proof of Theorem 17.1, we get

$$\begin{aligned} \|\varphi_\varepsilon c_{jk}\|_{M_p} &\le \|\varphi_\varepsilon c_{jk}\|_\infty^{1-t}\|\varphi_\varepsilon c_{jk}\|_{M_r}^t \\ &\le \|\varphi_\varepsilon c_{jk}\|_\infty^{1-t}\|\varphi_\varepsilon\|_{M_r}^t\|c_{jk}\|_{M_r}^t. \end{aligned}$$

By Theorem 17.1, $\|\varphi_\varepsilon\|_{M_r} \le 3\|S\|$, where $\|S\|$ is the norm of S on $L^p(\mathbf{R})$. It is clear that $\|\varphi_\varepsilon c_{jk}\|_\infty \to 0$ as $\varepsilon \to 0$. Thus, since

$$\|b-a\|_{[M_p]_{N\times N}} \le \delta_1 + \delta_2 + N\max_{j,k}\|\varphi_\varepsilon c_{jk}\|_{M_p},$$

we arrive at the conclusion that $\|b-a\|_{[M_p]_{N\times N}}$ is as small as desired if only first $\delta_1, \delta_2 > 0$ and then $\varepsilon > 0$ are chosen sufficiently small.

As $\det a_2(x_0) = 0$, there is a nonzero matrix $d \in \mathbf{C}^{N\times N}$ such that $a_2(x_0)d = 0$. As $d\varphi_{\varepsilon/2} \in [C_p(\dot{\mathbf{R}})]_{N\times N}$, it follows that

$$W(bd\varphi_{\varepsilon/2}) - W(b)W(d\varphi_{\varepsilon/2})$$

is compact (notice that $W(ba)-W(b)W(a)$ can be shown to be compact on $L^p(\mathbf{R}_+)$ whenever $b \in M_p$ and $a \in C_p(\dot{\mathbf{R}})$ or $b \in C_p(\dot{\mathbf{R}})$ and $a \in M_p$). If $x \in (x_0-\varepsilon,\, x_0+\varepsilon)$, then

$$b(x)d\varphi_{\varepsilon/2}(x) = a_2(x_0)d\varphi_{\varepsilon/2}(x) = 0 \cdot \varphi_{\varepsilon/2}(x) = 0,$$

while for $x \notin [x_0-\varepsilon,\, x_0+\varepsilon]$ we have

$$b(x)d\varphi_{\varepsilon/2}(x) = b(x)d \cdot 0 = 0.$$

Consequently, $bd\varphi_{\varepsilon/2}$ vanishes identically. Therefore $W(b)W(d\varphi_{\varepsilon/2})$ must be compact.

Now suppopse $W(a)$ is n-normal. Then $W(b)$ is also n-normal. By Lemma 18.5, there is a compact operator K and an $\eta > 0$ such that

$$\|W(b)f\| + \|Kf\| \le \eta\|f\| \text{ for all } f \in L^p_N(\mathbf{R}_+). \tag{19.6}$$

Replacing f by $W(d\varphi_{\varepsilon/2})g$, we see that

$$\|W(b)W(d\varphi_{\varepsilon/2})g\| + \|KW(d\varphi_{\varepsilon/2})g\| \ge \eta\|W(d\varphi_{\varepsilon/2})g\|$$

for all $g \in L^p_N(\mathbf{R}_+)$. Let $\{g_n\}$ be a bounded sequence in $L^p_N(\mathbf{R}_+)$. Since the operators $W(b)W(d\varphi_{\varepsilon/2})$ and $KW(d\varphi_{\varepsilon/2})$ are compact, there is a subsequence $\{g_{n_k}\}$ such that $W(b)W(d\varphi_{\varepsilon/2})g_{n_k}$ and $KW(d\varphi_{\varepsilon/2})g_{n_k}$ are convergent. From (19.6) we see that $\{W(d\varphi_{\varepsilon/2})g_{n_k}\}$ is also convergent. This implies that $W(d\varphi_{\varepsilon/2})$ must be compact. As a nonzero Wiener-Hopf operator is never compact, we have a contradiction. This shows that $W(a)$ cannot be n-normal. Consideration of adjoints shows that $W(a)$ is also not d-normal. □

19.3 Invertibility of the Almost Periodic Representatives

Theorem 2.28 again remains literally true for Wiener-Hopf operators $W(a)$ on $L^p(\mathbf{R}_+)$ $(1 < p < \infty)$ with symbols $a \in AP_p \setminus \{0\}$. The only difference in the proof is that Theorem 2.7 and Proposition 1.4(a) used in the proof of Corollary 2.8 must be replaced by Theorems 19.5 and 19.1(b) and that instead of (2.27) we now must use the estimate

$$\|e^c - 1\|_{M_p} \le e^{\|c\|_{M_p}} - 1 < 1.$$

The $L^p_N(\mathbf{R}_+)$ analogue of Proposition 18.1 is obviously also valid provided we require that a has a right AP_p factorization, i.e., a representation

$$a(x) = a_-(x)\,\mathrm{diag}\,(e^{i\lambda_1 x}, \dots, e^{i\lambda_N x})\,a_+(x)$$

with $\lambda_j \in \mathbf{R}$ and $a_\pm \in G[AP_p^\pm]_{N\times N}$.

Our next concern is the $L^p_N(\mathbf{R}_+)$ analogue of Theorems 10.3 and 18.10. Thus, let now $\mathcal{A}$ and $\mathcal{F}$ stand for the smallest closed subalgebras of $\mathcal{B}(L^p_N(\mathbf{R}))$ which contain the operators $\chi_\pm I$ and the sets $\{W^0(a) : a \in [AP_p]_{N\times N}\}$ and $\{W^0(a) : a \in [SAP_p]_{N\times N}\}$, respectively. Again it turns out that $\mathcal{F}$ contains all compact operators on $L^p_N(\mathbf{R}_+)$.

Theorem 19.6. *The maps*

$$\mu_r : \chi_\pm I \mapsto \chi_\pm I,\ W^0(a) \mapsto W^0(a_r) \quad (a \in [SAP_p]_{N\times N}),$$
$$\mu_l : \chi_\pm I \mapsto \chi_\pm I,\ W^0(a) \mapsto W^0(a_l) \quad (a \in [SAP_p]_{N\times N})$$

extend to (continuous) Banach algebra homomorphisms of $\mathcal{F}$ onto $\mathcal{A}$ such that

$$\|\mu_r(A)\| \leq \|A\|_{\text{ess}}, \ \|\mu_l(A)\| \leq \|A\|_{\text{ess}}$$

for all $A \in \mathcal{F}$.

Proof. It suffices to consider the case $N = 1$. We prove the theorem only for μ_r; the proof for μ_l is analogous.

Proceeding as in the proof of Lemma 10.1, we see that if $h_n \to +\infty$, then $e_{-h_n}I$ converges weakly to zero on $L^p(\mathbf{R})$. This implies that

$$e_{h_n}Ke_{-h_n}I \to 0 \ \text{strongly} \tag{19.7}$$

for every compact operator K on $L^p(\mathbf{R})$.

Let now $A = \sum_j \prod_k A_{jk}$, the sum and the products finite, where A_{jk} is either of the form $\chi_\pm I$ or of the form (10.5) where a^l_{jk}, a^r_{jk} are almost periodic polynomials and $a^0_{jk} \in C_p(\dot{\mathbf{R}})$ has compact support in $\mathbf{R}$. Moreover, we can also assume that $u(x) = 0$ for all $x < -x_0$ and $u(x) = 1$ for all $x > x_0$. We define

$$\mu_r(A) := \sum_j \prod_k \mu_r(A_{jk})$$

with $\mu_r(\chi_\pm I) = \chi_\pm I$ and $\mu_r(W^0(a_{jk})) = W^0(a^r_{jk})$. The theorem will follow as soon as we have shown that

$$\mu_r(A) \leq \|A + K\| \ \text{for every} \ K \in \mathcal{K}(L^p(\mathbf{R})). \tag{19.8}$$

By Lemma 10.2, there is a sequence $h_n \to +\infty$ such that

$$\|(a^r_{jk})_{h_n} - a^r_{jk}\|_\infty \to 0, \quad \|((a^r_{jk})')_{h_n} - (a^r_{jk})'\|_\infty \to 0, \tag{19.9}$$

the prime denoting the derivative. We prove that

$$e_{h_n}(A + K)e_{-h_n}I \to \mu_r(A) \ \text{strongly}$$

for every $K \in \mathcal{K}(L^p(\mathbf{R}))$. Clearly, this gives (19.8). Taking into account (19.7) and the fact that

$$e_{h_n}\chi_\pm e_{-h_n}I = \chi_\pm I \to \chi_\pm I = \mu_r(\chi_\pm I),$$

we are left with proving that

$$e_{h_n}W^0(a_{jk})e_{-h_n}I \to W^0(a^r_{jk}) \ \text{strongly}.$$

In what follows, we omit the subscripts j, k. Due to (10.3), we have to show that

$$\|W^0(a_{h_n} - a_r)f\|_p \to 0 \ \text{for every} \ f \in L^p(\mathbf{R}), \tag{19.10}$$

where $a_r := a^r \,(:= a^r_{jk})$. The set of all $f \in L^2(\mathbf{R}) \cap L^p(\mathbf{R})$ for which Ff has compact support is dense in $L^p(\mathbf{R})$. Thus, it is sufficient to show (19.10) for such f. Let $\varphi \in C^\infty(\mathbf{R})$ be a function with compact support such that $\varphi|\operatorname{supp} Ff = 1$. Then

$$\left\|W^0(a_{h_n} - a_r)f\right\|_p = \left\|F^{-1}(a_{h_n} - a_r)\varphi Ff\right\|_p.$$

For sufficiently large h_n and $x \in \operatorname{supp}\varphi$, we have

$$\begin{aligned}\big(a_{h_n}(x) - a_r(x)\big)\varphi(x) &= \big(1 - u(x+h_n)\big)a_l(x+h_n)\varphi(x)\\ &\quad + u(x+h_n)a_r(x+h_n)\varphi(x) - a_r(x)\varphi(x) + a_0(x+h_n)\varphi(x)\\ &= \big(a_r(x+h_n) - a_r(x)\big)\varphi(x) =: b_n(x)\varphi(x),\end{aligned}$$

whence

$$\|F^{-1}(a_{h_n} - a_r)\varphi Ff\|_p = \|F^{-1}b_n\varphi Ff\|_p \le \|W^0(b_n\varphi)\|\,\|f\|_p.$$

By Stechkin's inequality (Theorem 17.1),

$$\|W^0(b_n\varphi)\| \le \|S\|\big(\|b_n\varphi\|_\infty + V_1(b_n\varphi)\big). \tag{19.11}$$

Since

$$\begin{aligned}V_1(b_n\varphi) &\le \|(b_n\varphi)'\|_\infty|\operatorname{supp}(b_n\varphi)|\\ &\le \big(\|b_n'\|_\infty\|\varphi\|_\infty + \|b_n\|_\infty\|\varphi'\|_\infty\big)|\operatorname{supp}\varphi|,\end{aligned}$$

we infer from (19.9) that the right-hand side of (19.11) goes to zero as $h_n \to \infty$. This completes the proof of (19.10). □

Corollary 19.7. *If $A \in \mathcal{F}$ is Fredholm on $L^p_N(\mathbf{R})$, then $\mu_l(A)$ and $\mu_r(A)$ are invertible on $L^p_N(\mathbf{R})$.*

Proof. This follows from Theorem 19.6 by the reasoning employed in the proof of Corollary 18.11. □

Corollary 19.8. *If $a \in [SAP_p]_{N\times N}$ and $W(a)$ is Fredholm on $L^p_N(\mathbf{R}_+)$, then $W(a_l)$ and $W(a_r)$ are invertible on $L^p_N(\mathbf{R}_+)$.*

Proof. Immediate from Corollary 19.7. □

19.4 Matrix-Valued APW Symbols

To tackle Wiener-Hopf operators with symbols in $APW_{N\times N} \subset [AP_p]_{N\times N}$, we will make use of the following key result.

Theorem 19.9 (Kurbatov). *If $a \in APW_{N\times N}$, then $W(a)$ is invertible on $L^p_N(\mathbf{R}_+)$ for some $p \in (1,\infty)$ if and only if $W(a)$ is invertible on $L^p_N(\mathbf{R}_+)$ for all $p \in (1,\infty)$.*

To prove this theorem, we need an auxiliary result. For $1 < p < \infty$, let $\mathcal{V}^W_{N\times N}$ be the set of all operators $A \in \mathcal{B}(L^p_N(\mathbf{R}))$ which can be written in the form

$$A = \sum_{\lambda\in\mathbf{R}} a_\lambda U_{-\lambda}, \quad a_\lambda \in L^\infty_{N\times N}(\mathbf{R}), \quad \sum_\lambda \|a_\lambda\|_\infty < \infty, \tag{19.12}$$

where $a_\lambda \neq 0$ for at most countably many $\lambda \in \mathbf{R}$ and $U_{-\lambda}$ is defined by $(U_{-\lambda}f)(x) = f(x+\lambda)$ for $x \in \mathbf{R}$. Obviously, $\mathcal{V}^W_{N\times N}$ is a Banach algebra with the norm

$$\|A\| := \sum_{\lambda\in\mathbf{R}} \|a_\lambda\|_\infty = \sum_{\lambda\in\mathbf{R}} \|a_\lambda I\|_{\mathcal{B}(L^p_N(\mathbf{R}))}.$$

Theorem 19.10. *The algebra $\mathcal{V}^W_{N\times N}$ is inverse closed in $\mathcal{B}(L^p_N(\mathbf{R}))$, i.e., if an operator in $\mathcal{V}^W_{N\times N}$ is invertible on $L^p_N(\mathbf{R})$, then the inverse operator also belongs to $\mathcal{V}^W_{N\times N}$.*

Proof. We proceed again as in the proofs of Theorems 9.6 and 18.2. Let A be of the form (19.12), assume A is invertible, put $E := \mathcal{B}(L^p_N(\mathbf{R}))$, and define

$$a : \mathbf{R}_B \to E, \; a(\chi) := \sum_\lambda \chi(\lambda) a_\lambda U_{-\lambda}.$$

From (19.12) we infer that $a \in W(\mathbf{R}_B, E)$. Let $e_h(\lambda) := e^{i\lambda h}$. Clearly,

$$e_h^{-1} U_{-\lambda} e_h I = e^{i\lambda h} U_{-\lambda}. \tag{19.13}$$

The operators $a_\lambda I$ commute with every $e_h I$, i.e.,

$$e_h a_\lambda I = a_\lambda e_h I, \quad h \in \mathbf{R}. \tag{19.14}$$

Using (19.13) and (19.14) we get

$$\begin{aligned} a(e_h) &= \sum_\lambda e_h(\lambda) a_\lambda U_{-\lambda} = \sum_\lambda a_\lambda e_h(\lambda) U_{-\lambda} \\ &= \sum_\lambda a_\lambda e_h^{-1} U_{-\lambda} e_h I = e_h^{-1}\Big(\sum_\lambda a_\lambda U_{-\lambda}\Big) e_h I = e_h^{-1} A e_h I. \end{aligned}$$

Therefore,

$$\|a^{-1}(e_h)\| = \|e_h^{-1} A^{-1} e_h I\| = \|A^{-1}\| \text{ for all } h \in \mathbf{R},$$

and since $\mathbf{R}$ is dense in $\mathbf{R}_B$, we can conclude that a is invertible in $C(\mathbf{R}_B, E)$. The Bochner-Phillips Theorem (Theorem 9.5) therefore implies that a^{-1} belongs to $W(\mathbf{R}_B, E)$:

$$a^{-1}(\chi) = \sum_\lambda \chi(\lambda) c_\lambda, \quad c_\lambda \in E, \quad \sum_\lambda \|c_\lambda\| < \infty,$$

and $c_\lambda \neq 0$ for at most countably many $\lambda \in \mathbf{R}$. In particular,

$$A^{-1} = a^{-1}(e_0) = \sum_\lambda e_0(\lambda) c_\lambda = \sum_\lambda c_\lambda.$$

It remains to show that there are $d_\lambda \in L^\infty_{N\times N}(\mathbf{R})$ such that $c_\lambda = d_\lambda U_{-\lambda}$ for all $\lambda \in \mathbf{R}$. By (19.13) and (19.14),

$$\begin{aligned} e_h^{-1} a(e_s) e_h I &= \sum_\lambda e_h^{-1} e^{i\lambda s} a_\lambda U_{-\lambda} e_h I \\ &= \sum_\lambda e^{i\lambda s} a_\lambda e_h^{-1} U_{-\lambda} e_h I = \sum_\lambda e^{i\lambda s} a_\lambda e^{ih\lambda} U_{-\lambda} = a(e_{s+h}), \end{aligned}$$

whence $e_h^{-1} a^{-1}(e_s) e_h I = a^{-1}(e_{s+h})$. Because $a^{-1}(e_\nu) = \sum e^{i\lambda\nu} c_\lambda$, we obtain

$$e_h^{-1}\Big(\sum_\lambda e^{i\lambda s} c_\lambda\Big) e_h I = \sum_\lambda e^{i\lambda s} e^{i\lambda h} c_\lambda,$$

and comparison of the coefficients of $e^{i\lambda s}$ gives

$$e_h^{-1} c_\lambda e_h I = e^{i\lambda h} c_\lambda.$$

Thus,

$$e_h^{-1} c_\lambda e_h U_\lambda = e^{i\lambda h} c_\lambda U_\lambda.$$

Again by (19.13), $e_h U_\lambda = e^{i\lambda h} U_\lambda e_h$, which implies that

$$e_h^{-1} c_\lambda U_\lambda e_h = c_\lambda U_\lambda.$$

Consequently, $d_\lambda = c_\lambda U_\lambda$ commutes with every $e_h I\,(h \in \mathbf{R})$. Hence (see [164]) $d_\lambda \in L^\infty_{N\times N}(\mathbf{R})$. It finally results that

$$A^{-1} = \sum_\lambda c_\lambda = \sum_\lambda d_\lambda U_{-\lambda} \in \mathcal{V}^W_{N\times N}. \qquad \square$$

Proof of Theorem 19.9. The invertibility of $W(a)$ on $L^p_N(\mathbf{R}_+)$ is equivalent to the invertibility of the operator $\chi_+ W^0(a) + \chi_- I$ on $L^p_N(\mathbf{R})$. We have

$$\begin{aligned} a &= \sum_\lambda a_\lambda e_\lambda, \quad a_\lambda \in \mathbf{C}_{N\times N}, \quad \sum \|a_\lambda\| < \infty, \\ W^0(e_\lambda) &= F^{-1} e_\lambda F = U_\lambda, \end{aligned}$$

and hence,

$$\chi_+ W^0(a) + \chi_- I = \sum_\lambda \chi_+ a_\lambda U_\lambda + \chi_- U_0 \in \mathcal{V}^W_{N\times N}.$$

Theorem 19.10 implies that $(\chi_+ W^0(a) + \chi_- I)^{-1} \in \mathcal{V}^W_{N\times N}$. Therefore

$$(\chi_+ W^0(a) + \chi_- I)^{-1} = \sum d_\lambda U_\lambda, \quad d_\lambda \in L^\infty_{N\times N}(\mathbf{R}), \quad \sum \|d_\lambda\|_\infty < \infty,$$

which shows that $(\chi_+ W^0(a) + \chi_- I)^{-1}$ is bounded on every $L^r_N(\mathbf{R})$, $1 < r < \infty$. This yields the invertibility of $\chi_+ W^0(a) + \chi_- I$ on every $L^r_N(\mathbf{R})$ and thus the invertibility of $W(a)$ on every $L^r_N(\mathbf{R}_+)$. □

Corollary 19.11. *For $a \in APW_{N\times N}$, the following are equivalent:*

(i) *$W(a)$ is Fredholm on $L^p_N(\mathbf{R}_+)$;*

(ii) *$W(a)$ is invertible on $L^p_N(\mathbf{R}_+)$;*

(iii) *a has a canonical right APW factorization.*

Proof. The implications (iii)⇒(ii)⇒(i) are trivial. If $W(a)$ is Fredholm on $L^p_N(\mathbf{R}_+)$, then Corollary 19.7 applied to $W^0(a)\chi_+ I + \chi_- I$ shows that $W(a)$ is invertible on $L^p_N(\mathbf{R}_+)$. From Theorem 19.9 and Corollary 9.8 we deduce that a has a canonical right APW factorization. This completes the proof of the implication (i)⇒(iii). □

19.5 The Theorem by Duduchava and Saginashvili

Put $H^\infty_{p,\pm} := H^\infty_\pm \cap M_p$. Notice that $APW^\pm \subset H^\infty_{p,\pm}$ for all $p \in (1, \infty)$. It is easily seen that $H^\infty_{p,\pm}$ are closed subalgebras of M_p.

Theorem 19.12. (a) *The sets $C_p(\dot{\mathbf{R}}) + H^\infty_{p,\pm}$ are closed subalgebras of M_p.*

(b) *If $a_\pm \in C_p(\dot{\mathbf{R}}) + H^\infty_{p,\pm}$ and $b \in M_p$, then*

$$W(a_- b a_+) - W(a_-)W(b)W(a_+)$$

is a compact operator on $L^p(\mathbf{R}_+)$.

(c) *If $a_\pm \in [C_p(\dot{\mathbf{R}}) + H^\infty_{p,\pm}]_{N\times N}$, then for $W(a_\pm)$ to be Fredholm on $L^p_N(\mathbf{R}_+)$ it is necessary and sufficient that* $\det a_\pm$ *be invertible in $C_p(\dot{\mathbf{R}}) + H^\infty_{p,\pm}$. In that case*

$$\operatorname{Ind} W(a_\pm) = -\operatorname{wind}(\det a_\pm),$$

where wind a *is defined via the harmonic extension of a as in Section* 2.5.

For operators on $l^p(\mathbf{Z}_+)$, full proofs are in [46]. These proofs can be carried over to the $L^p(\mathbf{R}_+)$ case without difficulty.

Here is the L^p analogue of Lemma 3.5.

Lemma 19.13. *If $h \in AP^\pm_p$ and $M(h) = 0$, then $hv \in C_p(\dot{\mathbf{R}}) + H^\infty_{p,\pm}$ for every $v \in C_p(\overline{\mathbf{R}})$.*

Proof. One can proceed as in the proof of Lemma 3.5. Assume $h = e_\alpha$ ($\alpha > 0$) and construct σ as there. The total variation $V_1(\sigma)$ can be estimated by

$$V_1(\sigma) \le \int_{-\infty}^{\infty} |\sigma'(x)|\,dx = \int_{-\infty}^{\infty} |\varphi(x)|\,dx \le M \int_{-\infty}^{\infty} \frac{dx}{1+|x|^3}\,dx < \infty,$$

and Theorem 17.1 therefore implies that $\sigma \in C_p(\overline{\mathbf{R}})$. Every function $v \in C_p(\overline{\mathbf{R}})$ can be represented in the form (3.10) where $v_0 \in C_p(\dot{\mathbf{R}})$ and $v_0(\infty) = 0$. It is easy to see that $e_\alpha v_0 \in C_p(\dot{\mathbf{R}})$, and since $e^{i\alpha x}\sigma(\alpha x)$ is a function in $H^\infty_+ \cap M_p$, it follows that $e_\alpha v$ belongs to $C_p(\dot{\mathbf{R}}) + H^\infty_{p,+}$. □

The L^p analogue of Theorem 10.9 is as follows.

Theorem 19.14. *Let $a, b \in G[SAP_p]_{N\times N}$ and suppose the almost periodic representatives a_l, b_l, a_r, b_r are connected by the relations*

$$\begin{aligned} a_l &= \varphi_l^- b_l \varphi_l^+ \quad \text{with } \varphi_l^\pm \in G[AP_p^\pm]_{N\times N}, \\ a_r &= \varphi_r^- b_r \varphi_r^+ \quad \text{with } \varphi_r^\pm \in G[AP_p^\pm]_{N\times N}, \\ M(\varphi_l^-) &= M(\varphi_r^-), \quad M(\varphi_l^+) = M(\varphi_r^+). \end{aligned}$$

Then $a = f_- b f_+$ with matrix functions

$$f_\pm \in G[SAP_p]_{N\times N} \cap G[C_p(\dot{\mathbf{R}}) + H^\infty_{p,\pm}]_{N\times N}.$$

Proof. The construction of the proof of Theorem 10.9 also works in the case at hand. We have only to approximate $\varphi_l^{\pm 1}, \varphi_r^{\pm 1}$ by almost periodic matrix polynomials in the norm of $[M_p]_{N\times N}$ and we must replace Lemma 3.5 by Lemma 19.13. □

From Example 16.19 we know that if $w = 1$, then $I_0(p, w) = (-1/p, 1/q)$, whence $\nu_0^-(p, w) = \nu_0^+(p, w) = 1/p$. Thus, the horn

$$\mathcal{H}(a(+\infty), a(-\infty);\ \nu_0^-(p, w),\ \nu_0^+(p, w))$$

degenerates to the circular arc $\mathcal{A}(a(+\infty), a(-\infty);\ 1/p)$. Figure 19.1 illustrates the following theorem.

Theorem 19.15 (Duduchava and Saginashvili). *Let $a \in SAP_p \setminus \{0\}$ and $1 < p < \infty$.*

(a) *If $a \notin GSAP$, then $W(a)$ is not normally solvable on $L^p(\mathbf{R}_+)$.*

(b) *If $a \in GSAP$ and $\kappa(a_l)\kappa(a_r) < 0$, then $W(a)$ is not normally solvable on $L^p(\mathbf{R}_+)$.*

(c) *If $a \in GSAP$, $\kappa(a_l)\kappa(a_r) \ge 0$, and $\kappa(a_l) + \kappa(a_r) > 0$, then $W(a)$ is properly n-normal and has trivial kernel on $L^p(\mathbf{R}_+)$.*

(d) *If $a \in GSAP$, $\kappa(a_l)\kappa(a_r) \ge 0$, and $\kappa(a_l) + \kappa(a_r) < 0$, then $W(a)$ is properly d-normal and its range coincides with $L^p(\mathbf{R}_+)$.*

(e) *If $a \in GSAP$, $\kappa(a_l) = \kappa(a_r) = 0$, and*

$$0 \notin \mathcal{A}(\mathbf{d}(a_r),\, \mathbf{d}(a_l);\, 1/p), \tag{19.15}$$

then $W(a)$ is Fredholm on $L^p(\mathbf{R}_+)$ and

$$\operatorname{Ind} W(a) = -\operatorname{ind} a + \frac{1}{p} - \left\{\frac{1}{p} + \frac{1}{2\pi}\arg\frac{\mathbf{d}(a_l)}{\mathbf{d}(a_r)}\right\}.$$

(f) *If $a \in GSAP$; $\kappa(a_l) = \kappa(a_r) = 0$, and (19.15) is violated, then $W(a)$ is not normally solvable on $L^p(\mathbf{R}_+)$.*

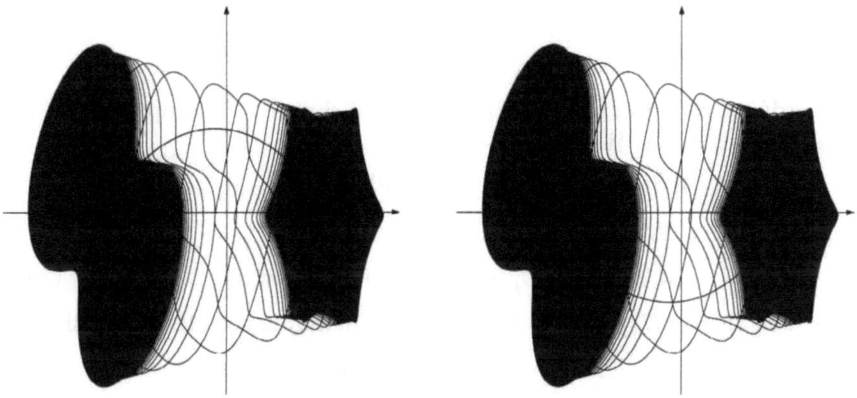

Figure 19.1: The pictures show the range of a function $a \in SAP$, the points at which $\kappa(a_l) \neq 0$ or $\kappa(a_r) \neq 0$, the arc $\mathcal{A}(\mathbf{d}(a_r), \mathbf{d}(a_l); 1/q)$ (left) and the arc $\mathcal{A}(\mathbf{d}(a_r), \mathbf{d}(a_l); 1/p)$ (right). The left picture and Theorem 18.8 imply that $T(a)$ is Fredholm on $H^p(\mathbf{R})$, the right picture in conjunction with Theorem 19.15 tells us that $W(a)$ is Fredholm on $L^p(\mathbf{R}_+)$.

Proof. By Proposition 19.4, invertibility in SAP of functions from SAP_p is equivalent to their invertibility in SAP_p. Part (a) therefore results from Theorems 19.1(b) and 19.5. So suppose $a \in GSAP_p$. Using Theorem 19.14, we see as in the proof of Theorem 3.9 that $W(a)$ is weakly Φ-equivalent to $W(b)$ where

$$b(x) := \big(1 - u(x)\big)\mathbf{d}(a_l)e^{i\kappa(a_l)x} + u(x)\mathbf{d}(a_r)e^{i\kappa(a_r)x} + b_0(x)$$

where $b_0 \in C_p(\dot{\mathbf{R}})$ is chosen so that $b_0(\infty) = 0$ and $b \in GSAP_p$. If $\kappa(a_l) = \kappa(a_r) = 0$, then $b \in C_p(\overline{\mathbf{R}})$, and Theorems 17.7 and 19.2 give parts (e) and (f). In the case where $\kappa(a_l) > 0$ and $\kappa(a_r) > 0$, we deduce from Lemma 19.13 that

$b \in C_p(\dot{\mathbf{R}}) + H^\infty_{p,+}$, and Theorem 19.12(b) implies that $W(b^{-1})$ is a left regularizer of $W(b)$. Corollary 19.8 and Theorem 2.28(b) (for $L^p(\mathbf{R}_+)$) show that $W(b)$ cannot be Fredholm. Hence, $W(b)$ is properly n-normal. If $\kappa(a_l) = 0$ and $\kappa(a_r) > 0$, we can proceed as in the proofs of Lemma 3.8 and Theorem 18.8 to deduce from Theorem 17.9 that $W(b)$ is n-normal. As $W(b)$ is not Fredholm due to Corollary 19.8 and Theorem 2.28(b) (for $L^p(\mathbf{R}_+)$), it follows again that $W(b)$ is properly n-normal. Theorem 19.1(a) finally tells us that $W(b)$ has trivial kernel, which completes the proof of part (c). Part (d) can be proved analogously.

We are so left with part (b). By Theorem 19.1(b) it suffices to show that $W(b)$ is not semi-Fredholm. Without loss of generality, assume $0 \notin \mathcal{A}(\mathbf{d}(a_r), \mathbf{d}(a_l); 1/p)$ (otherwise we can slightly perturb b without changing semi-Fredholmness). Then the argument of the proof of Lemma 3.6 shows that $W(b)$ is weakly Φ-equivalent to $W(c)$ where

$$c(x) := \alpha\chi_-(x)e^{i\lambda x} + \beta\chi_+(x)e^{i\mu x}$$

with $\alpha = \mathbf{d}(a_l)$, $\beta = \mathbf{d}(a_r)$, $\lambda = \kappa(a_l)$, $\mu = \kappa(a_r)$. Let $\mu > 0$. Suppose there is an $f_+ \in L^p(\mathbf{R}_+)$ such that $W(c)f_+ = 0$. Then

$$\alpha\chi_+ W^0(e_\lambda)W^0(\chi_-)f_+ + \beta\chi_+ W^0(e_\mu)W^0(\chi_+)f_+ = 0,$$

whence $\alpha U_\lambda P f_+ + \beta U_\mu Q f_+ = 0$ on $\mathbf{R}_+$. Since $Q = I - P$, we get

$$\alpha U_\lambda P f_+ - \beta U_\mu P f_+ + \beta U_\mu f_+ = 0 \text{ on } \mathbf{R}_+. \tag{19.16}$$

If $0 < x < \mu$, then $\beta(U_\mu f_+)(x) = \beta f_+(x - \mu) = 0$, and hence (19.16) implies that

$$\alpha(Pf_+)(x - \lambda) = \beta(Pf_+)(x - \mu) \text{ for } x \in (0, \mu). \tag{19.17}$$

The function Pf_+ belongs to $PL^p(\mathbf{R}) = H^p_+$. Since H^p_+ is translation invariant, both sides of (19.17) are functions in H^p_+. As these functions coincide on an interval, they must coincide almost everywhere by the Lusin-Privalov theorem. Thus, $\alpha U_\lambda P f_+ = \beta U_\mu P f_+$, and (19.16) therefore gives that $f_+ = 0$. This proves that $W(c)$ has a trivial kernel on $L^p(\mathbf{R}_+)$.

Now assume $W(b)$ is d-normal on $L^p(\mathbf{R}_+)$. Then $W(c)$ is also d-normal on $L^p(\mathbf{R}_+)$, and as $W(c)$ has a trivial kernel, it follows that $W(c)$ is Fredholm. This is impossible due to Corollary 19.8 and (the $L^p(\mathbf{R}_+)$ version of) Theorem 2.28(b). If $W(b)$ is n-normal on $L^p(\mathbf{R}_+)$, then $W(\overline{b})$ is d-normal on $L^q(\mathbf{R}_+)$ $(1/p + 1/q = 1)$. Repeating the above argument with $\overline{c}$ and q in place of c and p, we arrive again at a contradiction. Consequently, $W(b)$ cannot be semi-Fredholm. This completes the proof of part (b). □

19.6 Matrix-Valued SAP Symbols

We are now in a position to settle the Fredholm theory for operators with matrix-valued SAP_p symbols, provided the almost periodic representatives are a priorily known to have AP_p factorizations.

Theorem 19.16. *Let $a \in [SAP_p]_{N\times N}$ and suppose a_l and a_r have right AP_p factorizations. Then $W(a)$ is Fredholm on $L^p_N(\mathbf{R}_+)$ if and only if*

$$a \in GSAP_{N\times N}, \quad \kappa(a_l) = \kappa(a_r) = (0,\dots,0) \tag{19.18}$$

and one of the following two equivalent conditions is satisfied:

(i) $$\frac{1}{p} + \frac{1}{2\pi}\arg\xi_j\left(\mathbf{d}^{-1}(a_r)\mathbf{d}(a_l)\right) \notin \mathbf{Z}$$

for all eigenvalues $\xi_j = \xi_j(\mathbf{d}^{-1}(a_r)\mathbf{d}(a_l))$ of $\mathbf{d}^{-1}(a_r)\mathbf{d}(a_l)$;

(ii) $$\mathrm{sp}\left(\mathbf{d}^{-1}(a_r)\mathbf{d}(a_l)\right) \cap \Sigma[1/q, 1/q] = \emptyset,$$

where $\Sigma[1/q,1/q]$ is the ray $\{re^{2\pi i/q} : r \geq 0\}$.
If $W(a)$ is Fredholm, then

$$\mathrm{Ind}\, W(a) = -\,\mathrm{ind}\,(\det a) + \frac{N}{p} - \sum_{j=1}^{N}\left\{\frac{1}{p} + \frac{1}{2\pi}\arg\xi_j\right\}.$$

Proof. If $W(a)$ is Fredholm then $a \in G[SAP_p]_{N\times N}$ by virtue of Theorem 19.5, and $W(a_l), W(a_r)$ are invertible due to Corollary 19.8. This proves the necessity of (19.18).

Now suppose (19.18) holds. Using Theorems 17.10 and 19.14, one can show as in the proof of Theorem 16.22 that (i) (equivalently, (ii)) is necessary and sufficient for $W(a)$ to be Fredholm.

The index formula can be verified by an appropriate modification of the proof of Theorem 10.12. □

Corollary 19.17. *If $a \in SAPW_{N\times N}$, then $W(a)$ is Fredholm on $L^p_N(\mathbf{R}_+)$ if and only if a_l, a_r have right APW factorizations and* (19.18) *as well as one of the two equivalent conditions* (i),(ii) *of Theorem* 19.16 *are satisfied.*

Proof. Corollaries 19.8 and 19.11 imply that a_l, a_r automatically have canonical right *APW* (and thus AP_p) factorizations whenever $W(a)$ is Fredholm. The assertion is therefore immediate from Theorem 19.16. □

Notes

The definition of the classes AP_p and SAP_p used here was proposed in [130] and [131]; a slightly different definition of SAP_p was given earlier by Duduchava and Saginashvili [72]. It can be easily shown that both definitions lead to the same class of functions; see [133, Lemma 3.1 (ii)]. In the scalar case (Section 19.5), the Fredholm theory of Wiener-Hopf operators with SAP_p symbols was fully developed in [72]. A preliminary matrix case version, in which the AP_p factorability of the almost periodic representatives of the symbols was imposed as an a priori condition (culminating with Theorem 19.16), is in [130] and [131]. The exposition in Sections 19.3, 19.4, and 19.6 reflects the current state of the events, and is presented according to [122] and [133]. The major role in the reduction of the L^p case to the L^2 case is played by Kurbatov's Theorems 19.9 and 19.10, which can be found in his book [149]. This leads to the main result of this chapter - Corollary 19.17. Thus, the situation is as follows. The AP_p factorability condition on the almost periodic representatives of the symbol, imposed in Theorem 19.16, holds automatically in APW case, but is not necessary for Fredholmness in general. We will get around this difficulty when $p = 2$ in Chapter 21. The case $p \neq 2$ remains open at the present time.

Theorem 19.5 (a necessary condition for semi-Fredholmness) is in [130], [131]; see also [133]. These papers also contain semi-Fredholm criteria under the a priori assumptions of the AP_p factorability of the almost periodic representatives of the symbol.

The approach presented in this chapter allows us to consider Wiener-Hopf operators not only on L^p spaces, but also on spaces of Bessel potentials and, in particular, on Sobolev-Slobodetskii spaces. On top of that, equations on unions of several intervals, finite or infinite, may be tackled. In addition to already cited papers [128], [131], [132], [133], [211], we also mention [17], [18], [19], [54], [55], [137], [148]. This list is by no means complete.

Chapter 20

Hankel Operators on Besicovitch Spaces

The purpose of this section is to carry over a famous result by V. M. Adamyan, D. Z. Arov, and M. G. Krein to the norms of Hankel operators on the vector-valued Besicovitch spaces B_N^2.

20.1 B^1, B^2, B^∞

In analogy to the Besicovitch space B^2, one can define the Besicovitch spaces $B^p\,(1 \le p \le \infty)$. We define B^p as $L^p(\mathbf{R}_B, d\mu)$. Thus, B^∞ is the C^*-algebra of all essentially bounded and μ-measurable functions on $\mathbf{R}_B$ and the norm in B^∞ is given by

$$\|f\|_{B^\infty} = \operatorname*{ess\,sup}_{x \in \mathbf{R}_B} |f(x)|,$$

and $B^p\,(1 \le p < \infty)$ may also be viewed as the completion of the set of all almost periodic polynomials $f = \sum_\lambda f_\lambda e_\lambda \in AP^0$ with respect to the norm

$$\|f\|_{B^p} = \Big(\sum |f_\lambda|^p\Big)^{1/p} = \big(M(|f|^p)\big)^{1/p} = \left(\int_{\mathbf{R}_B} |f(x)|^p \, d\mu(x)\right)^{1/p}.$$

Let $B^p_{N\times N}$ be the linear space of all $(N \times N)$-matrix functions with entries in B^p.

We equip $B^1_{N\times N}$ and $B^2_{N\times N}$ with the norms

$$\|a\|_{B^1} := \|a\|_{B^1_{N\times N}} := \int_{\mathbf{R}_B} \operatorname{tr}\left(\left(a(x)a^*(x)\right)^{1/2}\right) d\mu(x), \tag{20.1}$$

$$\|a\|_{B^2} := \|a\|_{B^2_{N\times N}} := \left(\int_{\mathbf{R}_B} \operatorname{tr}\big(a(x)a^*(x)\big)\, d\mu(x)\right)^{1/2}, \tag{20.2}$$

where tr is the usual trace for matrices. Furthermore, the norm in $B^\infty_{N\times N}$ is defined by

$$\|a\|_{B^\infty} := \|a\|_{B^\infty_{N\times N}} := \operatorname*{ess\,sup}_{x\in\mathbf{R}_B} s_{\max}\big(a(x)\big)$$

where $s_{\max}(a(x))$ stands for the maximal singular value of the matrix $a(x)$. For $g \in B^\infty_{N\times N}$, the map Ψ_g given by

$$\Psi_g(a) = \int_{\mathbf{R}_B} \operatorname{tr}\big(a(x)g^*(x)\big)\, d\mu(x) \tag{20.3}$$

is a bounded linear functional on $B^1_{N\times N}$. Analogously to [25, Theorem 11(a) on p. 249]) one can show that the map $g \mapsto \Psi_g$ is an isometric isomorphism of $B^\infty_{N\times N}$ onto the dual space of $B^1_{N\times N}$.

Let $\mathbf{R}_+ := [0,\infty)$, $\mathbf{R}_- := (-\infty, 0]$, and for $p \in \{1,2,\infty\}$, put

$$\begin{aligned} \left[B^p_\pm\right]_{N\times N} &:= \left\{a \in B^p_{N\times N} : \Omega(a) \subset \mathbf{R}_\pm\right\}, \\ \left[\dot{B}^p_\pm\right]_{N\times N} &:= \left\{a \in B^p_{N\times N} : \Omega(a) \subset \mathbf{R}_\pm \setminus \{0\}\right\}. \end{aligned}$$

If $p = 1$ or $p = 2$, then $[B^p_\pm]_{N\times N}$ and $[\dot{B}^p_\pm]_{N\times N}$ can be shown to be the closure of the set $APW^\pm_{N\times N}$ and $[APW^\pm_0]_{N\times N}$ in $B^p_{N\times N}$, respectively, where $APW^\pm_0$ stand for the functions in $APW^\pm$ whose Bohr-Fourier spectrum does not contain the origin. Moreover, on identifying $(B^1_{N\times N})^*$ and $B^\infty_{N\times N}$, one can prove that

$$\left[B^\infty_\mp\right]_{N\times N} = \left(\left[\dot{B}^1_\pm\right]_{N\times N}\right)^\perp,$$

that is, $[B^\infty_\mp]_{N\times N}$ can be identified with the set of all $g \in B^\infty_{N\times N}$ for which

$$\int_{\mathbf{R}} \operatorname{tr}\big(a(x)g^*(x)\big)\, d\mu(x) = 0$$

for all $a \in [\dot{B}^1_\pm]_{N\times N}$. We also have the isometric isomorphisms

$$\left[\dot{B}^1_\pm\right]^*_{N\times N} \cong B^\infty_{N\times N}/\left[B^\infty_\mp\right]_{N\times N};$$

equivalently,

$$\|\Psi_g|[\dot{B}^1_\pm]_{N\times N}\| = \min\left\{\|g-h\|_{B^\infty} : h \in [B^\infty_\mp]_{N\times N}\right\} \tag{20.4}$$

for every $g \in B^\infty_{N\times N}$ (see, e.g., [84, Chapter IV, Section 1]).

20.2 Norms of Matrix Hankel Operators on B^2

Let $\tilde{P}, \tilde{P}_0$ and $\tilde{Q}, \tilde{Q}_0$ be the projections on B^2_N introduced in Section 9.2.

Theorem 20.1. *If $g \in B^\infty_{N\times N}$, then*

$$\|\tilde{P}_0 g\tilde{Q}\|_{\mathcal{B}(B^2_N)} = \|\tilde{P}g\tilde{Q}_0\|_{\mathcal{B}(B^2_N)} = \min\{\|g-h\|_{B^\infty} : h \in [B^\infty_-]_{N\times N}\}, \tag{20.5}$$
$$\|\tilde{Q}_0 g\tilde{P}\|_{\mathcal{B}(B^2_N)} = \|\tilde{Q}g\tilde{P}_0\|_{\mathcal{B}(B^2_N)} = \min\{\|g-h\|_{B^\infty} : h \in [B^\infty_+]_{N\times N}\}. \tag{20.6}$$

Proof. We consider the operator $\tilde{P}g\tilde{Q}_0$. By virtue of (20.4), it suffices to show that

$$\|\tilde{P}g\tilde{Q}_0\| = \|\Psi_g|[\dot{B}^1_\pm]_{N\times N}\|. \tag{20.7}$$

The space $B^2_{N\times N}$ is a Hilbert space with the inner product

$$(a,b) = \int_{\mathbf{R}_B} \operatorname{tr}\big(a(x)b^*(x)\big)\,d\mu(x),$$

and it is easily seen that

$$\|ab\|_{B^1} \le \|a\|_{B^2}\|b\|_{B^2} \ \text{ for } \ a,b \in B^2_{N\times N}.$$

Thus, the set

$$\{f = ab : a,b \in [B^2_+]_{N\times N},\ \|a\|_{B^2} < 1,\ \|b\|_{B^2} < 1\} \tag{20.8}$$

is contained in the open unit ball $\mathcal{V}$ of $[B^1_+]_{N\times N} \subset B^1_{N\times N}$. We claim that the set (20.8) is dense in $\mathcal{V}$.

It is obvious that the set $\{f \in APW^+_{N\times N} : \|f\|_{B^1} < 1\}$ is dense in $\mathcal{V}$. Fix an f in this set. There is a sufficiently small $\varepsilon > 0$ such that if we put $d := ff^* + \varepsilon I_N$, then

$$\|d^{1/2}\|_{B^1} = \lim_{T\to\infty} \frac{1}{2T}\int_{-T}^{T} \operatorname{tr}\big(f(x)f^*(x) + \varepsilon I_N\big)^{1/2}\,dx < 1. \tag{20.9}$$

The spectrum of d in $APW_{N\times N}$ coincides with the spectrum of d in $AP_{N\times N}$ and is a subset of $(0,\infty)$. Since $APW_{N\times N}$ is a completely symmetric ring (see [164, p. 299]), the matrix function d has a positive and selfadjoint square root $d^{1/2}$ in $APW_{N\times N}$ (see [164, p. 301]). Clearly, $d^{1/2} \in GAPW_{N\times N}$, and Corollary 9.15(b) applied to $(d^{1/2})^{-1}$ yields the factorization $d^{1/2} = aa^*$ with $a \in GAPW^+_{N\times N} \subset G[B^2_+]_{N\times N}$. By (20.9),

$$\begin{aligned}\|a\|^2_{B^2} &= \lim_{T\to\infty} \frac{1}{2T}\int_{-T}^{T} \operatorname{tr}\big(a(x)a^*(x)\big)\,dx \\ &= \lim_{T\to\infty} \frac{1}{2T}\int_{-T}^{T} \operatorname{tr} d^{1/2}(x)\,dx = \|d^{1/2}\|_{B^1} < 1.\end{aligned}$$

Put $b := a^{-1}f$. Because $b \in APW^+_{N\times N}$ and

$$\begin{aligned}
\|b\|^2_{B^2} &= \lim_{T\to\infty} \frac{1}{2T} \int_{-T}^{T} \operatorname{tr}\big(b(x)b^*(x)\big)\, dx \\
&= \lim_{T\to\infty} \frac{1}{2T} \int_{-T}^{T} \operatorname{tr}\Big(a^{-1}(x)f(x)f^*(x)\big(a^{-1}(x)\big)^*\Big)\, dx \\
&\le \lim_{T\to\infty} \frac{1}{2T} \int_{-T}^{T} \operatorname{tr}\Big(a^{-1}(x)d(x)\big(a^{-1}(x)\big)^*\Big)\, dx \\
&= \lim_{T\to\infty} \frac{1}{2T} \int_{-T}^{T} \operatorname{tr}\Big(a^{-1}(x)\big(a(x)a^*(x)\big)^2\big(a^{-1}(x)\big)^*\Big)\, dx \\
&= \lim_{T\to\infty} \frac{1}{2T} \int_{-T}^{T} \operatorname{tr}\big(a^*(x)a(x)\big)\, dx \\
&= \lim_{T\to\infty} \frac{1}{2T} \int_{-T}^{T} \operatorname{tr}\big(a(x)a^*(x)\big)\, dx = \|a\|^2_{B^2} < 1,
\end{aligned}$$

the matrix function b lies in the open unit ball of $[B^2_+]_{N\times N}$. This completes the proof of our claim that the set (20.8) is dense in $\mathcal{V}$.

Now it is easy to deduce that the collection of all products ab with $a \in [B^2_+]_{N\times N}$, $b \in [\dot{B}^2_+]_{N\times N}$, $\|a\|_{B^2} < 1$, $\|b\|_{B^2} < 1$ is dense in the unit ball $\{f \in [\dot{B}^1_+]_{N\times N} : \|f\|_{B^1} < 1\}$. We have

$$\begin{aligned}
&\|\Psi_g|[\dot{B}^1_+]_{N\times N}\| \\
&= \sup\left\{ \int_{\mathbf{R}_B} \operatorname{tr}\big(f(x)g^*(x)\big)\, d\mu(x) : f \in [\dot{B}^1_+]_{N\times N},\ \|f\|_{B^1} < 1 \right\}, \qquad (20.10)
\end{aligned}$$

and the supremum on the right may also be taken over the dense subset of all $f = f_1 f_2^*$ with $f_1 \in [B^2_+]_{N\times N}$, $f_2 \in [\dot{B}^2_-]_{N\times N}$, $\|f_1\|_{B^2} < 1$, $\|f_2\|_{B^2} < 1$. Furthermore,

$$\begin{aligned}
&\int_{\mathbf{R}_B} \operatorname{tr}\big(f_1(x)f_2^*(x)g^*(x)\big)\, d\mu(x) = \int_{\mathbf{R}_B} \operatorname{tr}\Big(\big(g(x)f_2(x)\big)^* f_1(x)\Big)\, d\mu(x) \\
&= \operatorname{tr}\left(\int_{\mathbf{R}_B} \big(g(x)f_2(x)\big)^* f_1(x)\, d\mu(x) \right). \qquad (20.11)
\end{aligned}$$

Let $f_j^{(1)}, \dots, f_j^{(N)}$ denote the columns of the matrix functions f_j $(j = 1, 2)$. The jk entry of the matrix $\int_{\mathbf{R}_B} \big(g(x)f_2(x)\big)^* f_1(x)d\mu(x)$ is the scalar product of $f_1^{(k)} \in B^2_N$ and $gf_2^{(j)} \in B^2_N$. Taking into account the equalities $f_1^{(k)} = \tilde{P}f_1^{(k)}$ and $f_2^{(k)} = \tilde{Q}_0 f_2^{(k)}$ we therefore get

$$\operatorname{tr}\left(\int_{\mathbf{R}_B} \big(g(x)f_2(x)\big)^* f_1(x)\, d\mu(x) \right) = \sum_{k=1}^{N} \big(f_1^{(k)}, gf_2^{(k)}\big) = \sum_{k=1}^{N} \big(f_1^{(k)}, \tilde{P}g\tilde{Q}_0 f_2^{(k)}\big).$$

Obviously, $\|f_j\|_{B^2} < 1$ if and only if $\sum_{k=1}^N \|f_j^{(k)}\|^2_{B^2_N} < 1$. Since the supremum of $\sum_{k=1}^N (f_1^{(k)}, \tilde{P}g\tilde{Q}_0 f_2^{(k)})$ over

$$f_1^{(k)} \in [B^2_+]_N,\ f_2^{(k)} \in [\dot{B}^2_-]_N,\ \sum_{k=1}^N \|f_j^{(k)}\|^2_{B^2_N} < 1 \quad (j = 1, 2)$$

equals $\|\tilde{P}g\tilde{Q}_0\|_{\mathcal{B}(B^2_N)}$, we arrive at the second equality in(20.5). The first equality follows from Theorem 9.3.

Equalities (20.6) can be verified analogously. □

20.3 Matrix Hankel Operators with *AP* Symbols

In Section 7.2, we introduced the Bochner-Fejér operators σ_γ. There we considered these operators on *AP*, while now we study these operators on B^∞. Thus, for $\gamma \in \Gamma := \mathcal{U} \times (0,1)$ and $a \in B^\infty = L^\infty(\mathbf{R}_B, d\mu)$, we put

$$(\sigma_\gamma a)(x) := (q_\gamma * a)(x) = \int_{\mathbf{R}_B} q_\gamma(x-\xi) a(\xi)\, d\mu(\xi),\ x \in \mathbf{R}_B.$$

Clearly, $\sigma_\gamma a$ is an almost periodic polynomial and

$$\|\sigma_\gamma a\|_{B^\infty} \le \|a\|_{B^\infty} \tag{20.12}$$

for all $a \in B^\infty$. Notice also that if $b \in AP_{N\times N}$ then

$$\|b\|_\infty := \|b\|_{L^\infty_{N\times N}(\mathbf{R})} = \|bI\|_{\mathcal{B}(L^2_N(\mathbf{R}))} = \|b\|_{B^\infty}. \tag{20.13}$$

Theorem 20.2. *If $g \in AP_{N\times N}$, then*

$$\|\tilde{P}_0 g\tilde{Q}\|_{\mathcal{B}(B^2_N)} = \|\tilde{P}g\tilde{Q}_0\|_{\mathcal{B}(B^2_N)} = \min_{h\in AP^-_{N\times N}} \|g-h\|_\infty = \inf_{h\in APW^-_{N\times N}} \|g-h\|_\infty, \tag{20.14}$$

$$\|\tilde{Q}_0 g\tilde{P}\|_{\mathcal{B}(B^2_N)} = \|\tilde{Q}g\tilde{P}_0\|_{\mathcal{B}(B^2_N)} = \min_{h\in AP^+_{N\times N}} \|g-h\|_\infty = \inf_{h\in APW^+_{N\times N}} \|g-h\|_\infty. \tag{20.15}$$

Proof. In view of Theorem 20.1, we are left to show that

$$\min\left\{\|g-h\|_{B^\infty} : h \in [B^\infty_\mp]_{N\times N}\right\} = \inf\left\{\|g-h\|_\infty : h \in APW^\mp_{N\times N}\right\}. \tag{20.16}$$

There is a matrix function $f \in [B^\infty_-]_{N\times N}$ such that

$$\operatorname{dist}_{B^\infty}\left(g, [B^\infty_-]_{N\times N}\right) = \|g-f\|_{B^\infty} = \operatorname*{ess\,sup}_{x\in\mathbf{R}_B} s_{\max}\left(g(x)-f(x)\right). \tag{20.17}$$

Put $g_\gamma = \sigma_\gamma g$, $f_\gamma = \sigma_\gamma f$, $s(x) = s_{\max}(f(x) - g(x))$. Clearly, $g_\gamma, f_\gamma \in AP_{N\times N}$. Since $s(x) \geq 0$ and $q_\gamma(x) \geq 0$, we obtain from (20.13), (20.17) and (20.12) that

$$\begin{aligned}
\|g_\gamma - f_\gamma\|_\infty &= \|g_\gamma - f_\gamma\|_{B^\infty} \\
&= \operatorname*{ess\,sup}_{x\in\mathbf{R}_B} s_{\max}\left(\int_{\mathbf{R}_B} q_\gamma(x-\xi)\big(g(\xi) - f(\xi)\big)\,d\mu(\xi)\right) \\
&\leq \operatorname*{ess\,sup}_{x\in\mathbf{R}_B} \int_{\mathbf{R}_B} q_\gamma(x-\xi) s_{\max}\big(g(\xi) - f(\xi)\big)\,d\mu(\xi) \\
&= \operatorname*{ess\,sup}_{x\in\mathbf{R}_B} (\sigma_\gamma s)(x) \leq \operatorname*{ess\,sup}_{x\in\mathbf{R}} s(x) = \|g - f\|_{B^\infty}
\end{aligned} \tag{20.18}$$

and

$$\begin{aligned}
\|g - g_\gamma\|_\infty &= \|g - g_\gamma\|_{B^\infty} = \operatorname*{ess\,sup}_{x\in\mathbf{R}_B} s_{\max}\big(g(x) - g_\gamma(x)\big) \\
&\leq \left(\operatorname*{ess\,sup}_{x\in\mathbf{R}_B} \operatorname{tr}\Big(\big(g(x) - g_\gamma(x)\big)\big(g(x) - g_\gamma(x)\big)^*\Big)\right)^{1/2} \\
&= \left(\operatorname*{ess\,sup}_{x\in\mathbf{R}_B} \sum_{i,j=1}^N |g_{ij}^{(\gamma)}(x)|^2\right)^{1/2} \\
&\leq \left(\sum_{i,j=1}^N \Big(\operatorname*{ess\,sup}_{x\in\mathbf{R}_B} |g_{ij}^{(\gamma)}(x)|\Big)^2\right)^{1/2},
\end{aligned} \tag{20.19}$$

where $g_{ij}^{(\gamma)}(x)$ are the entries of $g(x) - g_\gamma(x)$. We have $g_{ij}^{(\gamma)} \in AP$, and Theorem 7.5(c) implies that $\lim_\Gamma \|g_{ij}^{(\gamma)}\|_{AP} = 0$. Thus, (20.19) gives $\lim_\Gamma \|g - g_\gamma\|_\infty = 0$. Obviously, $AP \cap B_-^\infty = AP^-$. Since f_γ is an almost periodic polynomial and the Bohr-Fourier spectrum $\Omega(f_\gamma)$ is contained in $\Omega(f)$ for every $f \in [B_-^\infty]_{N\times N}$ (similarly to Theorem 7.5(b), where the case of AP functions was considered), we conclude that $f_\gamma \in [APW \cap B_-^\infty]_{N\times N} = APW_{N\times N}^-$. Therefore, by (20.18) and (20.17),

$$\begin{aligned}
\operatorname{dist}_{L^\infty}(g, AP_{N\times N}^-) &\leq \|g - f_\gamma\|_{B^\infty} \leq \|g_\gamma - f_\gamma\|_\infty + \|g - g_\gamma\|_\infty \\
&\leq \|g - f\|_{B^\infty} + \|g - g_\gamma\|_\infty = \operatorname{dist}_{B^\infty}\big(g, [B_-^\infty]_{N\times N}\big) + \|g - g_\gamma\|_\infty.
\end{aligned}$$

Passage to the limit $\lim_\Gamma$ now implies that

$$\operatorname{dist}_{L^\infty}(g, AP_{N\times N}^-) \leq \operatorname{dist}_{B^\infty}\big(g, [B_-^\infty]_{N\times N}\big).$$

The reverse inequality is obvious because $AP_{N\times N}^- \subset [B_-^\infty]_{N\times N}$. This completes the proof of the "$-$" version of (20.16) and thus of (20.14).

The proof of (20.15) is analogous. □

We are now in a position to establish an AP counterpart of Theorem 6.14.

Theorem 20.3. *If $a \in AP_{N\times N}$, then for the Hankel operator $H(a)$ on $L^2_N(\mathbf{R}_+)$ we have*

$$\|H(a)\| = \text{dist}_{L^\infty}(a, AP^-_{N\times N}). \tag{20.20}$$

Proof. Since $H(a) = \chi_+ W^0(a)\chi_- J$ and, by (2.10), $F\chi_+F^{-1} = P$ and $F\chi_-F^{-1} = Q$, we obtain

$$\begin{aligned}\|H(a)\|_{\mathcal{B}(L^2_N(\mathbf{R}_+))} &= \|\chi_+ W^0(a)\chi_- I\|_{\mathcal{B}(L^2_N(\mathbf{R}))}\\ &= \|F\chi_+ W^0(a)\chi_- F^{-1}\|_{\mathcal{B}(L^2_N(\mathbf{R}))} = \|PaQ\|_{\mathcal{B}(L^2_N(\mathbf{R}))}.\end{aligned} \tag{20.21}$$

For $a \in AP_{N\times N}$, we can use Theorem 9.3 to show that

$$\|PaQ\|_{\mathcal{B}(L^2_N(\mathbf{R}))} = \|\tilde{P}a\tilde{Q}_0\|_{\mathcal{B}(B^2_N)}. \tag{20.22}$$

Clearly, (20.20) results from (20.21), (20.22), and Theorem 20.2. □

20.4 Applications to Wiener-Hopf Operators

The following theorem is a consequence of Theorems 6.14, 6.15, and 20.3.

Theorem 20.4. *If $u \in AP_{N\times N}$ is unitary valued, then*

$$\begin{aligned} W(u) \text{ is invertible on } L^2_N(\mathbf{R}_+) &\Leftrightarrow \text{dist}_{L^\infty}(u, GAP^+_{N\times N}) < 1\\ &\Leftrightarrow \text{dist}_{L^\infty}(u, GAP^-_{N\times N}) < 1,\\ W(u) \text{ is left-invertible on } L^2_N(\mathbf{R}_+) &\Leftrightarrow \text{dist}_{L^\infty}(u, AP^+_{N\times N}) < 1,\\ W(u) \text{ is right-invertible on } L^2_N(\mathbf{R}_+) &\Leftrightarrow \text{dist}_{L^\infty}(u, AP^-_{N\times N}) < 1. \end{aligned}$$

□

We finally consider general (and not necessarily unitary valued) symbols in the algebra $APW_{N\times N}$.

Theorem 20.5. *For $a \in APW_{N\times N}$, the following are equivalent:*

(i) *$W(a)$ is invertible on $L^2_N(\mathbf{R}_+)$,*

(ii) *there is an $h \in GAP^+_{N\times N}$ such that $\|I - ah\|_\infty < 1$,*

(iii) *there is an $h \in GAP^-_{N\times N}$ such that $\|I - ha\|_\infty < 1$.*

Proof. If (ii) is satisfied then the operator $W(a)W(h)$ is invertible, because

$$\|I - W(a)W(h)\| = \|W(I - ah)\| = \|I - ah\|_\infty < 1.$$

Since $W(h)$ is invertible, this implies that $W(a)$ is also invertible. Conversely, if (i) holds then $a \in GL^\infty_{N\times N}(\mathbf{R})$ due to Theorem 6.5 and hence $a \in GAPW_{N\times N}$ by

Wieners's theorem. From the last inclusion we obtain that a^*a can be represented in the form $a^*a = b^*b$ with $b \in GAPW^+_{N\times N}$ (see Corollary 9.15(b)). The matrix function $u := ab^{-1} \in APW_{N\times N}$ is unitary valued:

$$u^*u = (b^{-1})^*(a^*a)b^{-1} = (b^{-1})^*(b^*b)b^{-1} = I,$$
$$uu^* = a(b^*b)^{-1}a^* = a(a^*a)^{-1}a^* = I.$$

Since $a = ub$ and $b \in GAPW^+_{N\times N}$, we deduce from (the matricial version of) Proposition 2.17 that $W(a) = W(u)W(b)$ and $W(b) \in G\mathcal{B}(L^2_N(\mathbf{R}_+))$. Consequently, the operators $W(u)$ and $W(u^*)$ are invertible together with $W(a)$. Now Theorem 20.4 tells us that there is a matrix function $c \in GAP^+_{N\times N}$ such that $\|u^* - c\|_\infty < 1$. Put $h = b^{-1}c$. Then $h \in GAP^+_{N\times N}$ and

$$\|I - ah\|_\infty = \|I - ubb^{-1}c\|_\infty = \|I - uc\|_\infty = \|u^* - c\|_\infty < 1,$$

which gives (ii).

The equivalence (i) $\Leftrightarrow$ (iii) follows from the equivalence (i) $\Leftrightarrow$ (ii) by passage to adjoints. □

Notes

Norms of Hankel operators on L^2 spaces were discussed in Chapter 6. A convenient reference for Besicovitch spaces is Chapter 1 of [171]. The transition from Hankel operators on L^2 to Hankel operators with matrix symbols in B^∞ on B^2 and its subspace AP was done in [121]; see also [126] for the statements of the results. Assertions (ii) and (iii) of Theorem 20.5 are equivalent to the existence of a canonical right APW factorization of a. This was proved in [212]. Thus, Theorem 20.5, when combined with Theorem 9.9, leads to an alternative proof of the Fredholm criterion in terms of factorization (Corollary 9.8). This approach is presented in [121].

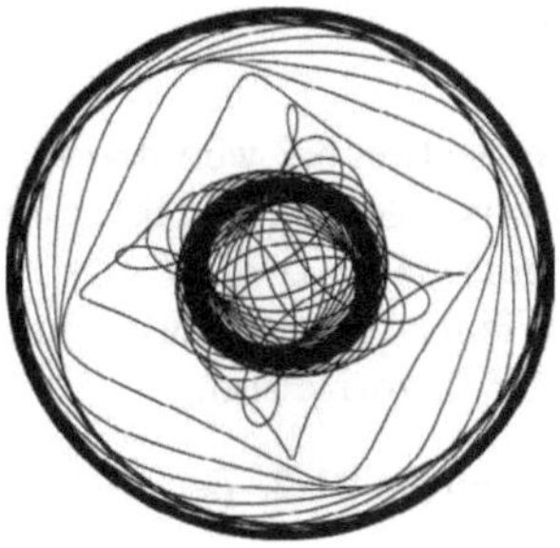

Chapter 21

Generalized AP Factorization

Simonenko's theorem tells us that the invertibility of the Wiener-Hopf operator $W(a)$ on $L^2_N(\mathbf{R}_+)$ is equivalent to the existence of a right Wiener-Hopf factorization of a. If a is continuous, then the factors of the Wiener-Hopf factorization need not be continuous. AP factorization is a factorization with "nice" factors, and we know from Section 9.4 that for $a \in APW_{N\times N}$ the existence of a canonical right AP factorization is equivalent to the invertibility of $W(a)$. In this chapter, we prove the AP analogue of Simonenko's theorem, that is, we introduce the notion of generalized AP factorization and then show that if $a \in AP_{N\times N}$, then $W(a)$ is invertible on $L^2_N(\mathbf{R}_+)$ if and only if a possesses a canonical generalized right AP factorization.

21.1 Definition of Generalized AP Factorization

In order to establish a criterion for the invertibility of $W(a)$ on $L^2_N(\mathbf{R}_+)$ with $a \in AP_{N\times N}$ and to study Fredholmness of $W(a)$ on $L^2_N(\mathbf{R}_+)$ with $a \in SAP_{N\times N}$ it is necessary to generalize the notion of AP factorization.

Definition 21.1. A *generalized right AP factorization* of a matrix function a in $GAP_{N\times N}$ is a representation

$$a = a_- d a_+ \tag{21.1}$$

where $d = \operatorname{diag}(e_{\lambda_1}, \dots, e_{\lambda_N})$ with $\lambda_1, \dots, \lambda_N \in \mathbf{R}$ and

$$a_- \in G[B^2_-]_{N\times N}, \quad a_+ \in G[B^2_+]_{N\times N}, \quad a_-\tilde{P}a_-^{-1}I \in \mathcal{B}(B^2_N). \qquad \square$$

In Section 6.2 we introduced Wiener-Hopf factorization for arbitrary matrix functions $a \in GL^\infty_{N\times N}(\mathbf{R})$. Generalized *AP* factorization is a special notion for matrix functions $a \in GAP_{N\times N} = G[C(\mathbf{R}_B)]_{N\times N}$. The role of the Hardy spaces $H^2_\pm$ in Wiener-Hopf factorization is played by the Besicovitch spaces $B^2_\pm$ in generalized *AP* factorization.

Theorem 21.2. *Let $a \in GAP_{N\times N}$ and suppose*

$$a = a_- d a_+, \quad d = \operatorname{diag}(e_{\lambda_1}, \dots, e_{\lambda_N}), \quad \lambda_1 \ge \dots \ge \lambda_N$$

and

$$a = b_- f b_+, \quad f = \operatorname{diag}(e_{\mu_1}, \dots, e_{\mu_N}), \quad \mu_1 \ge \dots \ge \mu_N$$

are generalized right AP factorizations. Then

$$b_- = a_- d h^{-1} d^{-1}, \quad f = d, \quad b_+ = h a_+$$

where $h = (h_{jk})_{j,k=1}^N$ is a matrix function in $G[B^1_+]_{N\times N}$ with nonzero and constant determinant, $h_{jk} = 0$ if $\lambda_k < \lambda_j$, and $\Omega(h_{jk}) \subset [0, \lambda_k - \lambda_j]$ if $\lambda_k \ge \lambda_j$.

Taking into account that $h_+ = b_+ a_+^{-1} \in [B^1_+]_{N\times N}$ and $h_- = b_-^{-1} a_- \in [B^1_-]_{N\times N}$, this can be proved by the arguments of the proof of Theorem 8.1.

As in Section 8.2, we call the real numbers λ_j the *right AP indices.* Notice that, by Theorem 21.2, the right *AP* indices exist and are well-defined for every matrix function in $GAP_{N\times N}$ which admits a generalized right *AP* factorization.

Corollary 21.3. *If $a \in GAP_{N\times N}$ has a generalized right AP factorization and all right AP indices are equal to each other, then the generalized right AP factorization of a is unique up to the transformation*

$$a_- \mapsto a_- h^{-1}, \quad a_+ \mapsto h a_+$$

with $h \in G\mathbf{C}^{N\times N}$. Moreover, in this case the matrix

$$\mathbf{d}(a) = M(a_-) M(a_+) \in G\mathbf{C}^{N\times N}$$

is well-defined. □

Lemma 21.4. *Suppose $a \in GAP_{N\times N}$ possesses a generalized right AP factorization with the right AP indices λ_j. Then*

(a) *$W(a)$ is invertible on $L^2_N(\mathbf{R}_+)$ if $\lambda_j = 0$ for all j;*

(b) *$W(a)$ is left-invertible but not invertible on $L^2_N(\mathbf{R}_+)$ if $\lambda_j \ge 0$ for all j and $\sum \lambda_j > 0$;*

(c) *$W(a)$ is right-invertible but not invertible on $L^2_N(\mathbf{R}_+)$ if $\lambda_j \le 0$ for all j and $\sum \lambda_j < 0$.*

Proof. The operator $W(a)$ is left (right)-invertible on $L^2_N(\mathbf{R}_+)$ if and only if the operator $FW(a)F^{-1} = PaP$ is left (right)-invertible on $[H^2_+]_N$, which is in turn equivalent to the left (right) invertibility of the operator $\tilde{W}(a) := \tilde{P}a\tilde{P}$ on $[B^2_+]_N$. Indeed, $W(a)$ is invertible from the left (resp. right) if and only if $W^*(a)W(a)$ (resp. $W(a)W^*(a)$) is invertible, and since the map $b \mapsto FbF^{-1}$ is a C^*-algebra isomorphism, we deduce from Theorem 9.3 that the invertibility of $W^*(a)W(a)$ (resp. $W(a)W^*(a)$) is equivalent to the invertibility of $\tilde{W}^*(a)\tilde{W}(a)$ (resp. $\tilde{W}(a)\tilde{W}^*(a)$), that is, to the left (resp. right) invertibility of $\tilde{W}(a)$.

Suppose $\lambda_j \le 0$ for all j. By Definition 21.1 the matrix function a is invertible in $AP_{N\times N}$ and the operator

$$R := a_+^{-1}d^{-1}\tilde{P}a_-^{-1}I = a^{-1}a_-\tilde{P}a_-^{-1}I$$

is bounded on B^2_N. Let $f \in APW^+_N$. Then $a_-^{-1}f \in B^2_N$, $\tilde{P}a_-^{-1}f \in [B^2_+]_N$, $a_+^{-1}d^{-1} \in [B^2_+]_{N\times N}$ and, hence, $Rf = a_+^{-1}d^{-1}\tilde{P}a_-^{-1}f \in [B^1_+]_N$. As $f \in B^2_N$ and $R \in \mathcal{B}(B^2_N)$, it follows that $Rf \in [B^1_+]_N \cap B^2_N = [B^2_+]_N$. Therefore

$$\begin{aligned} \tilde{P}a\tilde{P}Rf &= \tilde{P}a\tilde{P}a_+^{-1}d^{-1}\tilde{P}a_-^{-1}f = \tilde{P}aa_+^{-1}d^{-1}\tilde{P}a_-^{-1}f \\ &= \tilde{P}a_-\tilde{P}a_-^{-1}f = \tilde{P}f - \tilde{P}a_-\tilde{Q}_0a_-^{-1}f = f. \end{aligned} \tag{21.2}$$

The equality (21.2) is true for all $f \in APW^+_N$. Because APW^+_N is dense in $[B^2_+]_N$, equality (21.2) holds for all $f \in [B^2_+]_N$. Consequently, R is a right inverse of $\tilde{P}a\tilde{P}$ on $[B^2_+]_N$.

Analogously one can show that $\tilde{P}a\tilde{P}$ is left-invertible if $\lambda_j \ge 0$ for all j.

Finally, it is not difficult to see that $\tilde{P}a\tilde{P}$ cannot be invertible if all λ_j are nonnegative or nonpositive and at least one λ_j is nonzero. □

21.2 Two Lemmas

We denote by $\tilde{\mathcal{G}}_{N\times N}$ the set of all $a \in AP_{N\times N}$ for which $W(a)$ is invertible on $L^2_N(\mathbf{R}_+)$. Clearly, $\tilde{\mathcal{G}}_{N\times N}$ is an open subset of $AP_{N\times N}$. If $a \in \tilde{\mathcal{G}}_{N\times N}$, then, by Corollary 9.4, the operators

$$X_a := a\tilde{P} + \tilde{Q}_0 \quad \text{and} \quad Y_a := a\tilde{P}_0 + \tilde{Q}$$

are invertible on B^2_N. In what follows, the inverses of these operators are applied to matrices in $B^2_{N\times N}$ column-wise.

Lemma 21.5. *The maps*

$$a \mapsto M(X_a^{-1}I_N) \quad \text{and} \quad a \mapsto M(Y_a^{-1}a)$$

are continuous from $\tilde{\mathcal{G}}_{N\times N}$ *into* $\mathbf{C}^{N\times N}$.

Proof. Fix $c \in \tilde{\mathcal{G}}_{N\times N}$. Then Y_c is invertible on B_N^2. If $a \in AP_{N\times N}$ and $\|a-c\|_\infty < \|Y_c^{-1}\|^{-1}$, then Y_a is also invertible on B_N^2. Since $b := Y_a^{-1}a - Y_c^{-1}c \in B_{N\times N}^2$, we get

$$\begin{aligned} \|M(b)\|_\infty^2 &= \Big(s_{\max}\big(M(b)\big)\Big)^2 \le \operatorname{tr}\big(M(b)M(b^*)\big) \\ &\le \operatorname{tr} M(bb^*) = \operatorname{tr} \int_{\mathbf{R}_B} b(x)b^*(x)\,d\mu(x) \\ &= \int_{\mathbf{R}_B} \operatorname{tr}\big(b(x)b^*(x)\big)\,d\mu(x) = \|b\|_{B^2}^2. \end{aligned}$$

Consequently,

$$\|M(Y_a^{-1}a) - M(Y_c^{-1}c)\|_\infty \le \|(Y_a^{-1}a - Y_c^{-1}c)I_N\|_{B^2}.$$

Let b_{ij} be the entries of b and let b_k stand for the columns of b. We have

$$\begin{aligned} \|b\|_{B^2}^2 &= \int_{\mathbf{R}_B} \operatorname{tr}\big(b(x)b^*(x)\big)\,d\mu(x) \\ &= \int_{\mathbf{R}_B} \sum_{i,j=1}^{N} |b_{ij}(x)|^2 d\mu(x) = \sum_{k=1}^{N} \|b_k(x)\|_{B_N^2}^2, \end{aligned}$$

whence

$$\|(Y_a^{-1}a - Y_c^{-1}c)I_N\|_{B^2}^2 \le N\|Y_a^{-1}aI - Y_c^{-1}cI\|_{\mathcal{B}(B_{NtN}^2)}^2.$$

In summary,

$$\begin{aligned} &\|M(Y_a^{-1}a) - M(Y_c^{-1}c)\|_\infty \le \sqrt{N}\|Y_a^{-1}aI - Y_c^{-1}cI\|_{\mathcal{B}(B_N^2)} \\ &\le \sqrt{N}\big(\|Y_a^{-1} - Y_c^{-1}\|\,\|a\|_\infty + \|Y_c^{-1}\|\,\|a-c\|_\infty\big) \\ &\le \sqrt{N}\left(\frac{\|Y_c^{-1}\|^2\|a-c\|_\infty}{1-\|Y_c^{-1}\|\|a-c\|_\infty}(\|c\|_\infty + \|a-c\|_\infty) + \|Y_c^{-1}\|\,\|a-c\|_\infty\right) \\ &= \sqrt{N}\|a-c\|_\infty\|Y_c^{-1}\|(1+\|Y_c^{-1}\|\,\|c\|_\infty)(1-\|Y_c^{-1}\|\,\|a-c\|_\infty)^{-1}, \end{aligned}$$

which proves the continuity of the map $a \mapsto M(Y_a^{-1}a)$.

It can be shown similarly that the map $a \mapsto M(X_a^{-1}I_N)$ is also continuous. □

Lemma 21.6. *For every $a \in \tilde{\mathcal{G}}_{N\times N}$,*

$$M(X_a^{-1}I_N)M(Y_a^{-1}a) = M(Y_a^{-1}a)M(X_a^{-1}I_N) = I_N.$$

Proof. Choose $a_k \in APW_{N\times N}$ so that $\lim_{k\to\infty} \|a_k - a\|_\infty = 0$. The operators X_{a_k} and Y_{a_k} are invertible on B^2_N for all sufficiently large k, and Corollary 9.8 implies that $a_k = a_k^- a_k^+$ with $a_k^\pm \in GAPW^\pm_{N\times N}$ for these k. In particular, the matrices $\mathbf{d}(a_k) = M(a_k^-)M(a_k^+)$ are well-defined and invertible for all k large enough. We have

$$X_{a_k}^{-1} = (a_k^+)^{-1}\tilde{P}(a_k^-)^{-1}I + a_k^-\tilde{Q}_0(a_k^-)^{-1}I,$$
$$Y_{a_k}^{-1} = (a_k^+)^{-1}\tilde{P}_0(a_k^-)^{-1}I + a_k^-\tilde{Q}(a_k^-)^{-1}I,$$
$$M\big(a_k^-\tilde{Q}_0(a_k^-)^{-1}\big) = M\big((a_k^+)^{-1}\tilde{P}_0 a_k^+\big) = 0,$$

and thus,

$$M(X_{a_k}^{-1}I_N) = M\big((a_k^+)^{-1}\tilde{P}(a_k^-)^{-1}\big) = M(a_k^+)^{-1}M(a_k^-)^{-1} = \mathbf{d}(a_k)^{-1},$$
$$M(Y_{a_k}^{-1}a_k) = M(a_k^-\tilde{Q}a_k^+) = M(a_k^-)M(a_k^+) = \mathbf{d}(a_k).$$

This proves the assertion with a replaced by a_k. For general $a \in \tilde{\mathcal{G}}_{N\times N}$ the assertion now follows from Lemma 21.5. □

21.3 Invertibility of Matrix Operators with AP Symbols

A generalized right AP factorization will be called a *canonical generalized right AP factorization* if the right AP indices are all zero. Here is the main result of this section (compare with Corollary 9.8).

Theorem 21.7. *Let $a \in AP_{N\times N}$. Then a has a canonical generalized right AP factorization if and only if the operator $W(a)$ is invertible on $L^2_N(\mathbf{R}_+)$.*

Proof. The "only if" part is immediate from Lemma 21.4. We are therefore left with proving the "if" portion.

Suppose $W(a)$ is invertible. By Corollary 9.4 and the proof of Lemma 9.7, the operators $X_a = a\tilde{P} + \tilde{Q}_0$ and $Z_a = a^T\tilde{Q} + \tilde{P}_0$ are invertible on B^2_N. Consequently (again recall the proof of Lemma 9.7), the equations

$$(a\tilde{P} + \tilde{Q}_0)f = I_N,\ (a^T\tilde{Q} + \tilde{P}_0)g = I_N$$

have solutions $f, g \in B^2_{N\times N}$. Put

$$f_+ = \tilde{P}f,\quad f_- = I_N - \tilde{Q}_0 f,\quad g_-^T = \tilde{Q}g,\ g_+^T = I_N - \tilde{P}_0 g.$$

Then

$$af_+ = f_-,\quad g_-a = g_+, \tag{21.3}$$

whence

$$g_- f_- = g_-(af_+) = (g_- a) f_+ = g_+ f_+. \tag{21.4}$$

Since $g_- f_- \in [B^1_-]_{N\times N}$, $g_+ f_+ \in [B^1_+]_{N\times N}$, and $B^1_- \cap B^1_+ = \mathbf{C}$, we infer from (21.4) that

$$g_- f_- = g_+ f_+ = c \in \mathbf{C}^{N\times N}. \tag{21.5}$$

Taking into consideration that $M(g_+) = I_N$, we obtain

$$c = M(g_+) M(f_+) = M(f_+) = M(\tilde{P} f) = M(f) = M(X_a^{-1} I_N). \tag{21.6}$$

This in conjunction with Lemma 21.6 shows that $\det c \neq 0$, and (21.5) therefore yields that $f_\pm, g_\pm \in G[B^2_\pm]_{N\times N}$. From (21.3) we now get $a = a_- a_+$ with $a_- = f_- \in G[B^2_-]_{N\times N}$ and $a_+ = f_+^{-1} \in G[B^2_+]_{N\times N}$.

It remains to show that the operator $a_- \tilde{P} a_-^{-1} I$ is bounded on B^2_N. Without loss of generality assume that $\|a\|_\infty < 1$ (otherwise we can multiply X_a by $\gamma \tilde{P} + \tilde{Q}_0$ with an appropriate constant $\gamma \in \mathbf{C} \setminus \{0\}$). Set

$$R = a_+^{-1} \tilde{P} a_-^{-1} I + a_- \tilde{Q}_0 a_-^{-1} I.$$

We claim that

$$X_a R f = f \text{ for all } f \in B^2_N. \tag{21.7}$$

Let first $f = a_- f_0$ with $f_0 \in APW_N$. Then $f \in B^2_N$ and

$$\tilde{P} a_-^{-1} f \in APW_N^+, \quad \tilde{Q}_0 a_-^{-1} f \in [APW_0^-]_N,$$
$$a_+^{-1} \tilde{P} a_-^{-1} f \in [B^2_+]_N, \quad a_- \tilde{Q}_0 a_-^{-1} f \in [\dot{B}^2_-]_N.$$

This gives (21.7) for all $f \in a_- APW_N$. Thus, (21.7) will follow once we have shown that $a_- APW_N$ is dense in B^2_N.

Take any $\psi \in B^2_N \cap (a_- APW_N)^\perp$. For each column a_k^- of $a_- \in [B^2_-]_{N\times N}$ and every $\lambda \in \mathbf{R}$, we have

$$M(\psi^* a_k^- e_\lambda) = 0. \tag{21.8}$$

This implies that $\Omega(\psi^* a_k^-) = \emptyset$. Let σ_γ $(\gamma \in \Gamma)$ be the Bochner-Fejér operators (recall Section 7.2):

$$(\sigma_\gamma \varphi)(x) = (q_\gamma * \varphi)(x), \quad \varphi \in AP, \ x \in \mathbf{R}_B.$$

Since

$$\|q_\gamma * \varphi\|_{B^p} \leq \|q_\gamma\|_{B^1} \|\varphi\|_{B^p}$$

and $\|q_\gamma\|_{B^1} = M(q_\gamma) = 1$, we see that the operators σ_γ extend to bounded operators of norm at most 1 on B^p for every $p \in [1,\infty)$. Given $\varphi \in B^p$, choose $\varphi_k \in AP$ so that $\|\varphi - \varphi_k\|_{B^p} \to 0$. Because

$$\|\varphi_k\|_{B^p} \le \|\varphi_k\|_{AP}$$

and

$$\begin{aligned}\|\sigma_\gamma\varphi - \varphi\|_{B^p} &\le \|\sigma_\gamma(\varphi - \varphi_k)\|_{B^p} + \|\sigma_\gamma\varphi_k - \varphi_k\|_{B^p} + \|\varphi - \varphi_k\|_{B^p} \\ &\le 2\|\varphi - \varphi_k\|_{B^p} + \|\sigma_\gamma\varphi_k - \varphi_k\|_{AP},\end{aligned}$$

we deduce from Theorem 7.5(c) that $\lim_\Gamma \sigma_\gamma\varphi = \varphi$. As in the proof of Theorem 7.5(b) one can verify that $\Omega(\sigma_\gamma\varphi) \subset \Omega(\varphi)$. Thus, since $\psi^* a_k^- \in B^1$ and $\Omega(\psi^* a_k^-) = \emptyset$, we obtain that $\Omega(\sigma_\gamma(\psi^* a_k^-)) = \emptyset$ for all $\gamma \in \Gamma$. The almost periodic polynomial $\sigma_\gamma(\psi^* a_k^-)$ therefore vanishes identically for all $\gamma \in \Gamma$, and because $\lim_\Gamma \sigma_\gamma(\psi^* a_k^-) = \psi^* a_k^-$, we arrive at the conclusion that $\psi^* a_k^- = 0$ for all k. Hence $\psi^* a_- = 0$, and since a_- is invertible, it results that $\psi = 0$. This proves that $a_- APW_N$ is dense in B_N^2 and completes the proof of our claim (21.7).

We have shown that $X_a R = I$. Since X_a is invertible, we obtain that R must be bounded on B_N^2. Consequently, the operator

$$aR - aI = (I_N - a)a_-\tilde{P}a_-^{-1}I$$

is also bounded on B_N^2. As $\|a\|_\infty < 1$, we finally get the boundedness of $a_-\tilde{P}a_-^{-1}I$ on B_N^2. □

We note that (21.6) and Lemma 21.6 give

$$\begin{aligned}\mathbf{d}(a) &= M(a_-)M(a_+) = M(f_-)M(f_+^{-1}) = M(f_+^{-1}) \\ &= \left(M(X_a^{-1}I_N)\right)^{-1} = M(Y_a^{-1}a).\end{aligned} \tag{21.9}$$

Hence Theorem 21.7 and Lemma 21.5 imply the following.

Corollary 21.8. *The map $a \mapsto \mathbf{d}(a)$ given by (21.9) is continuous from $\tilde{\mathcal{G}}_{N\times N}$ onto $G\mathbf{C}^{N\times N}$.* □

Remark 21.9. Theorems 6.5 and 21.7 show that if $a \in AP_{N\times N}$ $(\subset L^\infty_{N\times N}(\mathbf{R}))$, then for $W(a)$ to be invertible on $L_N^2(\mathbf{R}_+)$ it is necessary and sufficient that, on the one hand,

$$a = a_- a_+$$

where $a_-^{\pm 1} \in [H_-^2(-1)]_{N\times N}$, $a_+^{\pm 1} \in [H_+^2(-1)]_{N\times N}$, and the operator $a_- S a_-^{-1} I$ is bounded on $L_N^2(\mathbf{R})$, and that, on the other hand,

$$a = b_- b_+$$

where $b_\pm \in G[B^2_\pm]_{N\times N}$ and $b_-\tilde{P}b_-^{-1}I$ is bounded on B^2_N. If $a \in APW_{N\times N}$, then the canonical right APW factorization $a = a_-a_+$ is simultaneously a canonical generalized right AP factorization and a canonical right Wiener-Hopf factorization (notice that $APW^\pm \subset B^2_\pm \cap H^2_\pm(-1)$). We do not know whether the last two factorizations coincide for $a \in AP_{N\times N} \setminus APW_{N\times N}$. □

By Theorem 21.7, the set $\tilde{\mathcal{G}}_{N\times N}$ is characterized as the collection of all matrix function $a \in GAP_{N\times N}$ which admit a canonical generalized right AP factorization.

Here is a straightforward consequence of Theorem 21.7 (compare with Proposition 8.23).

Corollary 21.10. *If $a \in GAP_{N\times N}$ has a canonical generalized right AP factorization, then all matrix functions in a sufficiently small $AP_{N\times N}$ neighborhood of a also admit a canonical generalized right AP factorization.* □

A check of the proof of Theorem 21.7 also shows that, under the normalization $M(a_-) = I_N$, the factors a_- and a_+ are stable in the topology of $B^2_{N\times N}$.

We finally remark that the sole invertibility of a matrix function in $AP_{N\times N}$ does not yet guarantee the existence of a generalized right AP factorization. This follows from Theorem 21.7, Corollary 9.8 and Theorem 8.17.

21.4 And Once More Matrix Operators with SAP Symbols

Armed with Theorem 21.7 and Corollary 21.8, we can reformulate the result of Section 18.8 as follows.

Theorem 21.11. *Let $a \in SAP_{N\times N}$. Then the operator $W(a)$ is Fredholm on $L^2_N(\mathbf{R}_+)$ if and only if*

(a) $a \in GSAP_{N\times N}$,

(b) *the almost periodic representatives a_l and a_r of a have a canonical generalized right AP factorization,*

(c) $\operatorname{sp}\left(\mathbf{d}^{-1}(a_r)\mathbf{d}(a_l)\right) \cap (-\infty, 0] = \emptyset$,

where $\operatorname{sp}\left(\mathbf{d}^{-1}(a_r)\mathbf{d}(a_l)\right)$ *stands for the set of the eigenvalues of the matrix* $\mathbf{d}^{-1}(a_r)\mathbf{d}(a_l) := (\mathbf{d}(a_r))^{-1}\mathbf{d}(a_l)$ *and* $\mathbf{d}(a_r)$, $\mathbf{d}(a_l)$ *are given by* (21.9).

If $W(a)$ is Fredholm, then $\operatorname{Ind} W(a)$ *is given by formulas* (10.31), (10.32). □

Combining Lemma 9.17 and Theorem 21.11 we get the following criterion.

Corollary 21.12. *If $a \in SAP_{N\times N}$, then for $W(a)$ to be n-normal (resp. d-normal) on $L^2_N(\mathbf{R}_+)$ it is necessary and sufficient that*

(a) $a \in GSAP_{N\times N}$,

(b) *the matrix functions*

$$\hat{a}_l = \begin{pmatrix} I_N & a_l \\ a_l^* & 0 \end{pmatrix}, \quad \hat{a}_r = \begin{pmatrix} I_N & a_r \\ a_r^* & 0 \end{pmatrix} \quad (\in GAP_{2N\times 2N})$$

$$\left(\textit{resp.}\quad \hat{a}_l = \begin{pmatrix} I_N & a_l^* \\ a_l & 0 \end{pmatrix}, \quad \hat{a}_r = \begin{pmatrix} I_N & a_r^* \\ a_r & 0 \end{pmatrix} \quad (\in GAP_{2N\times 2N})\right)$$

have canonical generalized right AP factorizations,

(c)
$$\mathrm{sp}\,\left(\mathbf{d}^{-1}(\hat{a}_r)\mathbf{d}(\hat{a}_l)\right) \cap (-\infty, 0] = \emptyset. \qquad \square$$

Here is an immediate consequence of Theorems 21.11 and 1.25.

Corollary 21.13. *If $a \in SAP_{N\times N}$ and $\lambda \in (0,\infty)$, then the operator $W_\lambda(a)$ is Fredholm on $L^2_N(0,\lambda)$ if and only if the matrix functions*

$$H_{a_l}(x) = \begin{pmatrix} e^{-i\lambda x} I_N & 0 \\ a_l(x) & e^{i\lambda x} I_N \end{pmatrix}, \quad H_{a_r}(x) = \begin{pmatrix} e^{-i\lambda x} I_N & 0 \\ a_r(x) & e^{i\lambda x} I_N \end{pmatrix}$$

have a canonical generalized right AP factorization and

$$\mathrm{sp}\,\left(\mathbf{d}^{-1}(H_{a_r})\mathbf{d}(H_{a_l})\right) \cap (-\infty, 0] = \emptyset.$$

In that case

$$\mathrm{Ind}\, W_\lambda(a) = -\sum_{k=1}^{2N} \left(\frac{1}{2} - \left\{\frac{1}{2} - \frac{1}{2\pi}\arg \xi_k\right\}\right),$$

where $\xi_1, \ldots, \xi_{2N} \in \mathbf{C}\setminus(-\infty, 0]$ are the eigenvalues of the matrix $\mathbf{d}^{-1}(H_{a_r})\mathbf{d}(H_{a_l})$ and $\{\cdot\}$ stands for the fractional part. Provided the arguments $\arg \xi_k$ are chosen in $(-\pi, \pi)$, we have

$$\mathrm{Ind}\, W_\lambda(a) = -\frac{1}{2\pi}\sum_{k=1}^{2N} \arg \xi_k. \qquad \square$$

Theorem 21.14. *Let $a \in SAP_{N\times N}$ and $\lambda \in (0,\infty)$. Then the operator $W_\lambda(a)$ is n-normal or d-normal on $L^2_N(0,\lambda)$ if and only if it is Fredholm.*

Proof. For the sake of definiteness, assume $W_\lambda(a)$ is n-normal. Put

$$H_a(x) = \begin{pmatrix} e^{-i\lambda x} I_N & 0 \\ a(x) & e^{i\lambda x} I_N \end{pmatrix}, \quad x \in \mathbf{R}.$$

By Theorem 1.25, the operator $W(H_a)$ is n-normal on $L^2_{2N}(\mathbf{R}_+)$. Corollary 10.5 therefore implies that $H_{a_l}, H_{a_r} \in GAP_{2N\times 2N}$ and that the operators $W(H_{a_l})$ and

$W(H_{a_r})$ are left-invertible. Since $\det H_{a_l} = \det H_{a_r} = 1$, we deduce from Corollary 10.8 that $W(H_{a_l})$ and $W(H_{a_r})$ are actually invertible.

We can slightly perturb a in $AP_{N\times N}$ so that $W(H_a)$ remains n-normal, $W(H_{a_l})$ and $W(H_{a_r})$ remain invertible, $\mathbf{d}(H_{a_l})$, $\mathbf{d}(H_{a_l})$ are not changed, and $\tilde{a}_l := H_{a_l}$, $\tilde{a}_r := H_{a_r}$ become almost periodic matrix polynomials that have canonical right APW factorizations $\tilde{a}_l = \tilde{a}_l^- \tilde{a}_l^+$ and $\tilde{a}_r = \tilde{a}_r^- \tilde{a}_r^+$ with $\tilde{a}_l^\pm$, $\tilde{a}_r^\pm$ in $AP^0_{2N\times 2N}$. Choose $\mu > 0$ so that

$$\Omega(\tilde{a}_l^-) \cap (-2\mu, 0) = \emptyset, \quad \Omega(\tilde{a}_l^+) \cap (0, 2\mu) = \emptyset,$$
$$\Omega(\tilde{a}_r^-) \cap (-2\mu, 0) = \emptyset, \quad \Omega(\tilde{a}_r^+) \cap (0, 2\mu) = \emptyset,$$

let $\tilde{f}_l, \tilde{f}_r \in [BV \cap C]_{2N\times 2N}$ be any matrix function such that $\operatorname{supp}\tilde{f}_l \subset [-\mu, 0]$, $\operatorname{supp}\tilde{f}_r \subset [-\mu, 0]$, and

$$\mathbf{d}(\tilde{a}_l)\hat{\chi}_- + I_{2N}\hat{\chi}_+ + \tilde{f}_l \in G[C(\overline{\mathbf{R}})]_{2N\times 2N},$$
$$\mathbf{d}(\tilde{a}_r)\hat{\chi}_- + I_{2N}\hat{\chi}_+ + \tilde{f}_r \in G[C(\overline{\mathbf{R}})]_{2N\times 2N}.$$

We consider the operator

$$\widetilde{R} := \tilde{a}\hat{P} + \hat{Q} + (1-u)W^0(\tilde{f}_l) + uW^0(\tilde{f}_r),$$

where $\tilde{a} := H_a$. By Lemma 18.16, the operator $\tilde{R}$ is Fredholm on $L^2_{2N}(\mathbf{R})$. As is known, the operator $W(H_a)$ is n-normal on $L^2_{2N}(\mathbf{R}_+)$ if and only the operator $R_{\tilde{a}} := \tilde{a}P + Q$ is n-normal and hence left-invertible on $L^2_{2N}(\mathbf{R})$. Put $\widetilde{V} := \widetilde{R}^{(-1)} R_{\tilde{a}}$ where $\widetilde{R}^{(-1)}$ is a regularizer of $\widetilde{R}$. Then $\widetilde{V}$ is left-invertible on $L^2_{2N}(\mathbf{R})$, and hence the coset $\widetilde{V} + J_\infty$ is left-invertible in Λ / J_∞, where Λ and J_∞ are defined in Section 18.7. Let

$$\widetilde{R}_{\tilde{a}} + J_\infty := \Big(\mathbf{d}(\tilde{a}_l)(1-u) + \mathbf{d}(\tilde{a}_r)u\Big)P + Q + J_\infty. \tag{21.10}$$

According to the proof of Lemma 18.19 (see (18.59), (18.64), (18.75), (18.76), and (18.77)), the cosets

$$\widetilde{V} + J_\infty = (\widetilde{R} + J_\infty)^{-1}(R_{\tilde{a}} + J_\infty)$$

and

$$\widetilde{V}_0 + J_\infty := (\widetilde{R} + J_\infty)^{-1}(\widetilde{R}_{\tilde{a}} + J_\infty)$$

are left-invertible only simultaneously. This implies that the cosets $\widetilde{V}_0 + J_\infty$ and hence (21.10) are also left-invertible. Then by Theorem 5.9(a),

$$\operatorname{sp}\Big(\mathbf{d}^{-1}(H_{a_r})\mathbf{d}(H_{a_l})\Big) \cap (-\infty, 0] = \emptyset.$$

Finally, from Corollary 21.13 we infer that $W(a)$ is Fredholm on $L^2_N(0,\lambda)$. □

Notes

Generalized AP factorization plays the same role for AP matrix functions as Wiener-Hopf factorization in L^p spaces plays for continuous functions. It is used as the next best thing to the factorization within the algebra AP when the latter does not exist. The notion of generalized AP factorization was introduced in [120], [122]. Except for Corollary 21.13 and Theorem 21.14, all results of this chapter were announced in these papers; full proofs were so far available only in [121]. Corollary 21.13 was established in [121]. Theorem 21.14 is new.

Chapter 22

Canonical Wiener-Hopf Factorization via Corona Problems

Explicit Wiener-Hopf factorization of matrix functions is a white whale of the theory of convolution type equations. Enormous effort has been spent on it, but the cases when it has been found are still scarce. The Portuguese transformation and its applications (Chapters 13, 14) can be thought of as results in this direction for triangular AP matrix functions based on the solution of certain corona problems. As it happens, there is an alternative approach to explicit factorization, also using corona problems, which allows us to consider more general 2×2 *matrix functions (not necessarily triangular and not necessarily AP). Assuming the existence of a non-trivial solution of the corresponding homogeneous Riemann-Hilbert problem in* $[H^\infty_\pm]_2$*, we get criteria for the existence of a canonical left* WH *factorization* $G = G_+G_-$ *in* $L^p(\mathbf{R}, w)$ $(1 < p < \infty,\ w \in A_p(\mathbf{R}))$ *and a bounded canonical left* WH *factorization in case* $G \in L^\infty_{2\times 2}(\mathbf{R})$ *and* $\det G = 1$*, as well as explicit formulas for* $G_\pm$ *in terms of the solutions of two associated corona problems (in the half-planes* $\mathbf{C}_+$ *and* $\mathbf{C}_-$*). A separation principle for constructing corona solutions from simpler ones is developed and corona solutions for a number of corona problems in* H^∞_+ *are obtained. Making use of these results we explicitly construct canonical left* WH *factorizations for triangular* 2×2 *matrix functions with diagonal entries* $e_{\pm\lambda}$ *and certain classes of* L^∞ *off diagonal entries.*

22.1 Canonical Left *WH* Factorization and Corona Problems

Throughout this section we suppose that $1 < p < \infty$ and $w \in A_p(\mathbf{R})$. We study the relationship between the existence of a canonical left WH factorization in $L^p(\mathbf{R}, w)$ for a matrix function $G \in L^\infty_{2\times 2}(\mathbf{R})$ with non-zero constant determinant and the solvability of two associated corona problems (in $\mathbf{C}_+$ and $\mathbf{C}_-$). Below, without loss of generality, we always assume $\det G = 1$. The solutions of the corona problems allow us to reduce the corresponding Riemann-Hilbert problem to a simpler one with a triangular matrix coefficient. Provided that there exists a non-trivial solution of the homogeneous Riemann-Hilbert problem in $[H^\infty_\mp]_2$, we get criteria for the existence of a canonical left WH factorization of G under some assumptions imposed below. We also show that the solvability of the corona problems mentioned above is necessary for the existence of a bounded canonical left WH factorization of G (in contrast to the case of canonical left WH factorization in $L^p(\mathbf{R}, w)$).

In this chapter we always identify functions belonging to Hardy spaces with their non-tangential limits on $\mathbf{R}$.

Definition 22.1. A matrix function $G \in L^\infty_{N\times N}(\mathbf{R})$ is said to admit a *bounded canonical left WH factorization* if

$$G = G_+ G_-$$

with $G_+^{\pm 1} \in [H^\infty_+]_{N\times N}$, $G_-^{\pm 1} \in [H^\infty_-]_{N\times N}$. □

According to Theorems 16.3(b) and 16.5, the operator

$$T_G = P + GQ \tag{22.1}$$

with coefficient $G \in L^\infty_{N\times N}(\mathbf{R})$ is invertible on $L^p_N(\mathbf{R}, w)$ ($1 < p < \infty$, $w \in A_p(\mathbf{R})$) if and only if G admits a canonical left WH factorization in $L^p(\mathbf{R}, w)$.

Usually, the operator T_G is considered together with the so-called Riemann-Hilbert (or, in some sources, just Riemann) boundary value problem in the Hardy space $[H^p_\pm(\mathbf{R}, w)]_N$ ($1 < p < \infty$, $w \in A_p(\mathbf{R})$): find two vector functions $\Phi^\pm$ in the spaces $[H^p_\pm(\mathbf{R}, w)]_N$ that satisfy the boundary condition

$$\Phi^+ = G\Phi^- + u, \tag{22.2}$$

on the real axis $\mathbf{R}$, where the vector function $u \in L^p_N(\mathbf{R}, w)$ is given. As is well-known (see, e.g., [37, p. 172] or [155, p. 46]), the operator T_G given by (22.1) is invertible on the space $L^p_N(\mathbf{R}, w)$ if and only if the corresponding Riemann-Hilbert boundary value problem (22.2) has a unique solution in $[H^p_\pm(\mathbf{R}, w)]_N$ for every $u \in L^p_N(\mathbf{R}, w)$.

The explicit factorization formula for triangular matrix functions with factorable diagonal entries (see, e.g., [155, Theorem 4.6]) implies the following simple result.

Lemma 22.2. *If $\varphi \in L^\infty(\mathbf{R})$, then the representation*

$$G := \begin{pmatrix} 1 & 0 \\ \varphi & 1 \end{pmatrix} = \begin{pmatrix} 1 & 0 \\ \widetilde{P}_+\varphi & 1 \end{pmatrix} \begin{pmatrix} 1 & 0 \\ \widetilde{P}_-\varphi & 1 \end{pmatrix},$$

where

$$(\widetilde{P}_\pm \varphi)(x) = \frac{\varphi(x)}{2} \pm \frac{x+i}{2\pi i} \int_{\mathbf{R}} \frac{(\tau+i)^{-1}\varphi(\tau)}{\tau - x}\, d\tau, \quad x \in \mathbf{R}, \tag{22.3}$$

is a canonical left WH factorization of the matrix function G in every space $L^p(\mathbf{R}, w)$ with $1 < p < \infty$, $w \in A_p(\mathbf{R})$.

Now consider an arbitrary matrix function

$$G = \begin{pmatrix} a & b \\ c & d \end{pmatrix} \tag{22.4}$$

with entries $a, b, c, d \in L^\infty(\mathbf{R})$ such that

$$\det G = ad - bc = 1. \tag{22.5}$$

We will study the canonical left WH factorability of the matrix function (22.4) under the following assumption (**A**):

(**A**) *there exist functions $f_+, g_+ \in H^\infty_+$ and $f_-, g_- \in H^\infty_-$ such that*

$$\begin{pmatrix} g_+ \\ f_+ \end{pmatrix} = \begin{pmatrix} a & b \\ c & d \end{pmatrix} \begin{pmatrix} g_- \\ f_- \end{pmatrix}. \tag{22.6}$$

In addition to (**A**), we require that

$$\inf_{z \in \mathbf{C}_+} \max\{|f_+(z)|, |g_+(z)|\} > 0, \tag{22.7}$$

$$\inf_{z \in \mathbf{C}_-} \max\{|f_-(z)|, |g_-(z)|\} > 0, \tag{22.8}$$

where $\mathbf{C}_\pm = \{z \in \mathbf{C} : \pm \operatorname{Im} z > 0\}$.

Theorem 12.11 implies that under conditions (22.7) and (22.8) the corona problems

$$f_+ \widetilde{f}_+ + g_+ \widetilde{g}_+ = 1, \quad f_- \widetilde{f}_- + g_- \widetilde{g}_- = 1 \tag{22.9}$$

have solutions $\widetilde{f}_+, \widetilde{g}_+ \in H^\infty_+$ and $\widetilde{f}_-, \widetilde{g}_- \in H^\infty_-$, respectively.

Theorem 22.3. *Under the assumptions* (**A**), (22.7), (22.8) *the matrix function* (22.4) *admits a canonical left WH factorization $G = G_+ G_-$ in every $L^p(\mathbf{R}, w)$* ($1 < p < \infty$, *w in $A_p(\mathbf{R})$*). *Here*

$$G_+ = \begin{pmatrix} \widetilde{f}_+ + g_+ F_+ & g_+ \\ -\widetilde{g}_+ + f_+ F_+ & f_+ \end{pmatrix}, \tag{22.10}$$

$$G_- = \begin{pmatrix} f_- & -g_- \\ \widetilde{g}_- + f_- F_- & \widetilde{f}_- - g_- F_- \end{pmatrix}, \tag{22.11}$$

where $\widetilde{f}_\pm, \widetilde{g}_\pm \in H^\infty_\pm$ are solutions of the corona problems (22.9) *and*

$$F_\pm = \widetilde{P}_\pm(\widetilde{g}_+ a \widetilde{f}_- - \widetilde{g}_+ b \widetilde{g}_- + \widetilde{f}_+ c \widetilde{f}_- - \widetilde{f}_+ d \widetilde{g}_-). \tag{22.12}$$

We have $F_\pm \in H^p_\pm(\mathbf{R}, w)$ for every $1 < p < \infty$ and every $w \in A_p(\mathbf{R})$.

Proof. If assumptions (**A**), (22.7), (22.8) are satisfied, then G admits a left canonical WH factorization in $L^p(\mathbf{R}, w)$ exactly if the matrix function $B_+ G B_-^{-1}$ with

$$B_\pm = \begin{pmatrix} f_\pm & -g_\pm \\ \widetilde{g}_\pm & \widetilde{f}_\pm \end{pmatrix}, \quad B_\pm^{-1} = \begin{pmatrix} \widetilde{f}_\pm & g_\pm \\ -\widetilde{g}_\pm & f_\pm \end{pmatrix}. \tag{22.13}$$

has such a factorization. By virtue of assumption (**A**) and (22.9), we have

$$\begin{aligned} B_+ G B_-^{-1} &= \begin{pmatrix} f_+ & -g_+ \\ \widetilde{g}_+ & \widetilde{f}_+ \end{pmatrix} \begin{pmatrix} a & b \\ c & d \end{pmatrix} \begin{pmatrix} \widetilde{f}_- & g_- \\ -\widetilde{g}_- & f_- \end{pmatrix} \\ &= \begin{pmatrix} f_+ & -g_+ \\ \widetilde{g}_+ & \widetilde{f}_+ \end{pmatrix} \begin{pmatrix} a \widetilde{f}_- - b \widetilde{g}_- & g_+ \\ c \widetilde{f}_- - d \widetilde{g}_- & f_+ \end{pmatrix} = \begin{pmatrix} y & 0 \\ F & 1 \end{pmatrix} \end{aligned}$$

where

$$F = \widetilde{g}_+ a \widetilde{f}_- - \widetilde{g}_+ b \widetilde{g}_- + \widetilde{f}_+ c \widetilde{f}_- - \widetilde{f}_+ d \widetilde{g}_- \tag{22.14}$$

and

$$y = (f_+ a - g_+ c) \widetilde{f}_- + (-f_+ b + g_+ d) \widetilde{g}_-.$$

On the other hand, $y = \det\, (B_+ \, G \, B_-^{-1}) = 1$ because $\det\, G = \det\, B_\pm = 1$ due to (22.5) and (22.9). Therefore

$$B_+ G B_-^{-1} = \begin{pmatrix} 1 & 0 \\ F & 1 \end{pmatrix}. \tag{22.15}$$

Since $F \in L^\infty(\mathbf{R})$, it follows from Lemma 22.2 that

$$\begin{pmatrix} 1 & 0 \\ F & 1 \end{pmatrix} = \begin{pmatrix} 1 & 0 \\ F_+ & 1 \end{pmatrix} \begin{pmatrix} 1 & 0 \\ F_- & 1 \end{pmatrix}$$

is a canonical left WH factorization in every $L^p(\mathbf{R}, w)$. Thus,

$$G_+ = B_+^{-1} \begin{pmatrix} 1 & 0 \\ F_+ & 1 \end{pmatrix}, \quad G_- = \begin{pmatrix} 1 & 0 \\ F_- & 1 \end{pmatrix} B_- \tag{22.16}$$

are factors of a canonical left WH factorization of G, $G = G_+ G_-$, in every $L^p(\mathbf{R}, w)$. It is easily seen that (22.16) coincides with (22.10), (22.11). □

Conditions (22.7) and (22.8) are in general not necessary for G to have a canonical left WH factorization: $f_+(z)$ and $g_+(z)$ (or $f_-(z)$ and $g_-(z)$) can approach zero simultaneously as z tends to $\mathbf{R}$; see [20]. The next result shows that joint zeros in $\mathbf{C}_+$ (or $\mathbf{C}_-$) are not possible.

Lemma 22.4. *Under assumption* (**A**), *the homogeneous Riemann-Hilbert problem* (22.2) *has a non-trivial solution in* $H^p_\pm(\mathbf{R}, w)$ *if*
(1) *the functions* f_+ *and* g_+ *have a common zero in* $\mathbf{C}_+$, *or*
(2) *the functions* f_- *and* g_- *have a common zero in* $\mathbf{C}_-$.

Proof. Let f_+ and g_+ have a common zero $z_0 \in \mathbf{C}_+$. Then by (22.6),

$$\begin{pmatrix} \Phi_1^+ \\ \Phi_2^+ \end{pmatrix} = \begin{pmatrix} a & b \\ c & d \end{pmatrix} \begin{pmatrix} \Phi_1^- \\ \Phi_2^- \end{pmatrix}$$

where $\Phi_1^\pm(z) = (z - z_0)^{-1} g_\pm(z)$, $\Phi_2^\pm(z) = (z - z_0)^{-1} f_\pm(z)$. It remains to observe that $\Phi_1^\pm, \Phi_2^\pm \in H^p_\pm(\mathbf{R}, w)$. Case (2) can be treated analogously. □

We now introduce an additional condition (**B**). Under this condition, (22.7) and (22.8) become necessary for the operator T_G to be invertible. Namely, assume that

(**B**) *either there exist functions* $\varphi_+, \psi_+ \in H^\infty_+$ *such that*

$$\inf_{z \in \mathbf{C}_+} \max\{|\varphi_+(z)|, |\psi_+(z)|\} > 0 \tag{22.17}$$

and

$$\operatorname*{ess\,inf}_{x \in \mathbf{R}} |\varphi_+(x)\, f_+(x) + \psi_+(x)\, g_+(x)| > 0, \tag{22.18}$$

or there exist functions $\varphi_-, \psi_- \in H^\infty_-$ *such that*

$$\inf_{z \in \mathbf{C}_-} \max\{|\varphi_-(z)|, |\psi_-(z)|\} > 0 \tag{22.19}$$

and

$$\operatorname*{ess\,inf}_{x\in\mathbf{R}} |\varphi_-(x)\, f_-(x) + \psi_-(x)\, g_-(x)| > 0. \tag{22.20}$$

Of course, condition (**B**) holds if at least one of the functions $f_\pm$, $g_\pm$ is bounded away from zero on **R**. (For example, if f_+ is bounded away from zero on **R**, then (22.18) is fulfilled by $\varphi_+ = 1$, $\psi_+ = 0$.)

Theorem 22.5. *Under assumptions* (**A**) *and* (**B**) *the following statements are equivalent:*

(a) *the matrix function* (22.4) *admits a canonical left WH factorization* $G = G_+ G_-$ *in every* $L^p(\mathbf{R}, w)$ *with* $1 < p < \infty$ *and* $w \in A_p(\mathbf{R})$,

(b) *the matrix function* (22.4) *admits a canonical left WH factorization* $G = G_+ G_-$ *in at least one of the spaces* $L^p(\mathbf{R}, w)$ *with* $1 < p < \infty$ *and* $w \in A_p(\mathbf{R})$,

(c) *conditions* (22.7) *and* (22.8) *hold.*

If one (and therefore all) of the statements (a), (b), (c) *is true, then the factors* $G_\pm$ *can be given by* (22.10), (22.11).

Proof. The implication (c) $\Rightarrow$ (a) is Theorem 22.3, and the implication (a) $\Rightarrow$ (b) is trivial. It remains to show that (b) $\Rightarrow$ (c).

To this end, suppose that the matrix function (22.4) admits a canonical left WH factorization in $L^p(\mathbf{R}, w)$ but, for example,

$$\inf_{z\in\mathbf{C}_+} \max\{|f_+(z)|, |g_+(z)|\} = 0. \tag{22.21}$$

For definiteness, assume that (22.19), (22.20) hold; the case where (22.17), (22.18) are satisfied can be treated similarly. By (22.19), the corona problem $\varphi_-\widetilde{\varphi}_- + \psi_-\widetilde{\psi}_- = 1$ is solvable in H^∞_- and hence the matrix functions

$$D_- = \begin{pmatrix} \psi_- & \varphi_- \\ -\widetilde{\varphi}_- & \widetilde{\psi}_- \end{pmatrix}, \quad D_-^{-1} = \begin{pmatrix} \widetilde{\psi}_- & -\varphi_- \\ \widetilde{\varphi}_- & \psi_- \end{pmatrix}$$

belong to $[H^\infty_-]_{2\times 2}$. Thus, the matrix function $\widetilde{G} = G\, D_-^{-1}$ also admits a canonical left WH factorization in $L^p(\mathbf{R}, w)$ and

$$\widetilde{G}\begin{pmatrix} \widehat{g}_- \\ \widehat{f}_- \end{pmatrix} = \begin{pmatrix} g_+ \\ f_+ \end{pmatrix} \tag{22.22}$$

where

$$\begin{pmatrix} \widehat{g}_- \\ \widehat{f}_- \end{pmatrix} = D_-\begin{pmatrix} g_- \\ f_- \end{pmatrix} = \begin{pmatrix} \psi_- g_- + \varphi_- f_- \\ -\widetilde{\varphi}_- g_- + \widetilde{\psi}_- f_- \end{pmatrix},$$

$f_+, g_+ \in H_+^\infty$, $\widehat{f}_-, \widehat{g}_- \in H_-^\infty$, and $1/\widehat{g}_- \in L^\infty(\mathbf{R})$ by virtue of (22.20). Let

$$\widetilde{G} = \begin{pmatrix} \widetilde{a} & \widetilde{b} \\ \widetilde{c} & \widetilde{d} \end{pmatrix}.$$

According to (22.22) and (22.20),

$$\widetilde{a} = \frac{g_+ - \widetilde{b}\,\widehat{f}_-}{\widehat{g}_-}, \quad \widetilde{c} = \frac{f_+ - \widetilde{d}\,\widehat{f}_-}{\widehat{g}_-}. \tag{22.23}$$

In view of the stability of the property to have a canonical left WH factorization (which follows from Theorem 16.5), there exists an $\varepsilon > 0$ such that for every $\widehat{f}_+, \widehat{g}_+ \in H_+^\infty$ subject to the constraints

$$\|f_+ - \widehat{f}_+\|_\infty < \varepsilon, \quad \|g_+ - \widehat{g}_+\|_\infty < \varepsilon \tag{22.24}$$

the perturbed matrix function

$$\widehat{G} = \begin{pmatrix} \widehat{a} & \widetilde{b} \\ \widehat{c} & \widetilde{d} \end{pmatrix} \tag{22.25}$$

with

$$\widehat{a} = \frac{\widehat{g}_+ - \widetilde{b}\,\widehat{f}_-}{\widehat{g}_-}, \quad \widehat{c} = \frac{\widehat{f}_+ - \widetilde{d}\,\widehat{f}_-}{\widehat{g}_-}, \tag{22.26}$$

admits a canonical left WH factorization in $L^p(\mathbf{R}, w)$ simultaneously with $\widetilde{G}$ (recall (22.23) and (22.26)). Finally, from (22.25), (22.26) we get the equality

$$\widehat{G}\begin{pmatrix} \widehat{g}_- \\ \widehat{f}_- \end{pmatrix} = \begin{pmatrix} \widehat{g}_+ \\ \widehat{f}_+ \end{pmatrix}. \tag{22.27}$$

By (22.21), there is a point $z_0 \in \mathbf{C}_+$ such that

$$|f_+(z_0)| < \varepsilon, \quad |g_+(z_0)| < \varepsilon.$$

Therefore the functions $\widehat{f}_+ = f_+ - f_+(z_0)$ and $\widehat{g}_+ = g_+ - g_+(z_0)$ satisfy (22.24) and have the common zero $z_0 \in \mathbf{C}_+$. By Lemma 22.4, the homogeneous Riemann-Hilbert problem

$$\begin{pmatrix} \Phi_1^+ \\ \Phi_2^+ \end{pmatrix} = \widehat{G}\begin{pmatrix} \Phi_1^- \\ \Phi_2^- \end{pmatrix}$$

has the non-trivial solution

$$\Phi_1^\pm(z) = \widehat{g}_\pm(z)/(z - z_0), \quad \Phi_2^\pm(z) = \widehat{f}_\pm(z)/(z - z_0) \tag{22.28}$$

in $H^p_\pm(\mathbf{R}, w)$. This contradicts the invertibility of the operator $T_{\widehat{G}} = P + \widehat{G}\, Q$ and completes the proof. □

In the case of *bounded* canonical left WH factorization we get a stronger result.

Theorem 22.6. *Let $a, b, c, d \in L^\infty(\mathbf{R})$ and $ad - bc = 1$. Then the matrix function* (22.4) *admits a bounded canonical left WH factorization $G = G_+ G_-$ if and only if there exist functions $f_\pm, g_\pm \in H^\infty_\pm$ satisfying* (22.6) to (22.8) *and the functions $F_\pm$ given by* (22.12) *are in $H^\infty_\mp$. In that case the factors $G_\pm$ can be given by* (22.10), (22.11).

Proof. Sufficiency. Since assumption (**A**) and (22.7), (22.8) hold we infer from Theorem 22.3 that the matrix function (22.4) admits a canonical left WH factorization $G = G_+ G_-$ in every $L^p(\mathbf{R}, w)$ with $1 < p < \infty$, $w \in A_p(\mathbf{R})$ and that $G_\pm$ can be given by (22.10), (22.11). Since $F_\pm \in H^\infty_\mp$, we see from (22.10), (22.11) that $G_\pm, G_\pm^{-1} \in [H^\infty_\pm]_{2\times 2}$ and thus the canonical left WH factorization $G = G_+ G_-$ is actually bounded.

Necessity. Let $G = G_+ G_-$ be a bounded canonical left WH factorization. Then $\det G_\pm \neq 0$ on $\mathbf{R}$, $\det G_+^{-1} = \det G_-$, $\det G_+^{-1} \in H^\infty_+$, and $\det G_- \in H^\infty_-$. Consequently,

$$\det G_+^{-1} = \det G_- = \text{const} \neq 0.$$

We may without loss of generality assume that

$$\det G_\pm = 1 \tag{22.29}$$

(otherwise we can make use of transformation $G_+ \mapsto G_+ c^{-1}$, $G_- \mapsto c\, G_-$ with the constant $c = (\det G_+)^{1/2} \neq 0$). Writing

$$G_+ = \begin{pmatrix} \widetilde{f}_+ & g_+ \\ -\widetilde{g}_+ & f_+ \end{pmatrix}, \quad G_-^{-1} = \begin{pmatrix} \widetilde{f}_- & g_- \\ -\widetilde{g}_- & f_- \end{pmatrix},$$

where $f_\pm$, $\widetilde{f}_\pm$, $g_\pm$, $\widetilde{g}_\pm \in H^\infty_\pm$, we get (22.6) from the equality $G_+ = G\, G_-^{-1}$. Equalities (22.29) imply (22.9), and hence (22.7), (22.8) hold. Finally, since $G_+^{-1} G\, G_-^{-1} = I$ we obtain that

$$F = \widetilde{g}_+\, a \widetilde{f}_- - \widetilde{g}_+\, b \widetilde{g}_- + \widetilde{f}_+\, c \widetilde{f}_- - \widetilde{f}_+\, d \widetilde{g}_- = 0,$$

whence $F_\pm = 0 \in H^\infty_\mp$. □

The proof of the necessity part of Theorem 22.6 shows that if G possesses a bounded canonical left WH factorization, then there exist solutions $\widetilde{f}_+, \widetilde{g}_+ \in H^\infty_+$

and $\widetilde{f}_-, \widetilde{g}_- \in H^\infty_-$ of the corona problems (22.9) such that the transformation

$$G \mapsto B_+ G B_-^{-1},$$

with $B_\pm$ given by (22.13), reduces G to the identity 2×2 matrix. On the other hand, for arbitrary corona solutions $\widetilde{f}_\pm, \widetilde{g}_\pm \in H^\infty_\pm$ such a transformation always reduces G to a *triangular* matrix function of the form $\begin{pmatrix} 1 & 0 \\ F & 1 \end{pmatrix}$.

22.2 Separation Principle for Corona Problems

In this section we present a separation principle for two-element corona problems. In case the corona data are represented by products of functions,

$$f = f_1 \, f_2 \dots f_n, \qquad g = g_1 \, g_2 \dots g_m, \tag{22.30}$$

this principle allows us to construct corona solutions to this problem in terms of the corona solutions to the (simpler) problems whose corona data run through the set of all pairs

$$(f_k, g_i), \qquad k = 1, 2, \dots, n, \qquad i = 1, 2, \dots, m,$$

of the initial corona data factors.

In what follows we will use the notation (f_k, g_i) for the two-element corona problem

$$f_k \, \widetilde{f}_{k,i} + g_i \, \widetilde{g}_{i,k} = 1$$

with the data f_k, g_i. The corresponding pair of solutions will be denoted by $(\widetilde{f}_{k,i}, \widetilde{g}_{i,k})$.

We first consider the corona problem with the data (22.30) for $m = 1$.

Lemma 22.7. *Let $f_1, f_2, \dots, f_n, g \in H^\infty_+$ and let $(\widetilde{f}_k, \widetilde{g}_k)$ be a solution in H^∞_+ to the corona problem (f_k, g), $k = 1, 2, \dots, n$. Then*

$$\left(\widetilde{f}_1 \, \widetilde{f}_2 \dots \widetilde{f}_n, \quad \widetilde{g}_1 + \sum_{k=1}^{n-1} \widetilde{g}_{k+1} \, (f_1 \, \widetilde{f}_1) \, (f_2 \, \widetilde{f}_2) \dots (f_k \, \widetilde{f}_k) \right) \tag{22.31}$$

is a solution in H^∞_+ to the corona problem

$$(f_1 \, f_2 \dots f_n, \; g), \qquad n = 2, 3, \dots .$$

Proof. Since $f_k \widetilde{f}_k + g\,\widetilde{g}_k = 1$ for $k = n, n-1, \dots, 1$, we get

$$\begin{aligned}
I_{n,1} &:= f_1 f_2 \dots f_n \widetilde{f}_1 \widetilde{f}_2 \dots \widetilde{f}_n \\
&\quad + g \left(\widetilde{g}_1 + \sum_{k=1}^{n-1} \widetilde{g}_{k+1} (f_1 \widetilde{f}_1)(f_2 \widetilde{f}_2) \dots (f_k \widetilde{f}_k) \right) \\
&= f_1 f_2 \dots f_{n-1} \widetilde{f}_1 \widetilde{f}_2 \dots \widetilde{f}_{n-1} (f_n \widetilde{f}_n + g\,\widetilde{g}_n) \\
&\quad + g \left(\widetilde{g}_1 + \sum_{k=1}^{n-2} \widetilde{g}_{k+1} (f_1 \widetilde{f}_1) \dots (f_k \widetilde{f}_k) \right) \\
&= I_{n-1,1} = \dots = I_{1,1} = 1.
\end{aligned}$$

□

The case of arbitrary m in (22.30) can be disposed of by induction on m, with Lemma 22.7 serving as the base of induction (see [20, Theorem 4.2] for details).

Theorem 22.8. *Let $f_k, g_i \in H_+^\infty$ and $f = f_1 f_2 \dots f_n$, $g = g_1 g_2 \dots g_m$. If $(\widetilde{f}_{k,i}, \widetilde{g}_{i,k})$ is a solution in H_+^∞ to the corona problem (f_k, g_i), then*

$$\begin{aligned}
\widetilde{f} &= \widetilde{f}_{1,1} \widetilde{f}_{2,1} \dots \widetilde{f}_{n,1} \\
&\quad + \sum_{i=1}^{m-1} \sum_{1 \le k_1 \le \dots \le k_i \le n} \widetilde{F}_{n,k_1,\dots,k_i} (g_1 \widetilde{g}_{1,k_1})(g_2 \widetilde{g}_{2,k_2}) \dots (g_i \widetilde{g}_{i,k_i}), \\
\widetilde{g} &= \widetilde{g}_{1,1} \widetilde{g}_{2,1} \dots \widetilde{g}_{m,1} \\
&\quad + \sum_{k=1}^{n-1} \sum_{1 \le i_1 \le \dots \le i_k \le m} \widetilde{G}_{m,i_1,\dots,i_k} (f_1 \widetilde{f}_{1,i_1})(f_2 \widetilde{f}_{2,i_2}) \dots (f_k \widetilde{f}_{k,i_k}),
\end{aligned}$$

where

$$\begin{aligned}
\widetilde{F}_{n,k_1,\dots,k_i} &= \left(\prod_{k=1}^{k_1-1} \widetilde{f}_{k,1} \right) \left(\prod_{k=k_1}^{k_2-1} \widetilde{f}_{k,2} \right) \dots \left(\prod_{k=k_{i-1}}^{k_i-1} \widetilde{f}_{k,i} \right) \left(\prod_{k=k_i}^{n} \widetilde{f}_{k,i+1} \right), \\
\widetilde{G}_{m,i_1,\dots,i_k} &= \left(\prod_{i=1}^{i_1-1} \widetilde{g}_{i,1} \right) \left(\prod_{i=i_1}^{i_2-1} \widetilde{g}_{i,2} \right) \dots \left(\prod_{i=i_{k-1}}^{i_k-1} \widetilde{g}_{i,k} \right) \left(\prod_{i=i_k}^{m} \widetilde{g}_{i,k+1} \right),
\end{aligned}$$

is a solution in H_+^∞ to the corona problem $f\,\widetilde{f} + g\,\widetilde{g} = 1$. □

The next result deals with corona data containing non-integral powers of functions. It allows to construct a corona solution for the data

$$f^\alpha, g^\beta, \quad \alpha, \beta \ge 0, \quad \alpha + \beta > 0 \tag{22.32}$$

provided a corona solution for the data f, g is known. Of course, some additional conditions on f, g, eliminating the non-uniqueness of roots, are needed.

Corollary 22.9. *Let $f, g \in H_+^\infty$ and suppose each of the sets $\mathbf{C}\setminus f(\mathbf{C}_+)$, $\mathbf{C}\setminus g(\mathbf{C}_+)$ contains a simple continuous curve connecting 0 and ∞. Let $(\widetilde{f}, \widetilde{g})$ be a solution in H_+^∞ to the corona problem $(f,\ g)$. Then the pair*

$$\left(f^{[\alpha]+1-\alpha}\, \widetilde{f}^{[\alpha]+1} \sum_{i=0}^{[\beta]} \binom{[\alpha]+i}{i} (g\widetilde{g})^i, \quad g^{[\beta]+1-\beta}\, \widetilde{g}^{[\beta]+1} \sum_{k=0}^{[\alpha]} \binom{[\beta]+k}{k} (f\widetilde{f})^k \right),$$

where $[x]$ is the greatest integer less than $x \in \mathbf{R}$, is a solution in H_+^∞ to the corona problem with the data (22.32).

Proof. If α and β are natural numbers, the assertion follows from Theorem 22.8 and the observation that the number of all tuples $(k_1, \cdots, k_i)$ satisfying the relation $1 \le k_1 \le \cdots \le k_i \le n$ is equal to $\binom{n+i-1}{i}$. The case of arbitrary non-negative α, β can be reduced directly to the case where $\alpha, \beta \in \mathbf{N}$. □

We remark that if any of the numbers α, β is an integer, the condition on the corresponding set $\mathbf{C}\setminus f(\mathbf{C}_+)$ or $\mathbf{C}\setminus g(\mathbf{C}_+)$ can be dropped.

Let $f_k, g_i \in H_+^\infty$, $\alpha_k > 0$, $\beta_i > 0$, and suppose that for each non-integral number α_k, β_i the corresponding set $\mathbf{C}\setminus f_k(\mathbf{C}_+)$, $\mathbf{C}\setminus g_i(\mathbf{C}_+)$ contains a simple continuous curve connecting 0 and ∞. Combining Theorem 22.8 and Corollary 22.9 one can construct corona solutions for the corona data

$$f = \prod_{k=1}^{n} f_k^{\alpha_k}, \quad g = \prod_{i=1}^{m} g_i^{\beta_i}$$

in terms of solutions $\widetilde{f}_{k,i}\,, \widetilde{g}_{i,k} \in H_+^\infty$ to the corona problems $(f_k\,, g_i)$. We confine ourselves to the case $m = 1$ and $\beta_1 = 1$.

Corollary 22.10. *Let $f_1, f_2, \ldots, f_n,\ g \in H_+^\infty$, let $(\widetilde{f}_k,\ \widetilde{g}_k)$ be a solution in H_+^∞ to the corona problem $(f_k,\ g)$, $k = 1, 2, \ldots, n$, and let $f = \prod_{k=1}^n f_k^{\alpha_k}$, where $\alpha_k \ge 0$ and the set $\mathbf{C}\setminus f_k(\mathbf{C}_+)$ contains a simple continuous curve connecting 0 and ∞ in case α_k is not an integer. Then the functions*

$$\begin{aligned}
\widetilde{f} &= \prod_{k=1}^{n} \left(f_k^{[\alpha_k]+1-\alpha_k}\, \widetilde{f}_k^{[\alpha_k]+1} \right), \\
\widetilde{g} &= \widetilde{g}_1 \sum_{s=0}^{[\alpha_1]} (f_1\, \widetilde{f}_1)^s + \sum_{k=1}^{n-1} \left(\widetilde{g}_{k+1} \sum_{s=0}^{[\alpha_{k+1}]} (f_{k+1}\, \widetilde{f}_{k+1})^s \right) \prod_{j=1}^{k} (f_j\, \widetilde{f}_j)^{[\alpha_j]+1}
\end{aligned}$$

are solutions in H_+^∞ to the corona problem (f, g).

Proof. By Corollary 22.9, the functions

$$\widehat{f}_k = f_k^{[\alpha_k]+1-\alpha_k}\, \widetilde{f}_k^{[\alpha_k]+1}, \quad \widehat{g}_k = \widetilde{g}_k \sum_{s=0}^{[\alpha_k]} (f_k\, \widetilde{f}_k)^s$$

are solutions in H_+^∞ to the corona problem $(f_k^{\alpha_k}, g)$ for $k = 1, 2, \ldots, n$. To obtain the asserted formulas for $\widetilde{f}$ and $\widetilde{g}$ it remains to replace in (22.31) the functions f_k, $\widetilde{f}_k$, and $\widetilde{g}_k$ by $f_k^{\alpha_k}$, $\widehat{f}_k$, and $\widehat{g}_k$, respectively. □

22.3 Explicit Formulas for Basic Corona Solutions

Theorem 22.8 and Corollary 22.10 allow us to construct corona solutions in terms of solutions of simpler corona problems as from bricks. In this section we consider a number of such (in a sense canonical) corona problems. We construct explicit corona solutions in H_+^∞ in case one of the functions in the corona data is arbitrary or "almost" arbitrary in H_+^∞, while the other is a periodic binomial $e_\mu - c$ ($\mu > 0, c \in \mathbf{C}$) and therefore has a periodic set of zeros in $\mathbf{C}$. Of course, the separation principle of Section 22.2 then allows us to construct solutions to corona problems with the second function being a product of periodic binomials.

Thus, consider the corona problem (f, g) with the data

$$g \in H_+^\infty, \quad f = e_\mu - c, \quad \mu > 0,\ c \in \mathbf{C}.$$

The case $|c| > 1$ is trivial: the pair $\big((e_\mu - c)^{-1}, 0\big)$ is a corona solution in H_+^∞. So assume $0 < |c| < 1$ (we will come to the cases $|c| = 1$ and $c = 0$ later). Then all zeros of f lie in $\mathbf{C}_+$ and are given by

$$z_k = (\arg c + 2\pi k - i \log |c|)/\mu, \qquad k \in \mathbf{Z}. \tag{22.33}$$

We need the following interpolation result.

Lemma 22.11. *Let $c \in \mathbf{C}$, $0 < |c| < 1$, let z_k be given by (22.33), and let $\{a_k\}$ be a sequence in l_∞. Then a solution $\Psi \in H_+^\infty$ of the interpolation problem*

$$\Psi(z_k) = a_k, \qquad k \in \mathbf{Z}, \tag{22.34}$$

is given by

$$\Psi(z) = -\sum_{k \in \mathbf{Z}} \frac{a_k}{\mu^2 c^2} \left(\frac{e^{i\mu z} - c}{z - z_k} \right)^2. \tag{22.35}$$

Proof. We have

$$\left(\frac{e^{i\mu z} - c}{z - z_k} \right)^2 = \frac{\eta_k^2(z)}{(z - \overline{z}_k)^2},$$

where the functions

$$\eta_k(z) := (e^{i\mu z} - c)\, \frac{z - \overline{z}_k}{z - z_k}, \qquad k \in \mathbf{Z},$$

belong to H^∞_+ and are uniformly bounded. The estimate

$$\begin{aligned}
&\sup_{z\in\mathbf{C}_+}\sum_{k\in\mathbf{Z}}\frac{|a_k|}{|z-\overline{z}_k|^2}\\
&\le \sup_{k\in\mathbf{Z}}|a_k|\cdot\sup_{z\in\mathbf{C}_+}\sum_{k\in\mathbf{Z}}\frac{\mu^2}{(\mu\operatorname{Re}z-\arg c-2\pi k)^2+\log^2|c|}\\
&\le 2\mu^2\sup_{k\in\mathbf{Z}}|a_k|\cdot\left(\frac{1}{\log^2|c|}+\sum_{k=1}^{\infty}\frac{1}{(2\pi k)^2}\right)<\infty,
\end{aligned}$$

shows that the series on the right of (22.35) is absolutely convergent. Hence (22.35) is a well-defined function in H^∞_+ and, obviously, (22.34) holds. □

Theorem 22.12. *Let $0<|c|<1$, $\mu>0$, $f=e_\mu-c$, $g\in H^\infty_+$. Then the corona problem (f,g) is solvable in H^∞_+ if and only if*

$$\inf_{k\in\mathbf{Z}}|g(z_k)|>0, \tag{22.36}$$

where the z_k's, given by (22.33), are all the zeros of f. If (22.36) is satisfied, a pair of corona solutions is given by

$$\widetilde{f}(z)=\frac{1-g(z)\,\widetilde{g}(z)}{e^{i\mu z}-c},\qquad \widetilde{g}(z)=-\sum_{k\in\mathbf{Z}}\frac{1}{\mu^2\,c^2\,g(z_k)}\left(\frac{e^{i\mu z}-c}{z-z_k}\right)^2. \tag{22.37}$$

Proof. The necessity of (22.36) is obvious. Conversely, assume (22.36) holds and put $a_k=1/g(z_k)$ for $k\in\mathbf{Z}$. By (22.36), $\{a_k\}\in l_\infty$. By Lemma 22.11, the function $\widetilde{g}$ in (22.37) is a solution of the H^∞_+ interpolation problem

$$\widetilde{g}\,(z_k)=1/g(z_k),\quad k\in\mathbf{Z}. \tag{22.38}$$

Obviously, $f\widetilde{f}+g\widetilde{g}=1$. It remains to prove that the function

$$\widetilde{f}(z)=\left(1-g(z)\,\widetilde{g}(z)\right)/(e^{i\mu z}-c)$$

belongs to H^∞_+. It is clear from (22.38) that $\widetilde{f}$ is analytic in $\mathbf{C}_+$. To prove the boundedness of $\widetilde{f}$ we may estimate $\widetilde{f}$ from above separately in every disc $D_k=\{z:|z-z_k|<r\}$, $k\in\mathbf{Z}$, and on the set $Y=\mathbf{C}_+\setminus\bigcup_{k\in\mathbf{Z}}D_k$, where r is any fixed number such that $r<\mu^{-1}\min\{\pi,-\log|c|\}$.

We first estimate $|e^{i\mu z}-c|$ on Y from below. Due to periodicity, it is sufficient to consider this function only on

$$Y_0=\{z:\arg\,c-\pi\le\mu\operatorname{Re}z\le\arg\,c+\pi,\ \operatorname{Im}z>0\}\setminus D_0.$$

Obviously,

$$M := \min_{\varphi\in[0,2\pi]} |e^{i\mu r(\cos\varphi + i\sin\varphi)} - 1| > 0. \tag{22.39}$$

Since the function $f = e_\mu - c$ does not vanish on the closure of Y_0 on the Riemann sphere, the minimum modulus principle can be applied to get

$$\begin{aligned}\inf_{z\in Y_0} |e^{i\mu z} - c| \geq \min\Big\{ &\min_{z\in\partial D_0} |e^{i\mu z} - c|, \min_{-\pi\leq x\leq\pi} |e^{i(\arg c + x)} - c|, \\ &\inf_{y>0} |e^{i(\arg c\pm\pi)}e^{-y} - c|\Big\} = \min\{M|c|, 1 - |c|, |c|\} =: M_0 > 0.\end{aligned} \tag{22.40}$$

Let $\|\Phi\|_\infty = \sup_{z\in\mathbf{C}_+} |\Phi(z)|$ stand for the norm of $\Phi \in H^\infty_+$. Then due to (22.40),

$$\sup_{z\in Y} |\widetilde{f}(z)| \leq \frac{1}{M_0}(1 + \|g\|_\infty \|\widetilde{g}\|_\infty) < \infty. \tag{22.41}$$

Obviously, the functions

$$\Phi_k(z) = \frac{g(z) - g(z_k)}{z - z_k}, \qquad \Psi_k(z) = \frac{\widetilde{g}(z) - \widetilde{g}(z_k)}{z - z_k}, \qquad k \in \mathbf{Z} \tag{22.42}$$

are analytic in $\mathbf{C}_+$ and, by virtue of the maximum modulus principle,

$$\begin{aligned}\|\Phi_k\|_\infty \leq \; & \frac{2\|g\|_\infty}{\operatorname{Im} z_k} = \frac{2\mu\,\|g\|_\infty}{-\log|c|} &&=: M_1 < \infty, \\ \|\Psi_k\|_\infty \leq \; & 2\mu\,\|\widetilde{g}\|_\infty/(-\log|c|) &&=: M_2 < \infty,\end{aligned} \tag{22.43}$$

whence $\Phi_k, \Psi_k \in H^\infty_+$, $k \in \mathbf{Z}$. In view of (22.42) and (22.38)

$$\begin{aligned}1 - g(z)\,\widetilde{g}(z) &= 1 - (g(z_k) + \Phi_k(z)\,(z - z_k))\,\,(\widetilde{g}(z_k) + \Psi_k(z)\,(z - z_k)) \\ &= -\,(\Phi_k(z)\widetilde{g}(z_k) + \Psi_k(z)g(z_k))\,(z - z_k) + \Phi_k(z)\Psi_k(z)(z - z_k)^2.\end{aligned} \tag{22.44}$$

Hence, for every $k \in \mathbf{Z}$, the function

$$f_k(z) = (1 - g(z)\,\widetilde{g}(z))\,/(z - z_k)$$

is analytic in $\mathbf{C}_+$ and, by (22.43) and (22.44),

$$\sup_{z\in D_k} |f_k(z)| \leq M_1\|\widetilde{g}\|_\infty + M_2\|g\|_\infty + M_1\,M_2\,r. \tag{22.45}$$

In the same vein, the functions

$$h_k(z) = (e^{i\mu z} - c)/(z - z_k), \qquad k \in \mathbf{Z}$$

are analytic in $\mathbf{C}_+$ as well and, by the minimum modulus principle and (22.39),

$$\inf_{z\in D_k} |h_k(z)| = \min_{z\in\partial D_k} |h_k(z)| = M|c|r^{-1} > 0. \tag{22.46}$$

Consequently, by (22.45) and (22.46),

$$\begin{aligned}\sup_{z\in D_k} |\widetilde{f}(z)| &= \sup_{z\in D_k} |f_k(z)/h_k(z)| \\ &\le \frac{r}{M\,|c|}\,(M_1\,\|\widetilde{g}\|_\infty + M_2\,\|g\|_\infty + M_1\,M_2\,r) < \infty \qquad (22.47)\end{aligned}$$

for all $k \in \mathbf{Z}$. From (22.41) and (22.47) we finally see that $f \in H_+^\infty$. □

In the remaining cases $|c| = 1$ and $c = 0$ the results are less complete. The case $|c| = 1$ can be tackled with the help of Theorem 22.12 and a small linear change in the argument of the function g.

Corollary 22.13. *Theorem* 22.12 *remains valid in the case* $|c| = 1$ *if in addition the function* $g_\theta(z) = g(z - i\log\theta)$ *belongs to* H_+^∞ *for some* $\theta > 1$.

Proof. It is easily seen that the corona problem $(e_\mu - c, g)$ is solvable in H_+^∞ if so is the corona problem $(\theta^\mu\, e_\mu - c, g_\theta)$. Since the numbers

$$\widehat{z}_k = (\arg c + 2\pi k)/\mu + i\log\theta \in \mathbf{C}_+, \qquad k \in \mathbf{Z},$$

are the zeros of $\theta^\mu\, e_\mu - c$ and since $g_\theta(\widehat{z}_k) = g(z_k)$, we obtain from Theorem 22.12 that the pair $(\widehat{f}, \widehat{\Psi})$ constituted by

$$\widehat{f}(z) = \frac{1 - g_\theta(z)\,\widehat{\Psi}(z)}{\theta^\mu\, e^{i\mu z} - c}, \qquad \widehat{\Psi}(z) = -\sum_{k\in\mathbf{Z}} \frac{1}{\mu^2\, c^2\, g(z_k)} \left(\frac{\theta^\mu\, e^{i\mu z} - c}{z - \widehat{z}_k}\right)^2 \qquad (22.48)$$

is a solution in H_+^∞ to the corona problem $(\theta^\mu\, e_\mu - c,\ g_\theta)$. Hence the pair

$$\widetilde{f}(z) = \widehat{f}(z + i\log\theta), \qquad \widetilde{g}(z) = \widehat{\Psi}(z + i\log\theta) \qquad (22.49)$$

is a corona solution in H_+^∞ to the problem $(e_\mu - c, g)$. It remains to observe that (22.48) and (22.49) give (22.37). □

In the case $c = 0$, we consider the corona problem in a narrower class of functions. Let $W_{\mathbf{R}}$ denote the Banach algebra of all functions $f : \mathbf{R} \to \mathbf{C}$ of the form

$$f(x) = \sum_{\lambda\in\mathbf{R}} c_\lambda\, e^{i\lambda x} + \int_{\mathbf{R}} k(\lambda)\, e^{i\lambda x}\, d\lambda \qquad (22.50)$$

where $c_\lambda \neq 0$ for at most countably many λ and $k \in L_1(\mathbf{R})$. The norm in $W_{\mathbf{R}}$ is defined by

$$\|f\|_{W_{\mathbf{R}}} = \sum_{\lambda\in\mathbf{R}} |c_\lambda| + \int_{\mathbf{R}} |k(\lambda)|\, d\lambda$$

and the algebraic operations in $W_{\mathbf{R}}$ are defined pointwise (see, e. g., Chapter 7 of [92]). Also, for any $\mu \in [-\infty, +\infty]$, $\mu \neq 0$, we define the projection P_μ on $W_{\mathbf{R}}$ by the formula

$$(P_\mu f)(z) = \begin{cases} \sum\limits_{0 \leq \lambda < \mu} c_\lambda e^{i\lambda z} + \int_0^\mu k(\lambda) e^{i\lambda z} \, d\lambda & \text{for } z \in \mathbf{C}_+ \text{ if } \mu > 0, \\ \sum\limits_{\mu < \lambda \leq 0} c_\lambda e^{i\lambda z} + \int_\mu^0 k(\lambda) e^{i\lambda z} \, d\lambda & \text{for } z \in \mathbf{C}_- \text{ if } \mu < 0, \end{cases}$$

where f is given by (22.50). Put $W_{\mathbf{R}}^{\pm} = P_{\pm\infty} W_{\mathbf{R}}$. Clearly, $W_{\mathbf{R}}^{\pm}$ are Banach subalgebras of $W_{\mathbf{R}}$ and $W_{\mathbf{R}}^{\pm} \subset H_{\pm}^{\infty}$. For finite μ, $P_\mu f$ is an entire function and the above formulas for $P_\mu f$ are valid for all $z \in \mathbf{C}$. For every $f \in W_{\mathbf{R}}^{\pm}$ the limit $\lim\limits_{y \to \pm\infty} f(iy)$ exists; it will be denoted by $c_0(f)$.

The following lemma is a complete analogue of Lemma 13.12 and can be proved in the same way.

Lemma 22.14. *If $f \in W_{\mathbf{R}}$ and $c_0(f) = 0$, then the series $\sum_{k=0}^{\infty} P_{[0,\mu)}(f^k)$ converges in $W_{\mathbf{R}}$ for every $\mu \in (0, \infty)$.* □

Theorem 22.15. *The corona problem in $W_{\mathbf{R}}^{+}$ with the data e_μ, g is solvable if and only if $c_0(g) \neq 0$. If $c_0(g) \neq 0$, a pair of corona solutions is given by*

$$\begin{aligned} \widetilde{e}_\mu &= e_{-\mu} \, (1 - g\,\widetilde{g}), \\ \widetilde{g} &= (c_0(g))^{-1} \sum_{k=0}^{\infty} P_{[0,\mu)} \left((1 - (c_0(g))^{-1} g)^k \right). \end{aligned} \tag{22.51}$$

The proof is direct verification, similarly to the proof of Theorem 13.13.

Using Corollary 22.10 and the results of this section, it is possible to find solvability criteria and to construct explicit corona solutions in H_{+}^{∞} to the corona problem (f, h) with the corona data

$$f = f_0 \prod_{k=1}^{n} (e_{\lambda_k} - c_k)^{\alpha_k}, \qquad h = h_0 \, g,$$

where $c_k \in \mathbf{C}$, $\lambda_k > 0$, $\alpha_k > 0$ if $|c_k| \geq 1$, $\alpha_k \in \mathbf{N}$ if $|c_k| < 1$, $f_0^{\pm 1} \in APW^{+}$, $h_0^{\pm 1} \in H_{+}^{\infty}$, and

$$g \in H_{+}^{\infty} \ \text{ if } \ \prod_{k=1}^{n} |c_k| \, (1 - |c_k|) \neq 0,$$

$$g \in W_\gamma \ \text{ for some } \gamma > 0 \ \text{ if } \ \prod_{k=1}^{n} |c_k| \, (1 - |c_k|) = 0.$$

22.4 Canonical WH Factorization of Triangular Matrix Functions

In this section we consider applications of the results presented in Sections 22.1, 22.2, 22.3 to canonical left WH factorization problems for triangular 2×2 matrix functions of the form

$$G = \begin{pmatrix} e_\lambda & 0 \\ f & e_{-\lambda} \end{pmatrix}, \tag{22.52}$$

where $\lambda > 0$ and $f \in L^\infty(\mathbf{R})$.

Theorem 22.16. *Let $\lambda > 0$ and*

$$f(x) = \left(f_+(x)\, e^{i\lambda x} - f_-(x)\right) \Big/ \left(e^{i\mu_+ x} \prod_{k=1}^{n} \left(a_k\, e^{i\mu_k x} - b_k\right)^{\alpha_k} \right), \tag{22.53}$$

where $\mu_+ \geq 0$, $\mu_- := \mu_+ + \sum_{k=1}^n \mu_k \alpha_k - \lambda \leq 0$,

$$f_\pm \in H_\pm^\infty \;\; \textit{if} \;\; \mu_\pm \neq 0 \;\; \textit{and} \;\; f_\pm \in W_{\mathbf{R}}^\pm \;\; \textit{if} \;\; \mu_\pm = 0,$$

$$a_k, b_k \in \mathbf{C} \setminus \{0\}, \quad |a_k| \neq |b_k|, \quad \mu_k > 0, \quad \alpha_k \in \mathbf{N}.$$

Then the matrix function (22.52) *possesses a canonical left WH factorization $G = G_+ G_-$ in one (and therefore all) of the spaces $L^p(\mathbf{R}, w)$ with $p \in (1, \infty)$ and $w \in A_p(\mathbf{R})$ if and only if the following conditions hold:*

(i) $c_0(f_\pm) \neq 0$ *if* $\mu_\pm \neq 0$*;*

(ii) *for every* $k = 1, 2, \ldots, n$, $\inf_{j \in \mathbf{Z}} |f_\pm(z_{kj})| > 0$ *if* $\pm(|a_k| - |b_k|) > 0$, *where*

$$z_{kj} = \frac{1}{\mu_k} \left(\arg \frac{b_k}{a_k} + 2\pi j - i \log \left| \frac{b_k}{a_k} \right| \right), \quad j \in \mathbf{Z}. \tag{22.54}$$

If (i) *and* (ii) *hold, the factors $G_\pm$ can be given by* (22.10), (22.11) *with*

$$F_\pm = \widetilde{P}_\pm(\widetilde{f}_+ f \widetilde{f}_- + \widetilde{g}_+ e_\lambda \widetilde{f}_- - \widetilde{f}_+ e_{-\lambda} \widetilde{g}_-), \tag{22.55}$$

$$g_\pm = e_{\mu_\pm} \prod_{k=1}^{n} (g_k^\pm)^{\alpha_k}, \quad g_k^+ = a_k\, e_{\mu_k} - b_k, \quad g_k^- = a_k - b_k\, e_{-\mu_k},$$

$$\widetilde{f}_\pm = \widetilde{f_0^\pm} + (1 - f_\pm \widetilde{f_0^\pm}) \sum_{k=1}^{n} \prod_{j=1}^{k-1} (1 - f_\pm \widetilde{f_j^\pm})^{\alpha_j}\, \widetilde{f_k^\pm} \sum_{s=0}^{\alpha_k - 1} (1 - f_\pm \widetilde{f_k^\pm})^s, \tag{22.56}$$

$$\widetilde{g}_{\pm} = (1 - f_{\pm}\, \widetilde{f}_0^{\pm})\, e_{-\mu_{\pm}} \prod_{k=1}^{n} \left((1 - f_{\pm}\, \widetilde{f}_k^{\pm})/g_k^{\pm}\right)^{\alpha_k}, \tag{22.57}$$

$$\widetilde{f}_0^{\pm} = \begin{cases} (c_0(f_{\pm}))^{-1} \sum_{k=0}^{\infty} P_{\mu_{\pm}} \left((1 - (c_0(f_{\pm}))^{-1} f_{\pm})^k\right), & \mu_{\pm} \neq 0, \\ 0, & \mu_{\pm} = 0, \end{cases} \tag{22.58}$$

$$\widetilde{f}_k^{\pm}(x) = \begin{cases} -\sum_{j\in\mathbf{Z}} \dfrac{1}{\mu_k^2\, (c_k^{\pm})^2\, f_{\pm}(z_{kj})} \left(\dfrac{g_k^{\pm}(x)}{x - z_{kj}}\right)^2, & \pm(|a_k| - |b_k|) > 0, \\ 0, & \pm(|a_k| - |b_k|) < 0, \end{cases} \tag{22.59}$$

and $c_k^+ = b_k$, $c_k^- = a_k$.

Proof. Sufficiency. Obviously, the functions $f_{\pm}, g_{\pm} \in H_{\pm}^{\infty}$ satisfy condition (**A**), which in the case at hand takes the form

$$g_+ f = e_{\lambda} f_+ - f_-, \quad g_+ = e_{\lambda} g_-. \tag{22.60}$$

We claim that the corona problems $(f_{\pm}, g_{\pm})$ are solvable in $H_{\pm}^{\infty}$ and that the pairs $(\widetilde{f}_{\pm}, \widetilde{g}_{\pm})$ given by (22.56), (22.57) are solutions.

Indeed, consider, for example, the "+" case. By Corollary 22.10, the corona problem (f_+, g_+) is solvable in H_+^{∞} if this is true for the corona problems (f_+, g_k^+), $k = 0, 1, \ldots, n$, where $g_0^+ = e_{\mu_+}$. Put

$$\widetilde{g}_k^+ = (1 - f_+\, \widetilde{f}_k^+)/g_k^+, \quad k = 0, 1, \ldots, n, \tag{22.61}$$

where $\widetilde{f}_0^+$ and $\widetilde{f}_k^+$ $(k > 0)$ are given by (22.58) and (22.59), respectively. If $\mu_+ > 0$, then $M(f_+) \neq 0$ in view of (i), and hence, by Theorem 22.15, the pair $(\widetilde{f}_0^+, \widetilde{g}_0^+)$ is a solution in H_+^{∞} to the corona problem (f_+, g_0^+). If $\mu_+ = 0$, then $g_0^+ = 1$, and hence (22.58) and (22.61) give $\widetilde{f}_0^+ = 0$ and $\widetilde{g}_0^+ = 1$, which again yields $f_+\, \widetilde{f}_0^+ + g_0^+\, \widetilde{g}_0^+ = 1$. Now fix $k \in \{1, 2, \ldots, n\}$. If $|a_k| > |b_k|$, then (ii) and Theorem 22.12 imply that $(\widetilde{f}_k^+, \widetilde{g}_k^+)$ is a solution in H_+^{∞} to the corona problem (f_+, g_k^+). If $|a_k| < |b_k|$, then g_k^+ is invertible in H_+^{∞} (and even APW^+). By (22.59), $\widetilde{f}_k^+ = 0$, whence, in accordance with (22.61), $\widetilde{g}_k^+ = 1/g_k^+ \in APW^+$. Again $(\widetilde{f}_k^+, \widetilde{g}_k^+)$ is a corona solution in H_+^{∞} to (f_+, g_k^+).

Finally, by virtue of Corollary 22.10, the corona problem (f_+, g_+) is solvable in H_+^{∞} and the functions

$$\begin{aligned} \widetilde{f}_+ &= \widetilde{f}_0^+ + (g_0^+\, \widetilde{g}_0^+) \sum_{k=1}^{n} \prod_{j=1}^{k-1} (g_j\, \widetilde{g}_j^+)^{\alpha_j}\, \widetilde{f}_k^+ \sum_{s=0}^{\alpha_k - 1} (g_k^+\, \widetilde{g}_k^+)^s, \\ \widetilde{g}_+ &= \widetilde{g}_0^+ \prod_{k=1}^{n} (\widetilde{g}_k^+)^{\alpha_k} \end{aligned} \tag{22.62}$$

are corona solutions in $H^\infty_\mp$. On substituting (22.61) into (22.62) we get (22.56) and (22.57) in the "+" case.

Since condition (**A**) is satisfied, it follows from Theorem 22.3 that the representation $G = G_+ G_-$ with the factors $G_\pm$ given by (22.10), (22.11) is a canonical left WH factorization in every space $L^p(\mathbf{R}, w)$. At this point the sufficiency portion is proved and the formulas for $G_\pm$ are established.

Necessity. The functions

$$f_\pm, \quad g_+ = e_{\mu_+} \prod_{k=1}^{n} (a_k\, e_{\mu_k} - b_k)^{\alpha_k}, \quad g_- = g_+ e_{-\lambda}$$

satisfy both the assumptions (**A**) and (**B**) of Section 22.1. Indeed, (**A**) coincides with (22.60) and (**B**) follows from the inclusion $1/g_+ \in L^\infty(\mathbf{R})$. Therefore Theorem 22.5 implies that the corona problems (f_+, g_+) and (f_-, g_-) are solvable in H^∞_+ and H^∞_-, respectively, which immediately gives (i) and (ii). □

Analogously, combining Corollary 22.10 with the results of Section 22.3, one can prove the following criterion.

Theorem 22.17. *Let* $\lambda > 0$ *and* $f(x) = \left(f_+(x)\, e^{i\lambda x/2} - f_-(x)\, e^{-i\lambda x/2}\right)/g(x)$ *where*

$$\begin{aligned} &1/g \in L_\infty(\mathbf{R}), \quad g_\pm = g\, e_{\pm\lambda/2} \in W^\pm_{\mathbf{R}}, \\ &f_\pm = c_\pm\, \varphi_\pm\, e_{\pm\gamma_\pm} \prod_{k \in K_\pm} (a_k\, e_{\pm\mu_k} - b_k)^{\alpha_k}, \\ &\varphi_\pm, \varphi_\pm^{-1} \in H^\infty_\pm, \quad c_\pm \in \mathbf{C}, \quad \gamma_\pm \geq 0, \\ &K_\pm \textit{ are finite and disjoint subsets of } \mathbf{N}, \end{aligned}$$

and for $k \in K := K_+ \cup K_-$,

$$\begin{aligned} &a_k, b_k \in \mathbf{C}, \quad 0 < |b_k| \leq |a_k|, \quad \mu_k > 0, \\ &\alpha_k > 0 \textit{ if } |b_k| = |a_k|, \textit{ and } \alpha_k \in \mathbf{N} \textit{ if } |b_k| < |a_k|. \end{aligned}$$

Then the matrix function (22.52) *admits a canonical left WH factorization* $G = G_+ G_-$ *in one (and therefore all) of the spaces* $L^p(\mathbf{R}, w)$ ($p \in (1, \infty)$, $w \in A_p(\mathbf{R})$) *if and only if one of the following conditions holds:*

(i) $c_- = 0 \neq c_+$, $g_-^{-1} \in W^-_{\mathbf{R}}$,

$$c_0(g_+) \neq 0 \textit{ if } \gamma_+ > 0, \quad \inf_{j \in \mathbf{Z}} |g(z_{kj})| > 0 \textit{ for } k \in K_+; \tag{22.63}$$

(ii) $c_- \neq 0 = c_+$, $g_+^{-1} \in W^+_{\mathbf{R}}$,

$$c_0(g_-) \neq 0 \textit{ if } \gamma_- > 0, \quad \inf_{j \in \mathbf{Z}} |g(-z_{kj})| > 0 \textit{ for } k \in K_-; \tag{22.64}$$

(iii) $c_- \neq 0 \neq c_+$ *and* (22.63), (22.64) *hold, where* z_{kj} ($k \in K$, $j \in \mathbf{Z}$) *are given by* (22.54).

If one of these conditions holds, the factors $G_\pm$ can be given by (22.10), (22.11) *where $F_\pm$ are defined by* (22.55),

$$\widetilde{f}_\pm = \begin{cases} c_\pm^{-1}\varphi_\pm^{-1}(1 - g_\pm \widetilde{g}_0^\pm)\, e_{\mp\gamma_\pm} \prod_{k\in K_\pm} (f_k^\pm)^{[\alpha_k]+1-\alpha_k} (\widetilde{f}_k^\pm)^{[\alpha_k]+1}, & c_\pm \neq 0, \\ 0, & c_\pm = 0, \end{cases}$$

$$\widetilde{g}_\pm = \widetilde{g}_0^\pm + (1 - g_\pm \widetilde{g}_0^\pm) \sum_{k\in K_\pm} \prod_{j\in K_\pm, j<k} (1 - g_\pm \widetilde{g}_j^\pm)^{[\alpha_j]+1}\, \widetilde{g}_k^\pm \sum_{s=0}^{[\alpha_k]} (1 - g_\pm \widetilde{g}_k^\pm)^s$$

if $c_\pm \neq 0$, $\widetilde{g}_\pm = g_\pm^{-1}$ *if* $c_\pm = 0$,

$$f_k^\pm = a_k\, e_{\pm\mu_k} - b_k \;\; (k \in K_\pm), \quad \widetilde{f}_k^\pm = (1 - g_\pm \widetilde{g}_k^\pm)/f_k^\pm,$$

$$\widetilde{g}_0^\pm = \begin{cases} (c_0(g_\pm))^{-1} P_{\pm\gamma_\pm}\left((1 - (c_0(g_\pm))^{-1}(g_\pm)^k\right), & \gamma_\pm \neq 0, \\ 0, & \gamma_\pm = 0, \end{cases}$$

$$g_k^\pm(x) = -\sum_{j\in\mathbf{Z}} \frac{1}{\mu_k^2 b_k^2\, g_\pm(\pm z_{kj})} \left(\frac{a_k\, e^{\pm i\mu_k x} - b_k}{x \pm z_{kj}}\right)^2 \quad (k \neq 0). \qquad \square$$

Notes

There is an enormous literature devoted to explicit factorization in general, and most of it is beyond the scope of our book. This chapter is written on the basis of the papers [20], [22], and [23]. We would like to point out that, according to this approach, the factorization problem is divided into two subproblems. The first consists in finding at least one solution of a homogeneous Riemann-Hilbert problem in $[H_\mp^\infty]_2$. The second subproblem amounts to finding solutions (in $H_\mp^\infty$) of two associated corona problems. An alternative approach, using two explicit solutions of the same Riemann-Hilbert problem (if available), was systematically developed by L. A. Sakhnovich [193], [194].

In fact, [20] contains a description of much more general classes of 2×2 triangular matrix functions to which the approach of this chapter is applicable. In particular, the diagonal entries in (22.52) can be of the form $h^{\pm 1}$, where h is a certain inner function, not necessarily of the form e_λ. On the other hand, the results of Section 22.4 are slightly more general than those of [20].

Generalizations of the approach of this chapter to matrix functions of bigger size, motivated by the study of convolution type equations on unions of intervals, are in [21].

Chapter 23

Canonical APW Factorization via Corona Problems

Armed with the results on Wiener-Hopf factorization from the previous chapter, we now go back to canonical APW factorization. A new variation of the corona problem becomes useful in this setting, namely, the one in which the corona data and the corona solutions are APW functions with bounded Bohr-Fourier spectrum. In terms of these solutions, factorability criteria and factorization formulas are established for some triangular APW matrix functions with bounded Bohr-Fourier spectrum. Explicit results are obtained in the case where the corona data are periodic binomials. We then revisit the trinomial factorization problem and, in addition to the results of Chapter 15, we derive explicit factorization formulas.

23.1 General Results

For APW matrix functions, the factorability results of Section 22.1 can be simplified greatly. Namely, the following result holds.

Theorem 23.1. *Let G be the matrix function* (22.4) *with a, b, c, $d \in APW$ and $ad - bc = 1$. Then the following three assertions are equivalent:*

(i) *G admits a canonical left APW factorization $G = G_+ G_-$,*

(ii) *there exist functions $f_\pm$, $g_\pm \in H^\infty_\pm$ satisfying* (22.6) to (22.8),

(iii) *there exist functions $f_\pm$, $g_\pm \in APW^\pm$ satisfying* (22.6) to (22.8).

If one of these assertions is true, the factors $G_\pm \in APW^\pm$ *can be given by* (22.10), (22.11).

Proof. (iii) $\Rightarrow$ (ii) is obvious.

(ii) $\Rightarrow$ (i). By Theorem 22.3, condition (ii) implies that the matrix function G admits a canonical left WH factorization $G = G_+ G_-$ in every $L^p(\mathbf{R}), 1 < p < \infty$. Therefore the operator T_G is invertible on every space $L^p_2(\mathbf{R})$. This in conjunction with Corollary 18.4 gives (i).

(i) $\Rightarrow$ (iii). One simply can repeat the proof of the necessity portion of Theorem 22.6 (taking into consideration that now $f_\pm$, $\widetilde{f}_\pm$, $g_\pm$, $\widetilde{g}_\pm \in APW^\pm$).

To prove the asserted formulas for $G_\pm$, notice first that under the conditions of the theorem, G admits two factorizations: a canonical left WH factorization $G = G_+ G_-$ in every $L^p(\mathbf{R})$ and a canonical left APW factorization. Since, by Theorem 8.1, these two factorizations coincide up to an invertible constant matrix-valued multiple, formulas (22.10), (22.11) give factors $G_\pm$ in $APW^\pm_{2\times 2}$. □

Here is an immediate consequence of Theorem 23.1.

Corollary 23.2. *Let*

$$G = \begin{pmatrix} e_\lambda & 0 \\ f & e_{-\lambda} \end{pmatrix}, \tag{23.1}$$

where $\lambda > 0$ *and* $f \in APW$. *Then the following three assertions are equivalent:*

(i) *G admits a canonical left APW factorization* $G = G_+ G_-$,

(ii) *there exist functions* $f_\pm$, $g_\pm \in H^\infty_\pm$ *satisfying* (22.60) *and* (22.7) to (22.8),

(iii) *there exist functions* $f_\pm$, $g_\pm \in APW^\pm$ *satisfying* (22.60) *and* (22.7) to (22.8).

If one of these assertions is true, the factors $G_\pm \in APW^\pm_{2\times 2}$ *can be given by* (22.10), (22.11) *where* $\widetilde{f}_\pm, \widetilde{g}_\pm \in H^\infty_\mp$ *are solutions of the corona problems* (22.9), *and* (22.55) *holds.* □

Observe that the solutions $G_\pm$ are automatically almost periodic, even when $f_\pm$, $g_\pm$ are merely chosen in $H^\infty_\pm$. This gives us more flexibility when factoring APW matrix functions via corona problems.

From now on, let us consider $N \times N$ almost periodic matrix functions G with bounded Bohr-Fourier spectrum: $\Omega(G) \subset [-\lambda, \lambda]$. Due to Proposition 13.4, this condition may be imposed without loss of generality on all matrices G of the form (23.1).

Some preliminaries are needed.

Lemma 23.3. *Let $G \in GAP_{N\times N}$, $\Omega(G^{\pm 1}) \subset [-\lambda, \lambda]$, $0 \le \lambda < \infty$, and let G possess a canonical left AP factorization $G = G_+G_-$. Then $\Omega(G_+^{\pm 1}) \subset [0, \lambda]$ and $\Omega(G_-^{\pm 1}) \subset [-\lambda, 0]$.*

Proof. Since G has a canonical left AP factorization, we have $G_+^{\pm 1} \in AP_{N\times N}^+$ and $G_-^{\pm 1} \in AP_{N\times N}^-$ by definition. As $G_+ = GG_-^{-1}$ and $G_+^{-1} = G_-G^{-1}$, we see that $\Omega(G_+) \subset [0, \lambda]$ and $\Omega(G_+^{-1}) \subset [0, \lambda]$. Analogously, as $G_- = G_+^{-1}G$ and $G_-^{-1} = G^{-1}G_+$, it follows that $\Omega(G_-) \subset [-\lambda, 0]$ and $\Omega(G_-^{-1}) \subset [-\lambda, 0]$. □

For each $\varphi \in \mathbf{C} \setminus \{0\}$, we define the φ-transform as the algebraic isomorphism of the non-closed subalgebra

$$APW_{N\times N}^0 = \{G \in APW_{N\times N} : \Omega(G) \text{ is bounded}\}$$

onto itself that acts by the rule $G \mapsto G_\varphi$ where

$$G(x) = \sum_\nu G_\nu \, e^{i\nu x}, \tag{23.2}$$

$$G_\varphi(x) = G(x - i\,\log\varphi) = \sum_\nu G_\nu \, \varphi^\nu \, e^{i\nu x}, \tag{23.3}$$

and $\log\varphi$ denotes an arbitrary fixed value of logarithm.

Lemma 23.4. *Let $G \in APW_{N\times N}$ be invertible and $\Omega(G^{\pm 1}) \subset [-\lambda, \lambda]$, where $0 \le \lambda < \infty$. Then G possesses a canonical left APW factorization if and only if for each $\varphi \in \mathbf{C} \setminus \{0\}$ (equivalently, for some $\varphi \in \mathbf{C} \setminus \{0\}$), the matrix function G_φ defined by (23.3) has a canonical left APW factorization. In addition, $\mathbf{d}(G) = \mathbf{d}(G_\varphi)$.*

Proof. Let $G \in APW_{N\times N}$ possess a canonical left APW factorization $G = G_+\,G_-$. Then, by Lemma 23.3, $\Omega(G_+^{\pm 1}) \subset [0, \lambda]$ and $\Omega(G_-^{\pm 1}) \subset [-\lambda, 0]$. Since the φ-transform $G \mapsto G_\varphi$ is an algebraic isomorphism of $APW_{N\times N}^0$ onto itself, the canonical left APW factorization $G = G_+\,G_-$ yields, for each $\varphi \in \mathbf{C} \setminus \{0\}$, the canonical left APW factorization $G_\varphi = G_\varphi^+\,G_\varphi^-$. Obviously, $\mathbf{d}(G) = \mathbf{d}(G_\varphi)$. The reverse statement follows from considering the φ^{-1}-transform. □

Next we state a result about the solvability of the corona problem for functions with bounded Bohr-Fourier spectrum. For $\lambda \ge 0$, put

$$\begin{aligned} APW^+(\lambda) &= \{f \in APW^+ : \Omega(f) \subset [0, \lambda]\}, \\ APW^-(\lambda) &= \{f \in APW^- : \Omega(f) \subset [-\lambda, 0]\}. \end{aligned}$$

Theorem 23.5. *The corona problem with corona data f_+, $g_+ \in APW^+(\lambda)$ such that $M(g_+e_{-\lambda}) \ne 0$ is solvable in $APW^+(\lambda)$ if and only if*

$$\inf_{z\in\mathbf{C}} \max\{|f_+(z)|,\ |g_+(z)|\} > 0. \tag{23.4}$$

Observe that in (23.5) the inf is taken over the whole $\mathbf{C}$, as opposed to corona problems for functions with unbounded Bohr-Fourier spectrum, for which see (22.7).

Proof. Necessity. Let the corona problem be solvable in $APW^+(\lambda)$. Then there exist functions $\widetilde{f}_+$, $\widetilde{g}_+ \in APW^+(\lambda)$ such that

$$f_+ \widetilde{f}_+ + g_+ \widetilde{g}_+ = 1. \tag{23.5}$$

Since the Bohr-Fourier spectra of all functions in (23.5) are bounded, we may apply the φ-transform and, for each $\varphi \in \mathbf{C} \setminus \{0\}$, we get

$$f_\varphi^+ \widetilde{f}_\varphi^+ + g_\varphi^+ \widetilde{g}_\varphi^+ = 1.$$

Thus, for each $\varphi \in \mathbf{C} \setminus \{0\}$,

$$\inf_{z \in \mathbf{C}_+} \max \{|f_\varphi^+(z)|, |g_\varphi^+(z)|\} > 0. \tag{23.6}$$

Since

$$f_\varphi^+(z) = f_+(z - i \log \varphi), \quad g_\varphi^+(z) = g_+(z - i \log \varphi),$$

inequality (23.6) can be rewritten as (23.4).

Sufficiency. Let condition (23.4) hold. We want to show that, for corona data $f_+, g_+ \in APW^+(\lambda)$ subject to the condition $M(g_+ e_{-\lambda}) \neq 0$, the corona problem (23.5) is solvable in $APW^+(\lambda)$.

Since $g_+ \in APW^+(\lambda)$, the function $g_- := g_+ e_{-\lambda}$ belongs to APW^-. As $M(g_-) \neq 0$, it is possible to choose $\varphi \geq 1$ so that the function g_φ^- becomes invertible in APW^-. Consider the function

$$f_\varphi = f_\varphi^+ / g_\varphi^- - h_\varphi^- e_{-\lambda}, \tag{23.7}$$

where $h_\varphi^- \in APW^-$ is chosen so that $\Omega(f_\varphi) \subset [-\lambda, \lambda]$. Then

$$g_\varphi^+ f_\varphi = f_\varphi^+ e_\lambda - f_\varphi^-, \tag{23.8}$$

where $f_\varphi^- = g_\varphi^- h_\varphi^-$ and, moreover, due to (23.8), $f_\varphi^- \in APW^-(\lambda)$. It is clear that

$$f_\varphi^+, \, g_\varphi^+ \in APW^+, \quad f_\varphi^-, \, (g_\varphi^-)^{\pm 1} \in APW^-,$$

and with the help of (23.4) we obtain

$$\begin{aligned} \inf_{z \in \mathbf{C}_+} \max \{|f_\varphi^+(z)|, |g_\varphi^+(z)|\} &= \inf_{z \in \mathbf{C}_+} \max \{|f_+(z - i\log \varphi)|, |g_+(z - i\log \varphi)|\} \\ &\geq \inf_{z \in \mathbf{C}} \max\{|f_+(z)|, |g_+(z)|\} > 0. \end{aligned}$$

Therefore Corollary 23.2 implies that the matrix-valued function (23.1), with $f = f_\varphi$ defined by (23.7), has a canonical left *APW* factorization $G = G_+ G_-$ with

$$G_+ = \begin{pmatrix} \widetilde{f}_\infty^+ + g_\varphi^+ F_+ & g_\varphi^+ \\ -\widetilde{g}_\infty^+ + f_\varphi^+ F_+ & f_\varphi^+ \end{pmatrix}, \quad G_- = \begin{pmatrix} f_\varphi^- & -g_\varphi^- \\ (g_\varphi^-)^{-1} + f_\varphi^- F_- & -g_\varphi^- F_- \end{pmatrix}.$$

Here $\widetilde{f}^+_\infty$, $\widetilde{g}^+_\infty$ are corona solutions in H^∞_+ for the corona data f^+_φ, $g^+_\varphi \in APW^+(\lambda)$, the pair 0, $(g^-_\varphi)^{-1}$ is a corona solution in $APW^-(\subset H^\infty_-)$ for the corona data f^-_φ, $g^-_\varphi \in APW^-(\lambda)$, and

$$F_\pm = -\widetilde{P}_\pm(\widetilde{f}^+_\infty\,(g^-_\varphi)^{-1}\,e_{-\lambda}). \tag{23.9}$$

Obviously, the functions

$$\widetilde{f}^+_\varphi = \widetilde{f}^+_\infty + g^+_\varphi F_+, \quad \widetilde{g}^+_\varphi = \widetilde{g}^+_\infty - f^+_\varphi F_+$$

also give a corona solution in H^∞_+ for the corona data f^+_φ, g^+_φ. Moreover, since $G_+ \in APW^+_{2\times 2}$, it follows from the formulas for $G_\pm$ displayed above that $\widetilde{f}^+_\varphi$, $\widetilde{g}^+_\varphi \in APW^+$. Hence, by Lemma 23.3, $\Omega(\widetilde{f}^+_\varphi) \subset [0,\lambda]$ and $\Omega(\widetilde{g}^+_\varphi) \subset [0,\lambda]$. We so arrive at the conclusion that the functions $\widetilde{f}_+ = (\widetilde{f}^+_\varphi)_{\varphi^{-1}}$, $\widetilde{g}_+ = (\widetilde{g}^+_\varphi)_{\varphi^{-1}}$ belong to $APW^+(\lambda)$ and satisfy (23.5). □

We emphasize once again that Theorem 23.5 allows us to construct corona solutions in $APW^+(\lambda)$ via corona solutions in H^∞_+.

Corollary 23.6. *Under the conditions of Theorem* 23.5, *the corona problem* (23.5) *has the solutions* $\widetilde{f}_+, \widetilde{g}_+ \in APW^+(\lambda)$ *given by*

$$\widetilde{f}_+ = (\widetilde{f}^+_\infty + g^+_\varphi F_+)_{\varphi^{-1}}, \quad \widetilde{g}_+ = (\widetilde{g}^+_\infty - f^+_\varphi F_+)_{\varphi^{-1}}.$$

Here $\widetilde{f}^+_\infty$, $\widetilde{g}^+_\infty \in H^\infty_+$ *are corona solutions for the data* $f^+_\varphi = (f_+)_\varphi$, $g^+_\varphi = (g_+)_\varphi$ *in* $APW^+(\lambda)$, $\varphi \geq 1$ *is chosen so that the function* $g^-_\varphi = g^+_\varphi\,\varphi^{-\lambda}\,e_{-\lambda}$ *is invertible in* APW^-, *and* $F_\pm$ *are defined by* (23.9). □

Clearly, Theorem 23.5 and Corollary 23.6 have analogues for $APW^-(\lambda)$.

Taking into account Theorem 23.5, we get the following supplement to Corollary 23.2.

Theorem 23.7. *Let* G *be defined by* (23.1) *with* $f \in APW$ *and* $\Omega(f) \subset [-\lambda,\lambda]$, $0 < \lambda < \infty$. *Then the following assertions are equivalent:*

(i) *G has a canonical left APW factorization* $G = G_+ G_-$;

(ii) *there exist functions* $f_+, g_+ \in APW^+(\lambda)$, $f_- \in APW^-(\lambda)$ *such that*

$$g_+ f = e_\lambda f_+ - f_-, \quad M(g_+ e_{-\lambda}) \neq 0, \tag{23.10}$$
$$\inf_{z\in\mathbf{C}} \max\{|f_+(z)|,\ |g_+(z)|\} > 0; \tag{23.11}$$

(iii) *there exist functions* $f_-, g_- \in APW^-(\lambda)$, $f_+ \in APW^+(\lambda)$ *such that*

$$g_- f = f_+ - e_{-\lambda} f_-, \quad M(g_- e_\lambda) \neq 0,$$
$$\inf_{z\in\mathbf{C}} \max\{|f_-(z)|,\ |g_-(z)|\} > 0.$$

If assertion (ii) *is true, the factors* $G_\pm$ *can be given by*

$$G_+ = \begin{pmatrix} \widetilde{f}_+ & g_+ \\ -\widetilde{g}_+ & f_+ \end{pmatrix}, \quad G_- = \begin{pmatrix} f_- & -g_- \\ \left[(g_\varphi^-)^{-1}(1-\varphi^{-\lambda} e_{-\lambda} \widetilde{f}_\varphi^+ f_\varphi^-)\right]_{\varphi^{-1}} & \widetilde{f}_+ e_{-\lambda} \end{pmatrix}, \tag{23.12}$$

where $\widetilde{f}_+, \widetilde{g}_+ \in APW^+(\lambda)$ *are corona solutions for the corona data* f_+, g_+ *in* $APW^+(\lambda)$, $g_- := g_+ e_{-\lambda}$, *and* $\varphi \geq 1$ *is chosen so that* g_φ^- *is invertible in* APW^-. *If assertion* (iii) *is true, we have*

$$G_+ = \begin{pmatrix} \widetilde{f}_- e_\lambda & g_+ \\ -\left[(g_\varphi^+)^{-1}(1-\varphi^{\lambda} e_\lambda \widetilde{f}_\varphi^- f_\varphi^+)\right]_{\varphi^{-1}} & f_+ \end{pmatrix}, \quad G_- = \begin{pmatrix} f_- & -g_- \\ \widetilde{g}_- & \widetilde{f}_- \end{pmatrix}, \tag{23.13}$$

where $\widetilde{f}_-, \widetilde{g}_-$ *in* $APW^-(\lambda)$ *are corona solutions for the corona data* $f_-, g_- \in APW^-(\lambda)$, $g_+ := g_- e_\lambda$, *and* $\varphi \in (0,1]$ *is chosen so that* g_φ^+ *is invertible in* APW^+.

Proof. (i) $\Rightarrow$ (ii). We have

$$G_+ = G G_-^{-1}, \tag{23.14}$$

where $G_+^{\pm 1} \in APW_{2\times 2}^+$, $G_-^{\pm 1} \in APW_{2\times 2}^-$, and

$$G_+ = \begin{pmatrix} g_{11}^+ & g_{12}^+ \\ g_{21}^+ & g_{22}^+ \end{pmatrix}, \quad G_-^{-1} = \begin{pmatrix} g_{11}^- & g_{12}^- \\ g_{21}^- & g_{22}^- \end{pmatrix} \tag{23.15}$$

with $\Omega(g_{kj}^+) \subset [0,\lambda]$, $\Omega(g_{kj}^-) \subset [-\lambda, 0]$ $(k,j = 1,2)$ due to Lemma 23.3. By (23.14) and (23.15),

$$\begin{aligned} g_{21}^+ &= f\, g_{11}^- + e_{-\lambda}\, g_{21}^-, \quad & g_{11}^+ &= e_\lambda\, g_{11}^-, \\ g_{22}^+ &= f\, g_{12}^- + e_{-\lambda}\, g_{22}^-, \quad & g_{12}^+ &= e_\lambda\, g_{12}^-, \end{aligned} \tag{23.16}$$

where $M(g_{11}^-) \neq 0$ or $M(g_{12}^-) \neq 0$ because $\det M(G_-) \neq 0$. Let us assume that $M(g_{12}^-) \neq 0$. From (23.16) we then immediately get (23.10) with f_+, g_+ in $APW^+(\lambda)$ and f_- in $APW^-(\lambda)$ if put

$$g_+ := g_{12}^+, \quad f_\pm := g_{22}^\pm.$$

Setting $\widetilde{g}_+ := -g_{21}^+$, $\widetilde{f}_+ := g_{11}^+$ we infer from the equality $\det G_-^{-1} = 1$ that (23.5) holds. Since the corona problem (23.5) is solvable in $APW^+(\lambda)$, we get (23.11) in view of Theorem 23.5 .

(i) $\Rightarrow$ (iii). The proof is analogous.

(ii) $\Rightarrow$ (i). Since $G^{\pm 1} \in APW_{2\times 2}$ and $\Omega(G^{\pm 1}) \subset [-\lambda, \lambda]$, it follows from Lemma 23.4 that G has a canonical left APW factorization only simultaneously with

$$G_\varphi = \begin{pmatrix} \varphi^\lambda e_\lambda & 0 \\ f_\varphi & \varphi^{-\lambda} e_{-\lambda} \end{pmatrix}. \tag{23.17}$$

As $M(g_+ e_{-\lambda}) \neq 0$ we can choose $\varphi \geq 1$ so that $g_\varphi^- = g_\varphi^+ \varphi^\lambda e_\lambda$ becomes invertible in APW^-. Then applying φ-transform to (23.10) we get the identities

$$\begin{pmatrix} f_\varphi^+ & -g_\varphi^+ \\ \widetilde{g}_\varphi^+ & \widetilde{f}_\varphi^+ \end{pmatrix} G_\varphi \begin{pmatrix} 0 & g_\varphi^- \\ -(g_\varphi^-)^{-1} & f_\varphi^- \end{pmatrix}$$
$$= \begin{pmatrix} f_\varphi^+ & -g_\varphi^+ \\ \widetilde{g}_\varphi^+ & \widetilde{f}_\varphi^+ \end{pmatrix} \begin{pmatrix} 0 & g_\varphi^+ \\ -(g_\varphi^+)^{-1} & f_\varphi^+ \end{pmatrix} = \begin{pmatrix} 1 & 0 \\ F & 1 \end{pmatrix}$$

where

$$F = -\widetilde{f}_\varphi^+ (g_\varphi^+)^{-1} = -\widetilde{f}_\varphi^+ \varphi^{-\lambda} e_{-\lambda} (g_\varphi^-)^{-1} \in APW^-$$

because $\widetilde{f}_\varphi^+ \in APW^+(\lambda)$ and $(g_\varphi^-)^{-1} \in APW^-$. Hence

$$G_\varphi = G_\varphi^+ G_\varphi^- \tag{23.18}$$

with

$$G_\varphi^+ = \begin{pmatrix} f_\varphi^+ & -g_\varphi^+ \\ \widetilde{g}_\varphi^+ & \widetilde{f}_\varphi^+ \end{pmatrix}, \quad G_\varphi^- = \begin{pmatrix} f_\varphi^- & -g_\varphi^- \\ (g_\varphi^-)^{-1} + F f_\varphi^- & -F g_\varphi^- \end{pmatrix} \tag{23.19}$$

is a canonical left APW factorization of the matrix function (23.17). By Lemma 23.3, $G_\varphi^\pm \in APW^\pm(\lambda)$. Finally, applying the φ^{-1}-transform to (23.18) and (23.19) we obtain (i) with the factors (23.12).

(iii) $\Rightarrow$ (i). The proof is analogous. □

Note that the form of factors $G_\pm$ in (23.12) and (23.13) is simpler than in (22.10), (22.11) with $F_\pm$ given by (22.55).

23.2 Binomial Periodic Corona Data

Let $E(\xi)$ be the integer part of $\xi \in \mathbf{R}$ and let $\tilde{E}(\xi)$ denote the greatest integer that is less than ξ.

Theorem 23.8. *Let c, $s \in \mathbf{C}$, $0 < \mu \leq \lambda$, and*

$$f_+(x) = e^{i\mu x} - c \ , \ \ g_+(x) = e^{i\lambda x} - s. \tag{23.20}$$

Then the corona problem (f_+, g_+) is solvable in $APW^+(\lambda)$ if and only if

(a) $c^n \neq s^m$ *if* λ/μ *is rational and* $m, n \in \mathbf{N}$ *are relatively prime such that* $\lambda/\mu = m/n$,

(b) $|c|^\lambda \neq |s|^\mu$ *if* λ/μ *is irrational.*

Under these conditions, there exist corona solutions $\tilde{f}_+$, $\tilde{g}_+ \in APW^+$ *with the additional property* $\Omega(\tilde{f}_+) \subset [0, \lambda)$, $\Omega(\tilde{g}_+) \subset [0, \mu)$. *One of them is given by the formulas*

(i)
$$\tilde{f}_+(x) = -\frac{1}{c^n - s^m} \sum_{j=0}^{m-1} \sum_{k=k_{j-1}+1}^{k_j} c^{n-k-1}\, s^j\, e^{i(k\mu - j\lambda)x},$$
$$\tilde{g}_+(x) = \frac{1}{c^n - s^m} \sum_{j=0}^{m-1} c^{n-k_j-1} s^j e^{i((k_j+1)\mu-(j+1)\lambda)x}$$
with $k_j = \tilde{E}((j+1)\lambda/\mu)$, *if* $\lambda/\mu = n/m$ *is rational,*

(ii)
$$\tilde{f}_+(x) = -\sum_{j=0}^{\infty} \sum_{k=k_{j-1}+1}^{k_j} c^{-k-1}\, s^j\, e^{i(k\mu - j\lambda)x},$$
$$\tilde{g}_+(x) = \sum_{j=0}^{\infty} c^{-k_j-1}\, s^j\, e^{i((k_j+1)\mu-(j+1)\lambda)x}$$
with $k_j = \tilde{E}((j+1)\lambda/\mu)$ *and* $0 \notin \Omega(\tilde{g}_+)$ *if* λ/μ *is irrational and* $|c|^\lambda > |s|^\mu$,

(iii)
$$\tilde{f}_+(x) = \sum_{j=1}^{\infty} \sum_{\tau=\tau_j+1}^{\tau_j} c^{\tau-1}\, s^{-j}\, e^{i(-\tau\mu + j\lambda)x},$$
$$\tilde{g}_+(x) = -\sum_{j=1}^{\infty} c^{\tau_j} s^{-j}\, e^{i(-\tau_j\mu+(j-1)\lambda)x}$$
with $\tau_j = E((j-1)\lambda/\mu)$ *and* $0 \notin \Omega(\tilde{f}_+)$ *if* λ/μ *is irrational and* $|c|^\lambda < |s|^\mu$.

Proof. Suppose that the corona problem (f_+, g_+) has a solution in $APW^+(\lambda)$. Then condition (23.4) holds due to Theorem 23.5. Obviously, (23.4) implies conditions (a) and (b).

Conversely, if (a) and (b) hold, then in all the cases (i) to (iii) the functions $\tilde{f}_+$, $\tilde{g}_+$ are well-defined. Note that the series in (ii) (respectively, in (iii)) converge because of the inequality $|c|^\lambda > |s|^\mu$ (respectively, $|c|^\lambda < |s|^\mu$).

Direct computation shows that in all the cases (i) to (iii) the functions $\tilde{f}_+$, $\tilde{g}_+$ satisfy the corona identity $f_+\,\tilde{f}_+ + g_+\,\tilde{g}_+ = 1$. Let, for example, λ/μ be irrational and $|c|^\lambda > |s|^\mu$. Then

$$f_+(x)\tilde{f}_+(x) = \sum_{j=0}^{\infty} c^{-k_{j-1}-1} s^j e^{i((k_{j-1}+1)\mu - j\lambda)x} - \sum_{j=0}^{\infty} c^{-k_j-1} s^j e^{i((k_j+1)\mu - j\lambda)x},$$

$$g_+(x)\tilde{g}_+(x) = -\sum_{j=0}^{\infty} c^{-k_j-1} s^{j+1} e^{i((k_j+1)\mu-(j+1)\lambda)x} + \sum_{j=0}^{\infty} c^{-k_j-1} e^{i((k_j+1)\mu - j\lambda)x},$$

whence, because $k_{-1} = \tilde{E}(0) = -1$,

$$f_+(x)\tilde{f}_+(x) + g_+(x)\tilde{g}_+(x) = c^{-k_{-1}-1} e^{i(k_{-1}+1)\mu x} = 1.$$

Obviously, $0 \notin \Omega(\tilde{g}_+)$.

Cases (i) and (iii) can be considered analogously. Finally, $\Omega(\tilde{f}_+)$ and $\Omega(\tilde{g}_+)$ are contained in $[0,\lambda)$ and $[0,\mu)$, respectively, since $k_j\mu < (j+1)\lambda \le (k_j+1)\mu$ and hence

$$0 \le (k_{j-1}+1)\mu - j\lambda \le k_j\mu - j\lambda = k_j\mu - (j+1)\lambda + \lambda < \lambda,$$
$$0 \le (k_j+1)\mu - (j+1)\lambda = k_j\mu - (j+1)\lambda + \mu < \mu. \qquad \square$$

Notice that in the case (iii) the functions $\tilde{f}_+$, $\tilde{g}_+$ can be rewritten in the form

$$\tilde{f}_+(x) = \sum_{j=-\infty}^{-1} \sum_{k=k_{j-1}+1}^{k_j} c^{-k-1} s^j e^{i(k\mu - j\lambda)x},$$
$$\tilde{g}_+(x) = -\sum_{j=-\infty}^{-1} c^{-k_j-1} s^j\, e^{i((k_j+1)\mu-(j+1)\lambda)x}$$

because $E((j-1)\lambda/\mu) = -\tilde{E}((-j+1)\lambda/\mu) - 1$, that is, $\tau_j = -k_{-j} - 1$. This representation shares some symmetry with (ii).

23.3 Trinomial Matrix Functions

In this section we return to the matrix functions

$$G = \begin{pmatrix} e_\lambda & 0 \\ c_{-1}e_{-\nu} - c_0 + c_1 e_\delta & e_{-\lambda} \end{pmatrix} \tag{23.21}$$

with

$$c_{-1}c_0c_1 \ne 0 \quad \text{and} \quad 0 < \delta,\, \nu < \lambda = \nu + \delta. \tag{23.22}$$

A criterion for the existence of a canonical left AP factorization for such matrix functions was established in Theorem 15.5 for irrational ν/δ and in Theorem 14.14 for rational ν/δ. Here we give an alternative approach, leading not only to existence results but also to explicit formulas for the factors. This approach is based on the corona solutions found in Theorem 23.8. Recall that in Section 15.1 we presented only an algorithm for calculating those factors in the case of irrational ν/δ.

In accordance with Theorem 23.8, we define

$$k_j = \tilde{E}((j+1)\lambda/\delta), \quad \tau_j = E((j-1)\lambda/\delta), \quad j \in \mathbf{Z}. \tag{23.23}$$

Let also $\beta = \nu/\delta$ and

$$c_{\pm 1,0} = c_{\pm 1}c_0^{-1}, \quad s = c_{-1}c_1^{-1}. \tag{23.24}$$

If β is rational, we denote by n and m the relatively prime natural numbers such that $1 + \beta\,(= \lambda/\delta\,) = n/m$.

Theorem 23.9. *The matrix function G defined by* (23.21), (23.22) *has a canonical left APW factorization $G = G_+G_-$ if and only if either*

$$\beta \text{ is rational and } c_1^{n-m}c_{-1}^{m} \neq c_0^{n}, \tag{23.25}$$

or

$$\beta \text{ is irrational and } |c_1|^{\beta}|c_{-1}| \neq |c_0|^{\beta+1}. \tag{23.26}$$

Under these conditions the factors $G_\pm$ can be given by

$$G_+(x) = \begin{pmatrix} -\dfrac{c_0^{-1}}{c_{1,0}^{-n} - s^m} \displaystyle\sum_{j=0}^{m-1} \sum_{k=k_{j-1}+1}^{k_j} c_{1,0}^{-n+k} s^j e^{i(k\delta - j\lambda)x} & e^{i\lambda x} - s \\ -\dfrac{1}{c_{1,0}^{-n} - s^m} \displaystyle\sum_{j=0}^{m-1} c_{1,0}^{-n+k_j+1} s^j e^{i((k_j+1)\delta-(j+1)\lambda)x} & c_1 e^{i\delta x} - c_0 \end{pmatrix},$$

$$G_-(x) = \begin{pmatrix} -c_0 s + c_{-1} s e^{-i\nu x} & s e^{-i\lambda x} - 1 \\ G_-^{21}(x) & G_-^{22}(x) \end{pmatrix}$$

with

$$G_-^{21}(x) = 1 - \frac{1}{c_{1,0}^{-n} - s^m} \sum_{j=0}^{m-1} \sum_{k=k_{j-1}+2}^{k_j} c_{1,0}^{-n+k} s^{j+1} e^{i(k\delta-(j+1)\lambda)x},$$

$$G_-^{22}(x) = -c_0^{-1} \frac{1}{c_{1,0}^{-n} - s^m} \sum_{j=0}^{m-1} \sum_{k=k_{j-1}+1}^{k_j} c_{1,0}^{-n+k} s^j e^{i(k\delta-(j+1)\lambda)x}$$

if (23.25) *holds,*

$$G_+(x) = \begin{pmatrix} -c_0^{-1}\sum\limits_{j=0}^{\infty}\sum\limits_{k=k_{j-1}+1}^{k_j} c_{1,0}^{k} s^j e^{i(k\delta - j\lambda)x} & e^{i\lambda x} - s \\ -\sum\limits_{j=0}^{\infty} c_{1,0}^{k_j+1} s^j e^{i((k_j+1)\delta-(j+1)\lambda)x} & c_1 e^{i\delta x} - c_0 \end{pmatrix}, \tag{23.27}$$

$$G_-(x) = \begin{pmatrix} -c_0 s + c_{-1} s e^{-i\nu x} & s e^{-i\lambda x} - 1 \\ G_-^{21}(x) & G_-^{22}(x) \end{pmatrix} \tag{23.28}$$

with

$$G_-^{21}(x) = 1 - \sum_{j=0}^{\infty}\sum_{k=k_{j-1}+2}^{k_j} c_{1,0}^{k} s^{j+1} e^{i(k\delta-(j+1)\lambda)x},$$

$$G_-^{22}(x) = -c_0^{-1}\sum_{j=0}^{\infty}\sum_{k=k_{j-1}+1}^{k_j} c_{1,0}^{k} s^j e^{i(k\delta-(j+1)\lambda)x}$$

if β is irrational and $|c_1|^\beta|c_{-1}| < |c_0|^{\beta+1}$, and

$$G_+(x) = \begin{pmatrix} c_0^{-1}\sum\limits_{j=1}^{\infty}\sum\limits_{\tau=\tau_j+1}^{\tau_{j+1}} c_{1,0}^{-\tau} s^{-j} e^{i(-\tau\delta+j\lambda)x} & e^{i\lambda x} - s \\ \sum\limits_{j=1}^{\infty} c_{1,0}^{-\tau_j} s^{-j} e^{i(-\tau_j\delta+(j-1)\lambda)x} & c_1 e^{i\delta x} - c_0 \end{pmatrix}, \tag{23.29}$$

$$G_-(x) = \begin{pmatrix} -c_0 s + c_{-1} s e^{-i\nu x} & s e^{-i\lambda x} - 1 \\ G_-^{21}(x) & G_-^{22}(x) \end{pmatrix} \tag{23.30}$$

with

$$G_-^{21}(x) = 1 + \sum_{j=1}^{\infty}\sum_{\tau=\tau_j+1}^{\tau_{j+1}-1} c_{1,0}^{-\tau} s^{-j+1} e^{i(-\tau\delta+(j-1)\lambda)x},$$

$$G_-^{22}(x) = c_0^{-1}\sum_{j=1}^{\infty}\sum_{\tau=\tau_j+1}^{\tau_{j+1}} c_{1,0}^{-\tau} s^{-j} e^{i(-\tau\delta+(j-1)\lambda)x}$$

if β is irrational and $|c_1|^\beta|c_{-1}| > |c_0|^{\beta+1}$.

Proof. Obviously, the function $f := c_{-1}e_{-\nu} - c_0 + c_1 e_\delta$ satisfies the identity $g_+ f = e_\lambda f_+ - f_-$ where $g_+ = e_\lambda - s$, $f_+ = c_1(e_\delta - c_1^{-1}c_0)$, $f_- = s(c_{-1}e_{-\nu} - c_0)$, and s is given by (23.24). Thus $g_+, f_+ \in APW^+(\lambda)$, $f_- \in APW^-(\lambda)$ and

$M(g_+e_{-\lambda}) = -s \neq 0$, which gives (23.10). On the other hand, conditions (23.25) and (23.26) guarantee the solvability of the corona problem (f_+, g_+) in $APW^+(\lambda)$ (recall Theorem 23.8). According to Theorem 23.5, this implies (23.11). Thus, if (23.25) and (23.26) are satisfied, then the matrix function (23.21) has a canonical left *APW* factorization in view of Theorem 23.7 (case (ii)), and its factors $G_\pm$ are given by (23.12).

To calculate the entries in (23.12) we use corona solutions $\tilde{f}_+, \tilde{g}_+ \in APW^+(\lambda)$ for the corona data $f_+, g_+ \in APW^+(\lambda)$. Let k_j, τ_j be given by (23.23) and let $c_{\pm 1,0}$, s be given by (23.24). From Theorem 23.8 we get the following corona solutions $\tilde{f}_+$, $\tilde{g}_+ \in APW^+(\lambda)$. If λ/δ is rational, then

$$\tilde{f}_+(x) = -\frac{c_0^{-1}}{c_{1,0}^{-n} - s^m} \sum_{j=0}^{m-1} \sum_{k=k_{j-1}+1}^{k_j} c_{1,0}^{-n+k} s^j e^{i(k\delta - j\lambda)x},$$

$$\tilde{g}_+(x) = \frac{1}{c_{1,0}^{-n} - s^m} \sum_{j=0}^{m-1} c_{1,0}^{-n+k_j+1} s^j e^{i((k_j+1)\delta - (j+1)\lambda)x}.$$

If λ/δ is irrational and $|c_1|^\beta |c_{-1}| < |c_0|^{\beta+1}$ then

$$\tilde{f}_+(x) = -c_0^{-1} \sum_{j=0}^{\infty} \sum_{k=k_{j-1}+1}^{k_j} c_{1,0}^{k}\, s^j\, e^{i(k\delta - j\lambda)x},$$

$$\tilde{g}_+(x) = \sum_{j=0}^{\infty} c_{1,0}^{k_j+1}\, s^j\, e^{i((k_j+1)\delta - (j+1)\lambda)x}.$$

If λ/δ is irrational and $|c_1|^\beta |c_{-1}| > |c_0|^{\beta+1}$ then

$$\tilde{f}_+(x) = c_0^{-1} \sum_{j=1}^{\infty} \sum_{\tau=\tau_j+1}^{\tau_{j+1}} c_{1,0}^{-\tau}\, s^{-j}\, e^{i(-\tau\delta + j\lambda)x},$$

$$\tilde{g}_+(x) = -\sum_{j=1}^{\infty} c_{1,0}^{-\tau_j}\, s^{-j}\, e^{i(-\tau_j\delta + (j-1)\lambda)x}.$$

From here and (23.12) we obtain the desired formulas for G_+. We now pass to the computation of G_-.

Let first $|s| < 1$. Then the function $g_- = g_+ e_{-\lambda} = 1 - se_{-\lambda}$ is invertible in APW^- and we may use (23.12) with $\varphi = 1$. In this case one can check directly that $(g_-(x))^{-1}\,(1 - e^{-i\lambda x}\,\tilde{f}_+(x)\,f_-(x))$ equals

$$1 - \frac{1}{c_{1,0}^{-n} - s^m} \sum_{j=0}^{m-1} \sum_{k=k_{j-1}+2}^{k_j} c_{1,0}^{-n+k} s^{j+1} e^{i(k\delta - (j+1)\lambda)x}$$

if λ/δ is rational,

$$1-\sum_{j=0}^{\infty}\sum_{k=k_{j-1}+2}^{k_j} c_{1,0}^{k}\, s^{j+1}\, e^{i(k\delta-(j+1)\lambda)x}$$

if λ/δ is irrational and $|c_1|^{\beta}|c_{-1}| < |c_0|^{\beta+1}$,

$$1+\sum_{j=1}^{\infty}\sum_{\tau=\tau_j+1}^{\tau_{j+1}-1} c_{1,0}^{-\tau}\, s^{-j+1}\, e^{i(-\tau\delta+(j-1)\lambda)x}$$

if λ/δ is irrational and $|c_1|^{\beta}|c_{-1}| > |c_0|^{\beta+1}$. This yields the desired formulas for G_-, and thus the asserted factorization of G when $|s| < 1$.

The case $|s| \geq 1$ can be disposed of with the following trick. Choose $\varphi > 0$ so that $|c_{-1}c_1^{-1}| < \varphi^{\lambda}$ and define $\tilde{c}_{-1} = c_{-1}\varphi^{-\nu}$, $\tilde{c}_1 = c_1\varphi^{\delta}$. According to (23.3),

$$G_{\varphi} = \begin{pmatrix} \varphi^{\lambda}e_{\lambda} & 0 \\ \tilde{c}_{-1}e_{-\nu} - c_0 + \tilde{c}_1 e_{\delta} & \varphi^{-\lambda}e_{-\lambda} \end{pmatrix} = \begin{pmatrix} \varphi^{\lambda} & 0 \\ 0 & 1 \end{pmatrix} \widetilde{G} \begin{pmatrix} 1 & 0 \\ 0 & \varphi^{-\lambda} \end{pmatrix}, \tag{23.31}$$

where

$$\widetilde{G} = \begin{pmatrix} e_{\lambda} & 0 \\ \tilde{c}_{-1}e_{-\nu} - c_0 + \tilde{c}_1 e_{\delta} & e_{-\lambda} \end{pmatrix}.$$

Since $|\tilde{c}_{-1}\tilde{c}_1^{-1}| < 1$, the factors $\widetilde{G}_{\pm}$ of $\widetilde{G}$ can be computed by the formulas above (with $c_{\pm 1}$ changed to $\tilde{c}_{\pm 1}$ and $s = \tilde{c}_{-1}\tilde{c}_1^{-1}$). After that, using (23.31) and a subsequent φ^{-1}-transform, we arrive at the asserted factorization formulas for the original G. Note that the applicability of the φ^{-1}-transform to $\widetilde{G}_{\pm}$ is justified by the boundedness of the Bohr-Fourier spectra.

Finally, we move on to the necessity of conditions (23.25), (23.26) for the existence of a canonical left APW factorization of G. For irrational β, everything follows from Lemma 15.4. For rational β, the matrix function G has a canonical left APW factorization only simultaneously with

$$G_0(x) = \begin{pmatrix} e^{inx} & 0 \\ c_{-1}e^{-ikx} - c_0 + c_1 e^{imx} & e^{-inx} \end{pmatrix}, \tag{23.32}$$

where $k = n - m$, because $G_0(x) = G(mx/\delta)$. But Theorem 14.14(a) shows that the matrix-valued function (23.32) has a canonical AP-factorization if and only if

$$\det\,(c_{-1}\delta_{i+k,j} - c_0\delta_{i,j} + c_1\delta_{i,m+j})_{i,j=1}^{n} \neq 0, \tag{23.33}$$

where δ_{ij} is the Kronecker symbol, that is, if and only if $(-1)^n\,(c_0^n - c_1^k c_{-1}^m) \neq 0$, which completes the proof. □

We remark that the matrix function G defined by (23.21) is not at all AP factorable if $|c_1|^\beta\,|c_{-1}| \neq |c_0|^{1+\beta}$ and $\beta = \nu/\delta$ is irrational (Theorem 15.5). In the case of rational β, G is always AP factorable (Theorem 14.13). Thus, if $c_1^{n-m}c_{-1}^m = c_0^n$, then G is AP factorable with non-zero partial AP-indices.

Corollary 23.2, combined with Theorems 22.8 and 23.8, allows to construct explicit APW factorizations of the matrix functions (23.1) with more complicated f's. In particular, the following case can be tackled:

$$f(x) = \frac{\prod_{j=1}^{s} (c_j\, e^{i\nu_j x} - d_j)^{\beta_j}\, e^{i\lambda x} - \prod_{j=s+1}^{\tau} (c_j\, e^{-i\nu_j x} - d_j)^{\beta_j}}{e^{i\mu_+ x} \prod_{k=1}^{n} (a_k\, e^{i\mu_k x} - b_k)^{\alpha_k}}, \tag{23.34}$$

where $\lambda > 0$ and, for $j = 1, 2, \ldots, \tau$,

$$c_j, d_j \in \mathbf{C} \setminus \{0\}, \quad \nu_j > 0, \tag{23.35}$$

$$\beta_j > 0 \text{ if } |c_j| = |d_j| \text{ and } \beta_j \in \mathbf{N} \text{ if } |c_j| \neq |d_j|, \tag{23.36}$$

$$\mu_+ \geq 0 \text{ and } \mu_- := \mu_+ + \sum_{k=1}^{n} \mu_k\, \alpha_k - \lambda \leq 0, \tag{23.37}$$

$$a_k,\, b_k \in \mathbf{C} \setminus \{0\}, \quad |a_k| \neq |b_k|, \quad \mu_k > 0, \quad \alpha_k \in \mathbf{N}. \tag{23.38}$$

We confine ourselves to a canonical APW factorization criterion. Let

$$J_+ = \{1, 2, \ldots, s\}, \quad J_- = \{s+1, \ldots, \tau\},$$
$$K_\pm = \{k = 1, \ldots, n : \pm(|a_k| - |b_k|) > 0\}.$$

Theorem 23.10. *Let* $\lambda > 0$ *and let* f *be given by* (23.34) *to* (23.38). *Then the matrix function* (23.1) *admits a canonical APW factorization if and only if for every choice of* $\pm$ *and for every pair* $(k, j) \in K_\pm \times J_\pm$ *we have*

$(d_j/c_j)^{m_k} \neq \left(\dfrac{b_k}{a_k}\right)^{\pm n_j}$ *if* ν_j/μ_k *is rational with* $n_j, m_k \in \mathbf{N}$ *relatively prime such that* $\nu_j/\mu_k = n_j/m_k$;

$|d_j/c_j|^{\mu_k} \neq \left|\dfrac{b_k}{a_k}\right|^{\pm\nu_j}$ *if* ν_j/μ_k *is irrational.* □

Notes

This chapter, as the previous one, is based on the papers [20], [22], and [23].

Note that with the help of Theorem 23.9 one can construct explicit canonical AP factorizations for all the triangular 2×2 matrix functions considered in [24]. Theorem 23.10 generalizes Theorem 5.1 of [23] and corrects assertion (iii) of that theorem.

Bibliography

[1] V. M. Adamjan, D. Z. Arov, and M. G. Krein: Infinite Hankel block matrices and related extension problems. *Amer. Math. Soc. Transl.* **111** (1978), 133–156.

[2] V. Adukov: On invertibility of matrix Wiener-Hopf operator on discrete linearly ordered abelian group. *Integral Equations Operator Theory* **23** (1995), 373–386.

[3] A. Aktosun, M. Klaus, and C. van der Mee: Inverse wave scattering with discontinuous wavespeed. *J. Math. Phys.* **36** (1995), 2880–2928.

[4] A. B. Antonevich: *Linear Functional Equations. Operator Approach.* Birkhäuser, Basel and Boston 1996. Russian original: University Press, Minsk 1988.

[5] A. B. Antonevich and A. Lebedev: *Functional-Differential Equations. I. C^*-Theory.* Longman Scientific & Technical, Harlow 1994.

[6] R. Arens and I. Singer: Generalized analytic functions. *Trans. Amer. Math. Soc.* **81** (1956), 379–393.

[7] R. G. Babadzhanyan and V. S. Rabinovich: A system of integro-difference equations on the half line. *Akad. Nauk Armyan. SSR Dokl.* **81** (1985), 107–111 [Russian].

[8] R. G. Babadzhanyan and V. S. Rabinovich: On the factorization of almost periodic matrix functions. In: *Differential and Integral Equations, and Complex Analysis*, pp. 13–22, Kalmyk University Press, Elista 1986 [Russian].

[9] M. Bakonyi, L. Rodman, I. M. Spitkovsky, and H. J. Woerdeman: Positive extensions of matrix functions of two variables with support in an infinite band. *C. R. Acad. Sci. Paris, Sér.* I, *Math.* **323** (1996), 859–863.

[10] M. Bakonyi, L. Rodman, I. M. Spitkovsky, and H. J. Woerdeman: Positive matrix functions on the bitorus with prescibed coefficients in a band. *J. Fourier Analysis and Appl.* **5** (1999), 789–812.

[11] J. A. Ball, Yu. I. Karlovich, L. Rodman, and I. M. Spitkovsky: Sarason interpolation and Toeplitz corona theorem for almost periodic matrix functions. *Integral Equations Operator Theory* **32** (1998), 243–281.

[12] C. S. Ballantine: Products of positive definite matrices, III. *J. Algebra* **10** (1968), 174–182.

[13] C. S. Ballantine: Products of positive definite matrices, IV. *Linear Algebra Appl.* **3** (1970), 79–114.

[14] H. Bart, I. Gohberg, and M. A. Kaashoek: The coupling method for solving integral equations. *Oper. Theory Adv. Appl.* **12** (1984), 39–73.

[15] H. Bart and V. Tsekanovskii: Matricial coupling and equivalence after extension. *Oper. Theory Adv. Appl.* **59** (1992), 143–160.

[16] M. A. Bastos, A. Bravo, and Yu. I. Karlovich: Singular integral operators with piecewise almost periodic coefficients and Carleman shift. *Preprint* 16/2001, Instituto Superior Técnico, Lisbon 2001.

[17] M. A. Bastos and A. F. dos Santos: Convolution equations of the first kind on a finite interval in Sobolev spaces. *Integral Equations Operators Theory* **13** (1990), 638–659.

[18] M. A. Bastos and A. F. dos Santos: Wiener-Hopf operators with oscillating symbols and convolution operators on a union of intervals. *Integral Equations Operator Theory* **15** (1992), 920–941.

[19] M. A. Bastos, A. F. dos Santos, and R. Duduchava: Finite interval convolution operators on the Bessel potential spaces H_p^s. *Math. Nach.* **173** (1995), 49–63.

[20] M. A. Bastos, Yu. I. Karlovich, and A. F. dos Santos: Oscillatory Riemann-Hilbert problems and the corona problem. *Preprint* 1/99, Instituto Superior Técnico, Lisbon 1999.

[21] M. A. Bastos, Yu. I. Karlovich, and A. F. dos Santos: The invertibility of convolution type operators on a union of intervals and the corona theorem. *Integral Equations Operator Theory,* to appear.

[22] M. A. Bastos, Yu. I. Karlovich, A. F. dos Santos, and P. M. Tishin: The corona theorem and the existence of canonical factorization of triangular *AP*-matrix functions. *J. Math. Anal. Appl.* **223** (1998), 494–522.

[23] M. A. Bastos, Yu. I. Karlovich, A. F. dos Santos, and P. M. Tishin: The corona theorem and the canonical factorization of triangular *AP*-matrix functions - Effective criteria and explicit formulas. *J. Math. Anal. Appl.* **223** (1998), 523–550.

[24] M. A. Bastos, Yu. I. Karlovich, I. M. Spitkovsky, and P. M. Tishin: On a new algorithm for almost periodic factorization. *Oper. Theory Adv. Appl.* **103** (1998), 53–74.

[25] M. Sh. Birman and M. Z. Solomjak: *Spectral Theory of Selfadjoint Operators in Hilbert Space.* D. Reidel Publishing Co., Dordrecht 1987. Russian original: University Press, St. Petersburg 1980.

[26] C. J. Bishop, A. Böttcher, Yu. I. Karlovich, and I. M. Spitkovsky: Local spectra and index of singular integral operators with piecewise continuous coefficients on composed curves. *Math. Nachr.* **206** (1999), 5–83.

[27] S. Bochner and R. S. Phillips: Absolutely convergent Fourier expansion for non-commutative normed rings. *Ann. of Math.* **43** (1942), 409–418.

[28] A. Böttcher: On the corona theorem for almost periodic functions. *Integral Equations Operator Theory* **33** (1999), 253–272.

[29] A. Böttcher, S. M. Grudsky, and I. M. Spitkovsky: On the Fredholm indices of associated systems of Wiener-Hopf equations. *J. Integral Equations Appl.* **12** (2000), 1–29.

[30] A. Böttcher, S. M. Grudsky, and I. M. Spitkovsky: Matrix functions with arbitrarily prescribed left and right partial indices. *Integral Equations Operator Theory* **36** (2000), 71–91.

[31] A. Böttcher, S. M. Grudsky, and I. M. Spitkovsky: The spectrum is discontinuous on the manifold of Toeplitz operators. *Arch. Math.* **75** (2000), 46–52.

[32] A. Böttcher, S. M. Grudsky, and I. M. Spitkovsky: Toeplitz operators with frequency modulated semi-almost periodic symbols. *J. Fourier Analysis Appl.* **7** (2001), 523–535.

[33] A. Böttcher, S. M. Grudsky, and I .M. Spitkovsky: On the essential spectrum of Toeplitz operators with semi-almost periodic symbols. *Oper. Theory Adv. Appl.*, to appear.

[34] A. Böttcher and Yu. I. Karlovich: Toeplitz and singular integral operators on Carleson curves with logarithmic whirl points. *Integral Equations Operator Theory* **22** (1995), 127–161.

[35] A. Böttcher and Yu. I. Karlovich: Toeplitz and singular integral operators on general Carleson Jordan curves. *Oper. Theory Adv. Appl.* **90** (1996), 119–152.

[36] A. Böttcher and Yu. I. Karlovich: Submultiplicative functions and spectral theory of Toeplitz operators. *Integral Transforms and Special Functions* **4** (1996), 181–202.

[37] A. Böttcher and Yu. I. Karlovich: *Carleson Curves, Muckenhoupt Weights, and Toeplitz Operators.* Progress in Mathematics, Vol. 154, Birkhäuser, Basel 1997.

[38] A. Böttcher and Yu. I. Karlovich: Toeplitz operators with PC symbols on general Carleson Jordan curves with arbitrary Muckenhoupt weights. *Trans. Amer. Math. Soc.* **351** (1999), 3143–3196.

[39] A. Böttcher, Yu. I. Karlovich, and V. S. Rabinovich: Emergence, persitence, and disappearance of logarithmic spirals in the spectra of singular integral operators. *Integral Equations Operator Theory* **25** (1996), 406–444.

[40] A. Böttcher, Yu. I. Karlovich, and V. S. Rabinovich: Mellin pseudodifferential operators with slowly varying symbols and singular integrals on Carleson curves with Muckenhoupt weights. *Manuscripta Math.* **95** (1998), 363–376.

[41] A. Böttcher, Yu. I. Karlovich, and V. S. Rabinovich: The method of limit operators for one-dimensional singular integrals with slowly oscillating data. *J. Operator Theory* **43** (2000), 171–198.

[42] A. Böttcher, Yu. I. Karlovich, and V. S. Rabinovich: Singular integral operators with complex conjugation from the viewpoint of pseudodifferential operators. *Oper. Theory Adv. Appl.* **121** (2001), 36–59.

[43] A. Böttcher, Yu. I. Karlovich, and I. M. Spitkovsky: Toeplitz operators with semi-almost periodic symbols on spaces with Muckenhoupt weight. *Integral Equations Operator Theory* **18** (1994), 261–276.

[44] A. Böttcher, Yu. I. Karlovich, and I. M. Spitkovsky: Toeplitz operators with semi-almost periodic matrix symbols on Hardy spaces. *Acta. Appl. Math.* **65** (2001), 115–136.

[45] A. Böttcher, N. Krupnik, and B. Silbermann: A general look at local principles with special emphasis on the norm computation aspect. *Integral Equations Operator Theory* **11** (1988), 455–479.

[46] A. Böttcher and B. Silbermann: *Analysis of Toeplitz Operators.* Springer, Berlin, Heidelberg, New York 1990.

[47] A. Böttcher and B. Silbermann: *Introduction to Large Truncated Toeplitz Matrices.* Springer, New York 1999.

[48] A. Böttcher and I. M. Spitkovsky: Wiener-Hopf integral operators with PC symbols on spaces with Muckenhoupt weight. *Revista Matemática Iberoamericana* **9** (1993), 257–279.

[49] A. Böttcher and I. M. Spitkovsky: Pseudodifferential operators with heavy spectrum. *Integral Equations Operator Theory* **19** (1994), 251–269.

[50] A. Brown and P. R. Halmos: Algebraic properties of Toeplitz operators. *J. Reine Angew. Math.* **213** (1963/1964), 89–102.

[51] F. M. Callier and J. Winkin: The spectral factorization problem for multivariable distributed parameter systems. *Integral Equations Operator Theory* **34** (1999), 270–292.

[52] F. M. Callier and J. Winkin: Spectral factorization of a spectral density with arbitrary delays. In: *Open Problems in Mathematical Systems and Control Theory*, pp. 79-82, Springer, London 1999.

[53] L. Carleson: Interpolation by bounded analytic functions and the corona problem. *Annals of Math.* **76** (1962), 547–559.

[54] L. P. Castro and F.-O. Speck: A Fredholm study for convolution operators with piecewise continuous symbols on a union of a finite and a semi-infinite interval. *Appl. Anal.* **64** (1997), 171–187.

[55] L. P. Castro and F.-O. Speck: Relations between convolution type operators on intervals and on the half-line. *Integral Equations Operator Theory* **37** (2000), 169–207.

[56] G. N. Chebotarev: Partial indices of the Riemann boundary value problem with a triangular matrix of second order. *Uspehi Matem. Nauk* **11** (1956), 192–202 [Russian].

[57] K. F. Clancey and I. Gohberg: *Factorization of Matrix Functions and Singular Integral Operators.* Birkhäuser, Basel and Boston 1981.

[58] L. Coburn: Weyl's theorem for non-normal operators. *Michigan Math. J.* **13** (1966), 285–286.

[59] L. Coburn and R. G. Douglas: Translation operators on the half-line. *Proc. Nat. Acad. Sci. USA* **62** (1969), 1010–1013.

[60] C. Corduneanu: *Almost Periodic Functions.* J. Wiley & Sons, New York, London, Sydney 1968.

[61] R. E. Curto, P. S. Muhly, and Jingbo Xia: Toeplitz operators on flows. *J. Funct. Anal.* **93** (1990), 391–450.

[62] K. R. Davidson: *C^*-Algebras by Example.* Amer. Math. Soc., Providence, RI 1996.

[63] A. Devinatz: Toeplitz operators on H^2 spaces. *Trans. Amer. Math. Soc.* **112** (1964), 304–317.

[64] J. Dixmier: *C^*-Algebras.* North-Holland Publishing Co., Amsterdam 1977.

[65] R. G. Douglas: Toeplitz and Wiener-Hopf operators in $H^\infty + C$. *Bull. Amer. Math. Soc.* **74** (1968), 895–899.

[66] R. G. Douglas: *Banach Algebra Techniques in Operator Theory.* Second ed., Springer, New York 1998.

[67] R. G. Douglas and D. Sarason: Fredholm Toeplitz operators. *Proc. Amer. Math. Soc.* **26** (1970), 117–120.

[68] R. V. Duduchava: Wiener-Hopf integral operators with discontinuous symbols. *Soviet Math. Dokl.* **14** (1973), 1001–1005.

[69] R. V. Duduchava: Integral operators of convolution type with discontinuous coefficients. *Math. Nachr.* **79** (1977), 75–98 [Russian].

[70] R. V. Duduchava: *Integral Equations with Fixed Singularities.* Teubner, Leipzig 1979. Russian original: Trudy Tbiliss. Mat. Inst. Razmadze Akad. Nauk Gruzin. SSR, Vol. 60, Tiflis 1979.

[71] R. V. Duduchava: On algebras generated by convolutions and discontinuous functions. *Integral Equations Operator Theory* **10** (1987), 505–530.

[72] R. V. Duduchava and A. I. Saginashvili: Convolution integral equations on a half-line with semi-almost-periodic presymbols. *Differential Equations* **17** (1981), 207–216.

[73] P. L. Duren: *Theory of H^p Spaces.* Academic Press, New York and London 1970.

[74] D. R. Farenick and W. Y. Lee: Hyponormality and spectra of Toeplitz operators. *Trans. Amer. Math. Soc.* **348** (1996), 4153–4174.

[75] J. Favard: Sur les equations differentielles á coefficients presque-periodiques. *Acta Math.* **51** (1927), 31–81.

[76] I. Feldman, I. Gohberg, and N. Krupnik: On explicit factorization and applications. *Integral Equations Operator Theory* **21** (1995), 430–459.

[77] I. Feldman, I. Gohberg, and N. Krupnik: Convolution equations on finite intervals and factorization of matrix functions. *Integral Equations Operator Theory* **36** (2000), 201–211.

[78] I. Feldman and A. Markus: On some properties of factorization indices. *Integral Equations Operator Theory* **30** (1998), 326–337.

[79] T. Fink, S. Roch, and B. Silbermann: Two projection theorems and symbol calculus for operators with massive local spectra. *Math. Nachr.* **162** (1993), 167–185.

[80] F. D. Gakhov: *Boundary Value Problems.* Dover Publ. Inc., New York 1990. Russian original: Nauka, Moscow 1977.

[81] M. P. Ganin: On a Fredholm integral equation whose kernel depends on the difference of the arguments. *Izv. Vys. Uchebn. Zaved. Matematika* **2/1963** (1963), 31–43 [Russian].

[82] F. R. Gantmacher: *The Theory of Matrices.* Vol. 1, Chelsea Publishing Company, New York 1959 and AMS Chelsea Publishing, Providence, RI, 1998. Fourth Russian edition: Nauka, Moscow 1988.

[83] J. García-Cuerva and J. L. Rubio de Francia: *Weighted Norm Inequalities and Related Topics.* North-Holland Publishing Co., Amsterdam 1985.

[84] J. B. Garnett: *Bounded Analytic Functions.* Academic Press, New York 1981.

[85] I. Gelfand, D. Raikov, and G. Shilov: *Commutative Normed Rings.* Chelsea, Bronx, N.Y. 1964. Russian original: Fizmatgiz, Moscow 1960.

[86] M. Gelfand and I. M. Spitkovsky: Almost periodic factorization: Applicability of the division algorithm. *Voronezh Winter Mathematical Schools, Amer. Math. Soc. Transl.* **184** (1998), 97–109.

[87] I. Gohberg: On an application of the theory of normed rings to singular integral equations. *Uspehi Matem. Nauk* **7** (1952), 149–156 [Russian].

[88] I. Gohberg: On the number of solutions of a homogeneous singular integral equation with continuous coefficients. *Dokl. Akad. Nauk SSSR* **122** (1958), 327–330 [Russian].

[89] I. Gohberg: Toeplitz matrices composed of the Fourier coefficients of piecewise continuous functions. *Funktsional. Anal. i Prilozhen.* **1** (1967), 91–92 [Russian].

[90] I. Gohberg and I. Feldman: Wiener-Hopf integro-difference equations. *Soviet Math. Dokl.* **9** (1968), 1312–1316.

[91] I. Gohberg and I. Feldman: Integro-difference Wiener-Hopf equations. *Acta Sci. Math. Szeged* **30** (1969), 199–224 [Russian].

[92] I. Gohberg and I. Feldman: *Convolution Equations and Projection Methods for Their Solution.* Amer. Math. Soc. Transl. of Math. Monographs **41**, Providence, RI 1974. Russian original: Nauka, Moscow 1971.

[93] I. Gohberg, S. Goldberg, and M. A. Kaashoek: *Classes of Linear Operators.* Vol. I, Birkhäuser, Basel 1990.

[94] I. Gohberg and M. G. Krein: Systems of integral equations on a half-line with kernel depending upon the difference of the arguments. *Amer. Math. Soc. Transl.* **14** (1960), 217–287.

[95] I. Gohberg and N. Krupnik: On the algebra generated by Toeplitz matrices. *Funct. Anal. Appl.* **3** (1969), 119–127.

[96] I. Gohberg and N. Krupnik: The spectrum of singular integral operators in L_p spaces with weight. *Soviet Math. Dokl.* **10** (1969), 406–410.

[97] I. Gohberg and N. Krupnik: Systems of singular integral equations in weighted L^p spaces. *Soviet Math. Dokl.* **10** (1969), 688–691.

[98] I. Gohberg and N. Krupnik: Singular integral operators with piecewise continuous coefficients and their symbols. *Math. USSR-Izv.* **5** (1971), 955–979.

[99] I. Gohberg and N. Krupnik: *One-Dimensional Linear Singular Integral Equations.* Vols. 1 and 2, Birkhäuser, Basel 1992. Russian original: Shtiintsa, Kishinev 1973.

[100] I. Gohberg, N. Krupnik, and I. M. Spitkovsky: Banach algebras of singular integral operators with piecewise continuous coefficients. General contour and weight. *Integral Equations Operator Theory* **17** (1993), 322–337.

[101] S. M. Grudsky: Singular integral equations and the Riemann boundary value problem with an infinite index in the space $L_p(\Gamma, \omega)$. *Math. USSR-Izv.* **26** (1986), 53–76.

[102] R. Hagen, S. Roch, and B. Silbermann: *Spectral Theory of Approximation Methods for Convolution Equations.* Birkhäuser, Basel 1995.

[103] R. Hagen, S. Roch, and B. Silbermann: *C^*-Algebras and Numerical Analysis.* Marcel Dekker Inc., New York 2001.

[104] P. R. Halmos: *A Hilbert Space Problem Book.* Second ed., Springer, New York 1982.

[105] P. Hartman: On completely continuous Hankel matrices. *Proc. Amer. Math. Soc.* **9** (1958), 862–866.

[106] P. Hartman and A Wintner: The spectra of Toeplitz's matrices. *Amer. J. Math.* **76** (1954), 867–882.

[107] E. Hewitt and K. A. Ross: *Abstract Harmonic Analysis.* Vol. 1, Springer, Berlin, Göttingen, Heidelberg 1963.

[108] E. Hewitt and K. A. Ross: *Abstract Harmonic Analysis.* Vol. 2, Springer, Berlin, Göttingen, Heidelberg 1970.

[109] E. Hlawka, J. Schoissengeier, and R. Taschner: *Geometric and Analytic Number Theory.* Springer, Berlin 1991.

[110] K. Hoffman: *Banach Spaces of Analytic Functions.* Dover, New York 1988.

[111] L. Hörmander: Estimates for translation invariant operators in L^p spaces. *Acta Math.* **104** (1960), 93–140.

[112] R. A. Horn and C. R. Johnson: *Matrix Analysis.* Cambridge University Press, New York 1987.

[113] R. Hunt, B. Muckenhoupt, and R. Wheeden: Weighted norm inequalities for the conjugate function and Hilbert transform. *Trans. Amer. Math. Soc.* **176** (1973), 227–251.

[114] B. V. Hvedelidze: The method of Cauchy type integrals for discontinuous boundary value problems of the theory of holomorphic functions of one complex variable. *J. Sov. Math.* **7** (1977), 309–414.

[115] I. S. Hwang and W. Y. Lee: On the continuity of spectra of Toeplitz operators. *Arch. Math.* **70** (1998), 66–73.

[116] Ronghui Ji and Jingbo Xia: On the classification of commutator ideals. *J. Funct. Anal.* **78** (1988), 208–232.

[117] N. K. Karapetiants and S. G. Samko: Singular convolution operators with a discontinuous symbol. *Siberian Math. J.* **16** (1975), 35–48.

[118] N. K. Karapetiants and S. G. Samko: *Equations with Involutive Operators.* Birkhäuser, Boston 2001.

[119] Yu. I. Karlovich: The local trajectory method of studying invertibility in C^*-algebras of operators with discrete groups of shifts. *Soviet Math. Dokl.* **37** (1988), 407–412.

[120] Yu. I. Karlovich: Riemann and Haseman boundary value problems with oscillating coefficients. *Reports of the extended sessions of a seminar of the I. N. Vekua Institute for Applied Mathematics* **5** (1990), 86–89 [Russian].

[121] Yu. I. Karlovich: *Algebras of Convolution Type Operators with Discrete Groups of Shifts and Oscillating Coefficients.* Doctoral dissertation, Math. Inst. Georgian Acad. Sci., Tiflis 1991 [Russian].

[122] Yu. I. Karlovich: On the Haseman problem. *Demonstratio Math.* **26** (1993), 581–595.

[123] Yu. I. Karlovich: The continuous invertibility of functional operators in Banach spaces. *Dissertationes Mathematicae* **340** (1995), 115–136.

[124] Yu. I. Karlovich and V. G. Kravchenko: A general method of investigating operators of singular integral operator type with non-Carleman shift. *Soviet Math. Dokl.* **22** (1981), 10–14.

[125] Yu. I. Karlovich and V. G. Kravchenko: The algebra of singular integral operators with piecewise-continuous coefficients and piecewise-smooth shift on a composite contour. *Math. USSR-Izv.* **23** (1984), 307–352.

[126] Yu. I. Karlovich and G. S. Litvinchuk: Some classes of semi-Noetherian operators. *Soviet Math.-Izv. VUZ* **34** (1990), 1–15.

[127] Yu. I. Karlovich and E. Ramírez de Arellano: A shift invariant algebra of singular integral operators with oscillating coefficients. *Integral Equations Operator Theory* **39** (2001), 441–474.

[128] Yu. I. Karlovich and I. M. Spitkovsky: On the Noether property for certain singular integral operators with matrix coefficients of the class SAP and the systems of convolution equations on a finite interval connected with them. *Soviet Math. Dokl.* **27** (1983), 358–363.

[129] Yu. I. Karlovich and I. M. Spitkovsky: Factorization of almost periodic matrix functions and Fredholm theory of Toeplitz operators with semi almost periodic matrix symbols. In: *Linear and Complex Analysis Problem Book. 199 Research Problems*, pp. 279–282, Lecture Notes in Mathematics, Vol. 1043, Springer, Berlin 1984.

[130] Yu. I. Karlovich and I. M. Spitkovsky: *Factorization of Almost Periodic Matrix Functions and (Semi-)Fredholmness of Some Convolution Type Equations.* No. 4421-85 dep., VINITI, Moscow 1985 [Russian].

[131] Yu. I. Karlovich and I. M. Spitkovsky: On the theory of systems of equations of convolution type with semi-almost-periodic symbols in spaces of Bessel potential. *Soviet Math. Dokl.* **33** (1986), 145–149.

[132] Yu. I. Karlovich and I. M. Spitkovsky: Factorization of almost periodic matrix-valued functions and the Noether theory for certain classes of equations of convolution type. *Math. USSR-Izv.* **34** (1990), 281–316.

[133] Yu. I. Karlovich and I. M. Spitkovsky: (Semi)-Fredholmness of convolution operators on the spaces of Bessel potentials. *Oper. Theory Adv. Appl.* **71** (1994), 122–152.

[134] Yu. I. Karlovich and I. M. Spitkovsky: Almost periodic factorization: an analogue of Chebotarev's algorithm. *Contemporary Math.* **189** (1995), 327–352.

[135] Yu. I. Karlovich and I. M. Spitkovsky: Factorization of almost periodic matrix functions. *J. Math. Anal. Appl.* **193** (1995), 209–232.

[136] Yu. I. Karlovich and I. M. Spitkovsky: Semi-Fredholm properties of certain singular integral operators. *Oper. Theory Adv. Appl.* **90** (1996), 264–287.

[137] Yu. I. Karlovich and I. M. Spitkovsky: On the semi-Noethericity of singular integral operators with matrix semi-almost-periodic coefficients. *Doklady Math.* **57** (1998), 176–178.

[138] Yu. I. Karlovich, I. M. Spitkovsky, and R. A. Walker: Almost periodic factorization of block triangular matrix functions revisited. *Linear Algebra Appl.* **293** (1999), 199–232.

[139] A. B. Khevelev: *On the Invertibility and Fredholmness of Some Classes of Operators with Non-Carleman Shift.* Candidate dissertation, Rostov State University, Rostov-on-Don 1979 [Russian].

[140] G. Khuskivadze, V. Kokilashvili, and V. Paatashvili: Boundary value problems for analytic and harmonic functions in domains with nonsmooth boundaries. Applications to conformal mappings. *Mem. Differential Equations Math. Phys.* **14** (1998), 1–195.

[141] U. Köhler and B. Silbermann: Einige Ergebnisse über Φ_+-Operatoren in lokalkonvexen, topologischen Vektorräumen. *Math. Nachr.* **56** (1973), 145–153.

[142] P. Koosis: *Introduction to H^p Spaces.* Second ed., Cambridge University Press, Cambridge 1998.

[143] M. G. Krein: Integral equations on a half-line with a kernel depending upon the difference of the arguments. *Amer. Math. Soc. Transl.* **22** (1962), 163–288.

[144] S. G. Krein: *Linear Equations in Banach Spaces.* Birkhäuser, Boston 1982. Russian original: Nauka, Moscow 1956.

[145] N. Krupnik: The exact constant in I. B. Simonenko's theorem on an envelope of a family of operators of local type. *Functional Anal. Appl.* **20** (1986), 144–145.

[146] N. Krupnik: *Banach Algebras with Symbol and Singular Integral Operators.* Birkhäuser, Basel 1987. Russian original: Shtiintsa, Kishinev 1984.

[147] N. Krupnik and I. Feldman: Relations between factorization and invertibility of finite Toeplitz matrices. *Izvestiya Akad. Nauk Moldavskoi SSR. Seriya Fiziko-Tehnicheskih i Matematicheskih Nauk* **3/1985** (1985), 20–26 [Russian].

[148] A. Kuijper and I. M. Spitkovsky: On convolution equations with semi-almost periodic symbols on a finite interval. *Integral Equations Operator Theory* **16** (1993), 530–538.

[149] V. G. Kurbatov: *Functional-Differential Operators and Equations.* Kluwer Academic Publishers, Dordrecht 1999.

[150] B. V. Lange and V. S. Rabinovich: Noethericity of multidimensional discrete convolution operators. *Math. Notes* **37** (1985), 237–407.

[151] B. V. Lange and V. S. Rabinovich: Pseudodifferential operators on R^n and limit operators. *Math. USSR-Sb.* **57** (1987), 183–194.

[152] B. Ya. Levin: *Distribution of Zeros of Entire Functions.* Amer. Math. Soc., Providence, RI 1964. Russian original: GITTL, Moscow 1956.

[153] B. M. Levitan: *Almost Periodic Functions.* GITTL, Moscow 1953 [Russian].

[154] B. M. Levitan and V. V. Zhikov: *Almost Periodic Functions and Differential Equations.*, Cambridge University Press, Cambridge and New York 1982. Russian original: University Press, Moscow 1978.

[155] G. S. Litvinchuk and I. M. Spitkovsky: *Factorization of Measurable Matrix Functions.* Birkhäuser, Basel and Boston 1987.

[156] Yu. Lyubarskii and I. M. Spitkovsky: Sampling and interpolating for a lacunary spectrum. *Royal Society of Edinburgh, Proceedings* **126A** (1996), 77–87.

[157] A. Markus and I. Feldman: The index of an operator matrix. *Funct. Anal. Appl.* **11** (1977), 149–151.

[158] B. Muckenhoupt: Weighted norm inequalities for the Hardy maximal function. *Trans. Amer. Math. Soc.* **165** (1972), 207–226.

[159] É. Muhamadiev: The invertibility of partial differential operators of elliptic type. *Soviet Math. Dokl.* **13** (1972), 1122–1126.

[160] G. J. Murphy: *C^*-Algebras and Operator Theory.* Academic Press, Boston 1990.

[161] N. Muskhelishvili: *Singular Integral Equations.* Nordhoff, Groningen 1963. Extended Russian version: Nauka, Moscow 1968.

[162] F. L. Nazarov and S. Treil: The hunt for a Bellman function: applications to estimates for singular integral operators and to other classical problems of harmonic analysis. *St. Petersburg Math. J.* **8** (1997), 721–824.

[163] Z. Nehari: On bounded linear forms. *Ann. of Math.* **104** (1957), 570–578.

[164] M. A. Naimark: *Normed Rings.* Noordhoff, Groningen 1964. Russian original: GITTL, Moscow 1956.

[165] N. K. Nikolskii: *Treatise of the Shift Operator: Spectral Function Theory.* Springer, Berlin and New York 1986. Russian original: Nauka, Moscow 1980.

[166] V. Yu. Novokshenov: Equations in convolutions on a finite interval and factorization of elliptic matrices. *Math. Notes* **27** (1980), 449–455.

[167] V. A. Paatashvili: On the discontinuous problem of linear conjugation. *Soobshch. Akad. Nauk Gruzin. SSR* **36** (1964), 539–540 [Russian].

[168] L. Page: Bounded and compact vectorial Hankel operators. *Trans. Amer. Math. Soc.* **150** (1970), 529–539.

[169] B. V. Pal'cev: Convolution equations on a finite interval for a class of symbols having powerlike asymptotics at infinity. *Math. USSR-Izv.* **16** (1981), 291–356.

[170] B. V. Pal'cev: A generalization of the Wiener-Hopf method for convolution equations on a finite interval with symbols having powerlike asymptotics at infinity. *Math. USSR-Sb.* **41** (1982), 289–328.

[171] A. A. Pankov: *Bounded and Almost Periodic Solutions of Nonlinear Differential Operator Equations.* Kluwer, Dordrecht, Boston, London 1990. Russian original: Naukova Dumka, Kiev 1985.

[172] J. R. Partington: *An Introduction to Hankel Operators.* Cambridge University Press, Cambridge 1988.

[173] G. K. Pedersen: *C^*-Algebras and Their Automorphism Groups.* Academic Press, London 1979.

[174] H. R. Pousson: Systems of Toeplitz operators on H^2. *Proc. Amer. Math. Soc.* **19** (1968), 603–608.

[175] S. C. Power: *Hankel Operators on Hilbert Space.* Pitman, Boston 1982.

[176] S. Prössdorf: *Some Classes of Singular Equations.* North-Holland Publishing Co., Amsterdam 1978.

[177] D. Quint, L. Rodman, and I. M. Spitkovsky: New cases of almost periodic factorization of triangular matrix functions. *Michigan Math. J.* **45** (1998), 73–102.

[178] M. Rabindranathan: On the inversion of Toeplitz operators. *J. Math. Mech.* **19** (1969/1970), 195–206.

[179] V. S. Rabinovich: Algebras of singular integral operators on compound contours with nodes that are logarithmic whirl points. *Russ. Acad. Sci. Izv. Math.* **60** (1996), 1261–1292.

[180] V. S. Rabinovich: C^*-algebras of pseudodifferential operators and limit operators. In: *Operator Theory for Complex and Hypercomplex Analysis (Mexico City, 1994)*, pp. 169–183, Amer. Math. Soc., Providence, RI 1998.

[181] V. S. Rabinovich: Mellin pseudodifferential operators techniques in the theory of singular integral operators on some Carleson curves. *Oper. Theory Adv. Appl.* **102** (1998), 201–218.

[182] V. S. Rabinovich, S. Roch, and B. Silbermann: Fredholm theory and finite section method for band-dominated operators. *Integral Equations Operator Theory* **30** (1998), 452–495.

[183] S. Roch and B. Silbermann: *Algebras of Convolution Operators and Their Image in the Calkin Algebra.* Report R-Math-05/90, Akad. Wiss. DDR, Karl-Weierstrass Institut f. Mathematik, Berlin 1990.

[184] L. Rodman and I. M. Spitkovsky: Almost periodic factorization and corona theorem. *Indiana Univ. Math. J.* **47** (1998), 1243–1256.

[185] L. Rodman, I. M. Spitkovsky, and H. J. Woerdeman: Carathéodory-Toeplitz and Nehari problems for matrix valued almost periodic functions. *Trans. Amer. Math. Soc.* **350** (1998), 2185–2227.

[186] L. Rodman, I. M. Spitkovsky, and H. J. Woerdeman: Factorization of almost periodic matrix functions of several variables and Toeplitz operators. *Oper. Theory Adv. Appl.* **122** (2001), 385–416.

[187] L. Rodman, I. M. Spitkovsky, and H. J. Woerdeman: Contractive extension problems for matrix valued almost periodic functions of several variables. *J. Operator Theory*, to appear.

[188] W. Rudin: *Fourier Analysis on Groups.* John Wiley & Sons Inc., New York 1990.

[189] W. Rudin: *Functional Analysis.* Second ed., McGraw-Hill Inc., New York 1991.

[190] A. I. Saginashvili: Singular integral operators with coefficients having discontinuities of semi-almost periodic type. *Soobshch. Akad. Nauk Gruzin. SSR* **94** (1979), 289–291 [Russian].

[191] A. I. Saginashvili: Singular integral operators with semi-almost periodic discontinuities in the coefficients. *Soobshch. Akad. Nauk Gruzin. SSR* **95** (1979), 541–543 [Russian].

[192] A. I. Saginashvili: Singular integral equations with coefficients having discontinuities of semi-almost periodic type. *Trudy Tbiliss. Mat. Inst. Razmadze* **66** (1980), 84–95 [Russian].

[193] L. A. Sakhnovich: Problems of factorization and operator identities. *Russian Math. Surveys*, **1/1986**(1986), 1–64.

[194] L. A. Sakhnovich: *Integral Equations with Difference Kernels on Finite Intervals.* Birkhäuser, Basel 1996.

[195] D. Sarason: Generalized interpolation in H^∞. *Trans. Amer. Math. Soc.* **127** (1967), 179–203.

[196] D. Sarason: Approximation of piecewise continuous functions by quotients of bounded analytic functions. *Canad. J. Math.* **24** (1972), 642–657.

[197] D. Sarason: Toeplitz operators with semi-almost periodic symbols. *Duke Math. J.* **44** (1977), 357–364.

[198] R. Schneider: Integral equations with piecewise continuous coefficients in L^p spaces with weight. *J. Integral Equations* **9** (1985), 135–152.

[199] V. N. Semenyuta: On singular operator equations with shift on a circle. *Soviet Math. Dokl.* **18** (1977), 1572–1574.

[200] E. Shamir: Solutions of Riemann-Hilbert systems with piecewise continuous coefficients in L^p. *Soviet Math. Dokl.* **7** (1966), 530–533.

[201] M. A. Shubin: On the coincidence of spectrums of pseudodifferential almost periodic operators on $L_2(R^n)$ and $B_2(R^n)$. *Siberian Math. J.* **17** (1976), 158–170.

[202] M. A. Shubin: Almost periodic functions and partial differential operators. *Russian Math. Surveys* **33** (1978), 1–52.

[203] I. B. Simonenko: Riemann's boundary problem with a measurable coefficient. *Soviet Math. Dokl.* **1** (1960), 1295–1298.

[204] I. B. Simonenko: The Riemann boundary value problem for n pairs of functions with measurable coefficients and its application to the investigation of singular integrals in the spaces L^p with weight. *Izv. Akad. Nauk SSSR Ser. Mat.* **28** (1964), 277–306 [Russian].

[205] I. B. Simonenko: A new general method of investigating linear operator equations of singular integral equation type, I. *Izv. Akad. Nauk SSSR Ser. Mat.* **29** (1965), 567–586 [Russian].

[206] I. B. Simonenko: A new general method of investigating linear operator equations of singular integral equation type, II. *Izv. Akad. Nauk SSSR Ser. Mat.* **29** (1965), 757–782 [Russian].

[207] I. B. Simonenko: Some general questions of the theory of the Riemann boundary value problem. *Math. USSR-Izv.* **2**, (1968), 1091–1099.

[208] I. B. Simonenko: Global and local factorability of a measurable matrix-function, and the Noether property of the singular operator generated by it. *Soviet Math.-Izv. VUZ* **27**, no. 4 (1983), 99–106.

[209] I. B. Simonenko and Chin Ngok Min: *Local Approach to the Theory of One-Dimensional Singular Integral Equations with Piecewise Continuous Coefficients. Noetherity.* Rostov University Press, Rostov-on-Don 1986 [Russian].

[210] I. M. Spitkovsky: The multipliers that do not influence factorizability. *Soviet Math. Dokl.* **17** (1976), 1733–1738 (1977).

[211] I. M. Spitkovsky: Factorization of several classes of semi-almost periodic matrix functions and applications to systems of convolution equations. *Soviet Math.-Izv. VUZ* **27**, no. 4 (1983), 107–115.

[212] I. M. Spitkovsky: On the factorization of almost periodic matrix functions. *Math. Notes* **45** (1989), 482–488.

[213] I. M. Spitkovsky: Singular integral operators with *PC* symbols on the spaces with general weights. *J. Functional Analysis* **105** (1992), 129–143.

[214] I. M. Spitkovsky and P. M. Tishin: Factorization of new classes of almost-periodic matrix functions. *Reports of the extended sessions of a seminar of the I. N. Vekua Institute for Applied Mathematics* **3** (1989), 170–173 [Russian].

[215] I. M. Spitkovsky and H. J. Woerdeman: The Carathéodory-Toeplitz problem for almost periodic functions. *J. Functional Analysis* **115** (1993), 281–293.

[216] I. M. Spitkovsky and D. Yong: Almost periodic factorization of certain block triangular matrix functions. *Mathematics of Computation* **69** (2000), 1053–1070.

[217] S. B. Stechkin: On bilinear forms. *Dokl. Akad. Nauk SSSR* **71** (1950), 237–240 [Russian].

[218] E. Stein and G. Weiss: *Introduction to Fourier Analysis on Euclidean Spaces.* Princeton University Press, Princeton 1971.

[219] J.-I. Tanaka: Corona problem and flows. *J. Functional Analysis* **102** (1991), 360–378.

[220] P. M. Tishin: *Factorization and Formulas for the Partial Indices of Certain Classes of Matrix Functions.* Candidate dissertation, Odessa State University, Odessa 1990 [Russian].

[221] T. V. Tonev: *Big-Planes, Boundaries, and Function Algebras*, North-Holland Publ. Co., Amsterdam 1992.

[222] S. Treil and A. Volberg: Wavelets and the angle between past and future. *J. Funct. Anal.* **143** (1997), 269–308.

[223] A. Volberg: Matrix A_p weights via S-functions. *J. Amer. Math. Soc.* **10** (1997), 445–466.

[224] H. Widom: Inversion of Toeplitz matrices, II. *Illinois J. Math.* **4** (1960), 88–99.

[225] H. Widom: Inversion of Toeplitz matrices, III. *Notices Amer. Math. Soc.* **7** (1960), p. 63.

[226] H. Widom: Singular integral equations in L^p. *Trans. Amer. Math. Soc.* **97** (1960), 131–160.

[227] H. Widom: On the spectrum of Toeplitz matrices. *Pacif. J. Math.* **14** (1964), 365–375.

[228] H. Widom: Asymptotic behavior of block Toeplitz matrices and determinants, II. *Advances in Math.* **21** (1976), 1–29.

[229] H. Widom: Wiener-Hopf integral equations. In: *The Legacy of Norbert Wiener: A Centennial Symposium (Cambridge, MA 1994)*, pp. 391–405, Proc. Sympos. Pure Math. **60**, Amer. Math. Soc., Providence, RI 1997.

[230] N. Wiener: Tauberian theorems. *Ann. Math.* **33** (1932), 1–100.

[231] N. Wiener and E. Hopf: Über eine Klasse singulärer Integralgleichungen. *Sitzungsber. Preuss. Akad. Wiss. Berlin, Phys.-Math. Kl.* **30/32** (1931), 696–706.

[232] J. Xia: Conditional expectations and the corona problem of ergodic Hardy spaces. *J. Functional Analysis* **64** (1985), 251–274.

[233] J. Xia: The K-theory and the invertibility of almost periodic Toeplitz operators. *Integral Equations Operator Theory* **11** (1988), 267–286.

Index

Operator Theory - Advances and Applications

Previously published titles

OT 114: **Ramírez de Arellano, E. / Shapiro, M.V. / Tovar, L.M. / Vasilevski, N.L.** (Eds.) Complex Analysis and Related Topics (1999) ISBN 3-7643-6228-6

OT 113: **Havin, V.P. / Nikolski, N.K.** (Eds.) Complex Analysis, Operators, and Related Topics (1999) ISBN 3-7643-6214-6

OT 112: **Maz'ya, V. / Nazarov, S. / Plamenevskij, B.** Asymptotic Theory of Elliptic Boundary Value Problems in Singularly Pertubed Domains,Volume II (2000) ISBN 3-7643-6398-3

OT 111: **Maz'ya, V. / Nazarov, S. / Plamenevskij, B.** Asymptotic Theory of Elliptic Boundary Value Problems in Singularly Pertubed Domains, Volume I (2000)ISBN 3-7643-6397-5

OT 110: **Rossmann, J. / Takáč, P. / Wildenhain, G.** (Eds.) The Maz'ya Anniversary Collection,Volume 2 (1999) ISBN 3-7643-6202-2

OT 109: **Rossmann, J. / Takáč, P. / Wildenhain, G.** (Eds.) The Maz'ya Anniversary Collection, Volume 1 (1999) ISBN 3-7643-6201-4

OT 108: **Dittrich, J. / Exner, P. / Tater, M.** (Eds.) Mathematical Results in Quantum Mechanics (1999) ISBN 3-7643-6097-6

OT 107: **Sakhnovich, L.A.** Spectral Theory of Canonical Differential Systems (1999) ISBN 3-7643-6057-7

OT 106: **Dijksma, A. / Gohberg, I. / Kaashoek, M.A. / Mennicken, R.** (Eds.), Contributions to Operator Theory in Spaces with an Indefinite Metric. The Heinz Langer Anniversary Volume (1998) ISBN 3-7643-6003-8

OT 105: **Bagdasarov, S.**, Chebyshev Splines and Kolmogorov Inequalities (1998) ISBN 3-7643-5984-6

OT 104: **Bercovici, H. / Foias, C.** (Eds.), Nonselfadjoint Operator Algebras, Operator Theory, and Related Topics. The Carl M. Pearcy Anniversary Volume (1998) ISBN 3-7643-5954-4

OT 103: **Gohberg, I. / Mennicken, R. / Tretter, C.** (Eds.), Recent Progress in Operator Theory. International Workshop on Operator Theory and Applications, IWOTA 95, in Regensburg, July 31–August 4, 1995 (1998) ISBN 3-7643-5891-2

OT 102: **Gohberg, I. / Mennicken, R. / Tretter, C.** (Eds.), Differential and Integral Operators. International Workshop on Operator Theory and Applications, IWOTA 95, in Regensburg, July 31–August 4, 1995 (1998) ISBN 3-7643-5890-4

OT 101: **Eidelman, S.D. / Zhitarashu, N.V.**, Parabolic Boundary Value Problems (1998) ISBN 3-7643-2972-6

OT 100: **Foias, C.I. / Frazho, A.E. / Gohberg, I. / Kaashoek, M.A.** (Eds.), Metric Constrained Interpolation, Commutant Lifting and Systems (1998) ISBN 3-7643-5889-0

OT 99: **Ayerbe Toledano, J.M. / Dominguez Benavides, T. / López Acedo, G.** Measures of Noncompactness in Metric Fixed Point Theory (1997) ISBN 3-7643-5794-0

OT 98: **Gohberg, I. / Lyubich, Y.** (Eds.), New Results in Operator Theory and Its Applications. The Israel M. Glazman Memorial Volume (1997) ISBN 3-7643-5775-4

OT 97: **Gorbachuk, M.L. / Gorbachuk, V.I.**, M.G. Krein's Lectures on Entire Operators (1997) ISBN 3-7643-5704-5

OT 96: **Alpay, D. / Dijksma, A. / Rovnyak, J. / Snoo, H. de** (Eds.), Schur Functions, Operator Colligations, and Reproducing Kernel Pontryagin Spaces (1997) ISBN 3-7643-5763-0

OT 95: **Dym, H. / Fritzsche, B. / Katsnelson, V. / Kirstein, B.** (Eds.), Topics in Interpolation Theory (1997) ISBN 3-7643-5723-1

OT 94: **Kadets, M.I. / Kadets, V.M.**, Series in Banach Spaces. Conditional and Unconditional Convergence (1997) ISBN 3-7643-5401-1

OT 93: **Egorov, Y. / Schulze, B.W.**, Pseudo-Differential Operators, Singularities, Applications (1997) ISBN 3-7643-5484-4

OT 92: **Shklyar, A.Y.**, Complete Second Order Linear Differential Equations in Hilbert Spaces (1997) ISBN 3-7643-5377-5

OT 91: **Skubachevskii, A.L.**, Elliptic Functional Differential Equations and Applications (1997) ISBN 3-7643-5404-6

OT 90: **Böttcher, A. / Gohberg, I.** (Eds.), Singular Integral Operators and Related Topics. Joint German-Israeli Workshop, Tel Aviv, March 1-10, 1995 (1996) ISBN 3-7643-5466-6

OT 89: **Egorov, V. / Kondratiev, V.**, On Spectral Theory of Elliptic Operators (1996) ISBN 3-7643-5390-2

OT 88: **van Neerven, J.**, The Asymptotic Behaviour of Semigroups of Linear Operators (1996) ISBN 3-7643-5455-0

OT 87: **Gohberg, I. / Lancaster, P. / Shivakumar, P.N.** (Eds.), Recent Developments in Operator Theory and Its Applications. International Conference in Winnipeg, October 2–6, 1994 (1996) ISBN 3-7643-5413-5

OT 86: **Berezansky, Y.M. / Sheftel, Z.G. / Us, G.F.**, Functional Analysis. Vol. II (1996) ISBN 3-7643-5345-7

OT 85: **Berezansky, Y.M. / Sheftel, Z.G. / Us, G.F.**, Functional Analysis. Vol. I (1996) ISBN 3-7643-5344-9

OT 83: **Antonevich, A.**, Linear Functional Equations. Operator Approach (1996) ISBN 3-7643-2931-9

OT 82: **Constantinescu, T.**, Schur Parameters, Factorization and Dilation Problems (1996) ISBN 3-7643-5285-X

OT 81: **Upmeier, H.**, Toeplitz Operators and Index Theory in Several Complex Variables (1996) ISBN 3-7643-5282-5

OT 80: **Gohberg, I. / Langer, H.** (Eds.), Operator Theory and Boundary Eigenvalue Problems. International Workshop in Vienna, July 27-30, 1993 (1995) ISBN 3-7643-5275-2

OT 79: **Gohberg, I. / Kaashoek, M.A. / van Schagen, F.**, Partially Specified Matrices and Operators. Classification, Completion, Applications (1995) ISBN 3-7643-5259-0

OT 78: **Demuth, M. / Schulze, B.W.** (Eds.), Partial Differential Operators and Mathematical Physics. International Conference in Holzhau (Germany), July 3 – 9, 1994 (1995) ISBN 3-7643-5208-6

OT 77: **Lindenstrauss, J. / Milman, V.D.** (Eds.), Geometric Aspects of Functional Analysis. Israel Seminar (GAFA) 1992–94 (1995) ISBN 3-7643-5207-8

OT 76: **Krasnosel'skij, A.M.**, Asymptotics of Nonlinearities and Operator Equations (1995) ISBN 3-7643-5175-6

OT 75: **Huijsmans, C.B. / Kaashoek, M.A. / Luxemburg, W.A. / Pagter, B. de** (Eds.), Operator Theory in Function Spaces and Banach Lattices. The A.C. Zaanen Anniversary Volume (1995) ISBN 3-7643-5146-2

OT 74: **Hagen, R. / Roch, S. / Silbermann, B.**, Spectral Theory of Approximation Methods for Convolution Equations (1994) ISBN 3-7643-5112-8

OT 73: **Feintuch, A. / Gohberg, I.** (Eds.), Nonselfadjoint Operators and Related Topics. Workshop on Operator Theory and Its Applications, Beersheva, February 24–28, 1992 (1994) ISBN 3-7643-5097-0

OT 72: **Gohberg, I. / Sakhnovich, L.A.** (Eds.), Matrix and Operator Valued Functions. The Vladimir Petrovich Potapov Memorial Volume (1994) ISBN 3-7643-5091-1

OT 71: **Basor, E.L. / Gohberg, I.** (Eds.), Toeplitz Operators and Related Topics. The Harold Widom Anniversary Volume. Workshop on Toeplitz and Wiener-Hopf Operators, Santa Cruz, California, September 20–22, 1992 (1994) ISBN 3-7643-5068-7

OT 70: **Demuth, M. / Exner, P. / Neidhardt, H. / Zagrebnov, V.** (Eds.), Mathematical Results in Quantum Mechanics. International Conference in Blossin (Germany), May 17-21, 1993 (1994) ISBN 3-7643-5025-3

OT 69: **Ashyralyev, A. / Sobolevskii, P.E.**, Well-Posedness of Parabolic Difference Equations (1994) ISBN 3-7643-5024-5

OT 68: **Halanay, A. / Ionescu, V.**, Time-Varying Discrete Linear Systems. Input-Output Operators. Riccati Equations. Disturbance Attenuation (1994) ISBN 3-7643-5012-1

OT 67: **Govorov, N.V.**, Riemann's Boundary Problem with Infinite Index (1994) ISBN 3-7643-2999-8

OT 66: **Khatskevich, V. / Sholykhet, D.**, Differentiable Operators and Nonlinear Equations (1993) ISBN 3-7643-2929-7

OT 65: **Djrbashian, M.M.**, Harmonic Analysis and Boundary Value Problems in the Complex Domain (1993) ISBN 3-7643-2855-X

OT 64: **Gohberg, I.** (Ed.), New Aspects in Interpolation and Completion Theories (1993) ISBN 3-7643-2948-3

OT 63: **Gohberg, I. / Goldberg, S. / Kaashoek, M.A.**, Classes of Linear Operators. Vol. 2 (1993) ISBN 3-7643-2944-0

OT 62: **Furuta, T. / Gohberg, I. / Nakazi, T.** (Eds.), Contributions to Operator Theory and its Applications. The Tsuyoshi Ando Anniversary Volume (1993) ISBN 3-7643-2928-9

OT 61: **Gheondea, A. / Timotin, D. / Vasilescu, F.H.** (Eds.), Operator Extensions, Interpolation of Functions and Related Topics. 14th International Conference on Operator Theory Timisoara, June 1-5, 1992 (1993) ISBN 3-7643-2902-5

OT 60: **Kuchment, P.A.**, Floquet Theory for Partial Differential Equations (1993) ISBN 3-7643-2901-7

OT 59: **Ando, T. / Gohberg, I.** (Eds.), Operator Theory and Complex Analysis. Workshop on Operator Theory and Complex Analysis, Sapporo (Japan), June 1991 (1992) ISBN 3-7643-2824-X

OT 58: **Gohberg, I.** (Ed.), Continuous and Discrete Fourier Transforms, Extension Problems and Wiener-Hopf Equations (1992) ISBN 3-7643-2809-6

OT 57: **Demuth, M. / Gramsch, B. / Schulze, B.W.**, Operator Calculus and Spectral Theory. Symposium on Operator Calculus and Spectral Theory, Lambrecht (Germany), December 1991 (1992) ISBN 3-7643-2792-8

MIX
Papier aus verantwortungsvollen Quellen
Paper from responsible sources
FSC® C105338
FSC
www.fsc.org

If you have any concerns about our products,
you can contact us on
ProductSafety@springernature.com

In case Publisher is established outside the EU,
the EU authorized representative is:
Springer Nature Customer Service Center GmbH
Europaplatz 3, 69115 Heidelberg, Germany

Printed by Libri Plureos GmbH
in Hamburg, Germany